Exploring Stochastic Laws

Exploring Stochastic Laws

Festschrift in Honour of the 70th Birthday of Academician Vladimir Semenovich Korolyuk

Editors:

A.V. Skorokhod and Yu.V. Borovskikh

///VSP///

Utrecht, The Netherlands, 1995

VSP BV
P.O. Box 346
3700 AH Zeist
The Netherlands

First published in 1995

ISBN 90-6764-196-0

CIP-DATA KONINKLIJKE BIBLIOTHEEK, DEN HAAG

Exploring

Exploring stochastic laws : Festschrift in honour of the
70th birthday of academician Vladimir Semenovich Korolyuk
/ ed.-in-chief: A.V Skorokhod, Yu.V. Borovskikh. -
Utrecht : VSP
ISBN 90-6764-196-0 bound
NUGI 811
Subject headings: stochastic laws

Printed in The Netherlands by Koninklijke Wöhrmann BV, Zutphen.

VLADIMIR SEMENOVICH KOROLYUK

Exploring Stochastic Laws

"From
the intrinsic evidence
of his creation
the Great Architect of the Universe
now begins to appear
as a pure mathematician"

Sir James Hopwood Jeans (1877 – 1946)
[*The Mysterious Universe* (1930)]

CONTENTS

Exploring Stochastic Laws. pp. ix–x.
Eds. A.V. Skorokhod and Yu.V. Borovskikh

Preface

This collection of articles gives an opportunity for the friends and collegues, the pupils and the pupils of the pupils to congratulate Vladimir Korolyuk on his seventieth birthday.

Present day troubles and conflicts have broken off many contacts between countries and people from around the world. I believe there are people who would have liked to have contributed to this book, but who could not because of the lack of information and communication. And there are our old friends who we shall never forget, who are sadly not with us today.

I remember clearly our first meeting forty five years ago at a University track and field athletics competition. He carried out his performance quickly, efficiently and excellently and left us as soon as possible to work on some scientific problems. The qualities Korolyuk possesses, to formulate problems exactly, to find the appropriate methods, and to approach quietly but confidently to the solution were always inherent in his mathematical studies. He discovered his famous results in such a difficult branch of Probability, as asymptotic expansions for distributions in boundary problems of statistics, primarily because of his purposefulness and persistence. This was the start of a glorious journey to the summit of scientific excellence. In all of his studies Korolyuk always saw clear goals, the best ways to reach them, and had the ability to find or to create the tools for the achievement of these goals. Sometimes it seemed that all the possibilities had been tried and the goal was totally inaccessible, but Korolyuk persistently tried new methods until the problem was conquered. For example, in the problem of merging in the theory of semi-Markov processes, up until perturbations on the spectrum were used. Korolyuk was never the worshipper of one idea only. He took great interest in Cybernetics and applied Mathematics. Even in Probability he often changed the object of his attention and began to work in a new field with immutable success, introducing new methods.

His pupils were greatly influenced by him, and would often follow him. Many of them are now in various cities and countries around the world, with their own students. His ability to work in new fields is preserved in him even now, and his new article in this collection supports this.

I believe our long friendship allows me to say, "Dear Volodya, your friends and pupils are happy that we were beside you all that time. We are waiting for your further successes and victories on the path that you have been working on for many fruitful years. We wish you a happy seventieth birthday. Many happy returns of the day!"

June 20, 1995

Anatoly V. Skorokhod
Michigan State University

Exploring Stochastic Laws. pp. xi–xx.
Eds. A.V. Skorokhod and Yu.V. Borovskikh

Vladimir Semenovich Korolyuk: A Short Biography

Vladimir Semenovich Korolyuk was born in Kiev on 19th August 1925. It was in Kiev that he obtained his secondary education, which he finished at the beginning of the Great Patriotic War (1941-1945).

When he was seventeen he entered the Red Army, and in a year he graduated from Kharkov Military Aviation School and started to work as an instructor of practical training at the same school. In 1947 he joined Kiev Taras Shevchenko University as a third year student of the mechanics and mathematics faculty (for the first two years he graduated in absentia from Kharkov University whilst still in the Army).

It was while he was at Kiev University, that Korolyuk's scientific interests were developed. It was under the supervision of Academician of the Ukrainian Academy of Sciences, B.V. Gnedenko, that he started scientific research work. He investigated the domains of attraction of stable distributions. In collaboration with his teacher he found a convenient form for the conditions of attraction in terms of characteristic functions. After graduating from the University, he worked as a senior teacher at Artemovsk Teachers' Institute. In a year he started postgraduate studies at the Institute of Mathematics of the Ukrainian Academy of Sciences where, under the supervision of B.V. Gnedenko he investigated non-parametric statistical Kolmogorov–Smirnov criteria.

When his supervisor went on a long foreign business trip, V.S. Korolyuk continued his education as a postgraduate student at Moscow University where he attended the lectures of A.N. Kolmogorov, A.Ya. Khintchine, and A.A. Liapunov. Under the supervision of A.N. Kolmogorov he developed a new approach to the problems of asymptotic analysis of non-parametric statistics to conditional random walks in Bernoulli and Poisson schemes. Liapunov's ideas in programming were actively developed and promoted by V.S. Korolyuk in Kiev.

Korolyuk's scientific activity and enormous enthusiasm during his years as a postgraduate student at Moscow University can be characterized by the following interesting incident. At one of A.Ya. Khintchine's lectures on queueing theory it was proved that if a stationary flow has parameter λ, then its intensity $\mu \geq \lambda$. To the next lecture Korolyuk brought the proof of the statement about the coincidence of μ and λ, which was then named "Korolyuk's Theorem" and went down as a classical result in all the textbooks of queueing theory.

In 1954, V.S. Korolyuk defended his PhD thesis. Since then he worked at the Institute of Mathematics of the National Academy of Sciences of Ukraine. First as the Junior Scientific Researcher and then from 1956 as the Senior Scientific Researcher. At this time he actively helped A.N. Kolmogorov in the creation of the well-known journal "Probability Theory and Its Applications". From 1960 he was the Head of the Department of Probability Theory. In 1963 he defended his Doctor of Mathematical Sciences thesis, and then from 1966 he was appointed Deputy Director of the Institute responsible for scientific work.

In 1967 he was elected to the position of Associate member of the National Academy of Sciences. In 1976 he was elected as an Academician of the National Academy of Sciences of Ukraine.

In connection with the creation of the Computing Centre of the Academy of Sciences in Kiev, V.S. Korolyuk actively trained students in mathematical programming. In 1958 he created a special course of lectures on the automatization of programming, which was based on elements of a universal address language for the description of digital computing processes in digital computing machines, which were developed by him.

Together with B.V. Gnedenko and E.L. Yushchenko, V.S. Korolyuk wrote the textbook,"Elements of programming", which was published in 1961 (with a second edition in 1963). In 1964 this textbook was translated into Hungarian and German, and then in 1969 into French. He was involved in the creation of the universal computing machine "Kiev", where under his initiative index registers were introduced.

In 1966 V.S. Korolyuk wrote the methodical textbook for teachers "Mathematical machines and programming", and in co-authorship with L.A. Kaluzhnin, the scientific textbook "Algorithms and mathematical machines".

In 1978, as a member of a team of famous specialists, he was awarded the State Premium of Ukraine for the creation of "The Cybernetic Encyclopedia".

At the same time Korolyuk continued to develop the method of asymptotic analysis in boundary problems for consequent sums of independent random variables connected into a Markov chain. For distributions of sojourn times of homogeneous Markov one-and multidimensional random walks in curvilinear domains he created algorithms for the construction of asymptotic decompositions containing regular terms and boundary layers, taking into account the effects of jumps over the boundaries.

The factorization methods for the investigation of the distributions of extremal deviations, for the time and the value of the first jump, and other functionals were developed side by side with asymptotic methods for the investigation of boundary functionals of processes with independent increments. For boundary functionals of semi-continuous processes with independent increments, V.S. Korolyuk developed a method of potential, which allows us to solve all of the most important types of boundary problems that appear in queueing theory, reliability theory, storage theory and other applied trends of Probability Theory.

In 1965, under the supervision of V.S. Korolyuk, intensive investigations on the theory of semi-Markov processes and their applications were carried out.

Towards the end of the seventies V.S. Korolyuk's attention was attracted to problems of mathematical statistics. In this area, his investigations were devoted to the problem of approximating distributions. The scope of his interests included asymptotic problems which appear in the theory of U-statistics, theory of range statistics and random permanents.

For V.S. Korolyuk deep analysis of applied problems, and the creation of new methods of investigations are characteristic of his work, and were mostly developed in his creation of mathematical reliability theory. Analysis of a great number of problems and the methods of their investigations made it possible for V.S. Korolyuk and his followers to put foundation to a new direction — analytical theory of the reliability of regenerative systems. V.S. Korolyuk proposed to apply Markov renewal processes to reliability theory.

The success of the new approaches to the solutions of applied problems of reliability theory was due to the application of new methods of harmonious and functional analysis.

Korolyuk's continuous mathematical investigations led him by necessity to step up his study of random evolutions — a direction in the theory of stochastic processes aimed at the investigation of complicated dynamic processes effected by internal and external random perturbations.

His fruitful scientific and organizing work has always been combined with teaching and social activity. In 1954 he began to deliver lectures at Kiev Taras Shevchenko University, and in 1965 he became a Professor of the Department of Probability Theory and Mathematical Statistics. V.S. Korolyuk regularly delivers lectures to help students improve their qualifications, and special courses of lectures for graduate students at the Mechanics and Mathematics Faculty of Kiev Taras Shevchenko University. On these special courses all trends of his investigations are reflected. This gives an opportunity for up and coming specialists to get acquainted with modern scientific achievements, and naturally to learn from Korolyuk's great mathematical experience.

V.S. Korolyuk has many disciples and students in various scientific disciplines related to probability theory and mathematical statistics, theory of random processes, cybernetics, queueing theory, reliability theory and others. In general every Kiev mathematician may consider himself to be one of the students of academician V.S.Korolyuk's students. More than forty mathematicians have obtained their PhD and Professors Degree under his supervision (among them are well known Doctors of Mathematics: N. Andreev, R.V. Bobryk, Yu.V. Borovskikh, N.S. Bratiĭchuk, V.S. Demianchuk, D.V. Gusak, B.P. Kredencer, B.G. Marchenko, A.F. Turbin, A.V. Swishchuk, A.P. Veliky). His complete list of publications contains more than three hundred titles, which includes more then twenty monographs.

Even now at seventy V.S. Korolyuk is full of creative resources, ideas, and plans. He actively engages in research work, often delivering lectures at scientific conferences and at universities in Italy, Spain, the Netherlands, Switzerland, Germany, Sweden and others. He is a member of the editorial boards of the "Ukrainian Mathematical Journal" and "Applied Stochastic Models and Data Analysis", and many other mathematical publications. He is regularly involved in scientific conferences around the world, often as chairman.

V.S. Korolyuk is laureate of Krylov's Premium of the Ukrainian Academy of Sciences. In 1995 academician V.S. Korolyuk was awarded N.N. Bogolyubov's Premium of the National Academy of Sciences of Ukraine for the development of fundamental Bogolyubov methods.

We wish him many more years of extraordinary success in his research work and in his other widespread interests.

21 June, 1995 **Editors**

Some works of V.S. Korolyuk from the last ten years

1. Korolyuk, V.S. and Swishchuk, A.V. (1990). *Applied Problems of the theory of random evolutions.* Kiev, Znannya. 30 p. (In Russian).

2. Korolyuk, V.S. and Swishchuk, A.V. (1992). *Semi-Markov random evolutions.* Kiev. Naukova Dumka Publishing. 256 p. (In Russian).

3. Korolyuk, V.S. and Swishchuk, A.V. (1995). *Semi-Markov random evolutions.* Kluwer Acad. Publish. 310 p.

4. Korolyuk, V.S. and Swishchuk, A.V. (1995). *Evolution of systems in random media.* USA, Fl., CRC Press. 255 p.

5. Korolyuk, V.S. and Swishchuk, A.V. (1986). CLT for semi-Markov random evolutions. Ukrain. Math. J. **38**, 330–334. (In Russian).

6. Korolyuk, V.S. and Swishchuk, A.V. (1987). CLT in phase merging scheme for semi-Markov random evolutions. Ukrain. Math. J. **39**, 314–319. (In Russian).

7. Korolyuk, V.S. and Swishchuk, A.V. (1989). Phase merging of inhomogeneous semi-Markov random evolutions. Ukrain. Math. J. **41** 163–170. (In Russian).

8. Korolyuk, V.S. and Swishchuk, A.V. (1989). Weak convergence of semi-Markov random evolutions in averaging scheme. Th. Prob. and Math. Stat. **42,** 80–90. (In Russian).

9. Korolyuk, V.S. and Swishchuk, A.V. (1989). CLT for inhomogeneous semi-Markov random evolutions. Ukrain. Math. J. **41**, 1064-1070. (In Russian).

10. Korolyuk, V.S. and Swishchuk, A.V. (1989). Limit representation of semi-Markov random evolution in series scheme. Ukrain. Math. J. **41**, 1476–1482. (In Russian).

11. Korolyuk, V.S. and Swishchuk, A.V. (1989). Weak convergence of semi-Markov random evolutions (martingale approach). In: *Pr. Th. Math. Stat., (VSP Science Press),* 1–9.

12. Korolyuk, V.S. and Swishchuk, A.V. (1995). *Evolutionary stochastic systems in random media.* Kiev. Naukova Dumka Publ. 260 p. (In Russian).

13. Korolyuk, V.S. and Swishchuk, A.V. (1984). Phase merging of semi-Markov random evolutions. In: *Intern. Conf. on stoch. optimiz. "Stoch. optimiz.". Abstracts.* Kiev, Naukova dumka, **1**, 127–129.

14. Korolyuk, V.S. and Swishchuk, A.V. (1985). Phase merging of semi-Markov random evolutions and storage theory. In: *Abstr. IV Vilnius conf. on Pr. Th. Math. Stat.* Vilnius, Mosklas, **2**, 62–64. (In Russian).

15. Korolyuk, V.S. and Swishchuk, A.V. (1988). Semi-Markov random evolutions. Phase merging and applications. In: *Proceed. V Internat. summer school on Th. Prob. and Math. Stat.* Varna 1985, 116–131.

16. Korolyuk, V.S. and Strok, V.V. (1992). Kravchuk Polynomials in algebraic code theory. *Abstracts of communications of the International Conference*

devoted to the memory of M. P. Kravchuk, Kiev, Institute of Mathematics of the Ukrainian Academy of Sciences, 96.

17. Korolyuk, V.S. (1993). Stochastic Kravchuk Polynomials. *Dokl. Acad. Nauk. Ukraine* **6**, 34–37.

18. Borovskikh, Yu.V. and Korolyuk, V.S. (1987). Limiting representation of *U*-statistics by the multiple Wiener-Ito integral. *Dokl. Acad. Nauk Ukrain.SSR* **A**(2), 16–18. (In Russian).

19. Borovskikh, Yu.V. and Korolyuk, V.S. (1987). Weak convergence of *U*-statistics to a multiple stochastic integral in Hilbert space. *Dokl. Akad. Nauk Ukrain.SSR* **A**(4), 9–13. (In Russian).

20. Borovskikh, Yu.V. and Korolyuk, V.S. (1988). Algebraic decomposition and asymptotics of the permanent of a random matrix. In: *Asymptotic and Applied Problems in the Theory of Random Evolution,* Institute of Math. Ukrain. Acad. of Sci., Kiev, 68–74. (In Russian).

21. Borovskikh, Yu.V. and Korolyuk, V.S. (1988). Asymptotic theory of *U*-statistics. *Ukr. Math. J.* **40**(2), 142–154.

22. Borovskikh, Yu.V. and Korolyuk, V.S. (1988). Rate of convergence of degenerate von Mises' functionals. *Teor Ver. i Prim.* **33**(1), 136–146; English translation: *Theory Probab. Appl.* **33**(1), 125–135.

23. Borovskikh, Yu.V. and Korolyuk, V.S. (1988). *Martingale Approximation.* Naukova Dumka, Kiev. (In Russian).

24. Borovskikh, Yu.V. and Korolyuk, V.S. (1990). Rate of convergence in the Central Limit Theorem for *UH*-statistics. *Ukr. Math. J.* **42**(8), 930–936.

25. Borovskikh, Yu.V. and Korolyuk, V.S. (1990). *U*-statistics of a bootstrap sample. *Ukr. Math. J.* **42**(10), 1245–1248.

26. Borovskikh, Yu.V. and Korolyuk, V.S. (1990). *UH*-statistics. *Sov. Math. Dokl.* **40**(2), 432–435.

27. Borovskikh, Yu.V. and Korolyuk, V.S. (1990). *UH*-statistics with increasing degrees. *Dokl. Akad. Nauk Ukrain.SSR* **A**(3), 3–7. (In Russian).

28. Borovskikh, Yu.V. and Korolyuk, V.S. (1990). Random symmetric polynomials with increasing degree. *Dokl. Akad. Nauk Ukrain.SSR*, **A**(4), 10–13. (In Russian).

29. Borovskikh, Yu.V. and Korolyuk, V.S. (1990). Elementary symmetric bootstrap polynomials. *Dokl. Akad. Nauk Ukrain.SSR* **A**(5), 9–12. (In Russian).

30. Borovskikh, Yu.V. and Korolyuk, V.S. (1991). *U-statistics with increasing degrees.* Preprint 91. 3, Institute of Math. Ukr. Acad. of Sci., Kiev. (In Russian).

31. Borovskikh, Yu.V. and Korolyuk, V.S. (1991). *Random permanents and symmetric statistics.* Preprint 91. 12, Institute of Math. Ukr. Acad. of Sci., Kiev. (In Russian).

32. Borovskikh, Yu.V. and Korolyuk, V.S. (1991). *U*-statistics with increasing degrees. *Sov. Math. Dokl.* **42**(2), 502–506.

33. Borovskikh, Yu.V. and Korolyuk, V.S. (1991). *U*-statistics of increasing degrees with asymptotic Poisson distribution. *Dokl. Akad. Nauk Ukrain.SSR* **A**(4), 29–32. (In Russian).

34. Borovskikh, Yu.V. and Korolyuk, V.S. (1991). *U*-statistics of increasing degrees with asymptotic Poisson distribution. *Report BS-R9120 August. Stichting Mathematical Centrum,* Amsterdam, 1–4.

35. Borovskikh, Yu.V. and Korolyuk, V.S. (1991). Random permanents and symmetric statistics. *VI USSR-JAPAN Symposium on Probability Theory and Mathematical Statistics. Abstr. of Communic.,* Kiev, p. 83.

36. Borovskikh, Yu.V. and Korolyuk, V.S. (1992). Random permanents and symmetric statistics. *Proceedings of the Sixth USSR-JAPAN Symposium on Probability Theory and Mathematical Statistics,* Kiev. World Scientific, 176–187.

37. Borovskikh, Yu.V. and Korolyuk, V.S. (1993). *Random permanents.* Institute of Mathematics, Kiev. (In Russian).

38. Borovskikh, Yu.V. and Korolyuk, V.S. (1993). Random permanents. *Sixth International Vilnius conference on Probability Theory and Mathematical Statistics,* Vilnius. Abstracts of communications, v.1(A-L). Institute of Mathematics and Informatics, 200–201.

39. Borovskikh, Yu.V. and Korolyuk, V.S. (1994). Random permanents and symmetric statistics. *Acta Applicandae Mathematicae* **31**, 1–63.

40. Borovskikh, Yu.V. and Korolyuk, V.S. (1994). *Theory of U-statistics.* Kluwer Academic Publishers. Dordrecht.

41. Borovskikh, Yu.V. and Korolyuk, V.S. (1988). *Approximation of nondegenerate U-statistics.* Preprint No. 88., Institute of Mathematics, Ukrain. Acad. of Sci., Kiev. (In Russian).

42. Borovskikh, Yu.V. and Korolyuk, V.S. (1988). A nonuniform estimate of the rate of convergence for *UH*-statistics. *Dokl. Akad. Nauk Ukrain.SSR,* **A**(10), 20–22. (In Russian).

43. Borovskikh, Yu.V. and Korolyuk, V.S. (1988). *Approximation of multisample U-statistics.* Preprint No. 88., Institute of Mathematics, Ukrain. Acad. of Sci., Kiev. (In Russian).

44. Borovskikh, Yu.V. and Korolyuk, V.S. (1988). Von Mises' statistics with values in Hilbert space. *Dokl. Akad. Nauk Ukrain.SSR* **A**(11), 15–18. (In Russian).

45. Borovskikh, Yu.V. and Korolyuk, V.S. (1989). The compact law of iterated logarithm for *UH*-statistics. *Ukrain. Mat. Zh.* **41**(5), 672–675. (In Russian).

46. Borovskikh, Yu.V. and Korolyuk, V.S. (1989). Approximation of distributions of *UH*-statistics. *Abstacts of 5th Internat. Vilnius Conf. in Probability Theory and Mathematical Statistics,* Vol. 3, Institute of Mathematics and Cybernetics, Lithuanian Academy of Sci., Vilnius, 1989, 319–320. (In Russian).

47. Borovskikh, Yu.V. and Korolyuk, V.S. (1990). *U*-statistics with increasing degrees. *Dokl. Akad. Nauk SSSR* **314**, 547–551. (In Russian). English translation: *Sov. Math. Dokl.* **42**(2), 502–506.

48. Borovskikh, Yu.V. and Korolyuk, V.S. (1991). Weak convergence of *UH*-statistics to stable laws. *Teor Ver.i Prim.* **39**(3), 597–600. (In Russian).

49. Borovskikh, Yu.V. and Korolyuk, V.S. (1986). The convergence rate for *U*-statistics. *Abstracts of 1st World Congr. of Bernoulli Society,* **2**, Tashkent, 1986, p.789.

50. Korolyuk, V.S. and Borovskikh, Yu.V. (1985). Approximation of non-degenerate *U*-statistics. *Teor Ver. i Prim.* **30** (3), 417–426. (In Russian). English translation: *Theory Probab. Appl.* **30**, (1985).

51. Korolyuk, V.S. and Borovskikh, Yu.V. (1985). Expansions for *U*-statistics and von Mises' functionals. *Ukr. Mat. Zh.* **37**(4), 450–456. English translation: *Ukr. Math. J.* **37**(4), 358–364.

52. Korolyuk, V.S. and Borovskikh, Yu.V. (1987). Central limit theorem for *U*-statistics in Hilbert space. *Dokl. Akad. Nauk Ukrain.SSR* **A**(3), 16–19. (In Russian).

53. Korolyuk, V.S. and Borovskikh, Yu.V. (1987). Functional limit theorem for the mean square deviation of the density estimate. In: *Asymptotic Methods in the Investigation of Stochastic Models,* Institute of Mathematics, Ukrain. Acad. of Sci., Kiev, 1987, 81–84. (In Russian).

54. Korolyuk, V.S. and Borovskikh, Yu.V. (1987). Asymptotic theory of *U*-statistics. *Ukr. Mat. Zh.* **39**(2). 169–182.

55. Korolyuk, V.S. and Borovskikh, Yu.V. (1988). Many-Dimensional *U*-statistical empirical process. *Dokl. Akad. Nauk Ukrain.SSR* **A**2, 12–15. (In Russian).

56. Korolyuk, V.S. and Borovskikh, Yu.V. (1988). *Symmetric Statistics in the Analysis of Experimental Data.* Obshchestvo "Znanie", Kiev. (In Russian).

57. Korolyuk, V.S. and Borovskikh, Yu.V. (1988). Glivenko Theorem for multi-sample *U*-statistical empirical distribution function. In: *Analytic Methods in the Investigation of Stochastic Models,* Institute of Mathematics, Ukrain. Acad. of Sci., Kiev, 60–62. (In Russian).

58. Korolyuk, V.S. and Borovskikh, Yu.V. (1989). The law of iterated logarithms for *U*-statistics. *Ukr. Mat. Zh.* **41**(1), 97–100.

59. Korolyuk, V.S. and Borovskikh, Yu.V. (1990). Rate of convergence in the central limit theorem for *UH*-statistics. *Teor. Ver. Mat. Statist.* **43**, 72–78 English translation: *Theor. Probab. Math. Statist.* **43**, 79–85.

60. Korolyuk, V.S. and Borovskikh, Yu.V. (1990). An estimate of the closeness of the moments of *UH*-statistics under the conditions of the central limit theorem. *Teor. Ver. Mat. Statist.* **42**, 48–53. English translation: *Theor. Probab. Math. Statist.* **42**, 55–60.

61. Королюк, В.С. и Боровских, Ю.В. (1989). *Теория U-статистик.* Киев, Наук. думка, 384 с.

62. Королюк, В.С. (1987). Эволюция систем в полумарковской случайной среде. *Кибернетика.* **5**, 106–109.

63. Королюк, В.С. (1989). *Стохастические модели систем.* Киев, Наук. думка, 208 с.

64. Королюк, В.С. и Свищук, А.В. Центральная предельная теорема для полумарковских случайных эволюций. *Укр. мат. журн.* **38**(3), 330–334.

65. Королюк, В.С., Свищук, А.В. и Королюк, В.В. (1987). Центральная предельная теорема в схеме фазового укркпнения для полумарковских случайных эволюций. *Там же.* **39**(3), 314–319.

66. Королюк В. С., Свищук А. В. (1989). Фазовое усреднение неоднородных полумарковских случайных эволюций. *Там же.* **41**(2), 163–170.

67. Королюк, В.С. и Свищук, А.В. (1990). Центральная предельная теорема для неоднородных полумарковских случайных эволюций. *Там же.* **8**, 1064–1070.

68. Королюк, В.С. и Свищук, А.В. (1990). Предельное представление полумарковских случайных эволюций в схеме серий. *Там же.* **11**, 1476–1482.

69. Королюк, В.С. и Свищук, А.В. (1990). Слабая сходимость полумарковских случайных эволюций в схеме усреднения. *Теория вероятности и мат. статистика.* **42**, 53–64.

70. Королюк, В.С. и Свищук, А.В. (1990). *Прикладные задачи теории случайных эволюций.* Киев, О-во "Знание" УССР, 32 с.

71. Королюк, В.С. и Боровских, Ю.В. О нормальной аппроксимации в банаховом пространстве L^P.

72. Korolyuk, V.S. and Swishchuk, A.V. (1985). Semi-Markov random evolutions. Phase Agregations and Applications. *Intern. Summer School on Probab. Th. Math. Stat., Varna, Gold Sands, Sofia 1988*, 116–131.

73. Korolyuk, V.S. and Swishchuk, A.V. (1990). Weak convergence of semi-Markov random evolutions (martingale approach). *Prob. Th. Math. Stat.* **1**, 1–9.

74. Korolyuk, V.S. (1990). CLT for semi-Markov random evolutions. *Comp. Math. Applic.* **19**(1), 83–88.

75. Korolyuk, V.S. (1991). Stability of autonomous dynamical systems with rapid Markov switchings. *Ukrainian Math. Journ.*, **33**(2), 1176–1181. (In Russian).

76. Korolyuk, V.S. (1991). *Averaging and stability of dynamical systems with rapid Markov switchings.*Preprint. University of UMEÅ, Sweden.

Exploring Stochastic Laws. pp. xxi–xxvii.
Eds. A.V. Skorokhod and Yu.V. Borovskikh

Memories of my youth

IGOR NIKOLAEVICH KOVALENKO

Presidium of the National Academy of Sciences of Ukraine

Received May 5, 1995

When Vladimir Semenovich initially became one of my University lecturers, he was not yet a full Professor or Doctor of Mathematics. In the first half of the fifties he obtained very interesting results about random Boolean-valued walks, which is a hard combinatorial subject, working in collaboration with our teacher B.V. Gnedenko. But his first results regarding maximal deviations of random walks, which can be interpreted as both combinatorial and in terms of difference parabolic schemes, naturally needed further refinement, i.e. the construction of more precise asymptotics in a small parameter scheme. V.S. Korolyuk's initial approach consisted in a direct decomposition in a small parameter scheme. This approach did not give the desired result, because of the complexity of the nature of this problem. Therefore V.S. Korolyuk undertook the challenging task of untangling asymptotics. Progress was not smooth. He once said, 'I have to discover the true nature of things, I must see how asymptotics are structured.' This led to the application of the method of sequentially exhausting, originated by Vishik and Lusternik (in their work only on differential equations) and later to the development (together with A.F. Turbin) of the theory of perturbations of Markov operators on a spectrum. But we, the students of Kiev Taras Shevchenko University, first got acquainted with V.S. Korolyuk as a teacher of computer programming. He was a pioneer of computing techniques, and the author of address programming (together with E.L. Yushchenko and B.V. Gnedenko). He was so fascinated with this subject that he did not care whether his students understood his lectures. Words flew like bullets from a machine gun, and the chalk from the blackboard crumbled into dust; usually it was difficult to follow his lectures. But when he supervised the individual term papers and diploma projects he was very patient. Usually, in a laconic way he explained the method of the solution to the problem. A thinking student received almost all of what was necessary to solve their problem, but was left with a lot to think about. Stormy, hectic and frantic scenes near the blackboard can always be recalled, when Vladimir Semenovich was outraged with his students who did not understand things that he thought were obvious.

Many interesting scientists worked at the Department of Probability Theory at Kiev Taras Shevchenko University: B.V. Gnedenko (Head of the Department for many years), I.I. Gikhman, V.S. Korolyuk, V.S. Michalevich (but not for a long time), L.A. Kaluzhnin, M.I. Yadrenko and others. Those who seriously developed enthusiasm for cybernetics grouped around Lev Arkadievich Kaluzhnin – a world famous and very versatile scientist. Vladimir Semenovich had original ideas and opinions about algorithms, electronic computing machines and methods of computation.

"Cybernetic friends" with L.A. Kaluzhnin at the head usually walked through forests and over hills having scientific discussions. Young science students were gladly welcomed to this environment. Victor Michailovich Glushkov, as I remember, did not participate in these walks, but scientific news and information were shared with him. Victor Michailovich evaluated the contribution of V.S. Korolyuk highly.

V.S. Korolyuk played a leading role in the creation of the Computing Centre of the Academy of Sciences which was reborn as the giant Institute of Cybernetics. Vladimir Semenovich provided assistance and influenced the development of computer programming, computing methods, mathematical models of real phenomena right up to radar-location. It seems that the development of the theory of Ukrainian cybernetics has not been truthfully described yet (in the Soviet Union history was not always recorded truthfully in detail).

In 1957, B.V. Gnedenko was Director of The Institute of Mathematics of The Ukrainian Academy of Sciences and Head of the Department of Probability Theory there, and at the same time Head of the Department of Probability Theory of Taras Shevchenko University. His deputies were V.S. Korolyuk at the Institute and I.I. Gikhman at the University. In 1961, after the departure of Boris Vladimirovich Gnedenko to Moscow they inherited these positions. Getting ready for leaving for Moscow B.V. Gnedenko said, 'I must entrust the Department to Korolyuk'.

I worked at this Department till 1962 when I moved to Moscow, not to work with B.V. Gnedenko at Moscow Lomonosov University, but to a defence organization, where all the employees were promised flats.

Boris Vladimirovich attended his Department once a week. On this day, which was obligatory for attendance, the workshop of the Department took place, and expert consultations for the postgraduates were given. Vladimir Semenovich Korolyuk, although he was not formally my supervisor helped me a lot. Many other postgraduates also felt his help greatly. Because of the initiative of B.V. Gnedenko, we studied in particular, queueing systems with time restrictions (literal English meaning is "systems with impatient customers"). This problem was set forth for postgraduates S.M. Brody and M.G. Farinich, who were one year ahead of me in their postgraduate studies. Brody solved two problems: service with bounded waiting time and service with bounded stay of a "customer" in a system in a $M/G/1$ scheme. As a first year postgraduate, I was not given any specific problem by my supervisor B.V. Gnedenko, and so I tried my abilities in different problems, selecting the more interesting ones. I took some problems from V.S. Michalevich undergoing on the postgraduate training by A.N. Kolmogorov. V.S. Korolyuk observed me rushing about without any definite direction and advised,'You should take problems with restrictions and invent the process which will unite all these problems'. He said nothing more. But this was sufficient for me, and in the same evening I found such a process. This incident characterizes the style of V.S. Korolyuk.

Problems associated with random walks led to the problem of the factorization of the kernel of an integral convolution equation. The question of the probabilistic interpretation of the factorization components arose in relation to this. This problem was given to me, and I obtained such an interpretation. The result of my work was published, and was even taken note of at a USSR Conference chaired by A.N. Kolmogorov. At that time D.V. Gusak, a postgraduate student of V.S. Korolyuk, started to study this topic seriously.

Another fact was that Vladimir Semenovich advised A.A. Borovkov to arm himself with the saddle-point method. And, being an analyst with great determination, A.A. Borovkov obtained by means of this method many important results associated with random walks. Vladimir Semenovich devoted a lot of his attention to queueing theory, or in the literal translation from Russian, "theory of mass customer service". In truth, he preferred to change the words "mass customer" and to say "theory of service *systems*". Indeed, this term is more precise. I remember, he advised me to study the famous article by Sir David Kendall, introducing into science the new notion "imbedded Markov processes". Possibly, Vladimir Semenovich was the first who realized in the middle of the fifties the importance of semi-Markov random processes, both in queueing and in the reliability theory of systems. He devoted great attention to all kinds (every reasonable possibility) of interpretations of semi-Markov processes and their functionals. V.S. Korolyuk was a teacher of the talented mathematician A.F. Turbin who perceived many of his ideas.

Kiev's mathematical school traditionally investigated small parameter methods and applied them whenever it was possible. The scientific school of nonlinear oscillations was created by N.N. Bogoliubov and Yu.A. Mitropolskii, and has obtained world fame. Vladimir Semenovich was an initiator of the investigations of asymptotic merging of phase states of semi-Markov random processes in a small parameter scheme. One day at the Institute of Mathematics a humorous wall newspaper was published. One of the caricatures was the following. Three men were carrying an enormous block with the inscription 'ε': two were ahead (these were local Professors) and only Vladimir Korolyuk with his legs giving way was behind. The caption to this caricature was: "What a heavy parameter! The load is hard although its volume is very small. But with the efforts of the two Professors it seems to be finished off".

Vladimir Korolyuk took hold of the hardest asymptotic problems. The formulas that came out were very complicated. It was easy to make a mistake. He often said,'If a resultant expression is simple then I will not make a mistake. This is my criterion'.

V.S. Korolyuk knows how to support, to hearten, and sometimes to drive a young scientist. I remember that B.V. Gnedenko was irreconcilable with someone's failure to attend his workshop and even to arrive on time. At that time I was married, and my older daughter was born. As all postgraduates, I did not lose a chance to earn some money on the side in addition to my scholarship (68 rubles per month). There was a woman teacher in one of the Universities who had maternity leave, and I took her practical course upon myself. (This activity was called "whimpering". It was said,'Here he comes to whimper'. To tell the truth, 'whimpering' was second only to the "training of dolts and blockheads" i.e. private tutoring). Once during the workshop's time at our Department I had to whimper. Boris Vladimirovich Gnedenko was very angry, as every other person in his place would have been. Vladimir Korolyuk protected me. In some time, I obtained results which were to the liking of B.V. Gnedenko, and so the misdeed was smoothed over.

During this time Vladimir Korolyuk actively took up social duties. He was the Head of the Trade Union of the Institute of Mathematics. He conducted staff meetings of the trade union bureau in a very business like way, and was intolerant towards verbosity. Vladimir Semenovich loved to have dinner at the restaurant "Cocktail Hall". Then the food was of very good quality, but in Kiev's restaurants you had to wait for an hour to be served with dinner. Vladimir Semenovich, at one time acted in the following way: he booked a dinner at the "Cocktail Hall", went to the Institute, conducted a staff meeting of the trade union bureau which lasted for forty minutes, and then came back to the restaurant exactly in time to eat his dinner.

Vladimir Semenovich reads a lot. He is interested in fiction, social and political journalism and many others. I think, he does not like long detailed descriptions, and so he tries to grab the main idea with a few sentences which he then considers to be a valuable find. I can still remember his witty explanation of "Parkinson's law".

Being a man of many interests and abilities, Vladimir Korolyuk always went in for tennis. At one time he was "ranked first" at the National Academy of Sciences in tennis.

Vladimir Korolyuk supported many mathematicians. In good time, such talented mathematicians as I.I. Ežov, B.B. Anisimov, and D.S. Silvestrov were supported by him a lot.

In Kiev, in the fifties and sixties, there were many scientific schools, groups and groupings. To put it mildly, all of them were unfriendly to each other. B.V. Gnedenko is a very kindhearted and responsive person, but he ended up in conflict with most of these groups, at least in the last period of his life in Kiev. His move to Moscow was not very peaceful. One of the people involved recalled in a private talk that, 'We trapped him in a cage with only one window as an exit. He had no way out except to go to Moscow'.

I can recollect how I was accepted to the position of Senior Research Fellow. I finished my postgraduate studies half a year ahead of schedule and obtained my PhD degree. The new director at The Institute of Mathematics Yu.A. Mitropolskii had a friendly attitude towards me from the very beginning, and Vladimir Korolyuk, himself wrote a character reference for me. (This old and tattered sheet of paper is still carefully kept by me now). The confirmation of my assignment took place at the Presidium of the Academy of Sciences.

V.S.Korolyuk (Deputy Director) and Yu.A.Mitropolskii (Director of Kiev's Institute of Mathematics)

Leading specialists of Kiev Probability School
from right to left:
Front row – **V.S. Korolyuk, A.V. Skorokhod, I.I. Gikhman, I.I. Ežov**
2nd row – **M.M. Diesperova, M.I.Portenko, M.I. Yadrenko, A.Ya. Dorogovtsev, G.N. Sytaya, D.V. Gusak**
3rd row – **A.F. Turbin, D.S. Selivestrov, A.A. Anisimov, R.V. Boiko, G.F. Butsan**

One of the former Vice-Presidents of the Academy examined the candidates to this position. He mounted the Presidium chair and we were allowed to enter the hall in turn and stood near the door at a distance of 15 metres from him, and were then questioned:

'Who is your supervisor?'

'Academician Gnedenko.'

'Did he really supervise your work?'

'Yes, he did supervise it.'

'Ha-ha!'

'And what is the subject of your dissertation?'

'It is ...'

'And what do you study there?'

'I generalize such results.'

'Well, then is the work compilation?' ...

This arch-academician did not understand what generalization means in mathematics; I could feel his unfriendly attitude towards B.V. Gnedenko.

I cannot remember one of Gnedenko's students becoming his adversary or behaving in an unfriendly way towards him. Vladimir Korolyuk was exemplary in this situation: he maintained both a deep respect for his teacher and for the traditions of the school. And on top of this Vladimir Korolyuk had a good relationship with his teacher's opponents. Generally, Korolyuk is distinguished by his ability to establish contacts with people, recognizing their best features. At the same time, I must emphasize that Vladimir Semenovich is a man of principle who believes in standing up for the truth, and will argue vehemently for it.

I described my impression of V.S. Korolyuk, relating to a short time interval, approximately from 1955 to 1962. In my so-called permanent memory, pictures of scientific and human contacts with him at that time appear in my mind most distinctly, although 30 years have passed. Memories of him delivering a lecture in programming; of him supervising the newly introduced practical professional training of the fourth year students at the University and, together with us, shaking in the bus (which was said to have been used previously for a funeral) on the road to the Computing Center in Feofania. I have memories of him at the USSR Congress of Mathematicians in Moscow in 1956 (there were present some of us students, as representatives of The Students Scientific Society). I remember when we swam together with Kolmogorov and Michalevich in the Dnieper. I remember him at the USSR Conference in Erevan, Tbilisi, and Uzhgorod. (Famous mathematician Sir David Kendall from Cambridge has the photograph of the Tbilisi Conference, where in front of the auditorium there are Kolmogorov, Gnedenko and Korolyuk). I have memories of him with his "comrades in arms" I.I. Gikhman, A.V. Skorokhod, M.I. Yadrenko, and others. I remember Vladimir Semenovich in a cybernetic group with L.A. Kaluzhnin at the head. And I have many other pictures connected with this marvellous man treasured in my "permanent" memory.

To his good health, happiness and future success!

May 5, 1995

Igor N. Kovalenko

Exploring Stochastic Laws. pp. xxix–xxxv.
Eds. A.V. Skorokhod and Yu.V. Borovskikh

"Forward, and only forward!"

M.M. DIESPEROVA*

Institute of Mathematics, National Academy of Sciences of Ukraine, Tereschenkivska St. 3, 252601 Kiev-4, Ukraine

Received May 20, 1995

At the beginning of the nineteen fifties Boris Vladimirovich Gnedenko selected some very gifted students, who formed the basis of "Gnedenko's Probability School". In 1953 he was invited to Germany, and in order to preserve his "school", he asked his teacher and friend A.N. Kolmogorov to take care of his alumni V.S. Korolyuk, V.S. Michalevich, and A.V. Skorokhod. All of them were assigned to M.V. Lomonosov University in Moscow. The postgraduate studies of V.S. Korolyuk and V.S. Michalevich were supervised directly by A.N. Kolmogorov, and A.V. Skorokhod's studies were passed over to E.B. Dynkin, who of course, has never regreted it for a single day. B.V. Gnedenko treated his students as if they were brothers, and never left them to the mercy of fate.

Very often Andrei Nikolaevich Kolmogorov invited his postgraduate students to his country home. When it was winter long skiing races were organized, and when it was summer, swimming competitions took place. It gave an opportunity for all the postgraduates to become familiar with one another in an informal environment. At that time Kolmogorov's postgraduates were Yu.V. Prokhorov, and L.N. Bolshev. Kolmogorov supervised the postdoctoral studies of S.H. Sirazhdinov, and V.M. Zolotarev was preparing for his postgraduate studies. They were all very extraordinary people and "Kolmogorov's brotherhood", as it was called, was retained throughout their lives.

Andrei Nikolaevich loved classical music, especially Bach, and he endeavored to initiate his students into this mystery. It was a tradition practiced for years. Gnedenko loved classical music as well. When students assembled at his house he would often switch on the record player, and together they would listen to the music of Tchaikovsky, Saint-Saënc and Bach. When Gnedenko's alumni came to Moscow, the new building of Moscow University was shown to them, and very often, Kirill

* Translated by O.A. Cherkashina (Adamenko)

Kondrashin's symphony orchestra performed music in the Great Assembly Hall in the University. Of course such concerts were attended by Kolmogorov's students regularly.

Participants of the Third USSR Conference in Probability Theory and Mathematical Statistics, 28th Sept-2nd Oct, 1953
from right to left:
Front row – **E.L. Yushchenko** (first), **B.V. Gnedenko** (third), **E.B. Dynkin** (seventh), **S.H. Sirazhdinov** (eighth);
2nd row – **M.G Krein**, **N.V. Smirnov**, **A.N. Kolmogorov**, **A.M. Yaglom**;
top row – **Yu.V. Linnik** (third), **G.M. Mania**, **I.I. Gikhman**;
left upper corner – **V.S. Michalevich**, **A.V. Skorokhod**, **V.S. Korolyuk**

A.N. Kolmogorov often invited V.S. Korolyuk and V.S. Michalevich to his dacha in Komarovka and sometimes they would ski distances of twenty kilometres. The young men would get tired, but A.N. Kolmogorov would not. He told his postgraduates that they had to come to his dacha and to ski with him many times again. Reports and new results were usually discussed in Komarovka.

Andrei Nikolaevich tried to unite the theoretical knowledge of his students with practice. He told them at his workshop that Probability Theory and Mathematical Statistics was the most progressive science with unexpected forks and an extensive scope of applications.

He asked Korolyuk and Michalevich to give advice to two engineers who designed a new lithographic machine. For the calculations they used some statistical methods but reached an impasse. Korolyuk and Michalevich found a mistake and proposed another, more successful way for the calculations.

At the end of Korolyuk's postgraduate studies Andrei Nikolaevich Kolmogorov discussed with him the prospects of his future direction in mathematics. Kolmogorov believed that Korolyuk had all the right abilities to start work on introducing probability theory into computer mathematics and to become the best specialist in this field in the Soviet Union. Later, A.N. Kolmogorov accepted V.S. Korolyuk for postdoctoral studies. He evaluated the abilities of V.S. Korolyuk very highly.

When V.S. Korolyuk, V.S. Michalevich and A.V. Skorokhod lived at Moscow University, their funds were very limited, and so V.S. Korolyuk was the 'treasurer' of this group. Usually, at the start of the month they paid for their dinners at the Professor's dining hall for the whole month. The rest of the money was given to them by V.S. Korolyuk according to their needs. The three of them lived as a commune, where they were all on duty for two days a week cooking breakfast in the morning and boiling tea in the evening.

V.S.Michalevich, A.V.Skorokhod (behind); A.G.Kostjuchenko, V.S.Korolyuk
("Two V's and two A's")

These 'three pilgrims' travelled to Moscow, which was the scientific and cultural center of the USSR, where not only the best scientific institutions worked but where also some of the world's best theatres and museums were located. The bright lights of Moscow attracted many young and energetic people, including these three young, talented and innocent men. They planned to take full advantage of these cultural and spiritual treasures.

As I remember, the defence of Korolyuk's thesis for his Doctor of Mathematics degree took place on 29th of June, 1954. Andrei Nikolaevich came to Kiev to see his defence. He addressed the meeting with a very laudatory speech about the dissertation and expressed certainty about the future scientific success of the candidate for degree. Vladimir Semenovich lived at that time with his mother and family in a small house on Vladimirskaya Street. After the defence of his thesis all his colleagues went to the house to celebrate. Tables were set up in the garden, and very jubilantly we congratulated Vladimir with the success of his defence. The next day we all went to Desna together to observe the eclipse of the sun, 'in honour' of Vladimir Korolyuk.

At this time, interest in mathematical statistics had only touched the surface. Andrei Nikolaevich Kolmogorov endeavoured to create an engineering-mathematical symbiosis. His students went to Gorky automobile factory, and had discussions with engineers there. They then attempted to solve problems of acceptance testing in a new way. After the defence of his thesis Vladimir Korolyuk remained as the Head of Probability Theory of Kiev Taras Shevchenko University. He arranged for the cooperation of a large mechanical engineering factory in Dnepropetrovsk, and developed statistical methods of data processing for this factory.

At that time Vladimir Korolyuk's motto was, "Forward, and only forward!" I can recall that in February of 1956 we went to Vorokhta in the Carpathian Mountains. We climbed midway up Goverla mountain, and reached a shepherd's house. It was still a long way to the top, and a slight snowstorm had just begun to start. Visibility was getting bad, but Vladimir Semenovich exclaimed, "Is it possible to be so close to the top and not to reach it?! Never! Forward, and only forward!" So, he together with V.M. Zolotarev skied up to the top, and V.S. Michalevich and I stayed at the shepherd's house. I was afraid for their lives, but eventually they came back very tired and satisfied with their achievement. Later they confessed that they were also afraid, because visibility was bad and the cornice along which they had climbed had precipices and deep abysses on both sides.

Vladimir Korolyuk was called 'an admirer of asymptotics' by his friends, because at that time he was enthusiastic about very many subjects, including Freud, programming and sport. A.V. Skorokhod once said in one of his epigrams:

"Adoration and delight
Leave his body with excite,
His inner soul does not forget
Doubt, sorrow and regret,
For every second things clutter his brain –
Breakfast, Russell, sex and Krein!"

In 1956, B.V. Gnedenko offered Victor Michailovich Glushkov the position of Head of the Laboratory of Computing Mathematics at the Institute of Mathematics of the Ukrainian Academy of Sciences. At that time Boris Vladimirovich was the Director of the Institute of Mathematics (Ishlinsky had moved to Moscow already). Victor Michailovich had many new ideas regarding the development of computing mathematics and computers. Together they often discussed how, from the Laboratory of Computing Mathematics, a new research institute, the Computing Center of the Academy of Sciences could be created. Boris Vladimirovich Gnedenko who was very busy as the Director of the Institute of Mathematics, had many foreign business trips, and did not like administrative work. That is why V.M. Glushkov carried on his shoulders the weight of organizing the Computing Center. He was the perfect organizer, very energetic and purposeful, and, in addition, a brilliant mathematician who was always ready to come out with a whole cascade of new ideas.

When it was revealed that when asymptotic decomposition is applied directly to Laplace transforms it does not give a desirable result to the problem of maximal deviation of random walks, Korolyuk became carried away for some time in programming. This happened in 1956-57.

The Laboratory of Computing Methods at the Institute of Mathematics was located in Feofania (outskirts of Kiev), and Vladimir Korolyuk with his family lived there in two rooms (if it is possible to call them 'rooms') of a small summer cottage with heating by stove, without many modern conveniences. The Head of the laboratory was V.M. Glushkov. Workshops at the laboratory took place very often. Vladimir Korolyuk taught students at the University and did his best to make his schedule suitable for him to attend Glushkov's workshops. He succeeded in programming so quickly that V.M. Glushkov invited him to create a Computing Centre at the Academy of Sciences and tempted him by various financial benefits suggesting that Korolyuk create his own Department at the Computing Centre. Most likely, this was a moral question. We had to give to Korolyuk what he was due, because he had always remained faithful to his teacher B.V.Gnedenko.

In the summer of 1957 B.V. Gnedenko, V.M. Glushkov, L.A. Kaluzhnin, V.S. Korolyuk, V.S. Michalevich and A.V. Skorokhod went to rest on the Desna river, an inflow of the Dnieper, upstream from Chernigov. Next to the Desna they pitched a soldiers' tent and put a huge log which served as a table, with the slogan, "Never set foot on the table". They then started to fish. In connection to this there was another slogan, "The only fish will be sausage". Fishing was great. The most successful was Michalevich. Generally, he was an avid fisherman and hunter. He did his best to initiate Korolyuk and Skorokhod into these activities. Although Korolyuk is the kind of person that easily gets carried away, he never become a fisherman or a hunter.

A.V. Skorokhod, V.S. Korolyuk and I.I.Gikhman

At that time, by the Desna, the words "Institute of Cybernetics" were mentioned for the very first time. To tell the truth, Boris Vladimirovich Gnedenko wanted to be the Director of the Computing Centre which was created by V.M. Glushkov, but Glushkov did not allow him. It was his brainchild, his creature and he was not going to refuse the post. That is why the question of who will remain at the Institute of Mathematics, and who will go to the Computing Center, changed from a practical to a moral question. Victor Michailovich Glushkov attempted to persuade V.S. Korolyuk to move to the Computing Center, and in some respects this was in Korolyuk's interests. When he understood that his proper place was at the Institute of Mathematics, with Gnedenko he quit programming and became seriously occupied with the investigation of the true nature of asymptotics which brought him brilliant results, and to the defence of his second thesis, Doctor Dissertation, and then to the creation of his own scientific school.

At that time, by the Desna, these disagreements between Gnedenko and Glushkov were only embryonic. The company was mainly harmonious and cheerful. They discussed many questions, mathematical and political, literary and musical. Boris Vladimirovich Gnedenko knew classical music very well, Victor Michailovich Glushkov knew many poems including Dante Alighieri. Michalevich glittered with wit, Skorokhod sparkled with intellect and Korolyuk possessed inexhaustible enthusiasm and energy. Lev Arkadievich told about his summer cottage in the suburbs of Berlin, near some beautiful lakes and about his yachting adventures. Of course, we did not have a yacht, but we had a boat which we went fishing on. Lev Arkadievich, who had almost got used to his new life in Ukraine was entirely absorbed into this group.

Being young spiritually, Vladimir Korolyuk is always open for friendly relationships, which is why there are always many interesting people around him. He gives to other people something which is very special and unique. He gives them a different vision of the world, and he makes them see the many fascinating things of life, which are easily missed by the normal person, leading a normal life.

Let Vladimir Semenovich Korolyuk continue to make his friends happy with the treasures of his soul and the new ideas of his mind for many many years to come.

"Forward, and only forward!"
"Вперед, и только вперед!"

PAPERS

Exploring Stochastic Laws. pp. 3–12.
Eds. A.V. Skorokhod and Yu.V. Borovskikh

Convergence of branching processes with several types of particles to Jirina processes

S.A. ALIEV[1] AND V.M. SHURENKOV

[1] *Azerbaijan, Baku, pr. Azadlyk 163, apt.37*

Received January 5, 1995

Abstract. In this paper there are formulated and proved three theorems about limit behaviour of finite-dimensional distributions of Markov branching processes with a finite number of types.

Let $\vec{\xi}_n(\varepsilon) = (\xi_n^1(\varepsilon), \dots \xi_n^d(\varepsilon))$, $n = 0, 1, 2, \dots$ be a Markov branching process with discrete time, d types of particles and generating function $\vec{F}_\varepsilon(\vec{s}) = (F_\varepsilon^1(\vec{s}), \dots, F_\varepsilon^d(\vec{s}))$, where ε is a series parameter, $\vec{s} = (s_1, \dots, s_d)$,

$$F_\varepsilon^j(\vec{s}) = E_j s_1^{\xi_1^1(\varepsilon)} \dots s_1^{\xi_1^d(\varepsilon)}.$$

Component $\xi_n^j(\varepsilon)$ is interpreted as the number of particles of j-th type in the n-th generation and E_j is the conditional mathematical expectation supposing that at the initial moment of time there was one particle of j-th type. We shall denote the average number of descendants of j-th type of one particle of i-th type by $a_{ij}(\varepsilon)$, i.e.

$$a_{ij}(\varepsilon) = E[\xi_1^j(\varepsilon) \mid \vec{\xi}_0(\varepsilon) = \vec{e}_i], \quad \vec{e}_i = (\underbrace{0, \dots, 0}_{i-1},\ 1,\ \underbrace{0, \dots, 0}_{d-i}).$$

Consider the matrix

$$A(\varepsilon) = \|a_{ij}(\varepsilon)\|_{i,j=1}^d$$

and suppose that the following decomposition holds

$$A(\varepsilon) = I + \varepsilon C + o(\varepsilon), \tag{1}$$

where $I = \|\delta_{ij}\|_{i,j=1}^d$ is an identity matrix and $C = \|c_{ij}\|_{i,j=1}^d$ is the matrix of some constant elements c_{ij}.

Choose normalizing constants b_ε^j such that $b_\varepsilon^j \to \infty$ when $\varepsilon \to 0$, and consider the family of random processes $\vec{\mu}_t(\varepsilon) = (\mu_t^1(\varepsilon), \dots, \mu_t^d(\varepsilon))$, where $\mu_t^j(\varepsilon) = \xi_{[t/\varepsilon]}^j(\varepsilon)/b_\varepsilon^j$, $\xi_0^j(\varepsilon) = [x^j, b_\varepsilon^j]$, $\vec{0} < \vec{x} = (x^1, \dots, x^d)$.

We introduce the definition and some properties of Jirina processes with several types of particles before formulating a theorem about the convergence of processes $\vec{\mu}_t(\varepsilon)$. This theorem is a multidimensional analogue of the corresponding result from [1].

Recall that a Markov process $\vec{\mu}_t = (\mu_t^1, \dots, \mu_t^d)$ is called an Jirina process with d types of particles if

$$E\left[e^{\langle \vec{\lambda}, \vec{\mu}_t \rangle} \mid \vec{\mu}_0 = \vec{x}\right] = e^{\langle \vec{x}, \vec{K}(t, \vec{\lambda}) \rangle},$$

where

$$\vec{0} < \vec{x} = (x^1, \dots, x^d),$$
$$\vec{0} > \vec{\lambda} = (\lambda^1, \dots, \lambda^d),$$
$$\langle \vec{x}, \vec{\lambda} \rangle = \sum_{i=1}^{d} x^i \lambda^i,$$
$$\vec{K}(t, \vec{\lambda}) = (K^1(t, \vec{\lambda}), \dots, K^d(t, \vec{\lambda})),$$

and $K^j(t, \vec{\lambda})$ for all $j = 1, \dots, d$ is the logarithm of a Laplace transformation of some infinitely divisible distribution. Vector $\vec{K}(t, \vec{\lambda})$ satisfies the following system of ordinary differential equations

$$\frac{\partial \vec{K}(t, \vec{\lambda})}{\partial t} = \vec{H}(\vec{K}(t, \vec{\lambda})), \quad \vec{K}(0, \vec{\lambda}) = \vec{\lambda}, \tag{2}$$

where

$$\vec{H}(\vec{\lambda}) = (H^1(\vec{\lambda}), \dots, H^d(\vec{\lambda}))$$

is a cumulant of an Jirina process with d types of particles.

These and others properties of Jirina processes with several types of particles can be found in [2].

Theorem 1. *Assume that decomposition* (1) *takes place. If the conditions*

$$\frac{1}{\varepsilon} b_\varepsilon^j [F_\varepsilon^j (e^{\lambda^1 / b_\varepsilon^1}, \dots, e^{\lambda^d / b_\varepsilon^d}) - e^{\lambda^j / b_\varepsilon^j}] \to H^j(\vec{\lambda}), \quad \varepsilon \to 0 \tag{3}$$

and

$$\left. \frac{\partial H^j(\vec{\lambda})}{\partial \lambda^i} \right|_{\vec{\lambda} = \vec{0}} = c_{ij},$$

are satisfied for $\vec{0} \geq \vec{\lambda} = (\lambda^1, \dots, \lambda^d)$, then finite-dimensional distributions of processes $\vec{\mu}_t(\varepsilon)$ weakly converge to finite-dimensional distributions of stochastically continuous Jirina process $\vec{\mu}_t = (\mu_t^1, \dots, \mu_t^d)$, $\vec{\mu}_0 = \vec{x}$ with the cumulant $\vec{H}(\vec{\lambda}) = (H^1(\vec{\lambda}), \dots, H^d(\vec{\lambda}))$, $\varepsilon \to 0$.

Proof. Let

$$H_\varepsilon^j(\vec{\lambda}) = \frac{1}{\varepsilon} b_\varepsilon^j \log \left\{ F_\varepsilon^j (e^{\lambda^1 / b_\varepsilon^1}, \dots, e^{\lambda^d / b_\varepsilon^d}) e^{-\lambda^j / b_\varepsilon^j} \right\}.$$

If $F^j_{k,\varepsilon}(\vec{s}) = E_j s_1^{\xi^1_k(\varepsilon)} \dots s_d^{\xi^d_k(\varepsilon)}$, then $\vec{F}_{k,\varepsilon}(\vec{s})$ is the k-th iteration of generating function $\vec{F}_\varepsilon(\vec{s})$, i.e. $\vec{F}_{k+1,\varepsilon}(\vec{s}) = \vec{F}_\varepsilon(\vec{F}_{k,\varepsilon}(\vec{s}))$. Therefore

$$b^j_\varepsilon \log F^j_{k+1,\varepsilon}(e^{\lambda^1/b^1_\varepsilon}, \dots, e^{\lambda^d/b^d_\varepsilon}) =$$

$$= \varepsilon H^j_\varepsilon \left(b^1_\varepsilon \log F^1_{k,\varepsilon}(e^{\lambda^1/b^1_\varepsilon}, \dots, e^{\lambda^d/b^d_\varepsilon}), \dots, b^d_\varepsilon \log F^d_{k,\varepsilon}(e^{\lambda^1/b^1_\varepsilon}, \dots, e^{\lambda^d/b^d_\varepsilon})\right) +$$

$$+ b^j_\varepsilon \log F^j_{k,\varepsilon}(e^{\lambda^1/b^1_\varepsilon}, \dots, e^{\lambda^d/b^d_\varepsilon}).$$

Put $k = [t/\varepsilon]$ and denote

$$K^j(t, \vec{\lambda}) = b^j_\varepsilon \log F^j_{[t/\varepsilon],\varepsilon}(e^{\lambda^1/b^1_\varepsilon}, \dots, e^{\lambda^d/b^d_\varepsilon}),$$

then, from the last equality, we obtain

$$K^j_\varepsilon(t+\varepsilon, \vec{\lambda}) - K^j_\varepsilon(t, \vec{\lambda}) = \varepsilon H^j_\varepsilon(\vec{K}_\varepsilon(i\varepsilon, \vec{\lambda})). \tag{4}$$

Taking $t = 0, \varepsilon, 2\varepsilon, \dots, (k-1)\varepsilon$ and summing the resulting inequalities, we obtain

$$K^j_\varepsilon(t, \vec{\lambda}) = \lambda^j + \varepsilon \sum_{i=0}^{[t/\varepsilon]-1} H^j_\varepsilon(\vec{K}_\varepsilon(i\varepsilon, \vec{\lambda})). \tag{5}$$

On the other hand, from differential equation (2) it follows that

$$\vec{K}(t, \vec{\lambda}) = \vec{\lambda} + \int_{[t/\varepsilon]\varepsilon}^{t} \vec{H}(\vec{K}(u, \vec{\lambda}))du + \int_0^{[t/\varepsilon]\varepsilon} \vec{H}(\vec{K}(u, \vec{\lambda}))du -$$

$$-\varepsilon \sum_{i=0}^{[t/\varepsilon]-1} \vec{H}(\vec{K}(i\varepsilon, \vec{\lambda})) + \varepsilon \sum_{i=0}^{[t/\varepsilon]-1} \vec{H}(\vec{K}(i\varepsilon, \vec{\lambda})).$$

Denote the norm of the sum of the first three terms by $\gamma_\varepsilon(t, \vec{\lambda})$. It is clear that $\gamma_\varepsilon(t, \vec{\lambda}) \to 0$ when $\varepsilon \to 0$ uniformly in t and λ. Denote $\Delta_\varepsilon(t, \vec{\lambda}) = |\vec{K}_\varepsilon(t, \vec{\lambda}) - \vec{K}(t, \vec{\lambda})|$.

It follows from the last two relations that

$$\Delta_\varepsilon(t, \vec{\lambda}) = \sum_{j=1}^{d} |K^j_\varepsilon(t, \vec{\lambda}) - K^j(t, \vec{\lambda})| \le \gamma_\varepsilon(t, \vec{\lambda}) +$$

$$+\varepsilon \sum_{j=1}^{d} \sum_{i=0}^{[t/\varepsilon]-1} |H^j(\vec{K}(i\varepsilon, \vec{\lambda})) - H^j_\varepsilon(\vec{K}(i\varepsilon, \vec{\lambda}))| +$$

$$+\varepsilon \sum_{j=1}^{d} \sum_{i=0}^{[t/\varepsilon]-1} |H^j_\varepsilon(\vec{K}(i\varepsilon, \vec{\lambda})) - H^j_\varepsilon(\vec{K}(i\varepsilon, \vec{\lambda}))|.$$

The second term on the right hand side of this formula does not exceed $t\delta_\varepsilon$, where $\delta_\varepsilon \to 0$ when $\varepsilon \to 0$, because relation (3) holds true uniformly in $\vec{\lambda}$ on every finite interval.

Further,

$$|H_\varepsilon^j(\vec{K}(i\varepsilon,\vec{\lambda}) - H_\varepsilon^j(\vec{K}_\varepsilon(i\varepsilon,\vec{\lambda}))| \leq \sum_i \left|\frac{\partial H_\varepsilon^j(\vec{\lambda}_\varepsilon)}{\partial\lambda^i}\right| \; |\vec{K}^i(i\varepsilon,\vec{\lambda}) - \vec{K}_\varepsilon^i(i\varepsilon,\vec{\lambda})|$$

where $\min\{K^j(i\varepsilon,\vec{\lambda}), K_\varepsilon^j(i\varepsilon,\vec{\lambda})\} \leq \lambda_\varepsilon^j \leq \max\{K^j(i\varepsilon,\vec{\lambda}), K_\varepsilon^j(i\varepsilon,\vec{\lambda})\}$.

Because on every finite interval of variation

$$\left|\frac{\partial H_\varepsilon^j(\vec{\lambda}_\varepsilon)}{\partial\lambda^i}\right| \leq \sup \|\vec{H}_\varepsilon'(\vec{\lambda})\| < \infty, \tag{6}$$

the right hand side of the last inequality does not exceed $c\Delta_\varepsilon(i\varepsilon,\vec{\lambda})$, $c = const$. Finally, we have

$$\Delta_\varepsilon(t,\vec{\lambda}) \leq \gamma_\varepsilon(t,\vec{\lambda}) + t\delta_\varepsilon + \varepsilon c \sum_{i=0}^{[t/\varepsilon]-1} \Delta_\varepsilon(i\varepsilon,\vec{\lambda}).$$

On the finite intervals of the variation of t and $\vec{\lambda}$ the sum $\gamma_\varepsilon(t,\vec{\lambda}) + t\delta_\varepsilon$ can be estimated from above by the value $\alpha_\varepsilon \to 0$. Therefore

$$\Delta_\varepsilon(t,\vec{\lambda}) \leq \alpha_\varepsilon + \varepsilon c \sum_{i=0}^{[t/\varepsilon]-1} \Delta_\varepsilon(i\varepsilon,\vec{\lambda}).$$

In just the same way as in [1], from this inequality we obtain

$$\Delta_\varepsilon(t,\vec{\lambda}) \leq \alpha_\varepsilon e^{ct} \to 0,$$

or, which is the same,

$$K_\varepsilon^j(t,\vec{\lambda}) \to K^j(t,\vec{\lambda}),$$

when $\varepsilon \to 0$.

Therefore, it is easy to see that

$$\left\{F_{[t/\varepsilon],\varepsilon}^j(e^{\lambda^1/b_\varepsilon^1},\dots,e^{\lambda^d/b_\varepsilon^d})\right\}^{[x^j b_\varepsilon^j]} \to e^{x^j K^j(t,\vec{\lambda})}. \tag{7}$$

The left hand side of (7) is equal to $E\{e^{\lambda^j \xi_{[t/\varepsilon]}^j(\varepsilon)/b_\varepsilon^j} \mid \xi_0^j(\varepsilon) = [x^j b_\varepsilon^j]\}$ and, consequently, relation (7) means that one-dimensional distributions of random processes $\vec{\mu}_t(\varepsilon)$ weakly converge to a one-dimensional stochastically continuous Jirina process with cumulant $\vec{H}(\vec{\lambda})$ when $\varepsilon \to 0$. Further, from the identity

$$\vec{F}_{k_1,\dots,k_r,\varepsilon}(\vec{s}^1,\dots,\vec{s}^r) = \vec{F}_{k_1,\varepsilon}(\vec{s}^1\vec{F}_{k_2-k_1,\dots,k_r-k_1}(\vec{s}^2,\dots,\vec{s}^r)),$$

where

$$\vec{s}\vec{F} = (s_1F^1,\dots,s_dF^d),$$

$$F_{k_1,\dots k_r,\varepsilon}(\vec{s}^1,\dots,\vec{s}^d) = E(\vec{s}^1)^{\vec{\xi}_{k_1}(\varepsilon)}\dots(\vec{s}^r)^{\vec{\xi}_{k_r}(\varepsilon)}, \quad k_1 < \dots < k_r,$$

the convergence of finite-dimensional distributions follows by induction.

Now we consider another extreme situation when

$$A(\varepsilon) \to A, \quad \varepsilon \to 0, \tag{8}$$

where matrix A is indecomposable and it's Perron's root is equal to 1. Denote by $\vec{u} = (u_1, \dots, u_d)$ and $\vec{v} = (v_1, \dots, v_d)$ the right and left, respectively, invariant vectors of matrix A with positive coordinates and normalizing conditions

$$\sum_{i=1}^{d} u_i = 1, \quad \sum_{i=1}^{d} v_i u_i = 1.$$

Choose normalizing $b_\varepsilon \to \infty$ when $\varepsilon \to 0$ and put

$$H_\varepsilon(\lambda) = \frac{1}{\varepsilon} b_\varepsilon \left[\sum_k v_k F_r^k(e^{\lambda u_1/b_\varepsilon}, \dots, e^{\lambda u_d/b_\varepsilon}) - e^{\lambda/b_\varepsilon} \right], \quad \lambda \leq 0.$$

Assume that when $\varepsilon \to 0$ the following decomposition takes place

$$\sum_{k,j} u_k a_{kj}(\varepsilon) u_j = 1 + \varepsilon c + o(\varepsilon), \tag{9}$$

where c is a constant.

Consider the family of random processes $\nu_t(\varepsilon) = (\nu_t^1(\varepsilon), \dots, \nu_t^d(\varepsilon))$, where $\nu_t^i(\varepsilon) = \xi_{[t/\varepsilon]}^i(\varepsilon)/b_\varepsilon$, $\xi_0^i(\varepsilon) = [x^i b_\varepsilon]$, $x^i \geq 0$, $\sum_i x^i > 0$.

Theorem 2. *Assume, that under conditions* (8) *and* (9),

$$H_\varepsilon(\lambda) \to H(\lambda), \quad H'(0) = c, \tag{10}$$

then finite-dimensional distributions of processes $\vec{\nu}_t(\varepsilon)$ weakly converge to finite-dimensional distributions of process $\mu_t \vec{v}$, where μ_t is a one-dimensional Jirina process with cumulant $H(\lambda)$ and $\mu_0 = \sum_{i=1}^d u_i x^i$.

To prove this Theorem we need the next lemma. Let $c_k(\varepsilon)$ be a sequence of matrices satisfying the following conditions

$$0 \leq c_k(\varepsilon) \leq A(\varepsilon), \tag{11}$$

$$\lim_{\substack{\varepsilon \to 0 \\ k \to \infty}} c_k(\varepsilon) = 0. \tag{12}$$

Consider the product

$$\begin{aligned} B_n(\varepsilon) &= \prod_{k=1}^{n} [A(\varepsilon) - c_k(\varepsilon)] \\ &= [A(\varepsilon) - c_n(\varepsilon)] \dots [A(\varepsilon) - c_1(\varepsilon)], \quad n = 1, 2, \dots \end{aligned}$$

Assume that $\vec{x} \geq 0$ is a vector such that $B_n(\varepsilon)\vec{x} \neq 0$ for all $n \geq 1$.

Lemma. *Under conditions* (8), (11) *and* (12)

$$\lim_{\substack{\varepsilon \to 0 \\ n \to \infty}} \frac{B_n(\varepsilon)\vec{x}}{\langle \vec{v} B_n(\varepsilon)\vec{x}\rangle} = \vec{u}.$$

Proof. Because $u_i,\ v_j > 0$, then, according to [3, p.193], it is possible to choose a sequence $\delta_n \to 0$, $\delta_n \geq 0$ such that

$$(1-\delta_n)D \leq A^n \leq (1+\delta_n)D,$$

$$D = \|u_i u_j\|_{i,j=1}^d.$$

Condition (8) implies that there exists

$$\beta(\varepsilon) \to 0, \quad \varepsilon \to 0,$$

such that

$$A - \beta(\varepsilon)D \leq A(\varepsilon) \leq A + \beta(\varepsilon)D.$$

Further, condition (12) implies that there exists a sequence

$$\gamma_k(\varepsilon) \to 0, \quad \varepsilon \to 0,\ k \to \infty,$$

such that

$$c_k(\varepsilon) \leq \gamma_k(\varepsilon)D.$$

From the last two inequalities it follows that

$$A - [\beta(\varepsilon) + \gamma_k(\varepsilon)]D \leq A(\varepsilon) - c_k(\varepsilon) \leq A + \beta(\varepsilon)D.$$

Denote $\alpha_k(\varepsilon) = \beta(\varepsilon) + \gamma_k(\varepsilon)$. By analogy to [3, p.193] we obtain that

$$\begin{aligned}\prod_{k=1}^n [A(\varepsilon) - c_k(\varepsilon)] &\geq \prod_{k=1}^n [A - \alpha_k(\varepsilon)D] \\ &\geq A^n - \sum_{k=1}^n \alpha_k(\varepsilon)D \\ &\geq [1 - \delta_n - \sum_{k=1}^n \alpha_k(\varepsilon)]D.\end{aligned}$$

On the other hand, since $AD = DA = D = D^2$, then

$$\begin{aligned}\prod_{k=1}^n [A(\varepsilon) - c_k(\varepsilon)] &\leq [A + \beta(\varepsilon)D]^n \\ &= A^n + [(1+\beta(\varepsilon))^n - 1]D \\ &\leq [\delta_n + (1+\beta(\varepsilon))^n]D.\end{aligned}$$

Fix $0 \le m \le n$. From the last two inequalities we obtain

$$\begin{aligned}[1-\delta_n - \sum_{k=n-m+1}^{n} \alpha_k(\varepsilon)]D &\le \prod_{k=n-m+1}^{n} [A(\varepsilon) - c_k(\varepsilon)] \\ &\le [\delta_m + (1+\beta(\varepsilon))^m]D.\end{aligned}$$

Denote $\vec{w} = B_{n-m}(\varepsilon)\vec{x}$. Apply this inequality to

$$B_n(\varepsilon)\vec{x} = [A(\varepsilon) - C_n(\varepsilon)]\dots[A(\varepsilon) - C_{n-m+1}(\varepsilon)]\vec{w}$$

and to $\vec{v}B_n(\varepsilon)\vec{x}$. Taking into account that

$$\frac{D\vec{w}}{\langle \vec{v}D\vec{w}\rangle} = \vec{u}$$

for arbitrary $\vec{w}$, we obtain

$$\begin{aligned}\frac{1-\delta_m - \sum_{k=n-m+1}^{n}\alpha_k(\varepsilon)}{\delta_m + (1+\beta(\varepsilon))^m}\vec{u} &\le \frac{B_n(\varepsilon)\vec{x}}{\langle \vec{v}B_n(\varepsilon)\vec{x}\rangle} \\ &\le \frac{\delta_m + (1+\beta(\varepsilon))^m}{1-\delta_m - \sum_{k=n-m+1}^{n}\alpha_k(\varepsilon)}\vec{u}.\end{aligned}$$

Taking the limit when $\varepsilon \to 0$, $n \to \infty$, and, then, when $m \to \infty$, we obtain the required equality.

Proof of Theorem 2. With the help of the proved lemma and repeating word for word the reasoning of [3, p.195], we obtain that

$$\lim_{\substack{\varepsilon\to 0\\ t\to\infty}} \frac{1-F_\varepsilon^j(t,\vec{s})}{\sum_k\{v_k[1-F_\varepsilon^k(t,\vec{s})]\}} = u^j \tag{13}$$

uniformly in $|s^j| \le 1$, $\vec{s} \ne \vec{1}$.

Further, to prove the Theorem it is sufficient to show that

$$\lim_{\varepsilon\to 0} b_\varepsilon \log F^j_{[t/\varepsilon],\varepsilon}(e^{\lambda^1/b_\varepsilon},\dots,e^{\lambda^d/b_\varepsilon}) = u^j K(t,\langle \vec{v},\vec{\lambda}\rangle), \tag{14}$$

for all $t > 0$, $\vec{\lambda} \le \vec{0}$, where $K(t,\lambda)$ is defined by function $H(\lambda)$ from (10) and differential equation

$$\begin{aligned}\frac{\partial K(t,\lambda)}{\partial t} &= H(K(t,\lambda)),\\ K(0,\lambda) &= \lambda.\end{aligned} \tag{15}$$

Denote the expression under the 'limit'-sign of (14) by $K_\varepsilon^i(t,\lambda)$ and put $K_\varepsilon(t,\vec{\lambda}) = \sum_i v_i K_\varepsilon^i(t,\vec{\lambda})$. From (13), it follows that

$$\lim_{\varepsilon\to 0}\frac{K_\varepsilon^i(t,\vec{\lambda})}{K_\varepsilon(t,\vec{\lambda})} = u^i \tag{16}$$

for every $t > 0$ uniformly in $\vec{\lambda} \leq \vec{0}$, $\vec{\lambda} \neq \vec{0}$.

Introduce the following notations

$$\begin{aligned} H_\varepsilon^j(\vec{\lambda}) &= \frac{1}{\varepsilon} b_\varepsilon \log \{F_\varepsilon^j(e^{\lambda^1/b_\varepsilon}, \ldots, e^{\lambda^d/b_\varepsilon}) e^{-\lambda^j/b_\varepsilon}\}, \\ H_\varepsilon^j(\lambda) &= H_\varepsilon^j(\lambda\vec{u}), \\ H_\varepsilon(\vec{\lambda}) &= \sum_{k=1}^d v_k H_\varepsilon^k(\vec{\lambda}), \\ H_\varepsilon(\lambda) &= H_\varepsilon(\lambda\vec{u}). \end{aligned}$$

Note that relations (4), (5) and (6) hold true in this situation as well. From (5), it is easy to obtain

$$K_\varepsilon(t, \vec{\lambda}) = \langle \vec{v}, \vec{\lambda} \rangle + \int_0^t H_\varepsilon(\vec{K}_\varepsilon(\tau, \vec{\lambda}))d\tau - \gamma_\varepsilon^1(t, \vec{\lambda}), \tag{17}$$

where

$$\gamma_\varepsilon^1(t, \vec{\lambda}) = H_\varepsilon(\vec{K}_\varepsilon(t, \vec{\lambda})) \left\{ t - \varepsilon \left[\frac{t}{\varepsilon}\right] \right\} \to 0, \quad \varepsilon \to 0,$$

uniformly in $t \geq 0$, $\lambda^1, \ldots, \lambda^d \leq 0$, on every finite interval of their variation.

Then, according to (16),

$$K_\varepsilon^i(t, \vec{\lambda}) = [u^i + \alpha_\varepsilon^i(t, \vec{\lambda})] K_\varepsilon(t, \vec{\lambda}), \tag{18}$$

where $\alpha_\varepsilon^i(t, \vec{\lambda}) \to 0$, $\varepsilon \to 0$, for every $t > 0$ uniformly in $\lambda^1, \ldots, \lambda^d \leq 0$ on finite intervals of their variation. Besides that, from the proof of relation (13) which, with evident changes, repeats the proof of Theorem 6.1.1 from [3, p.192], it is easy to see that the expression under the 'limit'-sign in (13) is bounded uniformly in $t > 0$, $|s^j| \leq 1$, $\vec{s} \neq \vec{1}$. From here it follows that $\alpha_\varepsilon^i(t, \vec{\lambda})$ in (18) is uniformly bounded in $t > 0$, $\lambda^1, \ldots, \lambda^d \leq 0$ on every finite interval of their variation. Therefore

$$\int_0^t H_\varepsilon(\vec{K}_\varepsilon(\tau, \vec{\lambda}))d\tau = \int_0^t H_\varepsilon(K_\varepsilon(\tau, \vec{\lambda}))d\tau + \gamma_\varepsilon^2(t, \vec{\lambda}), \tag{19}$$

where $\gamma_\varepsilon^2(t, \vec{\lambda}) \to 0$, $\varepsilon \to 0$, uniformly in $t \geq 0$, $\lambda^1, \ldots, \lambda^d \leq 0$, uniformly on every finite interval of their variance.

Finally, from (15) we find

$$K(t, \langle \vec{v}, \vec{\lambda} \rangle) = \langle \vec{v}, \vec{\lambda} \rangle + \int_0^t H(K(\tau, \langle \vec{v}, \vec{\lambda} \rangle))d\tau.$$

Denote $\Delta_\varepsilon(t) = |K_\varepsilon(t, \vec{\lambda}) - K(t, \langle \vec{v}, \vec{\lambda} \rangle)|$.

Comparing the last expression with (17) and (19) and similar reasonings to Theorem 1 show that the following inequality is true for some constant $c > 0$

$$\Delta_\varepsilon(t) \leq c \int_0^t \Delta_\varepsilon(\tau)d\tau + \gamma_\varepsilon(t, \vec{\lambda}), \tag{20}$$

where $0 \le \gamma_\varepsilon(t, \vec{\lambda}) \to 0$, $\varepsilon \to 0$, uniformly in $t \ge 0$ and $\lambda^1, \dots, \lambda^d \le 0$ on every finite interval of their variation. Inequality (20) leads to [4, p. 41]

$$\Delta_\varepsilon(t) \le c \int_0^t \gamma_\varepsilon(\tau, \vec{\lambda}) e^{c(t-\tau)} d\tau + \gamma_\varepsilon(t, \vec{\lambda}),$$

and, consequently, $\Delta_\varepsilon(t) \to 0$, $\varepsilon \to 0$. The proof can be accomplished in the same way as the proof of Theorem 1.

Combining the proofs of Theorems 1 and 2, it is easy to obtain the following result that generalizes them.

Assume matrix A has block diagonal form.

$$A = \operatorname{diag}\{A_1, A_2, \dots, A_r\},$$

where square matrices A_k, $k = 1, \dots, r$, are indecomposable, aperiodic, and their Perron's roots are equal to 1.

Denote the dimension of matrix A_k by d_k. Denote the right and left invariant vectors by $\vec{u}_k$ ad $\vec{v}_k$, respectively:

$$\sum_{i=1}^{d_k} v_k^i = 1, \quad \sum_{j=1}^{d_k} u_k^j v_k^j = 1, \quad k = 1, \dots, r, \quad \sum_{k=1}^{r} d_k = d.$$

Introduce into consideration operators U and V. Operator V associate d-dimensional vector $\vec{x} = (x^1, \dots, x^d)$ with r-dimensional vector $V\vec{x} = (y^1, \dots, y^r)$ with coordinates

$$y^k = \sum_{j=1}^{d_k} v_k^j x^{s_{k-1}+j},$$

where $s_0 = 0$, $s_k = d_1 + \dots + d_k$, $k = 0, 1, \dots, r$.

Operator U set the correspondence between r-dimensional vector $\vec{y} = (y^1, \dots, y^r)$ and d-dimensional vector $U\vec{y} = (x^1, \dots, x^d)$ with coordinates $x^j = u_k^{j-s_{k+1}}$ when $s_k < j \le s_{k+1}$, $k = 0, 1, \dots, r-1$.

Choose normalizing constants $b_k(\varepsilon) \to \infty$, $\varepsilon \to 0$, $k = 1, \dots, r$. Denote the diagonal matrix with elements $b_k(\varepsilon)$ on the main diagonal by $B(\varepsilon)$.

For r-dimensional vector $\vec{\lambda} = (\lambda^1, \dots, \lambda^r)$ with $\lambda^k \le 0$ set

$$\vec{H}_\varepsilon(\vec{\lambda}) = \varepsilon B(\varepsilon) V \vec{F}(e^{UB^{-1}(\varepsilon)\vec{\lambda}}),$$

where, for vector $\vec{x} = (x^1, \dots, x^d)$, vector $e^{\vec{x}}$ has the coordinates e^{x^i}, $i = 1, \dots, d$ by the definition.

Consider the family of random processes $\vec{\nu}_t(\varepsilon)$, where

$$\nu_t^i(\varepsilon) = \xi_{[t/\varepsilon]}^i(\varepsilon)/b_k(\varepsilon), \quad \xi_0^i(\varepsilon) = [x^i b_k(\varepsilon)],$$

when $s_{k-1} < i \le s, \quad k = 1, \dots r$.

Theorem 3. *If, within conditions* (8) *and* (9), $\vec{H}_\varepsilon(\vec{\lambda}) \to \vec{H}(\vec{\lambda})$ *and* $\vec{H}^1(0) = c$, *then the finite dimensional distributions of processes* $v_t^i(\varepsilon)$ *weakly converge to finite dimensional distributions of process* $\mu_t^k v^i$ *when* $s_{k-1} < i \le s_k$, $k = 1, \dots r$, *where* $\vec{\mu}_t = (\mu_t^1, \dots, \mu_t^r)$ *is an* r*-dimensional Jirina process with cumulant* $\vec{H}(\vec{\lambda})$ *and*

$$\mu_0^k = \sum_{j=1}^{d_k} x^{s_{k-1}+j} u_k^j.$$

References

1. **Aliev, S.A. and Shurenkov, V.M.** (1982). Transient occurrences and the convergence of Hamilton–Watson processes to the Jirina processes. *Theory Probab. and Appl.* **27**(3), 443–455. (In Russian).

2. **Ryžov, Yu.M. and Skorokhod, A.V.** (1970). Homogeneous branching processes with finite number of types and continuously changing mass. *Theory Probab. and Appl.* **15**(4), 722–726. (In Russian).

3. **Sevastjanov, B.A.** (1971). *Branching processes.* Nauka, Moscow. 436 p. (In Russian).

4. **Gikhman, I.I. and Skorokhod, A.V.** (1968). *Theory of Stochastic Differential Equations.* Naukova Dumka, Kiev, 354 p. (In Russian).

Exploring Stochastic Laws. pp. 13–40.
Eds. A.V. Skorokhod and Yu.V. Borovskikh

Diffusion approximation in switching stochastic models and its applications

VLADIMIR V. ANISIMOV*

Applied Statistics Department, Kiev University, Kiev, Ukraine

Received April 20, 1995

Abstract. A description of different classes of switching stochastic models (recurrent processes of semi-Markov type (*RPSM*))—simple and Markov cases, processes with semi-Markov switchings (*PSMS*) and switching processes (*SP*) is given. Limit theorems of the averaging principle and diffusion approximation types for *RPSM* and *PSMS* are considered in the case of "fast" switchings.

Applications of these results for asymptotic analysis of some interacting multi-component Markov systems and for studying the phenomena of self-organization on a macro-level and applications for the analysis of switching queueing systems and networks of Markov and non-Markov types in transient conditions and with a large number of calls are considered.

Introduction

The operation of a wide range of stochastic systems can be described in terms of random processes such that the character of their development varies spontaneously (switches) in some moments of time which are random functionals of the previous trajectory.

Such processes appear in the theory of queueing systems and networks, interactive multi-component systems, branching and migration processes, in the analysis of stochastic dynamic systems with random perturbations, random movements and other different applications.

To describe such models the author has introduced a special class of random processes with discrete interference of chance named *Switching Processes* (Anisimov (1975, 1977, 1978)).

Switching Processes (*SP*) are two-component processes $(x(t), \zeta(t))$, $t \geq 0$, taking their values in the space (X, R^r), for which there exists a sequence of times $t_1 < t_2 < \cdots$ such that on each interval $[t_k, t_{k+1})$ $x(t) = x(t_k)$. The behaviour of process

* The research described in this publication was made possible in part by Grant K3W100 from the Joint Fund of the Government of Ukraine and International Science Foundation

$\zeta(t)$ depends on the value $(x(t_k), \zeta(t_k))$ only. Times t_k are the switching times and $x(t)$ is the discrete switching component. Such processes can be described in terms of constructive characteristics (Anisimov (1975, 1977)). They are very suitable in analyzing and the asymptotic investigation of stochastic systems with "rare" or "fast" switchings (Anisimov (1978, 1988a, 1991–1995)).

SP are the natural generalization of such well-known classes of processes as Markov processes homogeneous on the 2-nd component (Ežov and Skorokhod (1969)), processes with independent increments and semi-Markov switchings (Anisimov (1973)), piecewise Markov aggregates (Buslenko et al. (1973)), Markov processes with semi-Markov interference of chance (Gikhman and Skorokhod (1973)) and Markov and semi-Markov evolutions (Kac (1956); Griego and Hersh (1969); Hersh (1974); Kertz (1978); Kurtz (1972, 1973); Papanicolaou and Hersh (1972); Pinsky (1971, 1975a,b, 1991); Korolyuk and Swishchuk (1986, 1991)).

For special classes of random evolutions (processes with independent increments and Markov and semi-Markov switchings), LLN and CLT-type theorems, were proved by many authors (Griego and Hersh (1969); Hersh and Papanicolaou (1972); Kurtz (1972, 1973); Kertz (1978); Pinsky (1975); Anisimov (1973, 1984, 1988a); Korolyuk and Turbin (1978); Watkins (1984); Korolyuk and Swishchuk (1986, 1992)). The results of averaging principle (*AP*) type for stochastic differential equations with independent Markov switchings (Khas'minskii (1968)) and for some Markov models in the case of feedback (Skorokhod (1987, 1992)) are very closely connected with this direction.

Limit theorems concerned with the convergence of one *SP* to another (in the class of *SP*) were proved in the case of "rare" switchings. These results allow us to construct a theory for the asymptotic consolidation (enlargement) of the states of non-homogeneous Markov and semi-Markov processes (Anisimov (1978, 1988a,b)).

In this paper we consider limit theorems of the averaging principle (*AP*) and diffusion approximation (*DA*) types in the case of fast switchings.

Theorems, concerned with the convergence of the trajectory of *SP* to a solution of some ordinary differential equation (*AP*), and the convergence of the normed difference to some diffusion process (*DA*), are proved for different important subclasses of *SP*: recurrent processes of a semi-Markov type (*RPSM*), (Anisimov and Aliev (1990); Anisimov (1992a)); non-homogeneous in time models and *RPSM* with an additional Markov component (Anisimov (1992c,d, 1993)); processes with semi-Markov switchings (Anisimov (1994, 1995)); and general *SP* with feedback between both components (Anisimov (1992b)).

The method of the proof uses a new approach which is based on the investigation of the asymptotic properties of a *RPSM*, theorems about the convergence of recurrent sequences with random coefficients to the solutions of stochastic differential equations, and the convergence of superpositions of random functions. It allows us to also study non-homogeneous in time models.

This class of theorems is the basis of a new approach in the investigation of transition phenomena for service processes in overloading queueing systems and networks of Markov and semi-Markov type; state-dependent branching processes with r types; asymptotically large number of particles and in a random medium; dynamic systems and movements with fast random switchings; and interactive migration processes. Some examples that illustrated AP and DA for these models are considered.

For Markov queueing systems and networks, different results were published in Anisimov (1992a), Anisimov and Lebedev (1992), and for non-Markov models, some results were obtained in Anisimov (1992c, 1993, 1995).

Another analytical approach which is based on the asymptotic analysis of integral Markov renewal equations and asymptotic inversion of perturbed linear operators was systematically developed in the works of Korolyuk and his pupils (Korolyuk and Turbin (1978, 1983); Korolyuk and Swishchuk (1988, 1992)). On the basis of this approach different results concerning the asymptotic consolidation of the states of Markov and semi-Markov systems (Korolyuk and Turbin (1978, 1983)), and AP and DA for semi-Markov random evolutions (Korolyuk and Swishchuk (1988, 1992)) are obtained.

Some results concerning AP and DA for overloading queueing systems with the help of the martingale approach are shown in (Basharin et al. (1989)).

This article consists of four parts. A description of different models of SP is given in Part 1. In Part 2 AP and DA are studied for the case of a simple $RPSM$. In Part 3 these results are applied for asymptotic analysis of some interactive multicomponent Markov systems and for studying the phenomena of self organization. Part 4 is devoted to the analysis of switching queueing systems and networks of Markov and non-Markov type with transient conditions and with a large number of requirements.

1. Switching stochastic models

In this part we consider some special models of switching stochastic systems which appear in different applications and will be used later for the problems of the asymptotic analysis of SP.

1.1. Simple case of RPSM

Let

$$F_k = \{(\xi_k(\alpha), \tau_k(\alpha)),\ \alpha \in R^r\}, \qquad k \geq 0$$

be jointly independent families of random variables taking values in the space $R^r \times$

$[0,\infty)$, and let S_0 be independent of F_k, $k \geq 0$ random variable in R^r.

We introduce the recurrent sequences:

$$t_0 = 0, \quad t_{k+1} = t_k + \tau_k(S_k), \quad S_{k+1} = S_k + \xi_k(S_k), \qquad k \geq 0 \tag{1.1}$$

and set

$$S(t) = S_k \quad \text{as } t_k \leq t < t_{k+1},\, t \geq 0. \tag{1.2}$$

Then process $S(t)$ forms a simple *RPSM*.

We assume that process $S(t)$ is regular, i.e., it has a finite number of jumps on each finite interval with probability 1.

In the case when distributions of the families F_k do not depend on the parameters α and k, times t_k, $k \geq 0$ form a recurrent flow and S(t) is a generalized renewal process. If as well distributions of the families F_k do not depend on the parameter k, process $S(t)$ is a homogeneous *SMP* (semi-Markov process). In particular when variables $\tau_k(\alpha)$ have exponential distributions, process $S(t)$ is a *MP* (Markov process).

1.2. RPSM with Markov switchings

Let

$$F_k = \{(\xi_k(x,\alpha), \tau_k(x,\alpha)),\ x \in X,\ \alpha \in R^r\}, \qquad k \geq 0$$

be jointly independent families of random variables taking values in $R^r \times [0,\infty)$, and let x_l, $l \geq 0$ be a *MP*, independent of F_k, $k \geq 0$, taking values in X, with (x_0, S_0) as its initial value. We put

$$t_0 = 0, \quad t_{k+1} = t_k + \tau_k(x_k, S_k), \quad S_{k+1} = S_k + \xi_k(x_k, S_k), \qquad k \geq 0, \tag{1.3}$$

$$S(t) = S_k, \quad x(t) = x_k \quad \text{as } t_k \leq t < t_{k+1},\ t \geq 0. \tag{1.4}$$

Then $(x(t), S(t))$ forms a *RPSM* with independent Markov switchings. When the distributions of variables $\tau_k(x,\alpha)$ do not depend on the parameters α and k, $x(t)$ is a *SMP*.

1.3. General case of RPSM

Let

$$F_k = \{(\xi_k(x,\alpha), \tau_k(x,\alpha), \beta_k(x,\alpha)),\ x \in X,\ \alpha \in R^r\}, \qquad k \geq 0$$

be jointly independent families of random variables taking values in the space $R^r \times [0,\infty) \times X$, with (x_0, S_0) as its initial value. We put

$$t_0 = 0, \quad t_{k+1} = t_k + \tau_k(x_k, S_k), \quad S_{k+1} = S_k + \xi_k(x_k, S_k),$$

$$x_{k+1} = \beta_k(x_k, S_k), \qquad k \geq 0, \tag{1.5}$$

$$S(t) = S_k, \quad x(t) = x_k \quad \text{as } t_k \leq t < t_{k+1}, \ t \geq 0. \tag{1.6}$$

Then the pair $(x(t), S(t))$, $t \geq 0$ forms a general *RPSM* with feedback between both components.

We assume that *RPSM* is regular, i.e., the component $x(t)$ has a finite number of jumps on each finite interval with probability 1.

In particular when distributions of the variables $\beta_k(x, \alpha)$ do not depend on the parameter α, the sequence x_k forms a *MP* in X and we obtain the previous case.

1.4. Processes with semi-Markov switchings (PSMS)

Let us consider a more general construction when we operate some random process in a semi-Markov environment.

Let

$$F_k = \{\zeta_k(t, x, \alpha), \ t \geq 0, \ x \in X, \ \alpha \in R^r\}, \qquad k \geq 0$$

be jointly independent parametric families of random variables, where (X, B_X) is a measurable space, $\zeta_k(t, x, \alpha)$ for each fixed k, x, α is a random process whose trajectory belongs to the Skorokhod space $D_\infty{}^r$, $x(t)$, $t \geq 0$ is independent of F_k, $k \geq 0$ right-continuous *SMP* in X, and S_0 is its initial value. We suppose that the variables introduced are measurable in the ordinary way on the pair (x, a) concerning σ-algebra $B_X \times B_{R^r}$.

We suppose also that $0 = t_0 < t_1 < \cdots$ are the times of the sequential jumps of $x(\cdot)$, $x_k = x(t_k)$, $k \geq 0$.

We construct a process with semi-Markov switchings in the following way:

Put

$$S_{k+1} = S_k + \xi_k,$$

where

$$\xi_k = \zeta_k(\tau_k, x_k, S_k), \quad \tau_k = t_{k+1} - t_k,$$

and

$$\zeta(t) = S_k + \zeta_k(t - t_k, x_k, S_k) \quad \text{as } t_k \leq t < t_{k+1}, \ t \geq 0. \tag{1.7}$$

Then the two-component process $(x(t), \zeta(t))$, $t \geq 0$ is a process with semi-Markov switchings (*PSMS*) and the component $x(t)$ itself is a *SMP*.

Let us introduce also the imbedded process

$$S(t) = S_k \quad \text{as } t_k \leq t < t_{k+1}. \tag{1.8}$$

It can be easily seen that the process $(x(t), S(t))$ forms a *RPSM* with independent Markov switchings.

Let from the construction above, $\{\zeta(t, x), \ t \geq 0,\}$ be the family of Markov processes and let $\zeta(t, x, \alpha)$ denote the process $\zeta(t, x)$ with initial value α. Then process $(x(t), \zeta(t))$ forms a Markov (when process $x(t)$ is Markov) or semi-Markov (when process $x(t)$ is semi-Markov) random evolution.

1.5. Switching processes (SP)

Let

$$F_k = \{((\zeta_k(t, x, \alpha), \tau_k(x, \alpha), \beta_k(x, \alpha)),\ t \geq 0, x \in X, \alpha \in R^r\}, \qquad k \geq 0$$

be jointly independent parametric families where (X, B_X) is a measurable space, $\zeta_k(t, x, \alpha)$ for each fixed k, x, α is a random process whose trajectory belongs to the Skorokhod space $D_\infty{}^r$ and $\tau_k(x, \alpha)$, $\beta_k(x, \alpha)$ are possibly dependent on $\zeta_k(\cdot, x, \alpha)$ random variables, $\beta_k(\cdot) \in X$. Let also (x_0, S_0) be independent of F_k, $k \geq 0$, with its initial vector in $X \times R^r$. We introduce the following recurrent sequences:

$$t_0 = 0, \quad t_{k+1} = t_k + \tau_k(x_k, S_k), \quad S_{k+1} = S_k + \xi_k(x_k, S_k),$$

$$x_{k+1} = \beta_k(x_k, S_k), \qquad k \geq 0, \tag{1.9}$$

where

$$\xi_k(x, \alpha) = \zeta_k(\tau_k(x, \alpha), x, \alpha).$$

We set

$$\zeta(t) = S_k + \zeta_k(t - t_k, x_k, S_k),$$

$$x(t) = x_k, \quad \text{as } t_k \leq t < t_{k+1},\ t \geq 0. \tag{1.10}$$

Then the two-component process $(x(t), \zeta(t))$, $t \geq 0$ is called a switching process (*SP*) (see Anisimov (1975, 1977)).

We also introduce an imbedded process

$$S(t) = S_k \quad \text{as } t_k \leq t < t_{k+1},\ t \geq 0. \tag{1.11}$$

As we can see, process $S(t)$ is *RPSM* (general case).

1.6. Other switching models

In terms of the different types of switching processes a wide classes of stochastic models can be described. For instance:

— sums of random variables defined on some Markov or semi-Markov random sequence;

— random movements switched by some random environment;

— dynamic systems and stochastic differential equations with coefficients switched by some Markov or semi-Markov process;

— branching and migration processes with state and age dependence, finite number of particles and in a random environment;

— Markov state dependent queueing systems and networks of the types $(M_Q/M_Q/\bar{m}/\bar{s})^r$ with vector input and service, impatient requirements etc.;

— non-Markov state dependent queueing systems and networks of the types $(SM_Q/M_Q/\bar{m}/\bar{s})^r$, $(G_Q/M_Q/\bar{m}/\bar{s})^r$ and $(G_{SM,Q}/M_{SM,Q}/\bar{m}/\bar{s})^r$ with input and service depending on the current values of queues and possibly on some Markov or semi-Markov environment.

2. Averaging principle and diffusion approximation for switching processes

In this part we study limit theorems for *RPSM* for the case of fast switchings. We consider the process on the interval $[0, nT]$ and the characteristics of the process dependent on the parameter of series n in such a way that the number of switchings tends, by probability, to infinity. It is the same when we consider the process on the interval $[0, T]$ in the scale of time nt (the switchings happen rapidly).

It has been proven that under natural assumptions, the normed trajectory of *RPSM* uniformly converges by probability to some function which is the solution of a non-stochastic differential equation. And the normed difference between the trajectory and this solution weakly converges in Skorokhod space D to some diffusion process.

2.1. Simple RPSM

First we consider a simple *RPSM* and give here two theorems of *AP* and *DA* types. In spite of the fact that it is the simplest case, these results nevertheless have very wide applications.

Let for each $n = 1, 2, \ldots$

$$F_{nk} = \{((\xi_{nk}(\alpha)), \tau_{nk}(\alpha)),\ \alpha \in R^r\}, \qquad k \geq 0$$

be jointly independent families of random variables taking their values in $R^r \times [0, \infty)$, the distributions of which do not depend on index k, and let S_{n0} be independent of F_{nk}, $k \geq 0$ with its initial value in R^r.

Put

$$\begin{aligned} t_{n0} &= 0, \quad t_{nk+1} = t_{nk} + \tau_{nk}(S_{nk}), \\ S_{n\,k+1} &= S_{nk} + \xi_{nk}(S_{nk}), \qquad k > 0, \\ S_n(t) &= S_{nk} \quad \text{as } t_{nk} \leq t < t_{nk+1},\ t > 0. \end{aligned} \tag{2.1}$$

Assume that there exist the functions

$$m_n(\alpha) = E\tau_{n1}(n\alpha),$$

$$b_n(\alpha) = E\xi_{n1}(n\alpha).$$

Theorem 2.1. (Averaging principle). *Suppose that for any* $N > 0$

$$\lim_{L\to\infty} \limsup_{n\to\infty} \sup_{|\alpha|<N} \Big\{ E\,\tau_{n1}(n\alpha)\,\chi(\tau_{n1}(n\alpha) > L) +$$

$$+ E\,|\xi_{n1}(n\alpha)|\,\chi(|\xi_{n1}(n\alpha)|) > L)\Big\} = 0 \qquad (2.2)$$

and as $\max(|\alpha_1|, |\alpha_2|) < N$,

$$|m_n(\alpha_1) - m_n(\alpha_2)| + |b_n(\alpha_1) - b_n(\alpha_2)| < C_N\,|\alpha_1 - \alpha_2| + \alpha_n(N), \qquad (2.3)$$

where C_N *are bounded constants,* $\alpha_n(N) \to 0$ *uniformly on* $|\alpha_1| < N$, $|\alpha_2| < N$, *and there exist functions* $m(a) > 0$ *and* $b(\alpha)$ *such that for any* $\alpha \in R^r$ *as* $n \to \infty$

$$m_n(\alpha) \to m(\alpha), \quad b_n(\alpha) \to b(\alpha) \qquad (2.4)$$

and

$$n^{-1} S_{n0} \xrightarrow{P} s_0. \qquad (2.5)$$

Then

$$\sup_{0\le t\le T} |n^{-1}\,S_n(nt) - s(t)| \xrightarrow{P} 0, \qquad (2.6)$$

where

$$s(0) = s_0, \quad \mathrm{d}s(t) = m(s(t))^{-1}\,b(s(t))\,\mathrm{d}t \qquad (2.7)$$

and T *is any positive value such that* $y(+\infty) > T$ *with probability* 1 *where*

$$y(t) = \int_0^t m(\eta(u))\,\mathrm{d}u, \qquad (2.8)$$

$$\eta(0) = S_0, \quad \mathrm{d}\eta(u) = b(\eta(u))\,\mathrm{d}u \qquad (2.9)$$

(*it is supposed that a solution of equation* (2.9) *exists on each interval and is unique*).

Now we consider the convergence of the process

$$\gamma_n(t) = \frac{1}{\sqrt{n}}\,(S_n(nt) - n\,s(t)), \qquad t \in [0, T]$$

to some diffusion process.

Denote

$$\tilde{b}_n(\alpha) = m_n(\alpha)^{-1}\,b_n(\alpha), \quad \tilde{b}(\alpha) = m(\alpha)^{-1}\,b(\alpha),$$

$$\rho_n(\alpha) = \xi_{n1}(n\alpha) - b_n(\alpha) - \tilde{b}(\alpha)\,(\tau_{n1}(n\alpha) - m_n(\alpha)),$$

$$D_n^2(\alpha) = E\,\rho_n(\alpha)\,\rho_n(\alpha)^*$$

(we denote the conjugate vector by the symbol *),

$$q_n(\alpha, z) = \sqrt{n}\left(\tilde{b}_n\left(\alpha + \frac{1}{\sqrt{n}} z\right) - \tilde{b}(\alpha)\right).$$

Theorem 2.2. (Diffusion approximation). *Let conditions* (2.3)–(2.5) *be satisfied where in* (2.3), $\sqrt{n}\,\alpha_n(N) \to 0$, *there exist continuous matrix-valued functions* $D^2(\alpha)$ *and* $Q(\alpha)$ *and vector-valued function* $g(a)$ *such that, uniformly in each bounded region as* $n \to \infty$,

$$D_n^2(\alpha) \to D^2(\alpha), \tag{2.10}$$

$$q_n(\alpha, z) \to Q(\alpha)\, z + g(a) \tag{2.11}$$

for any $z \in R^r$,

$$\gamma_n(0) \overset{w}{\Longrightarrow} \gamma_0, \tag{2.12}$$

and for any $N > 0$

$$\lim_{L\to\infty} \limsup_{n\to\infty} \sup_{|\alpha|<Nn} \Big\{ E\,\tau_{n1}^2(\alpha)\,\chi(\tau_{n1}(\alpha) > L) +$$

$$+ E\,|\xi_{n1}(\alpha)|^2\,\chi(|\xi_{n1}(a)|) > L) \Big\} = 0, \tag{2.13}$$

Then the sequence of the processes $\gamma_n(t)$ *J-converges on any interval* $[0,T]$ *such that* $y(+\infty) > T$ *to the diffusion process* $\gamma(t)$ *which satisfies the following stochastic differential equation:*

$$\mathrm{d}\gamma(t) = \{Q(s(t))\,\gamma(t) + g(s(t))\}\,\mathrm{d}t + D(s(t))\,m(s(t))^{-1/2}\,\mathrm{d}w(t), \tag{2.14}$$

$$\gamma(0) = \gamma_0,$$

where $s(\cdot)$ *satisfies equation* (2.7). (*J-convergence denotes the weak convergence of measures in Skorokhod space* D).

The proof of these two Theorems is given in (Anisimov and Lebedev (1992)). It can also be obtained from the results of (Anisimov and Aliev (1990)) when we put $V_n = 1,\ A_n = n$.

Let us consider an important case when process $S_n(t)$ is a homogeneous MP. Suppose that $S_n(t)$ is a regular stepwise process and there exist intensities of transition probabilities $q_n(\alpha, A),\ \alpha \in R^r,\ A \in B_R^r,\ \alpha \notin A$ such that

$$q_n(\alpha) = q_n(\alpha, R^r \setminus \{\alpha\}) < \infty$$

for any $\alpha \in R^r$.

We introduce the independent families $\{\xi_{nk}(\alpha),\ \alpha \in R^r\}$, $k \geq 0$ and $\{\tau_{nk}(\alpha),\ \alpha \in R\}$, $k \geq 0$ of random variables with values in R^r and $[0,\infty)$ respectively, such that $\tau_{nk}(n\alpha)$ has exponential distribution with parameter $q_n(\alpha)$ and

$$P\{\xi_{nk}(n\alpha) \in A\} = q_n(\alpha)^{-1} q_n(\alpha, A+\alpha), \qquad \alpha \notin A,$$

where $A+\alpha = \{z: z-\alpha \in A\}$. It is clear that *RPSM* which is defined by the families $(\xi_{nk}(\alpha), \tau_{nk}(\alpha))$ is equivalent to our MP $S_n(t)$. Denote $m_n(\alpha) = q_n(\alpha)^{-1}$, $D_n^2(\alpha) = E\,\xi_{n1}(n\alpha)\,\xi_{n1}(n\alpha)^*$ and keep the other notations.

Corollary 2.1. *If the conditions of Theorems* 2.1 *and* 2.2 *hold then relation* (2.6) *takes place and the sequence of processes* $\gamma_n(t)$ *weakly converges to the diffusion process* $\gamma(t)$ *satisfying equation* (2.14).

We remark that, in this case, conditions (2.2) and (2.13) for the variables $\tau_{n1}(\alpha)$ will be automatically fulfilled.

2.2. Processes with semi-Markov switchings

In this section we consider limit theorems for *PSMS* (see Part 1.4).

Let for each $n > 0$

$$F_{nk} = \{\zeta_{nk}(t,x,\alpha),\ t \geq 0,\ x \in X,\ \alpha \in R^r\}, \qquad k \geq 0$$

be jointly independent families of random processes in D_∞^r, $x_n(t)$, $t \geq 0$ be an independent of F_{nk}, $k \geq 0$ right-continuous *SMP* in X, S_{n0} is an initial value. We suppose also $0 = t_{n0} < t_{n1} < \cdots$ are the times of the sequential jumps of $x_n(\cdot)$, $x_{nk} = x_n(t_{nk})$, $k \geq 0$. We construct a *PSMS* $(x_n(t), \zeta_n(t))$, $t \geq 0$ according to formulae (1.7) and (1.8) in the following way:

$$S_{n\,k+1} = S_{nk} + \xi_{nk},$$

where

$$\xi_{nk} = \zeta_{nk}(\tau_{nk}, x_{nk}, S_{nk}), \qquad \tau_{nk} = t_{n\,k+1} - t_{nk},$$

and

$$\zeta_n(t) = S_{nk} + \zeta_{nk}(t - t_{nk}, x_{nk}, S_{nk}) \tag{2.15}$$

$$x_n(t) = x_{nk} \quad \text{as } t_{nk} \leq t < t_{nk+1},\ t \geq 0. \tag{2.16}$$

Then the process $(x_n(t), \zeta_n(t))$, $t \geq 0$ is a *PSMS*.

Let also $S_n(t)$, $t \geq 0$ be the corresponding imbedded *RPSM* with additional Markov switchings (see Part 1, 2., (1.3), (1.4)):

$$S_n(t) = S_{nk} \quad \text{as } t_{nk} \leq t < t_{nk+1}. \tag{2.17}$$

First we give the averaging principle for the switched component $\zeta_n(\cdot)$.

Suppose for simplicity that $x_n(\cdot)$ is a homogeneous *SMP* and distributions of the processes $\zeta_{nk}(\cdot)$ do not depend on index $k \geq 0$.

Let $\tau_n(x)$ be a sojourn time in the state x for *SMP* $x_n(\cdot)$. Denote for each $x \in X$, $\alpha \in R^r$

$$\xi_n(x,\alpha) = \zeta_{n1}(\tau_n(x), x, \alpha), \quad g_n(x,\alpha) = \sup_{t<\tau_n(x)} |\zeta_{n1}(t,x,\alpha)|.$$

Suppose that imbedded MP, $x_{nk},\ k \geq 0$ has at each $n \geq 0$ a stationary measure $\pi_n(A),\ A \in B_X$ and denote

$$m_n(x) = E\,\tau_{n1}(x), \quad b_n(x,\alpha) = E\,\xi_{n1}(x, n\alpha),$$

$$m_n = \int_X m_n(x)\,\pi_n(\mathrm{d}x), \quad b_n(\alpha) = \int_X b_n(x,\alpha)\,\pi_n(\mathrm{d}x),$$

$$\alpha_n(k) = \sup\Big\{|P\{x_{ni} \in A,\ x_{ni+k} \in B\} -$$

$$-P\{x_{ni} \in A\}\,P\{x_{ni+k} \in B\}|:\ A, B \in B_X,\ i \geq 0\Big\}.$$

Theorem 2.3. (Averaging Principle). *Suppose that there exists a sequence of integers* r_n *such that*

$$n^{-1} r_n \to O, \quad \sup_{k \geq r_n} \alpha_n(k) \to O, \tag{2.18}$$

for any $N > 0$

$$\lim_{n\to\infty} \sup_{|\alpha|<N} \sup_x n\,P\{n^{-1} g_n(x,\alpha) > \varepsilon\} = 0, \qquad \varepsilon > 0, \tag{2.19}$$

$$\lim_{L\to\infty} \limsup_{n\to\infty} \sup_{|\alpha|<N} \sup_x \Big\{E\tau_{n1}(x)\,\chi(\tau_{n1}(x) > L) +$$

$$+E\,|\xi_{n1}(x,n\alpha)|\,\chi(|\xi(x,n\alpha)| > L)\Big\} = 0 \tag{2.20}$$

and for any x *as* $\max(|\alpha_1|, |\alpha_2|) < N$

$$|b_n(x,\alpha_1) - b_n(x,\alpha_2)| < C_N\,|\alpha_1 - \alpha_2| + \alpha_n(N), \tag{2.21}$$

where C_N *is some constant,* $\alpha_n(N) \to 0$ *uniformly on* $|\alpha_1| < N$, $|\alpha_2| < N$*; there exists function* $b(\alpha)$ *such that for any* $\alpha \in R^r$

$$b_n(\alpha) \to b(\alpha), \tag{2.22}$$

$$m_n \to m > 0 \tag{2.23}$$

and

$$n^{-1} S_{n0} \xrightarrow{P} s_0. \tag{2.24}$$

Then for any $T > 0$

$$\sup_{0\le t\le T} |n^{-1}\zeta_n(nt) - s(t)| \xrightarrow{P} 0 \tag{2.25}$$

where $s(0) = s_0$,

$$ds(t) = m^{-1}\, b(s(t))\, dt. \tag{2.26}$$

(It is supposed that a solution of equation (2.26) exists on each interval and is unique).

Now we consider the convergence of the sequence of processes

$$\theta_n(t) = \frac{1}{\sqrt{n}}\,(\zeta_n(nt) - n\, s(t))$$

to some diffusion process.

Let us introduce the coefficient of uniformly strong mixing condition for the process x_{nk} :

$$\varphi_n(r) = \sup_{x,y,A} |P\{x_{nr} \in A/x_{n0} = x\} - P\{x_{nr} \in A/x_{n0} = y\}|.$$

Put

$$\tilde{b}_n(\alpha) = b_n(\alpha)\, m_n^{-1}, \quad \tilde{b}(\alpha) = b(\alpha) m^{-1},$$

$$\rho_{nk}(x,\alpha) = \xi_{nk}(x, n\alpha) - b_n(x,\alpha) - \tilde{b}(\alpha)(\tau_{nk}(x) - m_n(x)),$$

$$D_n(x,\alpha)^2 = E\, \rho_{n1}(x,\alpha)\, \rho_{n1}(x,\alpha)^*,$$

$$\gamma_n(x,\alpha) = b_n(x,\alpha) - b_n(\alpha) - \tilde{b}(\alpha)(m_n(x) - m_n).$$

Theorem 2.4. *Suppose that for some fixed $r > 0$, $q \in [0,1)$*

$$\varphi_n(r) \le q, \qquad n > 0, \tag{2.27}$$

the conditions (2.21) where $\sqrt{n}\,\alpha_n(N) \to 0$ and (2.22)-(2.24) hold, and for any $N > 0$ the following conditions are fulfilled:

1)

$$\lim_{n\to\infty} \sup_{|\alpha|<N} \sup_x n\, P\Big\{\frac{1}{\sqrt{n}}\, g_n(x,\alpha) > \varepsilon\Big\} = 0, \qquad \varepsilon > 0; \tag{2.28}$$

2)

$$\lim_{L\to\infty} \lim_{n\to\infty} \sup_{|\alpha|<N} \sup_x \Big\{ E\tau_{n1}(x)^2 + \chi(\tau_{n1}(x) > L) +$$

$$+ E\, |\xi_{n1}(x, n\alpha)|^2\, \chi(|\xi_{n1}(x, n\alpha)| > L)\Big\} = 0; \tag{2.29}$$

3)

$$|D_n(x,\alpha_1)^2 - D_n(x,\alpha_2)^2| \le C_N\,|\alpha_1-\alpha_2| + \alpha_n(N), \qquad (2.30)$$

$$\text{as}\ \ \max(|\alpha_1|,|\alpha_2|) < N$$

where $\alpha_n(N) \to 0$ *uniformly on* $|\alpha_1| < N$, $|\alpha_2| < N$;

4) *there exists function* $q(\alpha, z)$ *such that for any* N *in the domain* $|\alpha| < N$

$$|q(\alpha,z)| < C_N(1+|z|)$$

and uniformly by $|\alpha| < N$ *at each fixed* z

$$\sqrt{n}\left(\tilde{b}_n\left(\alpha + \frac{1}{\sqrt{n}} z\right) - \tilde{b}(\alpha)\right) \to q(\alpha, z), \qquad (2.31)$$

and there exist functions $D(\alpha)$ *and* $B(\alpha)$ *such that for any* $\alpha \in R^m$

$$D_n(\alpha)^2 = \int_X D_n(x,\alpha)\,\pi_n(\mathrm{d}x) \to D(\alpha)^2, \qquad (2.32)$$

$$B_n^{(1)}(\alpha)^2 + B_n^{(2)}(\alpha)^2 + (B_n^{(2)}(\alpha)^*)^2 \to B(\alpha)^2, \qquad (2.33)$$

where

$$B_n^{(1)}(\alpha)^2 = \int_X \gamma_n(x,\alpha)\,\gamma_n(x,\alpha)^*\,\pi_n(\mathrm{d}x),$$

$$B_n^{(2)}(\alpha)^2 = \sum_{k\ge 1} E\,\gamma_n(x_{n0},\alpha)\,\gamma_n(x_{nk},\alpha)^*,$$

$$P\{x_{n0}\in A\} = \pi_n(A), \qquad A \in B_X,$$

and also

$$\theta_n(0) \overset{w}{\Longrightarrow} \theta_0. \qquad (2.34)$$

Then the sequence of processes $\theta_n(t)$ *weakly converges in the space* D_T^r *(in the sense of weak convergence of measures) at each* $T > 0$ *to the diffusion process* $\theta(t)$ *which satisfies the following stochastic differential equation:*

$$\theta(0) = \theta_0$$

$$\mathrm{d}\theta(t) = q(s(t),\theta(t))\,\mathrm{d}t + m^{-\frac{1}{2}}\left(D(s(t))^2 + B(s(t))^2\right)^{\frac{1}{2}}\mathrm{d}w(t), \qquad (2.35)$$

where $w(t)$ *is a standard Wiener process in* R^r *and the solution of* (2.35) *exists and is unique.*

The proofs of Theorems 2.3 and 2.4 are given in (Anisimov (1994)). More general theorems concerning the convergence of *RPSM* with additional Markov component (variables $\tau_n(x)$ depend also on the parameter α) will appear in (Anisimov (1995)). *AP* for general *SP* with feedback between both components was given in (Anisimov (1992b)).

We mention here that the method of proof allows us to study non-homogeneous in time models also. It is possible because instead of parameter α we can consider parameter (α, t) and include the dependence on the current value of variable (S_{nk}, t_{nk}).

The results of Theorems 2.1–2.4 give us the theoretical background for studying the asymptotic behaviour of stochastic systems with switching structure. *AP* and *DA* type theorems for dynamic systems with fast switchings are studied on the base of these results in (Anisimov (1992, 1995). In the next parts we consider applications to other switching models.

3. Models of self-organization of multicomponent interacting Markov systems

First let us analyze in what sense the phenomena of self-organization could appear in stochastic switching systems. Consider equation (2.7) (or correspondingly (2.26)). Let s_* be some point such that

$$b(s_*) = 0 \tag{3.1}$$

and assume there exists some domain $D(s_*)$ such that if $s_0 \in D(s_*)$ then

$$s(t) \to s_* \quad \text{as } t \to \infty,$$

where $s(t)$ is a solution of equation (2.7) at $s(0) = s_0$.

It means that s_* is a stability point for equation (2.7). In this case taking into account relation (2.6) we obtain for any $\varepsilon > 0$

$$\lim_{t\to\infty} \limsup_{n\to\infty} P\left\{\left|\frac{1}{n} S_n(nt) - s_*(t)\right| > \varepsilon\right\} = 0. \tag{3.2}$$

This relation describes the situation of self-organization in switching systems.

Now we study a model of an asymptotically large number of interacting Markov systems with mutual influence between individual systems and some collective state of the whole super-system.

Let us consider n interacting Markov systems which develop in the following way. These systems are identical with a finite set of states $\{1, 2, \ldots, r\}$. The family of nonnegative functions $\lambda_{ij}(\bar{q})$, $i, j = \overline{1,r}$, $i \neq j$, $\bar{q} = (q_1, \ldots, q_r)$, $q_i \geq 0$, $\sum_{i=1}^{r} q_i = 1$ is given. Let $\nu_n(i,t)$ be the general number of systems which are in state i at time t. Denote

$$\bar{\nu}_n(t) = \left(\frac{1}{n}\,\nu_n(i,t),\; i = \overline{1,r}\right).$$

If at time t $\bar{\nu}_n(t) = \bar{q}$ some system is in state i, then this system on the interval $[t, t+h]$ independently of the other systems may pass to state j with probability

$$n^{-1}\,\lambda_{ij}(\bar{q})\,h + o(h), \qquad j \neq i.$$

Let us introduce column-vector

$$\tilde{b}(\bar{q}) = (-q_i\,\lambda_i(\bar{q}) + \sum_{k\neq i} q_k\,\lambda_{ki}(\bar{q}),\ i = \overline{1,r}) \tag{3.3}$$

where $\lambda_i(\bar{q}) = \sum_{k\neq i} \lambda_{ik}(\bar{q})$ and put

$$\lambda(\bar{q}) = \sum_{i=1}^{r} q_i\,\lambda_i(\bar{q}).$$

Theorem 3.1. *Suppose that $\bar{\nu}_n(0) = \bar{S}_0$ and functions $\lambda_{ij}(\bar{q})$ satisfy local Lipschitz conditions.*

Then

$$\sup_{t\leq T} |\bar{\nu}_n(nt) - \bar{s}(t)| \xrightarrow{P} 0 \tag{3.4}$$

where

$$\bar{s}(0) = \bar{s}_0, \quad d\bar{s}(t) = \tilde{b}(\bar{s}(t))\,dt \tag{3.5}$$

and T is any positive value such that $y(+\infty) > T$, where

$$\begin{aligned} y(t) &= \int_0^t \lambda(\bar{\eta}(u))^{-1}\,du, \\ \bar{\eta}(0) &= \bar{s}_0, \\ d\bar{\eta}(u) &= \tilde{b}(\bar{\eta}(u))\,\lambda(\bar{\eta}(u))^{-1}\,du. \end{aligned}$$

Proof. In our case $\bar{\nu}_n(t)$ is a homogeneous Markov process and it is easy to see that it can be described as a simple *RPSM* for which $\tau_{n1}(n\bar{q})$ has the exponential distribution with the parameter $\lambda(q)$ and

$$\xi_{n1}(n\bar{q}) = \frac{1}{n}\,\bar{e}_{ij} \quad \text{with probability } \frac{q_i\,\lambda_{ij}(q)}{\lambda(\bar{q})}, \qquad i,j = \overline{1,r},\ i \neq j \tag{3.6}$$

where $\bar{e}_{ij}$ is a column-vector for which the i-th component is equal to -1, j-th component is equal to 1 and other components are equal to 0. Thus we obtain that

$$\begin{aligned} E\,\tau_{n1}(n\bar{q}) &= \lambda(\bar{q})^{-1}, \\ E\xi_{n1}(n\bar{q}) &= \frac{1}{n}\,\tilde{b}(\bar{q})\,\lambda(\bar{q})^{-1}. \end{aligned}$$

and using Theorem 2.1 (Corollary 2.1) we prove our statement.

Now it is clear that for analyzing the phenomena of self-organization it is necessary to analyze the points of stability of equation (3.5).

These results can be used for analyzing the problems of self-organization for multi-component Markov models of the society suggested by W. Weidlich (1991,

1992) and W. Weidlich and G. Haag (1988) and give us a new approach in the investigation of "synergetic" phenomena in these models.

Now let us formulate the Theorem about diffusion approximation. By symbol $\bar{f}'(\bar{a})$ we denote a matrix of partial derivatives of vector-function $\bar{f}(\bar{a})$. Thus

$$\lim_{h\to 0} h^{-1}(\bar{f}(\bar{a}+h\bar{z}) - \bar{f}(\bar{a})) = \bar{f}'(\bar{a})\, z.$$

Let us introduce a matrix $B(\bar{q})^2 = \|b_{ij}(\bar{q})\|,\ i,j=\overline{1,r}$ where

$$\begin{aligned} b_{ij}(\bar{q}) &= -q_i\,\lambda_{ij}(\bar{q}) - q_j\,\lambda_{ji}(\bar{q}), \qquad i \neq j, \\ b_{ii}(\bar{q}) &= q_i\,\lambda_i(\bar{q}) - \sum_{k\neq i} q_k\,\lambda_{ki}(\bar{q}). \end{aligned}$$

Theorem 3.2. *Let the conditions of Theorem* 3.1 *be satisfied, assume continuous function* $Q(\bar{q}) = \tilde{b}'(\bar{q})$ *exists and let*

$$\sqrt{n}\,(\bar{\nu}_n(0) - s_0) \stackrel{w}{\Longrightarrow} \bar{\gamma}_0.$$

Then the sequence of processes

$$\bar{\gamma}_n(t) = \sqrt{n}\,(\bar{\nu}_n(nt) - \bar{s}(t))$$

on the interval $[0,T]$ *U-converges to the diffusion process* $\bar{\gamma}(t)$ *which satisfies the following stochastic differential equation*

$$\begin{aligned} \mathrm{d}\bar{\gamma}(t) &= Q(\bar{s}(t))\bar{\gamma}(t)\,\mathrm{d}t + B(\bar{s}(t))\,\mathrm{d}\bar{w}(t), \\ \bar{\gamma}(0) &= \bar{\gamma}_0 \end{aligned}$$

(here $\bar{w}(t)$ *is a standard Wiener process in* R^r*).*

Proof. It is easy to calculate that in our case

$$E\,\tau_{n1}(n\bar{q}) = \lambda(\bar{q})^{-1}, \qquad E\,\tau_{n1}(n\bar{q})^2 = \lambda(\bar{q})^{-2}$$

and

$$E\,\rho_n(\bar{q})\,\rho_n(\bar{q})^* = E\,\xi_{n1}(nq)\,\xi_{n1}(nq)^* = \frac{1}{n^2}\,D(\bar{q})^2\,\lambda(\bar{q})^{-1},$$

where

$$\rho_n(\bar{q}) = \xi_{n1}(n\bar{q}) - E\,\xi_{n1}(n\bar{q}) - \frac{1}{n}\,\tilde{b}(\bar{q})\,(\tau_{n1}(n\bar{q}) - \lambda(\bar{q})^{-1}).$$

Then by using Theorem 2.2 (Corollary 2.1) we obtain our statement.

We remark that if $\bar{q}_*$ is some stability point and $\bar{s}_0 \in D(\bar{q}_*)$ then the multidimensional distributions of the process $\bar{\gamma}(t)$ at large t are close to the distributions of stationary diffusion process $\tilde{\gamma}(t)$ satisfying the following stochastic differential equation

$$d\tilde{\gamma}(t) = Q(\bar{q}_*)\,\tilde{\gamma}(t)dt + B(\bar{q}_*)\,d\bar{w}(t), \quad \tilde{\gamma}(0) = \bar{\gamma}_* \tag{3.7}$$

where $\bar{\gamma}_*$ is a stationary point of equation (3.7). This result allows us to calculate the parameters of diffusion near the stability point.

Let us consider some examples.

Example 3.1. Let $r = 2$, $\lambda_{12}(\bar{q}) \equiv \alpha$, $\lambda_{21}(\bar{q}) \equiv \beta$. Then $\lambda_1(\bar{q}) \equiv \alpha$, $\lambda_2(\bar{q}) \equiv \beta$ and system (3.1) with function $\tilde{b}(\bar{q})$ from relation (3.3) has a form:

$$\begin{cases} -q_1\alpha + q_2\beta = 0 \\ -q_2\beta + q_1\alpha = 0 \end{cases} \tag{3.8}$$

with condition $q_1 + q_2 = 1$.

From (3.8) we can show that

$$q_{1*} = \beta\,(\alpha+\beta)^{-1}, \quad q_{2*} = \alpha\,(\alpha+\beta)^{-1}. \tag{3.9}$$

This solution as we can easily see is the point of stability for system (3.8). Denote $\tilde{b}(\bar{q}) = (b_1(\bar{q}), b_2(\bar{q}))$. Then in our example according to (3.3)

$$b_1(\bar{q}) = -q_1\alpha + q_2\beta = \beta - (\alpha+\beta)\,q_1, \quad b_2(\bar{q}) = -b_1(\bar{q}).$$

Then system (3.5) consists of one equation

$$ds_1(t) = (\beta - (\alpha+\beta)\,s_1(t))\,dt, \quad s_1(0) = s_0^{(1)}, \tag{3.10}$$

where $s_0^{(1)}$ is the 1-st component of the vector s_0. The solution of (3.10) has the form

$$s_1(t) = q_{1*} + (q_{1*} - s_0^{(1)})\,e^{-(\alpha+\beta)\,t},$$

where $q_{1*} = \beta\,(\alpha+\beta)^{-1}$. Correspondingly

$$s_2(t) = q_{2*} + (q_{2*} - s_0^{(2)})\,e^{-(\alpha+\beta)\,t}$$

where $q_{2*} = \alpha(\alpha+\beta)^{-1}$ and

$$s(t) \to q_* = (q_{1*}, q_{2*}).$$

Example 3.2. Consider now the general case of a system without feedback. Suppose that for any i, j, $i \neq j$, $\lambda_{ij}(\bar{q}) \equiv \lambda_{ij}$. It means that we have no influence between individual systems and the super-system. In this case it is easy to see that the solution of system (3.1) where function $\tilde{b}(\bar{q})$ is taken from relation (3.3), is a stationary probability distribution π for Markov process with state space $\{1,\ldots,r\}$

and intensities of transition probabilities λ_{ij}, $i \neq j$, and $s(t)$ corresponds to the vector of transition probabilities for this process on the interval $[0,t]$ with initial distribution s_0. In this case, when this process is irreducible, $s(t) \to \pi$.

Example 3.3. Let us consider the case of linear influence. Put

$$\lambda_{ij}(\bar{q}) = \lambda_{ij} q_i, \qquad i,j = \overline{1,r},\ i \neq j.$$

In this case, system (3.1) has the form:

$$\lambda_i q_i^2 = \sum_{k \neq i} q_k^2 \lambda_{kj}, \qquad i = \overline{1,r}, \tag{3.11}$$

where $\lambda_i = \sum_{k \neq i} \lambda_{ik}$.

Denote by $\pi = (\pi_1, \dots, \pi_r)$ a stationary distribution from Example 3.2. Then in this case the solution of (3.11) is unique and has the form $q_{*i} = \sqrt{\pi_i}$, $i = \overline{1,r}$.

Analogous results can be obtained for systems with discrete time.

4. Switching queueing systems and networks

In this part we consider models of queueing systems and networks which can be described in terms of *SP*. We say a switching system is such a queueing system for which the corresponding process of operation in time can be described in terms of *SP*.

Let us consider an example.

Example 4.1. *System* $SM/M_{SM,Q}/1/m$. Let $x(t)$, $t \geq 0$ be a right continuous *SMP* with a finite state space $\{1, 2, \dots, d\}$ given by the family of sojourn times $\{\tau(i),\ i = \overline{1,d}\}$ and transition probabilities $p(i,j)$, $i,j = \overline{1,d}$. For simplicity we suppose that the sojourn time $\tau_k(i)$ in state i does not depend on the state of the next step.

Let also the parametric family of intensities $\mu(i,l)$, $i = \overline{1,d}$, $l = 0, 1, 2, \dots, m$ be given. Denote by $t_1 < t_2 < \cdots$ the times of the consequential jumps for $x(t)$. Then we say that the input flow is semi-Markov if the calls enter the system one at a time at times t_k.

The system has one server and intensity of service $\mu(x(t), Q(t))$, where $Q(t)$ is the number of calls in the system at time t.

To describe process $(x(t), Q(t))$, $t \geq 0$ as a *PSMS* we introduce the jointly independent families of decreasing stepwise Markov processes $\{\eta_k(t,i,l),\ t \geq 0,\ i = \overline{1,d},\ l = \overline{1,m}\}$, $k \geq 0$, with values in $\{0, 1, 2, \dots\}$ such that for each i, l, k

$\eta_k(0,i,l) = l$. Transitions from any state j are possible only in state $j-1$ and

$$\begin{aligned} P\{\eta_k(t+h,i,l) &= j-1/\eta_k(t,i,l)=j\} \\ &= \mu(i,j)\,h + o(h). \end{aligned}$$

We suppose without loss of generality that $\mu(i,0) \equiv 0$. Each process $\eta_k(\cdot)$ is a process of pure death.

We introduce the families of processes $\zeta_k(t,x,l)$ in the following way:

$$\begin{aligned} \zeta_k(t,x,l) &= \eta_k(t,x,l) - l, \qquad t < \tau_k(x); \\ \zeta_k(\tau_k(x),x,l) &= \eta_k(\tau_k(x),x,l) - l + 1. \end{aligned}$$

Process $(x(t), Q(t))$, which is constructed by the introduced families $\zeta_k(\cdot)$ and $SMPx(\cdot)$, is *PSMS*.

By analogy it is possible to describe a semi-Markov queueing system $M_{SM.Q}/M_{SM.Q}/1/m$ which can be considered as system $M_Q/M_Q/1/m$ switched by some semi-Markov process. In the same way these results can be extended on Markov and semi-Markov queueing networks.

Now we consider the case of overloading systems. In this case the queue size tends to infinity and under some general conditions it is possible to prove *AP* and *DA* for the process describing the behaviour of the queue. The method of investigation is based on limit theorems (*AP* and *DA*) for *RPSM* or *PSMS* given in the part II.

4.1. Markov queueing networks

We consider the queueing network $(M_Q/M_Q/1/\infty)^r$ with group input and service which consists of r nodes with one server in each node and infinite number of places for waiting. Suppose that all calls are identical and the local characteristics of the network depend on the current values of queues in the nodes and the parameter of series n.

Denote the number of calls at the i-th node at time t by $S_n(i,t)$ and $S_n(t) = (S_n(i,t),\ i=\overline{1,r})$ is a column vector. Let the following be given:

1) nonnegative functions $\lambda_i(\alpha)$, $\mu_i(\alpha)$ and $\nu_i(\alpha)$, $i=\overline{1,r}$, where $\alpha = (\alpha^{(1)},\ldots,\alpha^{(r)})$ is a column vector in $[0,\infty)^r$;

2) families of integral random variables $\delta_i(\alpha)$, $\gamma_i(\alpha)$ with values in $\{0,1,\ldots\}$ and variables $\beta_i(\alpha)$ with values in $\{0,\pm1,\ldots\}$;

3) a family of transition probability matrices $P(\alpha) = \|p_{ij}(\alpha)\|$, $i=\overline{1,r}$, $j=\overline{1,r+1}$;

4) initial vector $Q_n(0)$.

The system works in the following way. If at time t $Q_n(t) = n\alpha$, then:

1) $\delta_i(\alpha)$ calls may enter at the i-th node with local intensity of entry $\lambda_i(\alpha)$, $i = \overline{1,r}$;

2) $\gamma_i(\alpha)$ calls may finish service in the i-th node with local intensity $\mu_i(\alpha)$ and all of them either with probability $p_{ij}(\alpha)$ pass to the j-th node, $j = \overline{1,r}$, or with probability $p_{ir+1}(\alpha)$ leave the network.

3) at the i-th node each customer in the queue independent of the others may transform into $\beta_i(\alpha)$ calls with local intensity $n^{-1}\nu_i(\alpha)$.

In our case process $Q_n(t)$ is a multidimensional Markov process.

Suppose that there exist the first and second moments of introduced random variables and denote

$$m_i(\alpha) = E\,\delta_i(\alpha), \quad g_i(\alpha) = E\,\gamma_i(\alpha), \quad e_i(\alpha) = E\,\beta_i(\alpha) - 1,$$

$$q(\alpha) = \sum_{i=1}^{r} (\lambda_i(\alpha) + \mu_i(\alpha) + \alpha^{(i)}\nu_i(\alpha)).$$

Put also

$$a_i(\alpha)^2 = E\,\delta_i(\alpha)^2, \quad c_i(\alpha)^2 = E\,\gamma_i(\alpha)^2, \quad d_i(\alpha)^2 = E\,(\beta_i(\alpha) - 1)^2, \qquad i = \overline{1,r}.$$

Let us introduce the following column vector-functions:

$$\begin{aligned} m(\alpha) &= (\lambda_1(\alpha)\,m_1(\alpha), \dots, \lambda_r(\alpha)\,m_r(\alpha)), \\ g(\alpha) &= (\mu_1(\alpha)\,g_1(\alpha), \dots, \mu_r(\alpha)\,g_r(\alpha)), \\ e(\alpha) &= (\alpha^{(1)}\nu_1(\alpha), \dots, \alpha^{(r)}\nu_r(\alpha)) \end{aligned}$$

and put

$$b(\alpha) = m(\alpha) - (I - P_0(\alpha)^*)\,g(\alpha) + e(\alpha),$$

where I is a unit matrix, $P_0(\alpha) = \|p_{ij}(\alpha)\|$, $i,j = \overline{1,r}$, and symbol "*" denotes the operation of transposition.

Let $G(\alpha) = b'(\alpha)$ be the matrix of partial derivatives for vector-function $b(\alpha)$ and let $B(\alpha)^2 = \|b_{ij}(\alpha)\|$, $i,j = \overline{1,r}$ be the matrix with the following elements:

$$\begin{aligned} b_{ij}(\alpha) &= -\mu_i(\alpha)\,p_{ij}(\alpha)\,c_i(\alpha)^2 - \mu_j(\alpha)\,p_{ji}(\alpha)\,c_j(\alpha)^2, \qquad i \neq j, \\ b_{ii}(\alpha) &= -2\mu_i(\alpha)\,p_{ii}(\alpha)\,c_i(\alpha)^2 + \lambda_i(\alpha)\,a_i(\alpha)^2 \\ &\quad + \mu_i(\alpha)\,c_i(\alpha)^2 + \sum_{k=1}^{r} \mu_k(\alpha)\,p_{ki}(\alpha)\,c_k(\alpha)^2 + \alpha^{(i)}\,\nu_i(\alpha)\,d_i(\alpha)^2, \qquad i = \overline{1,r}, \end{aligned}$$

Theorem 4.1. *Let the functions $\lambda_i(\alpha)$, $\mu_i(\alpha)$, $\nu_i(\alpha)$, $m_i(\alpha)$, $g_i(\alpha)$, $e_i(\alpha)$, $i = \overline{1,r}$, $P(\alpha)$ satisfy the local Lipschitz condition, uniformly on α in any bounded region* *in* $[0,\infty)^r$ *random variables* $\delta_i(\alpha)$, $\gamma_i(\alpha)$, $\beta_i(\alpha)$, $i = \overline{1,r}$ *are uniformly integrable,* $q(\alpha) > 0$, $\alpha \in [0,\infty)^r$,

$$n^{-1} Q_n(0) \xrightarrow{P} s_0, \tag{4.1}$$

and there exists $A > 0$ *such that the equation*

$$\mathrm{d}\eta(t) = b(\eta(t))\, q(\eta(t))^{-1}\, \mathrm{d}t, \quad \eta(0) = s_0$$

has unique solution $\eta(t)$ *such that* $\eta(t) > 0$ *on each component,* $t \in (0, A)$ *and*

$$\int_0^A q(\eta(t))^{-1}\, \mathrm{d}t > T.$$

Then

$$\sup_{0\le t\le T} |n^{-1} Q_n(nt) - s(t)| \xrightarrow{P} 0, \tag{4.2}$$

where

$$s(0) = s_0, \quad \mathrm{d}s(t) = b(s(t))\, \mathrm{d}t. \tag{4.3}$$

If in addition the functions $\lambda_i(\alpha)$, $\mu_i(\alpha)$, $\nu_i(\alpha)$, $m_i(\alpha)$, $g_i(\alpha)$, $e_i(\alpha)$, $i = \overline{1,r}$, $P(\alpha)$ *are continuously differentiable, uniformly on* α *in any bounded region in* $[0,\infty)^r$ *random variables* $\delta_i(\alpha)^2$, $\gamma_i(\alpha)^2$, $\beta_i(\alpha)^2$, $i = \overline{1,r}$ *are uniformly integrable and*

$$\frac{1}{\sqrt{n}} (Q_n(0) - n\, s_0) \overset{w}{\Longrightarrow} \gamma_0, \tag{4.4}$$

then the sequence of processes

$$\gamma_n(t) = \frac{1}{\sqrt{n}} (Q_n(nt) - n\, s(t))$$

$U-$*converges on interval* $[0,T]$ *to the multidimensional diffusion process* $\gamma(t)$ *such that*

$$\mathrm{d}\gamma(t) = G(s(t))\, \gamma(t)\, \mathrm{d}t + B(s(t))\, \mathrm{d}\omega(t), \quad \gamma(0) = \gamma_0. \tag{4.5}$$

where $\omega(t)$ *is the standard Wiener process in* R^r.

Proof. In our case process $Q_n(t)$ is a *MP* and we use Corollary 2.1 from part 2. Here, the variable $\tau_{n1}(n\alpha)$ has an exponential distribution with parameter $q(\alpha)$ and the variable $\xi_{n1}(n\alpha)$ does not depend on $\tau_{n1}(n\alpha)$ and

$$\xi_{n1}(n\alpha) = \begin{cases} \delta_i(\alpha)\, e_i & \text{with probability } \lambda_i(\alpha)\, q(\alpha)^{-1}, \\ \gamma_i(\alpha)\, (e_j - e_i) & \text{with probability } p_{ij}(\alpha)\, \mu_i(\alpha)\, q(\alpha)^{-1}, \\ -\gamma_i(\alpha)\, e_i & \text{with probability } p_{ir+1}(\alpha)\, \mu_i(\alpha)\, q(\alpha)^{-1}, \\ (\beta_i(\alpha) - 1)\, e_i & \text{with probability } \alpha^{(i)}\, \nu_i(\alpha)\, q(\alpha)^{-1}, \\ & i = \overline{1,r},\ j = \overline{1,r} \end{cases}$$

where e_i is a vector with the i-th component equal to 1 and the other components equal to 0. Calculating the characteristics of this variable and using the results of part 2 (Corollary 2.1) we obtain Theorem 4.1.

In conclusion we note that our *RPSM* is equivalent to the value of the queue only in the region $Q_n(t) \geq 0$. But the condition $\eta(t) > 0$, $t \in (0, A)$ together with relation (4.2) gives us that on each interval $[\alpha, \beta]$, $0 < \alpha$, $\beta < A$,

$$P\left\{Q_n(nt) > 0,\ \alpha \leq t \leq \beta\right\} \to 1$$

and $Q_n(t)$ is asymptotically equivalent to the value of the queue.

Let us consider an example, when $\lambda_i(\alpha) = \lambda_i$, $\mu_i(\alpha) = \mu_i\alpha_i$, $p_{ij}(\alpha) = p_{ij}$, $i = \overline{1,r}$, $j = \overline{0,r}$.

Then our network is equivalent to the network $(M/M/\infty)^r$. In this case equation (4.3) has the form

$$\mathrm{d}s(t) = (\lambda + (P^* - I)\, A\, s(t))\, \mathrm{d}t \tag{4.6}$$

where $\lambda = (\lambda_1, \ldots, \lambda_r)$, A—diagonal matrix with elements μ_i, $i = \overline{1,r}$.

Suppose that the matrix $P^* - I$ is invertible. Then iterating equation (4.6) we obtain the representation

$$s(t) = q_* + \exp\{(P^* - I)\, At\}\, (s_0 - q_*),$$

where

$$q_* = A^{-1}\, (I - P^*)^{-1}\, \lambda$$

(q_* is the stationary point).

Equation (4.5) has the form

$$\mathrm{d}\gamma(t) = (P^* - I)\, A\, \gamma(t)\, \mathrm{d}t + B(s(t))\, \mathrm{d}w(t). \tag{4.7}$$

If $s_0 = q_*$ then for all $t > 0$, $s(t) = q_*$, we have a quasi- stationary situation and stationary form for equation (4.7) with matrix of diffusion $B = B(q_*)$.

The general model of our system allows us to consider networks with impatient calls and unreliable servers. Many results devoted to Markov models were obtained in the book (Anisimov and Lebedev (1992)) and with the help of the martingale approach in the books (Korolyuk (1993); Basharin et al. (1989)).

4.2. Non-Markov queueing systems

Here we show how the results of part 2 can be applied for the analysis of non-Markov systems. Consider queueing system $SM/M_{SM,Q}/1/\infty$. This system is described in series scheme as it was done in Example 4.1. Let $x(t)$, $t \geq 0$ be a *SMP* with values in X such that the embedded MP x_k, $k \geq 0$ is uniformly ergodic with stationary measure $\pi(A)$, $A \in B_X$. Denote by $\{\tau(x),\ x \in X\}$ the sojourn time in state x. Let the family of non-negative functions $\mu(x,\alpha)$, $x \in X, \alpha \geq 0$, be given. There is one server and an infinite number of places for waiting. Calls enter the system one at a time at times of jumps $t_1 < t_2 < \cdots$ of the process $x(t)$. If a customer enters the system at time t_k and the number of calls in the system becomes equal to Q, then the intensity of service on the interval $[t_k, t_{k+1})$ is $\mu(x_k, \frac{1}{n}Q)$.

Let S_{n0} be the initial number of calls, and $Q_n(t)$ denote the number of calls in the system at time t.

We put

$$m(x) = E\,\tau(x); \quad m = \int_X m(x)\,\pi(\mathrm{d}x);$$

$$c(\alpha) = \int_X \mu(x,\alpha)\,m(x)\,\pi(\mathrm{d}x);$$

$$d^2(x) = D\,\tau(x); \quad d^2 = \int_X d^2(x)\,\pi(\mathrm{d}x);$$

$$e_1(\alpha) = \int_X \mu^2(x,\alpha)\,d^2(x)\,\pi(\mathrm{d}x); \quad e_2(\alpha) = \int_X \mu(x,\alpha)\,d^2(x)\,\pi(\mathrm{d}x);$$

$$D^2(\alpha) = c(\alpha) + e_1(\alpha) + 2\,(1 - c(\alpha))\,m^{-1}\,e_2(\alpha) + ((1 - c(\alpha))\,m^{-1})^2\,d^2;$$

$$b(\alpha) = (1 - c(\alpha))\,m^{-1};$$

$$g(x,\alpha) = 1 - m(x)\,m^{-1}(1 - c(\alpha) + \mu(x,\alpha)\,m).$$

Theorem 4.2. *Suppose that functions $\mu(x,\alpha)$ are locally Lipschitz with respect to α uniformly on $x \in X$, $m > 0$, the variables $\tau(x)$ are uniformly integrable, function $c(\alpha)$ has no more then linear growth and condition (4.1) is fulfilled. Then relation (4.2) holds, where*

$$\mathrm{d}s(t) = (1 - c(s(t))\,m^{-1}\,\mathrm{d}t, \qquad s(0) = s_0 \tag{4.8}$$

and T is any positive value such that $s(t) > 0$, $\quad t \in [0,T]$.

If in addition the variables $\tau(x)^2$ are uniformly integrable, function $c(\alpha)$, is continuously differentiable and condition (4.4) holds, then the sequence of processes

$$\gamma_n(t) = \frac{1}{\sqrt{n}}\,(Q_n(nt) - n\,s(t))$$

U-converges on interval $[0,T]$ to the multidimensional diffusion process $\gamma(t)$ such that

$$\mathrm{d}\gamma(t) = G(s(t))\,\gamma(t)\,\mathrm{d}t + m^{-1/2} \times (D^2(s(t)) + B^2(s(t)))^{1/2}\,\mathrm{d}w(t),$$

where $G(\alpha) = b'(\alpha)$ and

$$B^2(a) = E\left(g(x_0,\alpha)^2 + 2\sum_{k=1}^{\infty} g(x_0,\alpha)\,g(x_k,\alpha)\right)$$

where $P\{x_0 \in A\} = \pi(A)$, $A \in B_X$.

Proof. To prove this result we represent our system as above in terms of *PSMS*. Here times t_k are switching times and variable $\xi_{nk}(x, n\alpha)$ can be represented in the form

$$\xi_{n1}(x, n\alpha) = 1 - \Pi_{\mu(x,\alpha)}(\tau(x)),$$

where $\Pi_\lambda(t)$ is a Poisson process with the parameter λ. It is easy to see that $E\,\xi_{n1}(x, n\alpha) = 1 - \mu(x, \alpha)\, m(x)$. According to constructions in Theorems 2.3 and 2.4 it is easy to calculate other characteristics. We mention also that the process of changing of a queue is monotone on each interval $[t_k, t_{k+1})$. Then U-convergence *RPSM* to limit process automatically implies U-convergence of *PSMS* (there is no need to check conditions (2.19) and (2.20)). So under our assumptions all the conditions of Theorems 2.3 and 2.4 are fulfilled. That proves Theorem 4.2.

In particular when $\mu(x, \alpha) \equiv \alpha\mu$, our system is equivalent to system $SM/M/\infty$ with semi-Markov input and exponential service. In this case $c(\alpha) = \alpha\mu m$, $b(\alpha) = 1/m - \alpha\mu$ and from (4.8) we obtain

$$s(t) = (\mu m)^{-1} - ((\mu m)^{-1} - s_0)\, e^{-\mu t}.$$

Consequently in (4.9)

$$G(\alpha) = -\mu;$$

$$D(\alpha)^2 = \mu m\alpha + d^2/m^2;$$

$$g(x, \alpha) = 1 - m(x)/m.$$

The result of Theorem 4.2 also takes place when the intensity of service may be written in the form $\mu(x_k, n^{-1}\, Q(nt))$ (which may depend on the current value of the queue at time t).

In the same way it is possible to study the queueing networks of type $(SM/M_{SM,Q}/1/\infty)^r$ with semi-Markov input and service depending on the state of some *SMP* and current values of queues in the nodes and even more complicated systems of type $(M_{SM,Q}/M_{SM,Q}/1/\infty)^r$, intensity of input which also depends on the state of *SMP* and current values of queues.

Some other examples of non-Markov and even non-semi-Markov models of queueing systems of the types $G_Q/M_Q/1/\infty$, $SM_Q/M_Q/1/\infty$ and $(G_Q/M_Q/1/\infty)^r$ are considered in Anisimov (1993, 1995).

The results of parts 3 and 4 show that theorems of *AP* and *DA* types for *SP* give us a new approach in studying transition phenomena for wide classes of complex stochastic systems. For instance some examples for near-critical branching processes with a large number of particles were considered in (Anisimov (1991, 1993)).

References

Anisimov, V.V. (1973). Asymptotic consolidation of the states of random processes. *Cybernetics* **9**(3), 494–504.

Anisimov, V.V. (1975). Limit theorems for random processes and their application to discrete summation schemes, *Teoria Veroyatnostey i Primenen.* **20**(3), 692–694. English transl. in *Theor. Probab. Appl.* **20**.

Anisimov, V.V. (1977). Switching processes. *Cybernetics*, **13**(4), 590–595.

Anisimov, V.V. (1978). Limit theorems for switching processes and their applications. *Cybernetics* **14**(6), 917–929.

Anisimov, V.V. (1984). *Asymptotic Methods in Analysing Stochastic Systems.* Publ. Mecniereba, Tbilisi, 180 p. (In Russian).

Anisimov, V.V. (1988a). *Random Processes with Discrete Component. Limit Theorems.* Publ. Kiev Univ., 184 p. (In Russian).

Anisimov, V.V. (1988b). Limit theorems for switching processes. *Theory Probab. and Math. Statist.*, 37, 1–5.

Anisimov, V.V. (1991). Averaging principle for switching recurrent sequences. *Theor. Probability and Math. Statist.* **45**, 1–8.

Anisimov, V.V. (1992a). Limit theorems for Switching Processes. *Functional Analysis III, Proc. Postgrad. School and Conf., Dubrovnik, Yugoslavia, 1989; Var. publ. ser., Aarhus Univ.* 40, 235–262.

Anisimov, V.V. (1992b). Averaging principle for switching processes. *Theory Probab. and Math.Statist.* 46, 1–10.

Anisimov, V.V. (1992c). Switching processes: asymptotic theory and applications, *Proc. Int. Conf. "Stoch. Evolutionary Systems: Theory and Appl."* Katzively, Ukraine, (1992); New Trends in Probability and Statist., TVP Sci. Publ. 1–28, (1994).

Anisimov, V.V. (1992d). Asymptotic methods in the theory of switching processes. *Proc. of 2-nd Ukrainian-Hungarian Conf. (1992), VSP/TViMS, 1994*, 1–20.

Anisimov, V.V. (1993). Averaging principle for processes with fast switchings. *Random Oper. and Stoch. Eqv.* **1**(2), 151–160.

Anisimov, V.V. (1994). Limit theorems for processes with semi-Markov switchings and their applications, *Random Operators and Stochastic Equations* **2**(4), 333–352.

Anisimov, V.V. (1995). Switching processes: averaging principle, diffusion approximation and applications. *Acta Applicandae Mathematicae,* Kluwer, The Netherlands, 1–47. (To appear).

Anisimov, V.V. and Aliev, A.O. (1990). Limit theorems for recurrent processes of semi-Markov type. *Theory Probab. and Math. Statist.* 41, 7–13.

Anisimov, V.V. and Lebedev, E.A. (1992). *Stochastic Queueing Networks. Markov Models.* Publ. Kiev Univ., 206 p. (In Russian).

Basharin, G.P., Bocharov, P.P. and Kogan, Ja.A. (1989). *Analysis of Queues in Computing Networks.* Publ. Nauka, Moscow, 336 p. (In Russian).

Buslenko, N.P., Kalashnikov, V.V. and Kovalenko, I.N. (1973). *Lectures on the Theory of Complex Systems.* Publ. "Sov.Radio", Moscow, 440 p. (In Russian).

Ežov, I.I. and Skorokhod, A.V. (1969). Markov processes which are homogeneous in the second component. *Theor. Probab. Appl.* **14**, 679–692.

Gikhman, I.I. and Skorokhod, A.V. (1973). *Theory of Random Processes II.* Publ. "Nauka", Moscow, 640 p. (In Russian).

Griego, R. and Hersh, R. (1969). Random evolutions, Markov chains, systems of partial differential equations. *Proc. Nat. Acad. Sci.* 62, 305–308.

Hersh, R. (1974). Random evolutions: survey of results and problems. *Rocky Mount. J. Math.* **4**(3), 443–475.

Kac, M. (1956). A stochastic model related to the telegrapher's equation. Magnolia Petroleum Co. Colloq. Lectures repr. In: *Rocky Mount. J. Math.* 4, 497–510.

Kertz, R. (1978a). Limit theorems for semigroups with perturbed generators with applications to multi-scaled random evolutions. *TAMPS* 27, 215–233.

Kertz, R. (1978b). Random evolutions with underlying semi-Markov processes. *Publ. Res. Jnst. Math. Sci.* **14**, 589–614.

Khas'minskii, R.Z. (1968). About the averaging principle for ITO stochastic differential equations. *Cybernetics* **4**(3), 260–279.

Korolyuk, V.S. (1993). *Stochastic models of systems.* Publ. "Libed", Kiev, 135 p. (In Russian).

Korolyuk, V.S. and Swishchuk, A.V. (1986). Central limit theorem for semi-Markov random evolutions. *Ukrainian Math. J.* **38**, 330–335. (In Russian).

Korolyuk, V.S. and Swishchuk, A.V. (1988). Semi-Markov random evolutions. Phase aggregations and applications. *Intern. Summer. School on Probab. Theory Math. Stat., Varna, Gold Sands, 1985,* Sophia, 116–131.

Korolyuk, V.S. and Swishchuk, A.V. (1992). *Random evolutions.* Publ."Naukova Dumka", Kiev , 254 p. (In Russian); English transl., Kluwer Acad. Publ., 1994.

Korolyuk, V.S. and Turbin, A.F. (1978). *Mathematical foundations of phase consolidations of complex systems.* Publ. "Naukova Dumka", Kiev, 218 p. (In Russian).

Korolyuk, V.S. and Turbin, A.F. (1983). Limit theorems for Markov random evolutions in the scheme of asymptotic phase lumping. *Lecture Notes in Math.* **1021**, Springer, New York, 327–332.

Kurtz, T. (1972). A random Trotter product formula. *Proc. Amer. Math. Soc.* 35, 147 154.

Kurtz, T. (1973). A limit theorem for perturbed operator semigroups with applications to random evolutions. *J. Funct. Anal.* 12, 55–67.

Papanicolaou, G. and Hersh, R. (1972). Some limit theorems for stochastic equations and their applications. *Indiana Univ. Math. J.* 21, 815–840.

Pinsky, M. (1971). Multiplicative operator functionals of Markov processes. *BAMS* 77, 377–380.

Pinsky, M. (1975a). Multiplicative operator functionals and their asymptotical properties. *Adv. in Prob.* 3, 1–100.

Pinsky, M. (1975b). Random evolutions. *Lecture Notes in Math.* **451**, Springer, New York, 89–100.

Pinsky, M. (1991). *Lectures on Random Evolution.* World Scientific Press.

Skorokhod, A.V. (1987). *Asymptotic Methods in the Theory of Stochastic Differential Equations.* Publ. "Naukova Dumka", Kiev. 326p., English transl.: Amer. Math. Soc., Providence, RI, 1989.

Skorokhod, A.V. (1992). Dynamic systems with random influences. *Functional Analysis III,* Proc. Postgrad. School and Conf., Dubrovnik, Yugoslavia, 1989; Var. publ. ser., Aarhus Univ., 40, 193–234.

Watkins, J.C. (1984). A central limit problem in random evolutions. *Ann. Prob.* **12**(2), 480–513.

Weidlich, W. (1991). Physics and Social Science – the Approach of Synergetics. *Physics Reports* **204**, 1–163.

Weidlich, W. (1992). Modelling Concepts of Synergetics with Application to Transition between Totalitarian and Liberal Political Ideology. *Report on the Seminar Series on Catastrophic Phenomena in Soviet Society and Self-organized Behaviour of Social Systems.* Kiev, Ukraine.

Weidlich, W. and Haag, G (Eds.) (1988). *Interregional Migration-Dynamic theory and Comparative Analysis.* Springer, Berlin, Heidelberg, New York.

Exploring Stochastic Laws. pp. 41–54.
Eds. A.V. Skorokhod and Yu.V. Borovskikh

Approximation of generalized U-statistics

YURI V. BOROVSKIKH

Department of Applied Mathematics, Transport University, Moskovsky Prospekt 9, 190031, St Petersburg, Russia

Received February 3, 1995

Abstract. Multi-sample U-statistics are considered. Estimates of the rate of convergence in the central limit theorem are established.

1. Consider c independent sequences $\{X_{ji},\ \ i=1,2,\ldots\},\ \ j=1,\ldots,c$ independent random variables. All variables from the j-th sequence take values in a measurable space $(\mathcal{X}_j, \mathcal{A}_j)$ and have on it the same probability distribution P_j. Let $\Phi(x_{ji},\ i=1,\ \ldots,\ m_j,\ \ j=1,\ \ldots,\ c)$ be a measurable function which is symmetric with respect to m_j arguments of the j-th set.

Generalized U-statistics are defined (see [1, 2]) by

$$U = \prod_{j=1}^{c} \binom{n_j}{m_j}^{-1} \sum \Phi(X_{j\alpha},\ \ \alpha = i_{j1},\ \ldots,\ i_{jm_j},\ \ 1 < j < c), \tag{1}$$

where the sum is taken over all $1 \le i_{j1} < \ldots < i_{jm_j} \le n_j,\ \ m_j \le n_j,\ j=1,\ldots,c$.

The mapping Φ is called the *kernel* of U-statistics, the integer-valued vector $(m_1,\ldots,m_c)$ is called the *order of the kernel of* Φ, and number c – *order* of U-statistics.

Suppose $E\Phi = 0$ and $E\Phi^2 < \infty$. Under these conditions for each vector $(d_1,\ \ldots,\ d_c),\ \ 0 \le d_j \le m_j,\ 1 \le j \le c$ the functions are defined as

$$\Phi_{d_1\ldots d_c}(x_{ji},\ i=1,\ldots,d_j,\ 1 \le j \le c) =$$
$$= E\Phi(x_{j1},\ \ldots,\ x_{jd_j},\ x_{jd_{j+1}},\ \ldots, X_{jm_j},\ 1 \le j \le c),$$

so $\Phi_{0\ldots0} = 0,\ \ \Phi_{m_1\ldots m_c}(\ldots) = \Phi(\ldots)$ and in particular let

$$g_j(x) = E(\Phi(X_{11},\ \ldots,\ X_{1m_1};\ \ldots;\ X_{c1},\ \ldots,\ X_{cm_c})|\ X_{j1}=x),\quad j=1,\ \ldots,\ c.$$

Let

$$\eta_{d_1\ldots d_c} = E\Phi^2_{d_1\ldots d_c}(X_{ji},\ \ 1 \le i \le d_j,\ 1 \le j \le c),$$

$$\sigma_j^2 = Eg_j^2(X_{j1}), \quad j = 1, \ldots, c.$$

Then the variance $\sigma^2 = EU^2$ of U-statistics (1) is calculated by the Hoeffding formula

$$\sigma^2 = \prod_{j=1}^{c} \binom{n_j}{m_j}^{-1} \sum_{\substack{0 \le d_j \le m_j \\ 1 \le j \le c}} \prod_{j=1}^{c} \left\{ \binom{m_j}{d_j} \binom{n_j - m_j}{m_j - d_j} \right\} \eta_{d_1 \ldots d_c} \tag{2}$$

It is known that if $n = \min(n_1, \ldots, n_c) \to \infty$, then

$$\Delta = \sup_x |P(\sigma^{-1}U < x) - \Phi(x)| \to 0, \tag{3}$$

where $\Phi(x)$ is a limiting distribution function. Moreover, if $\max(\sigma_1, \ldots, \sigma_c) > 0$, then $\Phi(x)$ is a standard normal distribution function. If $\max(\sigma_1, \ldots, \sigma_c) = 0$, then the limiting distribution is concentrated, generally speaking, in an infinitely-dimensional space. More specifically the case with $c = 1$ was investigated in [2, 4].

In this paper we are interested in the problem of evaluating the rates of convergence in (3) for arbitrary $c \ge 1$. For $c = 1$ the question was studied in [2 – 7], for $c \ge 1$ in [2, 8–10]. Everywhere hereinafter we suppose that $\max(\sigma_1, \ldots, \sigma_c) > 0$. In this case U-statistics are called *nondegenerate* or the kernel Φ in (1) is called *nondegenerate*. For completeness let $\sigma_1 = \max(\sigma_1, \ldots, \sigma_c)$ and without loss of generality we assume that $n_1 \le n_2 \le \ldots \le n_c$. Let $m = \max(m_1, \ldots, m_c)$ and

$$\Gamma_c = \max_{1 \le j \le c} E|g_j(X_{j1})|^3.$$

2.

Theorem 1. *If the condition*

$$\max(\Gamma_c, E\Phi^2) < \infty \tag{4}$$

is satisfied, then for $n \ge m + 1$ the following inequality is true

$$\sup_x |P(\sigma^{-1}U < x) - \Phi(x)| \le C_\Phi n^{-1/2}, \tag{5}$$

where positive constant C_Φ is dependent on the kernel Φ only by $E\Phi^2$, $E|g_j(X_{j1})|^3$, σ_j, $1, \ldots, c$ and is estimated in the proof.

Proof. Let

$$\hat{U} = \sum_{j=1}^{c} \sum_{i=1}^{n_j} E(U|\ X_{ji}).$$

The value $\hat{U}$ is called the *projection* of U-statistics on the space of linear statistics. It is obvious that $\hat{U}$ can be represented as

$$\hat{U} = \sum_{j=1}^{c} \sum_{i=1}^{n_j} m_j n_j^{-1} g_j(X_{ji}).$$

Let

$$Y_{i_{11},\ldots,i_{cm_c}} = \Phi(X_{1i_{11}}, \ldots, X_{ci_{cm_c}}) - \sum_{j=1}^{c} \sum_{k=1}^{m_j} g_j(X_{ji_{jk}}).$$

These random variables satisfy the property of degeneracy:

$$EY_{i_{11},\ldots,i_{cm_c}} = 0, \quad E(Y_{i_{11},\ldots,i_{cm_c}} \mid X_{ji_{jk}}) = 0. \tag{6}$$

Further, let

$$\hat{\sigma}^2 = \sum_{j=1}^{c} m_j^2 n_j^{-1} \sigma_j^2, \quad \rho = \hat{\sigma}^{-1} \hat{U}, \quad \delta = \hat{\sigma}^{-1}(U - \hat{U}),$$

$$\delta_1 = \hat{\sigma}^{-1} \prod_{j=1}^{c} \binom{n_j}{m_j}^{-1} \sum Y_{i_{11},\ \ldots,\ i_{cm_c}},$$

where the sum is taken over all

$$1 \leq i_{j1} < \ldots < i_{jm_j} \leq k_j, \quad k_j = n_j - [\sqrt{n_j}], \quad j = 1, \ldots, c,$$

where $[\sqrt{n_j}]$ is an integral part of $\sqrt{n_j}$.

Denote

$$\Delta_1 = \sup_x |\Phi(x) - \Phi(x\hat{\sigma}\sigma^{-1})|,$$

$$\Delta_2 = \sup_x |\Phi(x) - \Phi(x + An^{-1/2})| + P(|\delta_1 - \delta| > An^{-1/2}),$$

$$\Delta_3 = \sup_x |P(\rho + \delta_1 < x) - \Phi(x)|,$$

where A is an arbitrary positive number.

It is obvious that

$$P(\sigma^{-1}U < x) = P(\rho + \delta < x\sigma\hat{\sigma}^{-1}),$$

$$\sup_x |P(\sigma^{-1}U < x) - \Phi(x)| \leq \sup_x |P(\rho + \delta < x) - \Phi(x)| + \Delta_1.$$

Further,

$$P(\rho + \delta < x) = P\{(\rho + \delta_1 < x + \delta_1 - \delta) \cap (|\delta_1 - \delta| \leq An^{-1/2})\} +$$

$$+ P\{(\rho + \delta_1 < x + \delta_1 < x + \delta_1 - \delta) \cap (|\delta_1 - \delta|) > An^{-1/2})\}.$$

From this, firstly we have

$$P(\rho + \delta < x) \geq P\{(\rho + \delta_1 < x + \delta_1 - \delta) \cap (|\delta_1 - \delta| < An^{-1/2})\} \geq$$

$$\geq P\{(\rho+\delta_1 < x - An^{-1/2}) \cap (|\delta-\delta_1| \leq An^{-1/2})\} \geq$$
$$\geq P(\rho+\delta_1 < x - An^{-1/2}) - P(|\delta-\delta_1| > An^{-1/2})$$

and secondly,

$$P(\rho+\delta < x) \leq P(\rho+\delta_1 < x + An^{-1/2}) + P(|\delta-\delta_1| > An^{-1/2}),$$

which gives the inequality

$$\sup_x |P(\rho+\delta < x) - \Phi(x)| \leq \Delta_2 + \Delta_3.$$

Thus,

$$\Delta \leq \Delta_1 + \Delta_2 + \Delta_3. \tag{7}$$

Estimate of Δ_1. From the properties of the normal distribution $\Phi(x)$ we have

$$\Delta_1 \leq (2\pi e)^{-1/2}(\sigma - \hat{\sigma})\hat{\sigma}^{-1},$$

where

$$\sigma^2 - \hat{\sigma}^{-2} = E(U - \hat{U})^2.$$

In connection with (2) we can write

$$E(U-\hat{U})^2 = \prod_{j=1}^{c} \binom{n_j}{m_j}^{-1} \sum \prod_{j=1}^{c} \left\{ \binom{m_j}{d_j} \binom{n_j - m_j}{m_j - d_j} \right\} y_{d_1,\ldots,d_c}, \tag{8}$$

where the sum is taken over all indexes

$$d_1 + \ \ldots \ + d_c \geq 2, \quad 1 \leq d_j \leq m_j, \quad j = 1, \ldots, c$$

$$y_{d_1,\ldots,d_c} = EY^2_{d_1,\ldots,d_c}(X_{ji}, \quad 1 \leq i \leq d_j, \ 1 \leq j \leq c)$$

and function $Y_{d_1,\ldots,d_c}$ is defined in the same manner as $\Phi_{d_1,\ldots,d_c}$. The expression on the right-hand side of (8) for $n \geq m$ can be evaluated firstly as

$$E(U-\hat{U})^2 \leq \sum_{d_1+\ldots+d_c \geq 2} \prod_{j=1}^{c} \prod_{s=1}^{d_j} \frac{d_j!}{n_j - m_j + s} \binom{m_j}{d_j}^2 y_{d_1,\ldots,d_c}$$

and then, taking into account that $y_{d_1,\ldots,d_c} \leq EY^2$, in the following way

$$E(U-\hat{U})^2 \leq m^c 2^{mc}(n-m+1)^{-2}EY^2.$$

As

$$(\sigma - \hat{\sigma})\hat{\sigma}^{-1} = \frac{\sigma^2 - \hat{\sigma}^2}{\hat{\sigma}(\sigma + \hat{\sigma})} \leq \frac{1}{2}\hat{\sigma}^{-2}E(U-\hat{U})^2,$$

then for all $n \geq m+1$

$$\Delta_1 \leq (2\pi e)^{-1/2}(m+1)^{c+2}2^{mc-1}a_c^{-1}n^{-1}EY^2, \tag{9}$$

where

$$a_c = \sum_{j=1}^{c} \lambda_j m_j^2 \sigma_j^2 \geq m_1^2 \sigma_1^2,$$

where

$$\lambda_j = \begin{cases} 1\,, & if \;\; j = 1, \\ \frac{n_1}{n_j}\,, & if \;\; j = 2, \ldots . \end{cases}$$

Estimate of Δ_2. The first term in Δ_2 is not greater than $A(2\pi)^{1/2}n^{-1/2}$ because of the properties of the distribution function Φ.

For the evaluation of the second item apply Chebyshev's inequality. Then

$$\Delta_2 \leq A(2\pi)^{1/2}n^{-1/2} + A^{-2}nE(\delta - \delta_1)^2. \tag{10}$$

Direct calculations show that

$$E(\delta - \delta_1)^2 = E\delta^2 - E\delta_1^2 - 2\hat{\sigma}^{-2} \prod_{j=1}^{c} \binom{n_j}{m_j}^{-2} \times$$

$$\times \sum\nolimits_1 \prod_{j=1}^{c} \left\{ \binom{n_j - k_j}{m_j - s_j} \binom{k_j}{s_j} \right\} \sum\nolimits_2 \prod_{j=1}^{c} \left\{ \binom{s_j}{r_j} \binom{k_j - s_j}{s_j - r_j} \right\} y_{r_1 \ldots r_c},$$

where in $\sum_1$ the sum is taken over indexes

$$\sum_{j=1}^{c} (m_j - s_j) \geq 1, \quad 1 \leq s_j \leq m_j, \;\; j = 1, \ldots, c,$$

and in $\sum_2$ for indexes

$$r_1 + \ldots + r_c \geq 2, \quad 1 \leq r_j \leq s_j, \;\; j = 1, \ldots, c.$$

From (8) for $n \geq m + 1$ we get

$$|E\delta^2 - E\delta_1^2| \leq n^{-3/2} 8c(m+1)^2 a_c^{-1} \prod_{j=1}^{c} \left(\sum_{d_j=1}^{m_j} d_j! \binom{m_j}{d_j}^2 \right) EY^2,$$

which together with previous considerations leads to the inequality

$$E(\delta - \delta_1)^2 \leq C_1 n^{-3/2}, \tag{11}$$

where

$$C_1 = 8c(m+1)^2 a_c^{-1} \prod_{j=1}^{c} \left(\sum_{d_j=1}^{m_j} d_j! \binom{m_j}{d_j}^2 \right) EY^2 +$$

$$+4(m+1)^2 a_c^{-1} \prod_{j=1}^{c} \left\{ \sum_{s_j=1}^{m_j} \binom{m_j}{s_j} 2^{s_j} \sum_{r_j=1}^{s_j} 2^{r_j} (s_j - r_j + 1) \binom{s_j}{r_j} \right\}.$$

Choosing $A = (2\sqrt{2\pi}C_1)^{1/3}$, from (10) and (11) we obtain

$$\Delta_2 \le \frac{3}{2}(C_1\pi^{-1})^{1/3}n^{-1/2}. \tag{12}$$

Estimate of Δ_3. Denote

$$\begin{aligned} \gamma_j &= E|g_j(X_{j1})|^3, \quad j = 1, \dots, c\,; \\ t_1 &= \sum_{j=1}^{c} m_j^3 n_j^{-2}\gamma_j, \\ t_2 &= 1 + \sum_{j=2}^{c}(m_j m_1^{-1})^2(\sigma_j\sigma_1^{-1})^2, \\ T &= \frac{1}{4}t_1^{-1}t_2^{-1}\hat{\sigma}^3. \end{aligned}$$

By the known Essen inequality (see [11])

$$\Delta_3 \le \frac{1}{\pi}\int_{-T}^{T}\Big|E\exp\{it(\rho+\delta_1)\} - \exp\{-t^2/2\}\Big|\,|t|^{-1}dt + \frac{24}{\pi\sqrt{2\pi}}T^{-1}. \tag{13}$$

For T, taken from the domain $|t| \le T$ (see [11])

$$\Big|E\exp\{it\rho\} - \exp\{-t^2/2\}\Big| \le 16\hat{\sigma}^3\sum_{j=1}^{c} m_j^3\gamma_j n_j^{-2}|t|^3\exp\{-t^3/3\}.$$

Then from (13) we have

$$\Delta_3 \le \frac{1}{\pi}\int_{-T}^{T}|E\exp\{it\rho\}(1-\exp\{it\delta_1\})|\,|t|^{-1}dt + \frac{1}{\pi}(6\sqrt{3\pi} + \frac{24}{\pi\sqrt{2\pi}})T^{-1}. \tag{14}$$

Consider the integrand function in (14). By Taylor's formula

$$E\exp\{it\rho\}(1-\exp\{it\delta_1\}) = -itE(\delta_1\exp\{it\rho\}) + t^2E(\delta_1^2\exp\{it\rho\}\theta_1) \tag{15}$$

where the random variable $|\theta_1| \le 1$.

According to definition

$$E(\delta_1\exp\{it\rho\}) = \hat{\sigma}^{-1}\prod_{j=1}^{c}\binom{n_j}{m_j}^{-1}\sum E(Y_{i_11},\dots,i_{cm_c}\exp(it\hat{\sigma}^{-1}\hat{U})),$$

where the sum is taken over all $1 \le i_{j1} < \dots < i_{jm_j} \le k_j, \quad j = 1, \dots, c$.

Let $\varphi_j(t)$ be a characteristic function of the random variable $g_j(X_{j1}), \quad j = 1, \dots, c$. By independence

$$E\left(Y_{i_{11},\dots,i_{cm_c}}\exp(it\hat{\sigma}^{-1}\hat{U})\right) =$$

$$= \prod_{j=1}^{c}\left(\varphi_j\left(\frac{tm_j}{\hat{\sigma}n_j}\right)\right)^{n_j-m_j} E\left(Y_{i_{11},\dots,i_{cm_c}}\exp\left(it\hat{\sigma}^{-1}\sum_{j=1}^{c}\sum_{k=1}^{m_j}m_jn_j^{-1}g_j(X_{ji_{jk}})\right)\right).$$

Therefore,

$$E(\delta_1 \exp\{it\rho\}) = \hat{\sigma}^{-1} \prod_{j=1}^{c} \left\{ \binom{k_j}{m_j} \binom{n_j}{m_j}^{-1} \left(\varphi_j \left(\frac{tm_j}{\hat{\sigma} n_j} \right) \right)^{n_j - m_j} \right\} \times$$

$$\times E \left(Y \exp \left(it\hat{\sigma}^{-1} \sum_{j=1}^{c} \frac{m_j}{n_j} \sum_{k=1}^{m_j} g_j(X_{jk}) \right) \right).$$

In the known formula

$$\prod_{s=1}^{m} (1 - z_s) = \sum_{k=0}^{m} (-1)^k \sum_{1 \le j_1 < \ldots < j_k \le m} \prod_{s=1}^{k} z_{j_s}$$

put

$$z_s = 1 - \exp\left(\frac{it}{\hat{\sigma}} \frac{m_j}{n_j} g_j(X_{js}) \right).$$

Then, taking into account property (6) we find the representation

$$E \left(Y \exp \left(it\hat{\sigma}^{-1} \sum_{j=1}^{c} \frac{m_j}{n_j} \sum_{k=1}^{m_j} g_j(X_{jk}) \right) \right) =$$

$$= \sum_{r=2}^{m_1 + \ldots + m_c} (-1)^r \sum_{r_1 + \ldots + r_c = r} \prod_{s=1}^{c} \binom{m_s}{r_s} \times$$

$$\times E \left(Y_{r_1, \ldots, r_c} \prod_{s=1}^{c} \prod_{j_s=1}^{r_s} \left(1 - \exp \left(\frac{it}{\hat{\sigma}} \frac{m_s}{n_s} g_s(X_{sj_s}) \right) \right) \right) = \pi_1,$$

where functions $Y_{r_1,\ldots,r_c}$ are defined in (8). From this representation by inequality $|\exp\{ix\} - 1| \le |x|$ we get

$$|\pi_1| \le \sum_{r=2}^{m_1 + \ldots + m_c} \sum_{r_1 + \ldots + r_c = r} \prod_{s=1}^{c} \left(\binom{m_s}{r_s} \left(\frac{m_s |t|}{\hat{\sigma} n_s} \right)^{r_s} \right) \times$$

$$\times E |Y_{r_1, \ldots, r_c} \prod_{s=1}^{c} \prod_{j_s=1}^{r_s} g_s(X_{sj_s})| = \pi_n(t).$$

The characteristic function $\varphi_j(tm_j/\hat{\sigma} n_j)$ can be evaluated by a standard method (see, for example [11]). Therefore in the domain $|t| < T$ for $n \ge m + 1$

$$|E(\delta_1 \exp\{it\rho\})| \le \hat{\sigma}^{-1} \prod_{j=1}^{c} \binom{k_j}{m_j} \binom{n_j}{m_j}^{-1} \pi_n(t) \exp\left(-\frac{t^2}{3(m+1)} \right) \tag{16}$$

Consider the second term on the right-hand side of (15). By reasoning of independence we obtain

$$E(\delta_1^2 \exp\{it\rho\}\theta_1) = \prod_{j=1}^{c} \left(\varphi_j \left(\frac{tm_j}{\hat{\sigma} n_j} \right) \right)^{n_j - k_j} E \left(\delta_1^2 \theta_1 \exp \left(\frac{ti}{\hat{\sigma}} \sum_{j=1}^{c} \frac{m_j}{n_j} \sum_{i=1}^{k_j} g_j(X_{ji}) \right) \right).$$

According to [11] for $|t| \le T$

$$\left|\prod_{j=1}^{c}\left(\varphi_j\left(\frac{tm_j}{\hat{\sigma}n_j}\right)\right)\right|^{n_j-n_k} \le \exp(-t^2 b_c),$$

where

$$b_c = 2\hat{\sigma}^{-2}\sum_{j=1}^{c} m_j^2 n_j^{-2}[\sqrt{n_j}]\sigma_j^2 - \frac{1}{6}t_1^{-1}t_2^{-1}\sum_{j=1}^{c} m_j^3 n_j^{-3}[\sqrt{n_j}]\gamma_j.$$

Analysis shows that for $n \ge m+1$

$$b_c \ge \frac{2\sqrt{3}-1}{6(\sqrt{3}+1)}t_2^{-1}n^{-1/2}.$$

Therefore inside the domain $|t| \le T$

$$|E(\delta_1^2\exp\{it\rho\}\theta_1)| \le (n-m+1)^{-2}\hat{\sigma}^{-2}EY^2\prod_{j=1}^{c}\left(\sum_{d_j=1}^{m_j} d_j!\binom{m_j}{d_j}^2\right)\exp(-t^2 b_c). \quad (17)$$

With the help of (15)–(17) the integral in (14) is easily evaluated and the following inequality can be written

$$\Delta_3 \le C_2 n^{-1/2} \quad (18)$$

with some positive constant C_2 dependent on the kernel only by EY^2, σ_j and γ_j.

Now (5) follows from (7), (9), (12) and (18) under condition (4).

Theorem 2. *If the conditions*

$$\Gamma_c < \infty \text{ and } E|\Phi|^{5/3} < \infty$$

are satisfied, then for $n \ge m+1$ *the following inequality is true*

$$\sup_x |P(\hat{\sigma}^{-1}U < x) - \Phi(x)| \le C(\Phi)n^{-1/2},$$

where the positive constant $C(\Phi)$ *depends on the kernel* Φ *only by* $E|\Phi|^{5/3}$, $E|g_j|^3$, σ_j, $j = 1, \dots, c$.

Proof. It is sufficient to apply the method of the proof of Theorem 1 and the considerations which are given in [3] for one-sample U-statistics.

3. Denote

$$h_c = \frac{1}{6}\sum_{j=1}^{c} m_j^3 n_j^{-2} E g_j^3(X_{j1})\left(\sum_{j=1}^{c} m_j^2 n_j^{-1}\sigma_j^2\right)^{-3/2} +$$

$$+E\left(Y\left(\sum_{j=1}^{c}\frac{m_j}{n_j}\sum_{k=1}^{m_j} g_j(X_{jk})\right)^2\right)\left(\sum_{j=1}^{c} m_j^2 n_j^{-1}\sigma_j^2\right)^{-3/2},$$

$$G_c(x) = \Phi(x) + \frac{1}{\sqrt{2\pi}} h_c(1 - x^2)\exp\{-x^2/2\}, \quad x \in R$$

$$g_c(t) = (1 + (it)^3 h_c)\exp\{-t^2/2\}, \quad t \in R.$$

Function $g_c(t)$ is the Fourier transformation of function $G_c(t)$.

Theorem 3. *Assume the Cramer property*

$$\lim_{|t|\to\infty} \sup |\varphi_j(t)| < 1, \quad j = 1, \dots, c \tag{19}$$

is satisfied and the moment conditions

$$\max(\max_{1\le j\le c} Eg_j^4(X_{j1}),\ \ E|\Phi|^3) < \infty \tag{20}$$

are fulfilled. Then for $n \to \infty$

$$\sup_x |P(\sigma^{-1}U < x) - G_c(x)| = O(n^{-3/5}). \tag{21}$$

Proof. Note that

$$\sup_x |P(\sigma^{-1}U < x) - G_c(x)| = \sup_x |P(\rho + \delta < x) - G_c(x\hat{\sigma}\sigma^{-1})|.$$

Further,

$$\begin{aligned} \sup_x |P(\rho + \delta < x) - G_c(x)| \ \le\ & \sup_x |G_c(x\hat{\sigma}\sigma^{-1}) - G_c(x)| + \\ & + \sup_x |G_c(x) - G_c(x + n^{-3/5})| \\ & + P(|\delta - \delta_1| > n^{-3/5}) \\ & + \sup_x |P(\rho + \delta_1 < x) - G_c(x)|. \end{aligned}$$

In this inequality the first three terms on the right-hand side have the evaluation $O(n^{-3/5})$ as $n \to \infty$: the first two by the properties of function $G_c(x)$ and the third one by Chebyshev's inequality and (11).

The fourth item we evaluate by the Esseen inequality:

$$\sup_x |P(\rho + \delta_1 < x) - G_c(x)| \le \frac{1}{\pi}\int_{-n}^{n} |E\exp\{it(\rho + \delta_1)\} - g_c(t)|\,|t|^{-1}dt + C_3 n^{-1}.$$

Consider the integral. First evaluate it in the following way

$$\int_{-n}^{n} |E\exp\{it(\rho + \delta_1)\} - g_c(t)|\,|t|^{-1}dt \le R_1 + R_2$$

with

$$\begin{aligned} R_1 &= \int_{-n}^{n} \left|E\exp\{it\rho\} - (1 + (it)^3 p_c)\exp\{-t^2/2\}\right| |t|^{-1}dt, \\ R_2 &= \int_{-n}^{n} \left|E\exp\{it(\rho + \delta_1)\} - E\exp\{it\rho\} - (it)^3 q_c \exp\{-t^2/2\}\right| |t|^{-1}dt, \end{aligned}$$

here

$$p_c = \frac{1}{6}\sum_{j=1}^{c} m_j^3 n_j^{-2} E g_j^3(X_{j1}) \left(\sum_{j=1}^{c} m_j^2 n_j^{-1}\sigma_j^2\right)^{-3/2},$$

$$q_c = \frac{1}{2}E\left(Y\left(\sum_{j=1}^{c}\frac{m_j}{n_j}\sum_{k=1}^{m_j} g_j(X_{jk})\right)^2\right)\left(\sum_{j=1}^{c} m_j^2 n_j^{-1}\sigma_j^2\right)^{-3/2}.$$

Under conditions (19) and (20) from estimates in the Central Limit Theorem it is known (see [11]) that $R_1 = O(n^{-1})$. At first evaluate the integrand in R_2. By Taylor's formula

$$E\exp\{it\rho\}(1-\exp\{it\delta_1\}) = -itE(\delta_1\exp\{it\rho\}) + \frac{1}{2}t^2E(\delta_1^2\exp\{it\rho\}) + t^3E(\delta_1^3\exp\{it\rho\}\theta_2),$$

where random variable $|\theta_2| \le 1$. The first and second mathematical expectations on the right-hand side are evaluated by formulas (16)–(17). For the third one from consideration of independence we can write the following representation

$$E(\delta_1^3\exp\{it\rho\}\theta_2) = \prod_{j=1}^{c}\left(\varphi_j(\frac{tm_j}{\hat\sigma n_j})\right)^{n_j-k_j} E\left(\delta_1^3\theta_2\exp\left(\frac{it}{\hat\sigma}\sum_{j=1}^{c}\frac{m_j}{n_j}\sum_{r=1}^{k_j} g_j(X_{jr})\right)\right).$$

Further,

$$\left|E(\delta_1^3\exp\{it\rho\}\theta_2)\right| \le E|\delta_1|^3\exp\{-t^2b_c\},$$

where by the inequality from [2]

$$E|\delta_1|^3 \le C_4 n^{-3/2}.$$

Mentioned estimates show that $R_2 = O(n^{-3/4})$ from which (21) follows.

4. Consider two generalized U-statistics $U^{(p)}$, $p = 1,2$ of the form (1) with the kernel $\Phi^{(p)}$ and order $(m_{1p},\dots,m_{cp})$, that is

$$U^{(p)} = \prod_{j=1}^{c}\binom{n_j}{m_{jp}}^{-1}\sum\Phi^{(p)}\left(X_{j\alpha},\alpha = i_{j1},\dots,i_{jm_{jp}},\quad 1\le j\le c\right) \tag{22}$$

where the sum is taken over all $1 \le i_{j1} < \dots < i_{jm_{jp}} \le n_j$, $j = 1,\dots,c$, and condition $E\Phi^{(p)} = 0$, $E(\Phi^{(p)})^2$, $p = 1,2$ is assumed.

Denote $\sigma_{ij} = E(U^{(i)}U^{(j)})$. It is known (see [1]) that

$$\sigma_{ij} = \prod_{\gamma=1}^{c}\binom{n_\gamma}{m_{\gamma j}}^{-1}\sum_{d_1=0}^{m_1(i,j)}\cdots\sum_{d_c=0}^{m_c(i,j)}\prod_{\gamma=1}^{c}\binom{m_{\gamma i}}{d_\gamma}\binom{n_\gamma - m_{\gamma i}}{m_{\gamma j}-d_\gamma}\eta_{d_1\dots d_c}^{(i,j)} \tag{23}$$

where $m_\gamma^{(i,j)} = \min(m_{\gamma i}, m_{\gamma j}), \quad i,j = 1,2.$

For $i = j$, formula (23) is reduced to (2). Let $\Phi_c(x_1,x_2) = P(\sigma_{11}^{-1}U^{(1)} < x_1,\ \sigma_{22}^{-1}U^{(2)} < x_2)$ and $\Phi(x_1,x_2)$ be two dimensional normal distributions with characteristic function

$$\varphi(t_1,t_2) = \exp\left(-\frac{1}{2}(t_1^2 + 2\rho_c t_1 t_2 + t_2^2)\right), \quad t_1, t_2 \in R$$

where

$$\rho_c = \hat{\sigma}_{11}^{-1}\hat{\sigma}_{22}^{-1}\sum_{j=1}^{c} m_{j1}m_{j2}n_j^{-1}\sigma_j^{(1,2)},$$

$$\sigma_j^{(1,2)} = E\left(g_j^{(1)}(X_{j1})g_j^{(2)}(X_{j1})\right).$$

It is proved in [2] that in the limiting relation

$$\sup_{x_1,x_2}|\Phi_c(x_1,x_2) - \Phi(x_1,x_2)| \to 0 \quad \text{as} \quad n \to \infty$$

the rates of convergence have an order $O(n^{-1/2})$ if $E\left|\Phi^{(p)}\right|^3 < \infty, \quad p = 1,2.$

Theorem 4. *Assume the conditions*

$$\max\left(\Gamma_c^{(p)},\quad E(\Phi^{(p)})^2\right) < \infty, \quad p = 1,2 \tag{24}$$

are satisfied.

Then

$$\sup_{x_1,x_2}|\Phi_c(x_1,x_2) - \Phi(x_1,x_2)| = O(n^{-1/2}) \quad as \quad n \to \infty \tag{25}$$

where index p denotes, as before, the number of U-statistics from (22).

Proof. First write the inequality

$$\sup_{x_1,x_2}|\Phi_c(x_1,x_2) - \Phi(x_1,x_2)| \le \sum_{j=1}^{4}\Delta_{1j}$$

where

$$\Delta_{11} = \sup_{x_1,x_2}\left|\Phi(x_1,x_2) - \Phi(x_1\frac{\sigma_{11}}{\hat{\sigma}_{11}}, x_2\frac{\sigma_{22}}{\hat{\sigma}_{22}})\right|,$$

$$\Delta_{12} = \sup_{x_1,x_2}|\Phi(x_1,x_2) - \Phi(x_1 + \hat{\sigma}_{11},\quad x_2 + \hat{\sigma}_{22})|,$$

$$\Delta_{13} = P(|\delta^{(1)} - \delta_1^{(1)}| > \hat{\sigma}_{11}) + P(|\delta^{(2)} - \delta_1^{(2)}| > \hat{\sigma}_{22}),$$

$$\Delta_{14} = \sup_{x_1,x_2}\left|P(\rho^{(1)} + \delta_1^{(1)} < x_1,\quad \rho^{(2)} + \delta_1^{(2)} < x_2) - \Phi(x_1,x_2)\right|.$$

From the properties of the normal distribution function

$$\Delta_{11} \leq \sup_{x_1}\left|\Phi(x_1)-\Phi(x_1)\frac{\sigma_{11}}{\hat\sigma_{11}}\right|+\sup_{x_2}\left|\Phi(x_2)-\Phi(x_2)\frac{\sigma_{22}}{\hat\sigma_{22}}\right|,$$

$$\Delta_{12} \leq \sup_{x_1}|\Phi(x_1)-\Phi(x_1+\hat\sigma_{11})|+\sup_{x_2}|\Phi(x_2)-\Phi(x_2+\hat\sigma_{22})|.$$

By the estimates used before we get for $n\to\infty$, $\Delta_{11}=O(n^{-1/2})$, $\Delta_{12}=O(n^{-1/2})$. For the estimation of Δ_{13} apply Chebyshev's inequality and (11). Then $\Delta_{13} = O(n^{-1/2})$. The value of Δ_{14} is considered with the help of Sadikova's inequality (see [12]) -

$$\Delta_{14} \leq \frac{2}{(2\pi)^2}\int_{-\epsilon\sqrt{n}}^{\epsilon\sqrt{n}}\int_{-\epsilon\sqrt{n}}^{\epsilon\sqrt{n}}|f(t_1,t_2)-h(t_1,t_2)||t_1t_2|^{-1}dt_1dt_2$$

$$+\,2\sup_{x_1}|P(\rho^{(1)}+\delta_1^{(1)}<x_1)-\Phi(x_1)|$$

$$+\,2\sup_{x_1}|P(\rho^{(2)}+\delta_1^{(2)}<x_2)-\Phi(x_2)|+C_5n^{-1/2},$$

where $\varepsilon>0$ is a sufficiently small number, C_5 – some positive constant,

$$f(t_1,t_2) = E\exp\{it_1(\rho^{(1)}+\delta_1^{(1)})+it_2(\rho^{(2)}+\delta_1^{(2)})\} -$$

$$-\,E\exp\{it_1(\rho^{(1)}+\delta_1^{(1)})\}E\exp\{it_2(\rho^{(2)}+\delta_1^{(2)})\},$$

$$h(t_1,t_2) = \varphi(t_1,t_2)-\varphi(0,t_2)\varphi(t_1,0).$$

In the last inequality the second and the third terms by (18) are evaluated as $O(n^{-1/2})$. Further,

$$|f(t_1,\ t_2)-h(t_1,\ t_2)|\leq$$

$$\leq|p(t_1,\ t_2)-h(t_1,\ t_2)|+|p(t_1,\ t_2)-f(t_1,\ t_2)|$$

where

$$p(t_1,\ t_2)=E\exp\{it_1\rho^{(1)}+it_2\rho^{(2)}\}-E\exp\{it_1\rho^{(1)}\}E\exp\{it_2\rho^{(2)}\}.$$

The integral from $|p(t_1,\ t_2)-h(t_1,\ t_2)|$ is evaluated like $O(n^{-1/2})$ (see [12]).

Thus

$$\Delta_{14}=\frac{1}{2\pi^2}\int_{-\epsilon\sqrt{n}}^{\epsilon\sqrt{n}}\int_{-\epsilon\sqrt{n}}^{\epsilon\sqrt{n}}|p(t_1,t_2)-f(t_1,t_2)||t_1t_2|^{-1}dt_1dt_2+O(n^{-1/2}).$$

The independence property functions $p(t_1,\ t_2)$, $f(t_1,\ t_2)$ allow the following representations

$$p(t_1,\ t_2) = \prod_{j=1}^{c}\left(\varphi_j\left(\frac{t_1 m_{j1}}{\hat{\sigma}_{11}n_j}, \frac{t_2 m_{j2}}{\hat{\sigma}_{22}n_j}\right)\right)^{n_j} - \prod_{j=1}^{c}\left(\varphi_j^{(1)}\left(\frac{t_1 m_{j1}}{\hat{\sigma}_{11}n_j}\right)\varphi_j^{(2)}\left(\frac{t_2 m_{j2}}{\hat{\sigma}_{22}n_j}\right)\right)^{n_j},$$

$$f(t_1,\ t_2) = \prod_{j=1}^{c}\left(\varphi_j\left(\frac{t_1 m_{j1}}{\hat{\sigma}_{11}n_j}, \frac{t_2 m_{j2}}{\hat{\sigma}_{22}n_j}\right)\right)^{n_j-k_j} E\exp\left\{it_1(\rho_1^{(1)} + \delta_1^{(1)}) + it_2(\rho_1^{(2)} + \delta_1^{(2)})\right\} -$$

$$- \prod_{j=1}^{c}\left(\varphi_j^{(1)}\left(\frac{t_1 m_{j1}}{\hat{\sigma}_{11}n_j}\right)\varphi_j^{(2)}\left(\frac{t_2 m_{j2}}{\hat{\sigma}_{22}n_j}\right)\right)^{n_j-k_j} E\exp\{it_1(\rho_1^{(1)}+\delta_1^{(1)})\}E\exp\{it_2(\rho_1^{(2)}+\delta_1^{(2)})\},$$

$$\varphi_j(t_1,\ t_2) = E\exp\{it_1 g_j^{(1)}(X_{j1}) + it_2 g_j^{(2)}(X_{j1})\},$$

$$\varphi_j^{(1)}(t_1) = \varphi_j(t_1, 0), \quad \varphi_j^{(2)}(t_2) = \varphi_j(0, t_2).$$

Directly from these representations and previous considerations connected with the evaluation of the integrand, we obtain $\Delta_{14} = O(n^{-1/2})$ under condition (24), which gives (25).

References

1. **Sugiura, N.** (1965). Multisample and multivariate non-parametric tests based on *U*-statistics and their asymptotic efficiencies. *Osaka J. Math.* **2**(4), 385–426.

2. **Borovskikh, Yu.V. and Korolyuk, V.S.** (1994). *Theory of U-statistics.* Kluwer Academic Publishers. Dordrecht.

3. **Borovskikh, Yu.V. and Korolyuk, V.S.** (1984). Approximation of non-degenerate *U*-statistics. *Theor. Probab. Appl.* **30**(3), 439–450.

4. **Rubin, H. and Vitale, R.A.** (1980). Asymptotic distribution of symmetric statistics. *Ann. Statist.* **8**(1), 165–170.

5. **Chan, Y.K. and Wierman, J.** (1977). On the Berry–Esseen theorem for *U*-statistics. *Ann. Probab.* **5**(1), 136–139.

6. **Callaert, H. and Janssen, P.** (1978). On the Berry–Esseen theorem for *U*-statistics. *Ann. Statist.* **6**(2), 417–421.

7. **Helmers, R. and Zwet van, W.R.** (1982). On the Berry–Esseen bound for *U*-statistics. In: *Statistical Theory and Related Topics III, Vol. 1,* S.S.Gupta and J.O.Berger (Eds). Academic Press, New York, pp. 497–512.

8. **Götze, F.** (1987). Approximations of multivariate U-statistics. *J. Multiv. Anal.* **22**(2), 212–229.

9. **Grams, W.E. and Serfling, R.J.** (1973). Convergence rates for U-statistics and related statistics. *Ann. Statist.* **1**(1), 153–160.

10. **Serfling, R.J.** (1980). *Approximation Theorems of Mathematical Statistics.* Wiley, New York, p.371.

11. **Petrov, V.V.** (1972). *Sums of independent random variables.* Nauka, Moscow, p.416. (In Russian).

12. **Sadikova, S.M.** (1966). Two-dimensional analogues of Esseen inequalities with application to the central limit theorem. *Theor. Probab. Appl.* **11**(3), 369–380. (In Russian).

Exploring Stochastic Laws. pp. 55–75.
Eds. A.V. Skorokhod and Yu.V. Borovskikh

Largest-fit selection of random sizes under a sum constraint: comparisons by weak convergence

FRANS A. BOSHUIZEN[1,*] AND ROBERT P. KERTZ[2,**]

[1] *Econometric Institute, Erasmus Universiteit Rotterdam, P.O. Box 1738, 3000 DR Rotterdam, The Netherlands*

[2] *School of Mathematics, Georgia Institute of Technology, Atlanta GA 30332, USA*

Received March 22, 1995

Abstract. Let $Y_1, Y_2, \ldots$ be a sequence of positive, independent, identically distributed random variables with distribution function $G(x)$ and denote the order statistics of $Y_1, \ldots, Y_n$ by $Y_{1,n} \leq \ldots \leq Y_{n,n}$. For positive numbers $(d_n)_{n\geq 1}$, the largest-fit off-line counting random variable $\hat{N}^e(d_n)$ is defined by $\hat{N}^e(d_n) = \max\{ j: 1 \leq j \leq n \text{ and } Y_{n,n} + \ldots + Y_{n-j+1,n} \leq d_n\}$ if this set is nonempty and $= 0$ otherwise. In a bin-packing context, the r.v. $\hat{N}^e(d_n)$ is the maximal number of objects with largest possible sizes (chosen from n objects sampled from distribution G) that can be packed into a bin of capacity d_n.

In the on-line situation the n objects with sizes $Y_1, \ldots, Y_n$ arrive one-by-one, and a decision maker must either accept or reject each object immediately, without recall. A 'good' on-line selection policy selects as few objects as possible with sizes as large as possible.

In this paper, quantitative comparisons are given between the off-line count $\hat{N}^e(d_n)$ and on-line counts $N_n^\tau(d_n)$ associated with 'good' policies τ. Specifically, for such policies τ, under appropriate conditions on the distribution function $G(x)$ and the constants $(d_n)_{n\geq 1}$, we find sequences of positive constants $(\beta_n)_{n\geq 1}$, $(\Delta_n)_{n\geq 1}$ and $(\Delta'_n)_{n\geq 1}$ such that

$$\left(\Delta_n\left(\frac{\hat{N}^e(d_n)}{\beta_n} - 1\right), \Delta'_n\left(\frac{N_n^\tau(d_n)}{\beta_n} - 1\right)\right) \Rightarrow (W, W'), \qquad n \to \infty,$$

for some nondegenerate r.v.'s W and W'.

1. Introduction

Let $Y_1, Y_2, \ldots$ be i.i.d. positive r.v.'s with distribution function $G(x)$. For positive numbers $(d_n)_{n\geq 1}$, the *largest fit counting r.v.* $\hat{N}^e(d_n) = \hat{N}_n^e(d_n)$ is defined by

$$\hat{N}^e(d_n) := \max\{ j: 1 \leq j \leq n \text{ and } Y_{n,n} + \cdots + Y_{n-j+1,n} \leq d_n\}$$

* The research of F.A. Boshuizen was partially supported by a Fulbright Grant

** The research of R.P. Kertz was partially supported by National Science Foundation Grant DMS 92-09586 if this set is nonempty and $= 0$ otherwise. In the bin-packing context, the r.v. $\hat{N}^e(d_n)$ is the maximal number of objects with largest possible sizes that can be packed into a bin of capacity d_n, chosen from n objects with random sizes sampled from distribution G. This is the largest count obtainable by a 'prophet' or individual using an off-line strategy, that is, a strategy which uses knowledge of all sizes in the sample, without any order restrictions on the sample values. Let τ^e denote this largest-fit off-line strategy. Note that, under this strategy, if the j-th largest item has been packed, and the $(j+1)$-th largest item does not fit in with the j largest one's, then the strategy stops with j items and does not proceed to pack items with smaller sizes. One also has the representation

$$\hat{N}^e(d_n) = \min\{j : 1 \le j \le n \text{ and } Y_{n,n} + \cdots + Y_{n-j+1,n} > d_n\} - 1$$

if this set is nonempty and $= n$ otherwise.

Define a *policy* τ to be a sequence of stopping times $(\tau_j)_{j\ge1}$ with respect to $Y_1, Y_2, \ldots$ with $1 \le \tau_1 < \tau_2 < \cdots$. For n objects, the counting r.v. associated with policy τ is defined by

$$N_n^\tau := \sum_{j\ge1} I(\tau_j \le n).$$

For any policy $\tau = (\tau_j)_{j\ge1}$ satisfying the sum constraint

$$\sum_{j\ge1} Y_{\tau_j} I(\tau_j \le n) \le d_n,$$

the r.v. N_n^τ has interpretation in a bin-packing context as the number of objects that are packed into a bin of size d_n, chosen sequentially under policy τ, without recall, as n objects with sizes sampled from distribution G appear in a given order. This N_n^τ is an 'on-line' count, a count under an on-line policy.

In this paper, quantitative comparisons are given between the off-line count $\hat{N}^e(d_n)$ and the on-line counts $N_n^\tau(d_n)$ associated with 'good' strategies τ for $(Y_j)_{j\ge1}$. Intuitively, a 'good' on-line policy for this setting is one which selects only items with large sizes and stops when the total-size sum reaches d_n. Difficulties arise if one seeks to translate these features into an on-line, expectation-based optimization problem. For example, the choice of optimization

$$\inf\left\{ EN_n^\tau : \tau \text{ is a policy for } (Y_j)_{j\ge1} \text{ such that } \sum_{j\ge1} Y_{\tau_j} I(\tau_j \le n) > d_n \right\}$$

rewards choice of large values of the Y_j's, but rules out policies $\tau = (\tau_j)_{j\ge1}$ with

$$\sum_{j\ge1} Y_{\tau_j} I(\tau_j \le n) \le d_n$$

on any part of the underlying probability space; and the choice of an optimization with a penalty feature in the optimization

$$\inf\left\{ E\hat{N}_n^\tau(d_n) : \tau \text{ is a policy for } (Y_j)_{j\ge1} \right\}$$

where

$$\hat{N}_n^{\tau}(d_n) = \begin{cases} \sum_{j=1}^{n} I(\tau_j \le n) & if \quad \sum_{j\ge 1} Y_{\tau_j} I(\tau_j \le n) \ge d_n \\ n & if \quad \sum_{j\ge 1} Y_{\tau_j} I(\tau_j \le n) < d_n, \end{cases}$$

again rewards choice of large values of the Y_j's, but imposes too great a penalty for the situation $\sum_{j\ge 1} Y_{\tau_j} I(\tau_j \le n) < d_n$. Explicit solution of such on-line expectation-based optimization problems also appears to be more difficult than for the comparable smallest-fit analogues in Coffman, Flatto and Weber (1987).

To place this problem, consider the following comparison. A manufacturer of a liquid chemical product anticipates production of an amount d_n of the product during the next business cycle. At the beginning of the period, the manufacturer knows there will be n customers (one order per customer) during the period. The manufacturer can fill orders partially if limited by her production level. The company requires that the manufacturer service as few orders as possible, with these being of largest possible order size, and should have total sum of order sizes attaining the anticipated production level d_n. If the manufacturer has knowledge of the sizes of all customer orders at the beginning of the business period, then she will service largest orders first and thereby fill $\hat{N}^e(d_n)$ full orders. If the manufacturer does not know the sizes of all customer orders at the beginning of the business period, but sees customer orders one-by-one as they arrive, and must either accept or reject each order immediately, without recall, then she must establish some criteria and choose a 'good' policy for handling orders under this criteria.

In this paper, the objective is to obtain asymptotic joint distributional comparisons of the off-line count $\hat{N}^e(d_n)$ and on-line counts $N_n^{\hat{\tau}}$ for 'good' policies $\hat{\tau}$. In analogy with the policies described in the smallest-fit problem discussed in Boshuizen and Kertz (1995), the specific policy in the analysis and the comparison criteria are defined as follows.

For positive constants $(\rho_n)_{n\ge 1}$ and $(d_n)_{n\ge 1}$ and n objects with sizes $Y_1, \dots, Y_n$ the *stopped threshold policy* $\hat{s}_n$ *with horizon* n accepts objects of size $\ge \rho_n$ and selects these only when the sum of the sizes of the selected objects is $\le d_n$; if the sum of the sizes $\ge \rho_n$ up to the present is $> d_n$, then the present object and all future objects are rejected. Thus $\hat{s}_n$ is the policy of stopping times $(\tau_j)_{j\ge 1}$ for $(Y_j)_{j\ge 1}$ defined by $\tau_1 = \min\{1 \le j \le n: Y_j \le d_n \text{ and } Y_j \ge \rho_n\}$ if this set is nonempty and $= \infty$ otherwise, and for $k = 2, 3, \dots, \tau_k = \min\{\tau_{k-1} < j \le n$: $Y_j \ge \rho_n$ and $\sum_{i=1}^{j} Y_i I(Y_i \ge \rho_n)) \le d_n\}$ if $\tau_{k-1} < \infty$ and this set is nonempty, and $= \infty$ otherwise. The counting r.v. associated with this policy at horizon n, denoted $N^{\hat{s}_n}(d_n) = N_n^{\hat{s}_n}(\rho_n, d_n) := N_n^{\hat{s}_n}$ has representation $N^{\hat{s}_n}(d_n) = \sum_{i=1}^{\hat{\nu}_n(d_n)} I(Y_i \ge \rho_n)$ where $\hat{\nu}_n(d_n) = \max\{j: 1 \le j \le n \text{ and } \sum_{i=1}^{j} Y_i I(Y_i \ge \rho_n) \le d_n\}$ if this set is nonempty and $= 0$ otherwise. A sequence of policies $(\hat{\tau}_n)_{n\ge 1}$ with $\hat{\tau}_n = (\hat{\tau}_{n,j})_{j\ge 1}$, satisfying the sum constraints $\sum_{j\ge 1} Y_{\hat{\tau}_{n,j}} I(\hat{\tau}_{n,j} \le n) \le d_n$ is said to be a *consistent approximater* of the largest-fit off-line strategy τ^e if there exists positive constants $(\beta_n)_{n\geq 1}$, $(\Delta_n)_{n\geq 1}$ and $(\Delta'_n)_{n\geq 1}$ for which

$$\left(\Delta_n\left(\frac{\hat{N}^e(d_n)}{\beta_n}-1\right),\Delta'_n\left(\frac{N^{\hat{\tau}_n}(d_n)}{\beta_n}-1\right)\right)\Rightarrow(W,W') \tag{1}$$

as $n\to\infty$, for some nondegenerate r.v.'s W and W'. In Theorems 2.1 and 3.1, it is shown that the sequence of stopped threshold policies $(\hat{s}_n)_{n\geq 1}$ is a consistent approximater of τ^e; and consequences of these results are given in Corollaries 2.6 and 3.4.

Observe that one could combine the theorems of this paper with the smallest-fit results of Boshuizen and Kertz (1995), if one allows negative X_i's in Boshuizen and Kertz (1995). However, the present arrangement is more natural for the given applications with positive r.v.'s. In fact, the theorems in this paper complement but do not duplicate the results in Boshuizen and Kertz (1995). Note, in particular, the interesting variations for asymptotic behaviors that occur for $G(x)$ in Case II for maxima, as identified in Theorems 2.1 and 3.1. The reader is also referred to the paper by Bruss and Robertson (1991) where expressions for the asymptotic behavior of $E\hat{N}^e(d_n)$ are obtained (under weaker assumption on the distribution $G(x)$).

Throughout the paper the following notation is used. For a nondecreasing function h on a subset S of $\mathbb{R}$, the left-continuous inverse of h is defined by $h^{\leftarrow}(s)=\inf\{x\in S: h(x)\geq s\}$. For a distribution function G, $r_G=\sup\{y: G(y)<1\}$ denotes the right end point of the support of G. The notations $o_P(1)$ and $O_P(1)$ are used to denote sequences of random variables which are respectively converging to zero in probability and bounded above and below by a finite constant uniformly for all n large. For two r.v.'s X and Y, $X\stackrel{d}{=}Y$ if the distributions of X and Y are same. For a real number x, $\lceil x\rceil$ denotes the smallest integer greater than x, and for two sequences $(m_n)_{n\geq 1}$ and $(l_n)_{n\geq 1}$ we write $m_n\approx l_n$ if $\lim_{n\to\infty} m_n/l_n=1$.

2. Normal convergence

In this section, settings are considered where appropriate normalizations of the counting r.v.'s $\hat{N}^e(d_n)$ and $N^{\hat{s}_n}(d_n)$ converge weakly (as $n\to\infty$) to a random pair of the form $(W_1, W_2+(W_3\wedge 0))$ where (W_1,W_2,W_3) has a (singular) multivariate normal distribution. The settings considered are $G(x)$ in Cases I, II with index $\alpha\geq 2$, and III for maxima (see the Appendix for background results on extreme value theory). Case II with index $0<\alpha<2$ will be treated in Section 3. Recall that for $G(x)$ in Case II for maxima, with index $\alpha>0$, the inverse function $G^{\leftarrow}$ has representation $G^{\leftarrow}(1-s)=s^{-a}L(s)$ for $L(s)$ slowly varying at zero and $a=1/\alpha$; and for $G(x)$ in Case III for maxima, with index $\alpha>0$, the inverse function $G^{\leftarrow}$ has representation

$G^{\leftarrow}(1-s) = r_G - s^{-a}L(s)$ for $L(s)$ slowly varying at zero and $a = -1/\alpha$. For G in Case I for maxima the function $c(s)$ is defined by $c(s) = s^{-1}\int_{1-s}^{1}(1-u)\,dG^{\leftarrow}(u)$.

For r.v.'s Y_i with $EY_i < \infty$, and for a given sequence of positive constants $(d_n)_{n\geq 1}$, define constants $(\beta_n)_{n\geq 1}$ and $(\rho_n)_{n\geq 1}$ by

$$d_n = n\int_0^{\beta_n/n} G^{\leftarrow}(1-s)\,ds \quad \text{and} \quad \rho_n = G^{\leftarrow}(1-(\beta_n/n)). \tag{2}$$

In the following settings of $G(x)$ in Cases I, II and III for maxima, it is assumed that the constants $(\beta_n)_{n\geq 1}$ satisfy $\beta_n/n \to 0$ as $n \to \infty$. This implies the following for the normalizing constants of Theorem 2.1: the location parameter $\beta_n \ll n$ and scaling parameters α_n, $\gamma_n \ll n$ as $n \to \infty$. This ensures that the total number of sizes selected goes to infinity, but stays small when compared to the horizon n, as $n \to \infty$, for the off-line largest fit strategy τ^e and for the on-line stopped threshold policy $\hat{s}_n$. Note, however, that in Cases I, II with $0 < a \leq 1$ and III, $\hat{\nu}_n(d_n)/n \to 1$ in probability as $n \to \infty$; in this sense, it is taking closer and closer to n observations for the stopped threshold policy $\hat{s}_n$ to achieve total sum of sizes near the level d_n as $n \to \infty$. Also, under the assumptions in the theorems of this paper, it follows that $\rho_n \ll d_n$, so the policies $\hat{s}_n$ are feasible, for all n large.

It is also assumed that

$$(i) \qquad \beta_n^{1/2}\left(\frac{1-G(\rho_n-)}{\beta_n/n} - 1\right) \to 0, \qquad n \to \infty, \tag{3}$$

and

(ii) $\alpha_n \to \infty$ as $n \to \infty$ where

$$\alpha_n \approx \begin{cases} \beta_n^{1/2}c(\beta_n/n)/G^{\leftarrow}(1-(\beta_n/n)) & \text{for } G(x) \text{ in Case I,} \\ \beta_n^{1/2}\big(r_G - G^{\leftarrow}(1-(\beta_n/n))\big)\big/r_G & \text{for } G(x) \text{ in Case III with } r_G > 0. \end{cases}$$

Condition $(3i)$ is imposed to ensure that the location parameters in the convergence of the two counting r.v. sequences of Theorem 2.1 can be taken to be the same β_n. In the case that $G(x)$ is a continuous d.f., it is immediate that $1 - G(\rho_n-) = \beta_n/n$; so condition $(3i)$ holds in this case. For d.f. $G(x)$ in Case I, II or III for maxima, it follows that

$$1 \leq (1-G(\rho_n-))/(\beta_n/n) \leq (1-G(\rho_n-))/(1-G(\rho_n)) \to 1, \qquad n \to \infty;$$

so the limit in $(3i)$ can be thought of as a condition on the rate of convergence of $(1-G(\rho_n-))/(\beta_n/n)$ to 1. In the Appendix we give an example of a d.f. $G(x)$ in Case II for maxima with $a = 1/3$ for which condition $(3i)$ fails. If d.f. $G(x)$ is in Case II with $a = 1/2$, one may replace $(3i)$ by the weaker assumption

$$\left(\frac{\beta_n}{\log\beta_n}\right)^{1/2}\left(\frac{1-G(\rho_n-)}{\beta_n/n} - 1\right) \to 0, \qquad n \to \infty \tag{3b}$$

whenever (3i) is assumed for this case in this section.

Condition (3ii) is imposed to ensure that the scaling parameters in the convergence of the off-line counting r.v. sequence in Theorem 2.1 are sufficiently large (see Example 2.5 (i) and (ii)). For $G(x)$ in Case II, no additional condition is needed in Theorem 2.1 to ensure that $\alpha_n \to \infty$ in this case.

Theorem 2.1. *Let $G(x)$ be in Case I, II with $0 < a \le 1/2$, or III for maxima with $G(0-) = 0$ and $0 < r_G \le \infty$, and let the positive constants $(d_n)_{n\ge1}$, $(\beta_n)_{n\ge1}$ and $(\rho_n)_{n\ge1}$ satisfy* (2). *Assume that $d_n/n \to 0$ and $\beta_n \to \infty$ as $n \to \infty$ and* (3) *holds. Then there exist positive constants $(\alpha_n)_{n\ge1}$ and $(\gamma_n)_{n\ge1}$ for which*

$$\left(\alpha_n^{-1}(\hat{N}^e(d_n) - \beta_n), \gamma_n^{-1}(N^{\hat{s}_n}(d_n) - \beta_n)\right) \Rightarrow (\hat{\mathcal{N}}^e, \hat{\mathcal{N}}^s)$$

where $(\hat{\mathcal{N}}^e, \hat{\mathcal{N}}^s) = (W_1, W_2 + (W_3 \wedge 0))$ and $\mathbf{W} = (W_1, W_2, W_3)$ is $N(\mathbf{0}, \Sigma_{\mathbf{W}})$-distributed.

For $G(x)$ in Case I, the constants $(\alpha_n)_{n\ge1}$ and $(\gamma_n)_{n\ge1}$ satisfy

$$\alpha_n \approx \beta_n^{1/2} c(\beta_n/n)/G^{\leftarrow}(1 - (\beta_n/n)) \quad \text{and} \quad \gamma_n \approx \beta_n^{1/2},$$

and

$$\Sigma_{\mathbf{W}} = \begin{pmatrix} 2 & -1 & 1 \\ -1 & 1 & -1 \\ 1 & -1 & 1 \end{pmatrix}.$$

For $G(x)$ in Case II, the constants $(\alpha_n)_{n\ge1}$ and $(\gamma_n)_{n\ge1}$ satisfy

$$\alpha_n = \gamma_n \approx \begin{cases} \beta_n^{1/2} & \text{for} \quad 0 < a < 1/2, \\ (\beta_n \log \beta_n)^{1/2} & \text{for} \quad a = 1/2, \end{cases}$$

and

$$\Sigma_{\mathbf{W}} = \begin{pmatrix} K_a^2 & \frac{-a}{1-a} & \frac{a}{1-2a} \\ \frac{-a}{1-a} & 1 & -1 \\ \frac{a}{1-2a} & -1 & \frac{(1-a)^2}{1-2a} \end{pmatrix}$$

for $0 < a < 1/2$, and

$$\Sigma_{\mathbf{W}} = \begin{pmatrix} 1 & 0 & 1/2 \\ 0 & 0 & 0 \\ 1/2 & 0 & 1/4 \end{pmatrix}$$

for $a = 1/2$, where $K_a^2 = \frac{2a^2}{(1-a)(1-2a)}$.

For $G(x)$ in Case III, the constants $(\alpha_n)_{n\ge1}$ and $(\gamma_n)_{n\ge1}$ satisfy

$$\alpha_n \approx \beta_n^{1/2}\left(r_G - G^{\leftarrow}(1 - (\beta_n/n))\right)/r_G \quad \text{and} \quad \gamma_n \approx \beta_n^{1/2}$$

and

$$\Sigma_{\mathbf{W}} = \begin{pmatrix} K_a^2 & \frac{a}{1-a} & \frac{-a}{1-a} \\ \frac{a}{1-a} & 1 & -1 \\ \frac{-a}{1-a} & -1 & 1 \end{pmatrix}.$$

In fact, for Cases I and III, $W_3 = -W_2$*; and for Case II,* $W_3 = (1-a)(W_1 - W_2)$ *for* $0 < a < 1/2$ *and* $W_3 = (1/2)W_1$ *for* $a = 1/2$.

Note that it is not true in general that $N^{\hat{s}_n}(d_n) \leq \hat{N}^e(d_n)$. However, Theorem 2.1 implies that $\lim_{n\to\infty} P(N^{\hat{s}_n}(d_n) \leq \hat{N}^e(d_n)) = 1$ in Cases I and III, since

$$\begin{aligned} P(N^{\hat{s}_n}(d_n) \leq \hat{N}^e(d_n)) &= P(\beta_n^{-1/2}(N^{\hat{s}_n}(d_n) - \beta_n) \\ &\leq \delta_n \alpha_n^{-1}(\hat{N}^e(d_n) - \beta_n)) \to P(\hat{\mathcal{N}}^s \leq 0) = 1 \end{aligned}$$

(for specific $\delta_n \to 0$) in Cases I and III. In Case II with $0 < a \leq 1/2$, Theorem 2.1 implies

$$\begin{aligned} P(N^{\hat{s}_n}(d_n) \leq \hat{N}^e(d_n)) &= P(\alpha_n^{-1}(N^{\hat{s}_n}(d_n) - \beta_n) \\ &\leq \alpha_n^{-1}(\hat{N}^e(d_n) - \beta_n)) \to P(\hat{\mathcal{N}}^s \leq \hat{\mathcal{N}}^e) = 1/2 \end{aligned}$$

The proof of Theorem 2.1 is based on a Brownian bridge approximation to the uniform empirical process. Before the theorem is proved, some lemmas concerning the Brownian bridge approximation are given. In the sequel we work on a probability space $(\Omega, \mathcal{A}, P)$ constructed by Csörgő et al. (1986) carrying an infinite sequence $U_1, U_2, \ldots$ of i.i.d. r.v.'s uniformly distributed on $(0,1)$ and a sequence of Brownian bridges $U_n(s)$, $0 \leq s \leq 1$, $n = 1, 2, \ldots$ such that for the uniform empirical process $\alpha_n(s) = n^{1/2}(E_n(s) - s)$, $0 \leq s \leq 1$, where $E_n(s) = n^{-1}\sum_{i=1}^n I(U_i \leq s)$

$$\sup_{1/n \leq s \leq 1-1/n} n^{\nu} \frac{|\alpha_n(s) - U_n(s)|}{(s(1-s))^{(1/2)-\nu}} = O_P(1), \qquad n \to \infty, \tag{4}$$

where ν is any fixed number such that $0 \leq \nu < 1/4$. This can be assumed without loss of generality. Note that for an G-distributed r.v. Y_i we have $Y_i \stackrel{\mathrm{d}}{=} G^{\leftarrow}(U_i)$, so on this space we use $Y_i = G^{\leftarrow}(U_i)$ for $i = 1, 2, \ldots$. Recall that the constants $(\beta_n)_{n\geq 1}$ and $(\rho_n)_{n\geq 1}$ are defined in equation (2). The proofs of the following two lemmas use standard arguments, for example, found in the papers by Csörgő and Mason (1986), Csörgő et al. (1991), and Lo (1989).

Lemma 2.2. *Let* $G(x)$ *be in Case I, II with index* $\alpha \geq 2$*, or III for maxima with* $G(0-) = 0$, $0 < r_G \leq \infty$*, let* $(\beta_n)_{n\geq 1}$ *satisfy* $\beta_n/n \to 0$ *and* $\beta_n \to \infty$*, and let* $(k_n)_{n\geq 1}$ *be any sequence of positive integers such that* $k_n \approx \beta_n$*. Then*

$$\frac{\sum_{i=1}^{k_n} Y_{n-i+1,n} - n\int_0^{k_n/n} G^{\leftarrow}(1-s)\,ds}{n^{1/2}A_n^e(k_n/n)} = -\frac{\int_{1-(\beta_n/n)}^{1-(1/n)} U_n(s)\,dG^{\leftarrow}(s)}{A_n^e(\beta_n/n)} + o_P(1)$$

where for $0 < t < 1$,

$$A_n^e(t) = \begin{cases} t^{1/2}c(t) & \text{in Case I,} \\ t^{(1/2)-a}L(t) & \text{in Case II with } 0 < a < 1/2 \\ & \quad \text{and in Case III,} \\ \left(\int_{1/n}^{t} u^{-1}L^2(u)\,du\right)^{1/2} & \text{in Case II with } a = 1/2. \end{cases}$$

Lemma 2.3. *Let* $G(x)$ *be in Case I, II with index* $\alpha \geq 2$, *or III for maxima with* $G(0-) = 0$ *and* $0 < r_G \leq \infty$, *and* $(\beta_n)_{n\geq 1}$ *and* $(\rho_n)_{n\geq 1}$ *satisfy* (2) *with* $\beta_n/n \to 0$ *and* $\beta_n \to \infty$ *and let* $\tau_n := 1 - G(\rho_n -)$ *for* $n \geq 1$. *Then*

$$(n\tau_n)^{-1/2}\Big(\sum_{i=1}^{n} I(Y_i \geq \rho_n) - n\tau_n\Big) = -\tau_n^{-1/2}\, U_n(1-\tau_n) + o_P(1)$$

and

$$\frac{\sum_{i=1}^{n} Y_i I(Y_i \geq \rho_n) - n\int_0^{\tau_n} G^{\leftarrow}(1-s)\,ds}{n^{1/2}A_n^s(\tau_n)}$$

$$= \begin{cases} -\tau_n^{-1/2}\, U_n(1-\tau_n) + o_P(1) & \text{in Cases I and III,} \\ -\tau_n^{-1/2}\, U_n(1-\tau_n) - \dfrac{\int_{1-\tau_n}^{1-(1/n)} U_n(s)\,dG^{\leftarrow}(s)}{A_n^s(\tau_n)} + o_P(1) & \\ & \text{in Case II with } 0 < a < 1/2, \\ -\dfrac{\int_{1-\tau_n}^{1-(1/n)} U_n(s)\,dG^{\leftarrow}(s)}{A_n^s(\tau_n)} + o_P(1) & \text{in Case II with } a = 1/2, \end{cases}$$

where for $0 < t < 1$,

$$A_n^s(t) = \begin{cases} t^{1/2}G^{\leftarrow}(1-t) & \text{in Cases I, II with } 0 < a < 1/2 \text{ and III,} \\ \left(\int_{1/n}^{t} u^{-1}L^2(u)\,du\right)^{1/2} & \text{in Case II with } a = 1/2. \end{cases}$$

If, in addition, condition (3*i*) holds as a hypothesis in Lemma 2.3, then the conclusions hold with $\tau_n = 1 - G(\rho_n-)$ replaced by β_n/n. Note that the norming constants $A_n^e(t)$ and $A_n^s(t)$ coincide in Case II but differ in Cases I and III, just as α_n and γ_n in Theorem 2.1.

The next lemma is also needed in the proof of Theorem 2.1. The proof of the lemma uses the representations of $G^{\leftarrow}(1-s)$ in each of the Cases I, II and III, and results on slowly varying functions. (See the Appendix.)

Lemma 2.4. *Let $(d_n)_{n\geq 1}$ and $(\beta_n)_{n\geq 1}$ be the sequences of positive constants defined in (2), let $k_n = \lceil \alpha_n \mu + \beta_n \rceil$, $-\infty < \mu < \infty$, and $j_n = \lceil n(1+\gamma_n \beta_n^{-1} z)\rceil$, $-\infty < z < 0$, and let the constants $(\alpha_n)_{n\geq 1}$ and $(\gamma_n)_{n\geq 1}$ be given as in Theorem 2.1. Assume condition (3ii) holds. Then*

$$\frac{d_n - n\int_0^{k_n/n} G^{\leftarrow}(1-s)\,ds}{n^{1/2}A_n^e(k_n/n)} = -\mu + o(1)$$

and

$$\frac{d_n - (j_n d_n/n)}{n^{1/2}A_n^s(\beta_n/n)} = \begin{cases} -z + o(1) & \text{in Case I and III,} \\ -z/(1-a) + o(1) & \text{in Case II with } 0 < a \leq 1/2, \end{cases}$$

where the functions $A_n^e(t)$ and $A_n^s(t)$ are given as in Lemmas 2.2 and 2.3, respectively.

Proof of Theorem 2.1. Consider first $G(x)$ in Case II with $0 < a < 1/2$. For the convergence analysis of $N^{\hat{s}_n}(d_n)$, observe that

$$\frac{N^{\hat{s}_n}(d_n) - \beta_n}{\beta_n^{1/2}} = \frac{\sum_{i=1}^{\hat{\nu}_n(d_n)} \Big(I(Y_i \geq \rho_n) - (1 - G(\rho_n -))\Big)}{\beta_n^{1/2}} + \beta_n^{1/2}\left(\frac{\hat{\nu}_n(d_n)}{n} - 1\right) + o_P(1)$$

where $\hat{\nu}_n(d_n) = \max\{j: 1 \leq j \leq n \text{ and } \sum_{i=1}^{j} Y_i I(Y_i \geq \rho_n) \leq d_n\}$ if this set is nonempty and $= 0$ otherwise; and obtain the following representations associated with this sum by using the argument in the proof of the Doeblin–Ascombe Central Limit Theorem as given in the book by Chow and Tcicher (1978), Theorem 1, page 317, the result that $\hat{\nu}_n(d_n)/n \to 1$ in probability, and Lemma 2.3:

$$\beta_n^{-1/2} \sum_{i=1}^{\hat{\nu}_n(d_n)} \Big(I(Y_i \geq \rho_n) - (1 - G(\rho_n -))\Big) = -(\beta_n/n)^{-1/2} U_n(1 - (\beta_n/n)) + o_P(1)$$

and

$$\frac{\sum_{i=1}^{j_n} Y_i I(Y_i \geq \rho_n) - (j_n d_n)/n}{n^{1/2}(\beta_n/n)^{1/2} G^{\leftarrow}(1 - (\beta_n/n))}$$

$$= -(\beta_n/n)^{-1/2} U_n(1 - (\beta_n/n)) - \frac{\int_{1-(\beta_n/n)}^{1-(1/n)} U_n(s)\,dG^{\leftarrow}(s)}{(\beta_n/n)^{1/2} G^{\leftarrow}(1 - (\beta_n/n))} + o_P(1)$$

where $j_n := \lceil n(1 + z_2/\beta_n^{1/2})\rceil$ with $-\infty < z_2 < 0$. Also define integers $k_n := \lceil \alpha_n \mu + \beta_n \rceil$ where the constants $(\alpha_n)_{n\geq 1}$ are defined in the statement of the theorem, and use the representations above and the convergence results from Lemma 2.4 to obtain the weak convergence

$$\begin{aligned}
& P\Bigg(\frac{\hat{N}^e(d_n)-\beta_n}{\alpha_n}\le\mu;\ \frac{\sum_{i=1}^{\hat{\nu}_n(d_n)}(I(Y_i\ge\rho_n)-(1-G(\rho_n-)))}{\beta_n^{1/2}}<z_1;\\
&\qquad\beta_n^{1/2}\Big(\frac{\hat{\nu}_n(d_n)}{n}-1\Big)\le z_2\Bigg)\\
&= P\Bigg(\sum_{i=1}^{k_n}Y_{n-i+1,n}>d_n;\ \frac{\sum_{i=1}^{\hat{\nu}_n(d_n)}(I(Y_i\ge\rho_n)-(1-G(\rho_n-)))}{\beta_n^{1/2}}<z_1;\\
&\qquad\sum_{i=1}^{j_n}Y_iI(Y_i\ge\rho_n)>d_n\Bigg)\\
&= P\Bigg(\frac{\int_{1-(\beta_n/n)}^{1-(1/n)}U_n(s)\,dG^{\leftarrow}(s)}{A_n^e(\beta_n/n)}<\mu;\ -(\beta_n/n)^{-1/2}U_n(1-(\beta_n/n))<z_1;\\
&\qquad(1-a)\Bigg((\beta_n/n)^{-1/2}U_n(1-(\beta_n/n))+\frac{\int_{1-(\beta_n/n)}^{1-(1/n)}U_n(s)\,dG^{\leftarrow}(s)}{A_n^s(\beta_n/n)}\Bigg)<z_2\Bigg)+o(1)\\
&= P(W_1<\mu;\ W_2<z_1;\ W_3<z_2)+o(1)
\end{aligned}$$

where $\mathbf{W}=(W_1,W_2,W_3)\overset{d}{=}N(0,\Sigma_{\mathbf{W}})$ and the covariance matrix $\Sigma_{\mathbf{W}}$ is defined in the assertion of the theorem.

Hence, for $-\infty<\mu<\infty$ and $-\infty<\nu<\infty$,

$$P\left(\frac{\hat{N}^e(d_n)-\beta_n}{\alpha_n}\le\mu;\ \frac{N^{\hat{s}_n}(d_n)-\beta_n}{\beta_n^{1/2}}\le\nu\right)=P(W_1\le\mu;W_2+(W_3\wedge 0)\le\nu)+o(1)$$

and the theorem is proved for $G(x)$ in Case II with $0<a<1/2$.

For the proof of the theorem in all other cases, use straightforward modifications of this argument, together with the Lemmas 2.2, 2.3 and 2.4.

Example 2.5. *(i)* For Exponential (1)-distribution G (so $G(x)$ is in Case I for maxima), and positive constants $(\beta_n)_{n\ge1}$ with $\beta_n\to\infty$, $\beta_n/n\to0$ and $\beta_n^{1/2}/\log(n/\beta_n)\to\infty$, and for $d_n=\beta_n-\beta_n\log(\beta_n/n)$ and $\rho_n=-\log(\beta_n/n)$, Theorem 2.1 gives

$$\left(\beta_n^{1/2}\log(n/\beta_n)\left(\frac{\hat{N}^e(d_n)}{\beta_n}-1\right),\beta_n^{1/2}\left(\frac{N^{\hat{s}_n}(d_n)}{\beta_n}-1\right)\right)\Rightarrow(W_1,W_2\wedge0)$$

where (W_1,W_2) is $N(\mathbf{0},\Sigma_{\mathbf{W}})$-distributed with $\Sigma_{\mathbf{W}}=\begin{pmatrix}2&-1\\-1&1\end{pmatrix}$.

(ii) For Uniform (0,1)-distribution G (so $G(x)$ is in Case III for maxima), and positive constants $(\beta_n)_{n\ge1}$ and $(d_n)_{n\ge1}$ satisfying $\beta_n/n\to0$ and $\beta_n^{3/2}/n\to\infty$, with $d_n=\beta_n(1-\beta_n/(2n))$ (and $\beta_n=d_n(1-(d_n/(2n))+o(d_n/n))$), and $\rho_n=1-(\beta_n/n)$,

Theorem 2.1 gives

$$\left(\beta_n^{1/2}(n/\beta_n)\left(\frac{\hat{N}^e(d_n)}{\beta_n}-1\right),\beta_n^{1/2}\left(\frac{N^{\hat{s}_n}(d_n)}{\beta_n}-1\right)\right) \Rightarrow (W_1, W_2 \wedge 0)$$

where (W_1, W_2) is $N(\mathbf{0}, \Sigma_{\mathbf{W}})$-distributed with $\Sigma_{\mathbf{W}} = \begin{pmatrix} 1/3 & -1/2 \\ -1/2 & 1 \end{pmatrix}$.

(*iii*) Let $G(x) = 1 - x^{-1/a}$ for $x \geq 1$, with $0 < a < 1/2$ (so $G(x)$ is in Case II for maxima with index $\alpha > 2$). For positive constants $(\beta_n)_{n\geq 1}$ and $(d_n)_{n\geq 1}$ satisfying $d_n/n \to 0$ and $d_n/n^a \to \infty$, and $\beta_n = ((1-a)d_n n^{-a})^{1/(1-a)}$, with $\rho_n = (n/\beta_n)^a$, Theorem 2.1 gives

$$\left(\beta_n^{1/2}\left(\frac{\hat{N}^e(d_n)}{\beta_n}-1\right),\beta_n^{1/2}\left(\frac{N^{\hat{s}_n}(d_n)}{\beta_n}-1\right)\right) \Rightarrow \left(W_1, W_2 + ((1-a)(W_1 - W_2)\wedge 0)\right)$$

where (W_1, W_2) is $N(\mathbf{0}, \Sigma_{\mathbf{W}})$-distributed with $\Sigma_{\mathbf{W}}$ the left-upper 2×2 matrix as in the corresponding part of Theorem 2.1.

(*iv*) For $G(x)$ as in (*iii*) with $a = 1/2$ (so $G(x)$ is in Case II for maxima with index $\alpha = 2$), and positive constants $(\beta_n)_{n\geq 1}$ and $(d_n)_{n\geq 1}$ satisfying $d_n/n \to 0$ and $d_n/\sqrt{n} \to \infty$, and $\beta_n = d_n^2/(4n)$ with $\rho_n = (n/\beta_n)^{1/2}$, Theorem 2.1 gives

$$\left(\left(\frac{\beta_n}{\log\beta_n}\right)^{1/2}\left(\frac{\hat{N}^e(d_n)}{\beta_n}-1\right),\left(\frac{\beta_n}{\log\beta_n}\right)^{1/2}\left(\frac{N^{\hat{s}_n}(d_n)}{\beta_n}-1\right)\right) \Rightarrow (W_1, W_3 \wedge 0)$$

where (W_1, W_3) is $N(\mathbf{0}, \Sigma_{\mathbf{W}})$-distributed with $\Sigma_{\mathbf{W}} = \begin{pmatrix} 1 & 1/2 \\ 1/2 & 1/4 \end{pmatrix}$.

The following corollary is an immediate consequence of Theorem 2.1.

Corollary 2.6. *Let $G(x)$ be in Case I, II with $0 < a \leq 1/2$, or III for maxima with $G(0-) = 0$ and $0 < r_G \leq \infty$, and let positive constants $(d_n)_{n\geq 1}$, $(\beta_n)_{n\geq 1}$ and $(\rho_n)_{n\geq 1}$ satisfy* (2) *and* (3) *as in Theorem* 2.1. *Then*

(*i*) *the sequence of stopped threshold policies $(\hat{s}_n)_{n\geq 1}$ is a consistent approximater of τ^e;*

(*ii*) *$\hat{N}^e(d_n)/N^{\hat{s}_n}(d_n) \to 1$ in probability as $n \to \infty$;*

(*iii*) *for $G(x)$ in Case II with $0 < a \leq 1/2$,*

$$\eta_n^{-1}(\hat{N}^e(d_n) - N^{\hat{s}_n}(d_n)) \Rightarrow \hat{\mathcal{N}}^e - \hat{\mathcal{N}}^s$$

with $\eta_n = \beta_n^{1/2}$ for $0 < a < 1/2$, and $= (\beta_n \log\beta_n)^{1/2}$ for $a = 1/2$, and

$$(N^{\hat{s}_n}(d_n) - \beta_n)/(\hat{N}^e(d_n) - \beta_n) \Rightarrow \hat{\mathcal{N}}^s/\hat{\mathcal{N}}^e$$

where $\hat{\mathcal{N}}^e$ and $\hat{\mathcal{N}}^s$ are given in Theorem 2.1.

3. Stable convergence

When the d.f. G has a 'fatter tail', that is, for G in Case II for maxima with index $0 < \alpha < 2$, the asymptotic distributional convergence of $\hat{N}^e(d_n)$ and $N^{\hat{s}_n}(d_n)$ toward stable distributions is analyzed in this section by using variations of the techniques of proof of Theorem 2.1. For G in Case II for maxima with $1/2 < a < \infty$, $a = 1/\alpha$, and for positive constants $(d_n)_{n\geq 1}$, define constants $(\beta_n)_{n\geq 1}$ and $(\rho_n)_{n\geq 1}$ by

$$\begin{aligned} d_n &= \begin{cases} n\int_0^{\beta_n/n} G^{\leftarrow}(1-s)\,ds & \text{for} \quad 1/2 < a < 1, \\ n\int_{1/n}^{\beta_n/n} G^{\leftarrow}(1-s)\,ds & \text{for} \quad 1 \leq a < \infty, \end{cases} \\ \rho_n &= G^{\leftarrow}(1-(\beta_n/n)). \end{aligned} \tag{5}$$

For $1/2 < a < \infty$, let $\Delta(a)$ denote a completely asymmetric stable r.v. with index $\alpha = 1/a$, that is, with characteristic function

$$\phi_{\alpha,\beta,\gamma,\theta}(t) = \exp\Big\{i\theta t - \gamma|t|^\alpha(1 - i\beta\,\mathrm{sgn}(t)\,\omega(t))\Big\}$$

where function $\omega(t) = \omega(t,\alpha)$, location parameter $\theta = \theta(\alpha)$ and scale parameter $\gamma = \gamma(\alpha)$ are given by $\omega(t) = \tan(\pi\alpha/2)$ if $\alpha \neq 1$, and $= -(2/\pi)\log|t|$ if $\alpha = 1$; $\theta = 0$ if $\alpha \neq 1$, and $= \int_0^\infty \left(\frac{\sin x}{x^2} - \frac{1}{x(1+x)}\right) dx$ if $\alpha = 1$; and $\gamma = \Gamma(1-\alpha)\cos(\pi\alpha/2)$ if $0 < \alpha < 1$, $= \pi/2$ if $\alpha = 1$, and $= \alpha(\alpha-1)^{-1}\Gamma(2-\alpha)\cos(\pi\alpha/2)$ if $1 < \alpha < 2$, and skewness parameter $\beta = 1$. (The r.v. $\Delta(a)$ can also be expressed in terms of a rate one Poisson process $N(t)$, $t \geq 0$: for example, $\Delta(a) = \int_0^\infty N(t)t^{-a-1}\,dt$ if $a > 1$; see Csörgő et al. (1986), p. 104). The assumption in this section which is the counterpart of condition (3*i*) and (3b) in Section 2 is the following:

for d.f. $G(x)$ in Case II with $1/2 < a < \infty$,

$$\beta_n^{1-a}\left(\frac{1 - G(\rho_n -)}{\beta_n/n} - 1\right) \to 0, \qquad n \to \infty. \tag{6}$$

Again, for a continuous d.f. $G(x)$, condition (6) is clearly satisfied. In the theorems of this section, Condition (6) is an additional assumption if $1/2 < a < 1$; for sequences $(\beta_n)_{n\geq 1}$ in the theorems, Condition (6) follows immediately from d.f. $G(x)$ being in Case II for maxima if $a \geq 1$.

Theorem 3.1. *Let $G(x)$ be in Case II for maxima with $1/2 < a \leq 1$ and with $G(0-) = 0$, and let the positive constants $(d_n)_{n\geq 1}$, $(\beta_n)_{n\geq 1}$ and $(\rho_n)_{n\geq 1}$ satisfy* (5). *Assume that $\beta_n \to \infty$ and $\beta_n/n \to 0$ as $n \to \infty$ and that* (6) *holds. Then there exist positive constants $(\alpha_n)_{n\geq 1}$ and $(\gamma_n)_{n\geq 1}$ for which*

$$\left(\alpha_n^{-1}(\hat{N}^e(d_n) - \beta_n), \gamma_n^{-1}(N^{\hat{s}_n}(d_n) - \beta_n)\right) \Rightarrow (\hat{\mathcal{N}}^e, \hat{\mathcal{N}}^s)$$

where $(\hat{\mathcal{N}}^e, \hat{\mathcal{N}}^s) = (R_1, R_2 \wedge 0)$ for some nondegenerate r.v.'s R_1 and R_2.

The constants $(\alpha_n)_{n\geq 1}$, $(\gamma_n)_{n\geq 1}$ *satisfy* $\alpha_n \approx \beta_n^a$ *for* $1/2 < a \leq 1$, *and* $\gamma_n \approx \beta_n^a$ *for* $1/2 < a < 1$ *and* $\approx \beta_n / \log \beta_n$ *for* $a = 1$; *and the pair* (R_1, R_2) *is given by*

$$(R_1, R_2) = \begin{cases} (-\Delta(a), -(1-a)\Delta(a)) & \textit{for} \quad 1/2 < a < 1, \\ (\exp(-\Delta(1)) - 1, -\Delta(1)) & \textit{for} \quad a = 1. \end{cases}$$

The following constants and functions are used to identify norming constants in the proof of Theorem 3.1:

$$A_n^e = n^a L(1/n), \quad B_n(s) = \begin{cases} n \int_0^s G^{\leftarrow}(1-u)\,du & \text{for} \quad 1/2 < a < 1, \\ n \int_{1/n}^s G^{\leftarrow}(1-u)\,du & \text{for} \quad a = 1, \\ 0 & \text{for} \quad 1 < a. \end{cases}$$

From Theorem 3 of Csörgő et al. (1986), it follows that for $G(x)$ in Case II for maxima with $1/2 < a < \infty$, and for any constants $(k_n)_{n\geq 1}$ with $1 \leq k_n \leq n$ and $k_n \approx \beta_n$

$$\begin{aligned} &(A_n^e)^{-1}\left(\sum_{i=1}^{k_n} Y_{n-i+1,n} - B_n(k_n/n)\right) \\ &= (A_n^e)^{-1}\left(\sum_{i=1}^{n} Y_i - B_n(1)\right) + o_P(1) \stackrel{\text{d}}{=} \Delta(a) + o_P(1) \end{aligned} \tag{7}$$

The statement and proof of the following lemma should be compared to that of Theorem 3 of Csörgő et al. (1986) and to Corollary 2 of Csörgő et al. (1991); in particular, contrast (7) and (8) in the case of $k_n = j_n = [nc]$ for $0 < c < 1$.

Lemma 3.2. *Let* $G(x)$ *be in Case II for maxima with* $1/2 < a < \infty$ *and* $G(0-) = 0$; *and let positive constants* $(d_n)_{n\geq 1}$, $(\beta_n)_{n\geq 1}$ *and* $(\rho_n)_{n\geq 1}$ *satisfy* (5) *with* $\beta_n \to \infty$ *and* $\beta_n/n \to 0$ *as* $n \to \infty$. *Let* $\tau_n := 1 - G(\rho_n-)$ *for* $n \geq 1$. *Then for any integers* $(j_n)_{n\geq 1}$ *with* $1 \leq j_n \leq n$ *and* $j_n \approx cn$ *for some constant* $0 < c \leq 1$, *it follows that*

$$(A_n^e)^{-1}\left(\sum_{i=1}^{j_n} Y_i I(Y_i \geq \rho_n) - (j_n/n)\, B_n(\tau_n)\right) \Rightarrow \begin{cases} c^a \Delta(a) & \textit{for} \quad a \neq 1, \\ c(\Delta(1) + \log c) & \textit{for} \quad a = 1. \end{cases} \tag{8}$$

In particular, for $j_n \approx n$,

$$\begin{aligned} &(A_n^e)^{-1}\left(\sum_{i=1}^{j_n} Y_i I(Y_i \geq \rho_n) - (j_n/n) B_n(\tau_n)\right) \\ &= (A_n^e)^{-1}\left(\sum_{i=1}^{n} Y_i - B_n(1)\right) + o_P(1) \stackrel{\text{d}}{=} \Delta(a) + o_P(1). \end{aligned} \tag{9}$$

Proof. For the proof of (8), first write

$$
\begin{aligned}
&(A_n^e)^{-1}\left(\sum_{i=1}^{j_n} Y_i I(Y_i \geq \rho_n) - (j_n/n)B_n(\tau_n)\right) \\
&= o(1) + \left(\frac{\sum_{i=1}^{j_n} Y_i - (j_n/n)B_n(1)}{j_n^a L(1/j_n)}\right)\left(\frac{j_n^a L(1/j_n)}{n^a L(1/n)}\right) \\
&\quad - \left((A_n^e)^{-1} n^{1/2}\tau_n^{(1/2)-a} L(\tau_n)\right)\left(\frac{\sum_{i=1}^{j_n} Y_i I(Y_i < \rho_n) - j_n \int_{\tau_n}^1 G^{\leftarrow}(1-s)\,ds}{n^{1/2}\tau_n^{(1/2)-a} L(\tau_n)}\right).
\end{aligned}
$$

Observe that the second expression on the right converges in distribution to $c^a\Delta(a)$ for $a \neq 1$ and to $c(\Delta(1) + \log c)$ for $a = 1$; and in the third expression

$$(A_n^e)^{-1} n^{1/2}\tau_n^{(1/2)-a} L(\tau_n) = o(1)$$

(e.g., use Lemma 2 in Csörgő, Horváth and Mason (1986)) and the quotient in parentheses converges to a $N(0,\sigma^2)$-distributed r.v., with σ^2 finite, since it equals

$$c^{1/2}\left(\tau_n^{-1/2} U_{j_n}(1-\tau_n) - \frac{\int_{1/n}^{1-\tau_n} U_{j_n}(s)\,dG^{\leftarrow}(s)}{\tau_n^{1/2} G^{\leftarrow}(1-\tau_n)}\right) + o_P(1).$$

To prove (9), use an argument analogous to that used for (8), together with the observation that

$$(A_n^e)^{-1}\left(\sum_{i=j_n+1}^{n} Y_i - \left(\frac{n-j_n}{n}\right) B_n(1)\right) = o_P(1).$$

If, in addition to the hypotheses of Lemma 3.2, it is true that d.f. satisfies condition (6), then the conclusion and proof of Lemma 3.2 hold with τ_n replaced by β_n/n.

Proof of Theorem 3.1. The verifications in the cases $1/2 < a < 1$ and $a = 1$ are analogous, so arguments are given here for $1/2 < a < 1$, and not for $a = 1$. First, let $(k_n)_{n\geq 1}$ and $(j_n)_{n\geq 1}$ be given by $k_n = \lceil \beta_n + \alpha_n\mu \rceil$ and $j_n = \lceil n(1 + \gamma_n\beta_n^{-1}\eta) \rceil$ for $-\infty < \mu < \infty$ and $-\infty < \eta < 0$, and use results on slowly varying functions (see the Appendix) to obtain

$$(A_n^e)^{-1}(d_n - B_n(k_n/n)) = -\mu + o(1)$$

and

$$(A_n^e)^{-1}(d_n - (j_n/n)B_n(\beta_n/n)) = -(1-a)^{-1}\eta + o(1).$$

Next, use that $\hat{\nu}_n(d_n)/n \to 1$ in probability (from (8)) to obtain that

$$\gamma_n^{-1} \sum_{i=1}^{\hat{\nu}_n(d_n)} \left(I(Y_i \geq \rho_n) - (1 - G(\rho_n -)) \right) = o_P(1).$$

From these results, (7) and (9), and (6), obtain for $-\infty < \mu < \infty$ and $-\infty < \eta < 0$ that

$$P(\alpha_n^{-1}(\hat{N}^e(d_n) - \beta_n) \leq \mu, \gamma_n^{-1}(N^{\hat{s}_n}(d_n) - \beta_n) \leq \eta)$$

$$= P(\hat{N}^e(d_n) \leq \alpha_n\mu + \beta_n, \hat{\nu}_n(d_n) \leq n(1 + \beta_n^{a-1}\eta)) + o(1)$$

$$= P\left(\sum_{i=1}^{k_n} Y_{n-i+1,n} > d_n, \sum_{i=1}^{j_n} Y_i I(Y_i \geq \rho_n) > d_n \right) + o(1)$$

$$= P\left((A_n^e)^{-1} \left(\sum_{i=1}^{n} Y_i - n \int_0^1 G^{\leftarrow}(1-u)\, du \right) > -\mu, \right.$$
$$\left. (A_n^e)^{-1} \left(\sum_{i=1}^{n} Y_i - n \int_0^1 G^{\leftarrow}(1-u)\, du \right) > -(1-a)^{-1}\eta \right) + o(1)$$

$$= P\left((-1)(A_n^e)^{-1} \left(\sum_{i=1}^{n} Y_i - n \int_0^1 G^{\leftarrow}(1-u)\, du \right) < \mu, \right.$$
$$\left. -(1-a)(A_n^e)^{-1} \left(\sum_{i=1}^{n} Y_i - n \int_0^1 G^{\leftarrow}(1-u)\, du \right) < \eta \right) + o(1)$$

and the conclusion follows.

Observe that in the case $a = 1$ of Theorem 3.1, the location parameter is β_n for both $\hat{N}^e(d_n)$ and $N^{\hat{s}_n}(d_n)$; and the scaling constants are β_n and $\beta_n / \log \beta_n$ for $\hat{N}^e(d_n)$ and $N^{\hat{s}_n}(d_n)$ respectively (with $\beta_n \gg \beta_n / \log \beta_n$ in this case). Note also that $\beta_n / n \to 0$, but in this case, $d_n / n \approx (\log \beta_n) L(\beta_n / n)$. Finally, in this case the distributional convergence of $\hat{N}^e(d_n)$ can be rewritten as $\beta_n^{-1}\hat{N}^e(d_n) \Rightarrow \exp(-\Delta(1))$.

Example 3.3. Let $G(x) = 1 - x^{-1/a}$ for $x \geq 1$, with $1/2 < a \leq 1$, so $G(x)$ is in Case II for maxima with index $1 \leq \alpha < 2$.
(*i*) For $1/2 < a < 1$, and positive constants $(d_n)_{n \geq 1}$ satisfying $d_n / n \to 0$ and $d_n / n^a \to \infty$, Theorem 3.1 gives

$$\left(\beta_n^{1-a} \left(\frac{\hat{N}^e(d_n)}{\beta_n} - 1 \right), \beta_n^{1-a} \left(\frac{N^{\hat{s}_n}(d_n)}{\beta_n} - 1 \right) \right) \Rightarrow \left(-\Delta(a), (-(1-a)\,\Delta(a)) \wedge 0 \right)$$

where $\Delta(a)$ is a r.v. with stable distribution of index $\alpha = 1/a$. Note that β_n can be written as $\beta_n = ((1-a)\, d_n n^{-a})^{1/(1-a)}$.

(*ii*) For $a = 1$ and positive constants $(d_n)_{n\geq 1}$ satisfying $d_n/n \to \infty$ and $(d_n/n) - \log n \to -\infty$, $\beta_n = \exp(d_n/n)$ (see (5)), Theorem 3.1 gives

$$\left(\frac{\hat{N}^e(d_n)}{\beta_n} - 1, (\log \beta_n)\left(\frac{N^{\hat{s}_n}(d_n)}{\beta_n} - 1\right)\right) \Rightarrow (\exp(-\Delta(1)) - 1, (-\Delta(1)) \wedge 0)$$

where $\Delta(1)$ is a r.v. with stable distribution of index $\alpha = 1$.

The next corollary is the 'stable' analogue of Corollary 2.6.

Corollary 3.4. *Let $G(x)$ be in Case II for maxima with $1/2 < a \leq 1$ and $G(0-) = 0$, and let positive constants $(d_n)_{n\geq 1}$, $(\beta_n)_{n\geq 1}$ and $(\rho_n)_{n\geq 1}$ satisfy the hypotheses of Theorem 3.1. Then*
(*i*) *the sequence of stopped threshold policies $(\hat{s}_n)_{n\geq 1}$ is a consistent approximater of τ^e;*
(*ii*) *$\hat{N}^e(d_n)/N^{\hat{s}_n}(d_n) \to 1$ in probability as $n \to \infty$;*
(*iii*) *for $G(x)$ in Case II with $1/2 < a < 1$,*

$$\beta_n^{-a}(\hat{N}^e(d_n) - N^{\hat{s}_n}(d_n)) \Rightarrow \hat{\mathcal{N}}^e - \hat{\mathcal{N}}^s,$$
$$(N^{\hat{s}_n}(d_n) - \beta_n)/(\hat{N}^e(d_n) - \beta_n) \Rightarrow \hat{\mathcal{N}}^s/\hat{\mathcal{N}}^e$$

where $\hat{\mathcal{N}}^e$ and $\hat{\mathcal{N}}^s$ are given as in Theorem 3.1.

For distributions $G(x)$ in Case II with $1 < a < \infty$ the following two results demonstrate that, in contrast to Cases I, II with $0 < a \leq 1$ and III, the stopped threshold policy $\hat{s}_n$ with threshold $\rho_n = G^{\leftarrow}(1 - (\beta_n/n))$ is *not* a consistent approximater of the off-line largest-fit strategy τ^e.

Theorem 3.5. *Let $G(x)$ be in Case II with $1 < a < \infty$ and with $G(0-) = 0$, and let positive constants $(d_n)_{n\geq 1}$ and $(\beta_n)_{n\geq 1}$ satisfy (5). Assume that $\beta_n \to \infty$ as $n \to \infty$. Then*

$$\hat{N}^e(d_n) \Rightarrow \hat{\mathcal{N}}^e := \sup\left\{k \geq 1 : \sum_{j=1}^{k} S_j^{-a} \leq 1/(a-1)\right\}$$

if this set is nonempty and $:= 0$ otherwise, where $S_j = T_1 + \cdots + T_j$ and $T_1, T_2, \ldots$ are standard exponentially distributed r.v.'s.

Proof. The theorem follows immediately from the convergence $d_n/(n^a L(1/n)) \to 1/(a-1)$ which can be obtained from results on slowly varying functions (see the Appendix), and the following result which can be found in Csörgő and Mason (1986), p. 974:

$$\frac{\sum_{j=1}^{k} Y_{n-j+1,n}}{n^a L(1/n)} \Rightarrow \sum_{j=1}^{k} S_j^{-a}, \qquad \text{for fixed } k \geq 1,$$

with $\{S_j\}$ as in the assertion of the theorem.

Theorem 3.6. *Let $G(x)$ be in Case II with $1 < a < \infty$ and with $G(0-) = 0$, and let the positive constants $(d_n)_{n\geq 1}$, $(\beta_n)_{n\geq 1}$ and $(\rho_n)_{n\geq 1}$ satisfy (5). Assume that $\beta_n \to \infty$ and $\beta_n/n \to 0$ as $n \to \infty$. Then*

$$\beta_n^{-1} N^{\hat{s}_n}(d_n) \Rightarrow \hat{\mathcal{N}}^s$$

where $\hat{\mathcal{N}}^s = \Big(((a-1)\Delta(a))^{-1/a} I(\Delta(a) > 0)\Big) \wedge 1$ and $\Delta(a)$ is a r.v. with stable distribution of index $\alpha = 1/a$.

Proof. First, use Chebyshev's inequality to obtain that for each $0 < \gamma \leq 1$,

$$[\gamma n]^{-1} \sum_{j=1}^{[\gamma n]} I(Y_j \geq \rho_n)\, n\beta_n^{-1} \to 1 \qquad \text{in probability as } n \to \infty. \tag{10}$$

Second, from properties of slowly varying functions, it follows that $d_n/(n^a L(1/n)) = (a-1)^{-1} + o(1)$, and hence from the definition of $\hat{\nu}_n(d_n)$, Lemma 3.2, and the assumption that $\beta_n \to \infty$ and $\beta_n/n \to 0$ we obtain

$$\hat{\nu}_n(d_n)/n \Rightarrow \hat{\mathcal{N}}^s \tag{11}$$

where $\hat{\mathcal{N}}^s$ is defined above.

Now, one can use (10) and (11) and an argument analogous to that used to prove Theorem 9.4.1 in Chow and Teicher (1978), Theorem 1 in Mogyoródi (1971), or Lemma 1 in Azlarov, Dzamiraev and Abdullaev (1986), to obtain

$$(\hat{\nu}_n(d_n))^{-1} \sum_{j=1}^{\hat{\nu}_n(d_n)} I(Y_j \geq \rho_n)\, n\beta_n^{-1} \to 1 \qquad \text{in probability as } n \to \infty.$$

Finally, use this convergence result, (11) and the representation

$$\beta_n^{-1} N^{\hat{s}_n}(d_n) = \left((\hat{\nu}_n(d_n))^{-1} \sum_{j=1}^{\hat{\nu}_n(d_n)} I(Y_j \geq \rho_n)\, n\beta_n^{-1} \right) (\hat{\nu}_n(d_n)/n)$$

to show that $\beta_n^{-1} N^{\hat{s}_n}(d_n) = (\hat{\nu}_n(d_n)/n) + o_P(1)$; and the conclusion follows.

4. Large capacities

The paper is concluded with the analogue of Theorems 2.1 and 3.1 in the 'large capacity' setting, in which it is assumed that the positive capacities $(d_n)_{n\geq 1}$ satisfy $n^{-1/2}(d_n - n\theta) \to 0$ as $n \to \infty$, for some $0 < \theta < EY_1$. Define constants τ and $\rho_n = \rho$ by $\theta = \int_0^\tau G^{\leftarrow}(1-s)\, ds$ and $\rho = G^{\leftarrow}(1-\tau)$. The proof of Theorem 4.1

goes along the same lines as the proof of Theorem 2.1 using standard approximation arguments, for example, given in Csörgő et al. (1986), and is therefore omitted here.

Theorem 4.1. *If $G(x)$ is continuous and strictly increasing on its support, $G(0-) = 0$ and $EY_1^2 < \infty$, then*

$$\left(n^{-1/2}(\hat{N}^e(d_n) - n\tau), n^{-1/2}(N^{\hat{s}_n}(d_n) - n\tau)\right) \Rightarrow (\hat{\mathcal{N}}^e, \hat{\mathcal{N}}^s)$$

where $(\hat{\mathcal{N}}^e, \hat{\mathcal{N}}^s) = (W_1, W_2 + (W_3 \wedge 0))$ and (W_1, W_2, W_3) is $N(\mathbf{0}, \Sigma(\theta))$-distributed. The covariance matrix $\Sigma(\theta)$ is a symmetric matrix $\Sigma(\theta) = (\Sigma(\theta)_{i,j})_{i,j=1,2,3}$ given by

$$\begin{aligned}
(\Sigma(\theta)_{1,j})_{j=1,2,3} &= \left(\sigma^2/\rho^2, (1-\tau)(\tau - (\theta/\rho)), \theta^{-1}\tau((\sigma^2/\rho) - (1-\tau)(\tau\rho - \theta))\right); \\
(\Sigma(\theta)_{2,j})_{j=2,3} &= (\tau(1-\tau), -\tau(1-\tau)); \\
\Sigma(\theta)_{3,3} &= \theta^{-2}\tau^2\left(\sigma^2 + 2\rho(1-\tau)\theta - \rho^2\tau(1-\tau)\right); \\
\sigma^2 = \sigma^2(\tau) &= \int_{1-\tau}^1 \int_{1-\tau}^1 (s \wedge t - st)\, dG^{\leftarrow}(s)\, dG^{\leftarrow}(t).
\end{aligned}$$

In fact, $W_3 = ((\tau\rho)/\theta)(W_1 - W_2)$.

The following corollary follows immediately.

Corollary 4.2. *If $G(x)$ is continuous and strictly increasing on its support, $G(0-) = 0$ and $EY_1^2 < \infty$, then*
(i) the sequence of stopped threshold policies $(\hat{s}_n)_{n\geq 1}$ is a consistent approximater of τ^e;
(ii) $\hat{N}^e(d_n)/N^{\hat{s}_n}(d_n) \to 1$ in probability as $n \to \infty$;
(iii) $(N^{\hat{s}_n}(d_n) - n\tau)/(\hat{N}^e(d_n) - n\tau) \Rightarrow \hat{\mathcal{N}}^s/\hat{\mathcal{N}}^e$;
(iv) $n^{-1/2}(\hat{N}^e(d_n) - N^{\hat{s}_n}(d_n)) \Rightarrow \hat{\mathcal{N}}^e - \hat{\mathcal{N}}^s$
where $(\hat{\mathcal{N}}^e, \hat{\mathcal{N}}^s)$ is given in Theorem 4.1.

Appendix

Recall the following definitions, relations, and results concerned with domains of attraction for maxima. An excellent monograph on extreme value theory is Resnick (1987).

Let Y_1, Y_2, ... be i.i.d. r.v.'s with d.f. G. $G(x)$ is said to be in the domain of attraction of d.f. $H(x)$ for maxima if there exists constants $(a_n)_{n\geq 1}$, $a_n > 0$ and $(b_n)_{n\geq 1}$, $b_n \in \mathbb{R}$, such that

$$P\left(a_n\left(\max_{1\leq i\leq n} Y_i - b_n\right) \leq x\right) \to H(x) \qquad \text{for all continuity points } x \text{ of } H. \quad (12)$$

In this setting we say G is in Case I, in Case II with index $\alpha > 0$ or in Case III with index $\alpha > 0$, if the limit d.f. H is respectively given by $\Lambda(x) = \exp(-e^{-x})$ for

$x \in \mathbb{R}$; by $\Phi_\alpha(x) = \exp(-x^{-\alpha})$ for $x > 0$; or by $\Psi_\alpha(x) = \exp(-(-x)^\alpha)$ for $x \le 0$. In the following paragraphs we list some properties in each of the Cases I, II, and III which are used in this paper.

- If $G(x)$ is in Case I for maxima, then constants $(a_n)_{n\ge1}$ and $(b_n)_{n\ge1}$ are given by
$$b_n = G^{\leftarrow}(1-1/n) \quad \text{and} \quad a_n^{-1} = g(b_n) \qquad \text{for } n \ge 1, \tag{13}$$
for example with $g(t) = \int_t^{r_G}(1-G(x))\,dx/(1-G(t))$. The function $G^{\leftarrow}(1-s)$ is slowly varying at zero. An auxiliary function useful in analysis is the function $c(s)$ defined on $[0,1]$ by
$$c(s) := s^{-1}\int_{1-s}^{1}(1-u)\,dG^{\leftarrow}(u) = s^{-1}\int_0^s G^{\leftarrow}(1-u)\,du - G^{\leftarrow}(1-s).$$
As proven by Lo (1989), the function $c(s)$ satisfies the following properties. There exists a finite constant k such that for $0 < s \le 1/2$,
$$G^{\leftarrow}(1-s) = k - c(s) + \int_s^1 u^{-1}c(u)\,du.$$
Note also that the function $\phi(s) := \int_0^s G^{\leftarrow}(1-u)\,du$ can be written as $\phi(s) = s\left(k + \int_s^1 u^{-1}c(u)\,du\right)$. Moreover, $c(s) > 0$; $c(s)$ is slowly varying at zero; $\lim_{s\downarrow0} c(s)/G^{\leftarrow}(1-s) = 0$; and if r_G is finite, then $\lim_{s\downarrow0} c(s) = 0$.

- If $G(x)$ is in Case II for maxima, with $\alpha > 0$, and $a = 1/\alpha$, then
$$a_n^{-1} = G^{\leftarrow}(1-1/n) \quad \text{and} \quad b_n = 0 \qquad \text{for } n \ge 1, \tag{14}$$
$r_G = \infty$ and $\lim_{n\to\infty} n(1-G(a_n^{-1}x)) = x^{-\alpha}$ for $x > 0$. The function $1-G(x)$ is regularly varying as $x \to \infty$ with index $-\alpha = -1/a$; and $G^{\leftarrow}(1-s) = s^{-a}L(s)$ where L is slowly varying at zero.

- If $G(x)$ is in Case III for maxima, with $\alpha > 0$, and $a = -1/\alpha$, then
$$a_n^{-1} = r_G - G^{\leftarrow}(1-1/n) \quad \text{and} \quad b_n = r_G \qquad \text{for } n \ge 1, \tag{15}$$
r_G is finite and $\lim_{n\to\infty} n(1-G(a_n^{-1}x + r_G)) = (-x)^\alpha$ for $x < 0$. The function $1 - G(r_G - x^{-1})$ is regularly varying as $x \to \infty$ with index $-\alpha = 1/a$; and $G^{\leftarrow}(1-s) = r_G - s^{-a}L(s)$ where L is slowly varying at zero.

We state two results concerning functions which are slowly varying at zero, which we frequently use in this paper.

- (Karamata's Theorem (see Csörgő and Mason (1986), Lemma 1.)) Let $L(x)$ be slowly varying at zero. If $\beta < 1$, then
$$\lim_{s\downarrow0}\int_0^s u^{-\beta}L(u)\,du\Big/\left(s^{1-\beta}L(s)\right) = \frac{1}{1-\beta}.$$

- (Karamata Representation (see e.g. Resnick (1987), Section 0.4).) For $0 < t < t_0$, $L(t)$ is slowly varying at zero if and only if $L(t) = c(t)\exp\left(\int_t^{t_0} \varepsilon(u)/u\,du\right)$ for some measurable functions $c : (0, t_0) \to \mathbb{R}_+$ and $\varepsilon: (0, t_0) \to \mathbb{R}$ satisfying $\lim_{t\downarrow 0} c(t) = c_0$ for some constant $c_0 > 0$ and $\lim_{t\to 0} \varepsilon(t) = 0$.

Finally, we give an example of a d.f. $G(x)$ which is in Case II for maxima with $a = 1/3$, and a sequence $(d_n)_{n\geq 1}$ for which condition (3*i*) fails. Other examples illustrating conditions (3b) and (6) can be constructed similarly. Let $G(x) = 1-[x]^{-3}$ for $x \geq 1$, where $[x]$ is the greatest integer less than or equal to x; thus $G(x) = \sum_{k=1}^{\infty}(1-k^{-3})I(k \leq x < k+1)$. It is clear that $G(x)$ is a d.f. in Case II for maxima with $\alpha = 3$ and $a = 1/3$. Let $B_k := 1 - k^{-3} = G(k)$ for $k = 1, 2, \ldots$; and observe that $G^{\leftarrow}(w) = \sum_{k=2}^{\infty} kI(B_{k-1} < w \leq B_k)$ for $0 < w < 1$. Let $k_n := [\log n]$ for $n = 1, 2, \ldots$, and let $(d_n)_{n\geq 1}$ be the positive integers satisfying

$$d_n/n = \int_{B_{k_n}}^1 G^{\leftarrow}(u)\,du = \sum_{l=k_n}^{\infty}(l+1)(B_{l+1} - B_l) = k_n^{-2} + \sum_{l=k_n}^{\infty} l^{-3}$$

for $n \geq 1$. Note that $d_n \approx (3/2)n/(\log n)^2$. These $(d_n)_{n\geq 1}$ were chosen so that the constants $(\beta_n)_{n\geq 1}$ of (2) are given by

$$\beta_n = n(1 - B_{k_n}) = nk_n^{-3}$$

for $n \geq 1$; note that $\beta_n \approx n/(\log n)^3$. Also the constants $(\rho_n)_{n\geq 1}$ of (2) satisfy

$$\rho_n = G^{\leftarrow}(1 - (\beta_n/n)) = G^{\leftarrow}(B_{k_n}) = k_n$$

for $n \geq 1$. Thus, it follows that for this example

$$\begin{aligned} \beta_n^{1/2}\left(\frac{1 - G(\rho_n -)}{\beta_n/n} - 1\right) &= n^{1/2}(1 - B_{k_n})^{1/2}\left(\frac{1 - B_{k_n-1}}{1 - B_{k_n}} - 1\right) \\ &\approx 3n^{1/2}k_n^{-5/2} \approx (1/5)(n/\log n)^{1/2}, \end{aligned}$$

which tends to infinity as $n \to \infty$.

Acknowledgment. The authors are grateful to the School of Mathematics of the Georgia Institute of Technology in Atlanta, the Econometric Institute of the Erasmus University Rotterdam and the Stieltjes Institute in The Netherlands for reciprocal invitations to pursue this research, and for their hospitality.

References

Azlarov, T.A., Dzamiraev, A.A. and Abdullaev, A.G. (1986). Remarks on limit theorems for sequences of random variables with random index. *Prob. Theory and Math. Stat.* **1**, 99–113.

Boshuizen, F.A. and Kertz, R.P. (1995). Smallest-fit selection of random sizes under a sum constraint: weak convergence comparisons, *preprint.*

Bruss, F.T. and Robertson, J.B. (1991). "Wald's lemma" for sums of order statistics of i.i.d. random variables. *Adv. Appl. Prob.* **23**, 612–623.

Chow, Y.S. and Teicher, H. (1978). *Probability Theory: Independence, Interchangeability, Martingales.* Springer-Verlag, New York.

Coffman, E.G., Flatto, L. and Weber, R.R. (1987). Optimal selection of stochastic intervals under a sums constraint. *Adv. Appl. Prob.* **19**, 454–473.

Csörgő, M., Csörgő, S., Horváth, L. and Mason, D.M. (1986). Weighted empirical and quantile processes. *Ann. Probab.* **14**, 31–85.

Csörgő, M., Csörgő, S. and Horváth, L. (1986). *An Asymptotic Theory for Empirical Reliability and Concentration Processes.* Lecture Notes in Statistics **33**. Springer-Verlag, Berlin.

Csörgő, M., Csörgő, S., Horváth, L. and Mason, D.M. (1986). Normal and stable convergence of integral functions of the empirical distribution function *Ann. Prob.* **14**, 86–118.

Csörgő, S. and Mason, D.M. (1986). The asymptotic distribution of sums of extreme values from a regularly varying distribution. *Ann. Probab.* **14**, 974–983.

Csörgő, S., Horváth, L. and Mason, D.M. (1986). What portion of the sample makes a partial sum asymptotically stable or normal? *Probab. Th. Rel. Fields* **72**, 1–16.

Csörgő, S., Haeusler, E. and Mason, D. (1991). The asymptotic distribution of extreme sums. *Ann. Probab.* **19**, 783–811.

Lo, G.S. (1989). A note on the asymptotic normality of extreme values. *J. Statist. Planning and Inference* **22**, 127–136.

Mogyoródi, J. (1971). A remark on limiting distributions for sums of a random number of independent random variables. *Rev. Roum. Math. Pures et Appl.* **16**, 552–557.

Resnick, S. (1987). *Extreme Values, Regular Variation, and Point Processes.* Springer-Verlag, New York.

Exploring Stochastic Laws. pp. 77–90.
Eds. A.V. Skorokhod and Yu.V. Borovskikh

Limit theorems for some characteristics of system $GI/GI/1$

N.S. BRATIĬCHUK

National Academy of Sciences of Ukraine, Institute of Mathematics, Tereschenkivska St. 3, 252601 Kiev-4, Ukraine
Institute of Mathematics Silesian Technical University, Poland

Received March 25, 1995

Abstract. A method based on the solution of a system of integral equations on a half axis is proposed to investigate queueing system $GI/GI/1$. Some results for such a system are stated.

1. Introduction

This paper describes a method based on the theory of integral equations on a half axis for the study of queueing system GI/GI/1. This system may be formally described by the sequence $\{\alpha_n, , \beta_n, , n \geq 0\}$, where α_n is the arrival time of the n-th customer. The system includes a single server which accepts the customers according to the discipline "first come - first serve". The service time of the n-th customer is β_n.

When a served customer leaves the system the server immediately begins to serve another customer. Unserved customers form a queue which can be unlimited. The random variables $\alpha_{n+1} - \alpha_n, \beta_n, n \geq 0$ are supposed to be totally independent and their distribution functions do not depend on n.

The traditional approach to the investigation of such systems consists of the study of random walk

$$S_n = \sum_{k=1}^{n}(x_k - y_k), \quad n \geq 0,\ S_0 = 0,\ x_n = \alpha_n - \alpha_{n-1},\ y_n = \beta_n$$

since, as it turn out, there exists a close relation between characteristics of the system and named random walk [1]. But when the system at time $t = 0$ contains more than one customer there are some difficulties in using this approach. At the same time it is well known that for the characteristics of systems $EI/GI/1$ and $GI/EI/1$ we can write the integral (or differential-integral) equations on the half axis using the fact that the arrival times for the first system or the departure times for the second are Markovian. For system $GI/GI/1$ it is not the case, but adding to the service process (process $\xi(t)$ below) additional components we can derive a new process which will have such times and hence, we can write the equations again (usually it is a system of equations) for the characteristics. The main difficulty is to find a solution of this system in a form acceptable for further analysis. In 1989 V.S. Korolyuk proposed a method for the solution. At first in the investigation of boundary functionals for random walks on the superposition of two renewal processes and then, in [2], this method was applied to the study of system $GI/GI/1$. This paper is a further development of this approach.

2. System of integral equations for characteristics

Denote

$$\alpha_n^1 = \alpha_n - \alpha_{n-1},$$

$$\alpha_n^2 = \beta_n \quad n \geq 1,$$

$$\mu_i(t) = \max\{n : \sum_{k=1}^{n} \alpha_k^i \leq t\}, \quad t \geq 0,$$

$$\xi(t) = \xi(0) + \mu_1(t) - \mu_2(t),$$

$$F_i(x) = P\{\alpha_n^i < x\}.$$

Let $\rho_n, \ n = 0, 1, \ldots, \rho_0 = 0$ be the jump-times of process $\xi(t)$ and accept the following notation: we say that $\rho_n, \ n \geq 1$ has type "1" if ρ_n is the jump time for process $\mu_1(t)$ and has type "2" if it is the jump - time for process $\mu_2(t)$. If both processes have jumps at time ρ_n we say that this time has type "1" too.

Let the event $A_n, n \geq 0$ mean that time ρ_n has type "1". Set

$$\kappa_n(\omega) = 2 - I\{A_n\},$$

where $I\{A\}$ is the characteristic function of the set A,

$$\nu(n, \omega) = \min\{k > 0 : \kappa_n \neq \kappa_{n+k}\}$$

and define the processes $\beta(t)$ and $\kappa(t)$ in the following way

$$\beta(t) = \rho_{n+\nu(n,\omega)} - t, \quad \kappa(t) = \kappa_n$$

for $\rho_n \leq t < \rho_{n+\nu(n,\omega)}, \ n \geq 0$.

Lemma 1.

$$(\xi_n, \kappa_n, \beta_n) = (\xi(\rho_n), \kappa_n(\rho_n), \beta(\rho_n)), n \geq 0,$$

is a Markov chain with state space $R \times \{1,2\} \times R^+$.

The proof of this lemma follows immediately from definitions of processes $\xi(t)$, $\kappa(t)$ and $\beta(t)$.

For system GI/GI/1 we denote the following:

τ — the busy period ,

$N(\tau)$ — the number of customers served during the busy period,

ν — the first idle time,

$$W_i(x,n) := \ W_i(x,n,u,\lambda,z) = E\{e^{-\lambda\tau - u\nu} z^{N(\tau)} | \xi_0 = n, \kappa_0 = i, \beta_0 = x\},$$

$$\lambda > 0, u > 0, |z| < 1.$$

Thus, for example, $W_1(x,n)$ is the joint characteristic function of the random variables $\tau, N(\tau), \nu$ provided that at time $t = 0$ the system contains n customers. The system starts at time $t = 0$ which is the arrival time of the customer and in time x the first served customer leaves the system. The function $W_2(x,n)$ has a similar meaning.

Using the full probability formula and Lemma 1 one can derive the following system of integral equations for functions $W_i(x,n)$ and $n \geq 1$

$$W_1(x,n) - \int_0^x e^{-\lambda y} W_1(x-y, n+1) dF_1(y) - z e^{-\lambda x} \int_x^\infty W_2(y-x, n-1) dF_1(y) -$$

$$= z e^{(u-\lambda)x} I\{n=1\} \int_x^\infty e^{-uy} dF_1(y),$$

$$W_2(x,n) - \int_x^\infty e^{-\lambda x} W_1(y-x, n+1) dF_2(y) - z \int_0^x e^{-\lambda y} W_2(x-y, n-1) dF_2(y) =$$

$$= z e^{-ux} I\{n=1\} \int_0^x e^{-(\lambda-u)y} dF_2(y).$$

For the functions

$$\hat{W}_1(x,n) = e^{-(u-\lambda)x} W_1(x, n+1),$$

$$\hat{W}_2(x,n) = e^{ux} W_2(x, n+1)$$

from this system we have for $n \geq 0$

$$\hat{W}_1(x,n) - \int_0^x e^{-uy}\hat{W}_1(x-y,n+1)dF_1(y) - z\int_x^\infty e^{-uy}\hat{W}_2(y-x,n-1)dF_1(y) =$$

$$= zI\{n=0\}\int_x^\infty e^{-uy}dF_1(y),$$

$$\hat{W}_2(x,n) - \int_x^\infty e^{-(\lambda-u)y}\hat{W}_1(y-x,n+1)dF_2(y) - z\int_0^x e^{-(\lambda-u)y}\hat{W}_2(x-y,n-1)dF_2(y) =$$

$$= zI\{n=0\}\int_0^x e^{-(\lambda-u)y}dF_2(y),$$

where it is assumed that $\hat{W}_2(x,-1)=0$.

Let $0 \le u < \lambda$ and

$$\hat{W}_2(n) = \lim_{x\to\infty}\hat{W}_2(x,n). \tag{1}$$

At first we assume that this limit exists. Then

$$\hat{W}_2(n) - z\hat{W}_2(n-1)f_2(\lambda-u) = zI\{n=0\}f_2(\lambda-u), \qquad n\ge 0,$$

and hence

$$\hat{W}_2(n) = (zf_2(\lambda-u))^{n+1}, \qquad n\ge 0. \tag{2}$$

Now it is easy to show that limit (1) exists for $0\le u<\lambda$. Indeed, since $P\{\tau+\nu\ge x/\xi_0=n, \kappa_0=2, \beta_0=x\}=1$ then

$$|\hat{W}_2(x,n)| \le 1,$$

and one can find the sequence $x_n \to \infty$ as $n\to\infty$ such that limit (1) exists if instead of x we take x_n. Hence formula (2) holds. Since the right hand side of this formula does not depend on sequence x_n it means that limit (1) exists.

Now we denote

$$V_1(x,n) = \hat{W}_1(x,n),$$

$$V_2(x,n) = \hat{W}_2(x,n) - (zf_2(\lambda-u))^{n+1}, \quad n\ge 0,$$

and for function $V_1(x,n), V_2(x,n)$ we have the following system of integral equations.

$$\hat{V}_1(x,n) - \int_0^x e^{-uy}\hat{V}_1(x-y,n+1)dF_1(y) - z\int_x^\infty e^{-uy}\hat{V}_2(y-x,n-1)dF_1(y) =$$

$$= z^{n+1}f_2^n(\lambda-u)\int_x^\infty e^{-uy}dF_1(y),$$

$$\hat{V}_2(x,n) - \int\limits_x^\infty e^{-(\lambda-u)y}\hat{V}_1(y-x,n+1)dF_2(y) - z\int\limits_0^x e^{-(\lambda-u)y}V_2(x-y,n-1)dF_2(y) =$$
$$= -z^{n+1}f_2^n(\lambda-u)\int\limits_x^\infty e^{-(\lambda-u)y}dF_2(y), \qquad n \geq 0. \tag{3}$$

We are going to investigate a system of this type in the next section.

3. A system of integral equations on a half axis

In this section our main task is to derive a representation for the solution of system (3) in which the functions on the right hand side belong to some space. We begin by defining this space.

For some fixed $a > 0$ we introduce the space W_a in the following way.

W_a is the set $(\phi_1(x,n), \phi_2(x,n))$, $n \geq 0, x \geq 0$, where the functions $\phi_i(x,n)$ are right hand continuous on x for any $n \geq 0$ and for any $\varepsilon > 0$

$$\int\limits_0^\infty e^{-\varepsilon x}\sup_n |\phi_1(x,n)|dx + \int\limits_0^\infty e^{(a-\varepsilon)x}\sup_n |\phi_2(x,n)|dx < \infty.$$

For some fixed $\lambda > 0$ consider in the space W_λ the following system of integral equations

$$\begin{aligned}
V_1(x,n) &- \int\limits_0^x V_1(x-y,n+1)dF_1(y) - \\
&- z\int\limits_x^\infty V_2(y-x,n-1)dF_1(y) = \phi_1(x,n), \\
V_2(x,n) &- \int\limits_x^\infty e^{-\lambda y}V_1(y-x,n+1)dF_2(y) - \\
&- z\int\limits_0^x e^{-\lambda y}V_2(x-y,n-1)dF_2(y) = \phi_2(x,n), \qquad n \geq 0, x \geq 0,
\end{aligned} \tag{4}$$

where $|z| \leq 1$.

Let

$$f_i(s) = \int\limits_0^\infty e^{-sx}dF_i(x), \quad \text{Re } s > 0,$$

$$\phi_+(s,n) = \int\limits_0^\infty e^{-sx}\phi_1(x,n)dx, \quad \text{Re } s > 0,$$

$$\phi_-(s,n) = \int_0^\infty e^{sx}\phi_2(x,n)dx, \qquad \text{Re } s < \lambda,$$

$$\psi_+(s,n,\lambda) = I_+[\phi_-(s,n)f_1(s) - \phi_+(s,n)f_2(\lambda - s)],$$

$$\hat{\phi}_+(s,n) = \sum_{k=0}^\infty f_1^n(s)\phi_+(s,n+k),$$

$$\hat{\psi}_+(s,n,\lambda) = \sum_{k=0}^\infty f_1^n(s)\psi_+(s,n+k,\lambda),$$

$$d(s,n,\lambda) = \hat{\phi}_+(s,n,\lambda)f_2(\lambda - s) + \hat{\psi}_+(s,n,\lambda),$$

where $I_\pm$ denotes the projectors defined by the correlations

$$I_\pm\left[\int_{-\infty}^\infty e^{-sx}dF(x)\right] = \pm\int_{-0}^{\pm\infty} e^{-sx}dF(x).$$

Theorem 1. *Let* $0 < \text{Re } s < \lambda$. *For the solution of system* (4) *the following representations hold*

$$\int_0^\infty e^{-sx}V_1(x,n)dx = \sum_{k=1}^{n-1} z^k I_+[f_2^{k-1}(\lambda - s)d(s,n-k,\lambda)] + \hat{\phi}_+(s,n)+$$

$$+z^n I_+\left[\frac{f_2^{n-1}(\lambda - s)}{f_+(s,\lambda,z)} I_+\left[\frac{d(s,0,\lambda)}{f_-(s,\lambda,z)}\right]\right], \quad n \geq 1, \tag{5}$$

$$\int_0^\infty e^{-sx}V_1(x,0)dx = \hat{\phi}_+(s,0) + \frac{zf_1(s)}{f_+(s,\lambda,z)} I_+\left[\frac{d(s,0,\lambda)}{f_-(s,\lambda,z)}\right], \tag{6}$$

$$\int_0^\infty e^{sx}V_2(x,n)dx = \sum_{k=0}^{n} z^k I_-[f_2^k(\lambda - s)d(s,n-k+1,\lambda)] + \sum_{k=0}^{n} z^k f_2^k(\lambda - s)\hat{\psi}_-(s,n-k)+$$

$$+z^n I_-\left[\frac{f_2^{n-1}(\lambda - s)}{f_+(s,\lambda,z)} I_+\left[\frac{d(s,0,\lambda)}{f_-(s,\lambda,z)}\right]\right], \quad n \geq 0. \tag{7}$$

where the functions $f_\pm(s,\lambda,z)$ *are the components of the canonical factorisation*

$$1 - zf_1(s)f_2(\lambda - s) = f_+(s,\lambda,z)f_-(s,\lambda,z), \qquad \text{Re } s = 0. \tag{8}$$

Remark 1. From the representation

$$1 - zf_1(s)f_2(\lambda - s) = 1 - zf_2(\lambda)\frac{f_1(s)f_2(\lambda - s)}{f_2(\lambda)} =$$

$$= 1 - zf_2(\lambda) \int_{-\infty}^{\infty} e^{-sx} dF(x, \lambda),$$

where $F(x, \lambda)$ is some distribution function on $(-\infty, \infty)$ for any $\lambda \geq 0$, and from well known results about canonical factorisation (see for example, [3]), it follows that factorisation (8) is possible and the functions $f_{+}^{\pm 1}(s, \lambda, z)$ and $f_{-}^{\pm 1}(s, \lambda, z)$ are regular and bounded in the half-plane $\pm\text{Re } s > 0$ and $f_{\pm}(\infty, \lambda, z) = 1$. Such a factorisation is unique.

Proof of Theorem 1.

Multiply the first equation in (4) by e^{-sx}, $\text{Re } s = a \in (0, \lambda)$ and the second by e^{sx}. Then integrate them over x from 0 to ∞. Now multiply these equations by $\mu^n, |\mu| < 1$ and sum them over n from 0 to ∞. After necessary transformations we obtain the following system,

$$\begin{aligned} v_+(s,\mu)(1-\mu^{-1}f_1(s)) \quad &- \quad z\mu f_1(s) v_-(s,\mu) = \\ &= \quad \Phi_+(s,\mu) - \mu^{-1} f_1(s) v_+(s) + \omega_-(s,\mu,z), \\ v_-(s,\mu)(1 - z\mu f_2(\lambda - s)) \quad &- \quad \mu^{-1} f_2(\lambda - s) v_+(s,\mu) = \\ &= \quad \Phi_-(s,\mu) - \mu^{-1} f_2(\lambda - s) v_+(s) + \omega_+(s,\mu,z), \end{aligned} \tag{9}$$

where

$$v_+(s,\mu) = \sum_{n=0}^{\infty} \mu^n \int_0^{\infty} e^{-sx} V_1(x,n) dx,$$

$$v_-(s,\mu) = \sum_{n=0}^{\infty} \mu^n \int_0^{\infty} e^{sx} V_2(x,n) dx,$$

$$\Phi_{\pm}(s,\mu) = \sum_{k=0}^{\infty} \mu^k \phi_{\pm}(s,k),$$

$$\omega_-(s,\mu,z) = -z \sum_{n=0}^{\infty} \mu^n \int_{-\infty}^{0} e^{-sx} \int_x^{\infty} V_2(y - x, n-1) dF_1(y) dx,$$

$$\omega_+(s,\mu,z) = -z \sum_{n=0}^{\infty} \mu^n \int_{-\infty}^{0} e^{sx} \int_x^{\infty} e^{-\lambda y} V_2(y - x, n-1) dF_2(y) dx,$$

$$v_+(s) = \int_0^{\infty} e^{-sx} V_1(x,0) dx.$$

Multiply the first equation in (9) by $(1 - z\mu f_2(\lambda - s))\mu^{-1}$, the second by $-\mu z(1 - \mu^{-1} f_1(s))$ and then sum them. Now we have

$$v_+(s,\mu)(1 - \mu^{-1} f_1(s))\mu^{-1} - \mu z v_-(s,\mu)(1 - z\mu f_2(\lambda - s)) = \Phi(s,\mu) +$$
$$+\mu^{-1}(1 - z\mu f_2(\lambda - s))\omega_-(s,\mu,z) - \mu z(1 - \mu^{-1} f_1(s))\omega_+(s,\mu,z) -$$
$$-v_+(s)(\mu^{-2} f_1(s) - z f_2(\lambda - s)), \tag{10}$$

where

$$\Phi(s,\mu) = \mu^{-1}(1 - z\mu f_2(\lambda - s))\Phi_+(s,\mu) - \mu z(1 - \mu^{-1} f_1(s))\Phi_-(s,\mu).$$

From (10) we have

$$\mu^{-1}(1 - z\mu f_2(\lambda - s))\omega_-(s,\mu,z) = -z\mu v_-(s,\mu)(1 - z\mu f_2(\lambda - s)) -$$
$$-I_-[\Phi(s,\mu)] - zI_-[v_+(s) f_2(\lambda - s)], \tag{11}$$

Now multiply the first equation in (9) by $\mu^{-1}(1 - z\mu f_2(\lambda - s))$, taking into consideration formula (11), and after necessary transformations we obtain

$$(\mu^{-1} v_+(s,\mu) + z\mu v_-(s,\mu))(1 - z\mu f_2(\lambda - s))(1 - \mu^{-1} f_1(s)) =$$
$$= \mu^{-1} I_+[(1 - z\mu f_2(\lambda - s))\Phi_+(s,\mu)] + +\mu z I_-[(1 - \mu^{-1} f_1(s))\Phi_-(s,\mu)] -$$
$$-\mu^{-2} f_1(s) v_+(s)(1 - z\mu f_2(\lambda - s)) - zI_-[v_+(s) f_2(\lambda - s)]. \tag{12}$$

The right hand side of this equation equals zero when $\mu = f_1(s)$. Hence

$$(1 - z f_1(s(f_2(\lambda - s))\frac{v_+(s)}{f_1(s)} = \frac{\Phi_+(s, f_1(s))}{f_1(s)} + z\hat{\psi}_+(s,0,\lambda) - zI_-[v_+(s) f_2(\lambda - s)] \tag{13}$$

or

$$\frac{v_+(s) - \Phi_+(s, f_1(s)}{f_1(s)} = zI_+[v_+(s) f_2(\lambda - s)] + z\hat{\psi}_+(s,0,\lambda). \tag{14}$$

From (14) it follows that function $f_1^{-1}(s)(v_+(s) - \Phi_+(s, f_1(s))$ is regular and bounded in the half-plain Re $s > 0$. Then from equation (13) written in the form

$$(1 - z f_1(s) f_2(\lambda - s))\frac{v_+(s) - \Phi_+(s, f_1(s))}{f_1(s)} = zd(s,0,\lambda) - zI_-[v_+(s) f_2(\lambda - s)]$$

it follows that

$$f_+(s,\lambda,z)\frac{v_+(s) - \Phi_+(s, f_1(s))}{f_1(s)} = zI_+\left[\frac{d(s,0,\lambda)}{f_-(s,\lambda,z)}\right].$$

is equivalent to (6).

In equation (12) the term $I_-[v_+(s)f_2(\lambda - s)]$ can be substituted according to equation (13) and then after the necessary transformations equation (12) takes the following form

$$(\mu^{-1}v_+(s,\mu) + z\mu v_-(s,\mu))(1 - z\mu f_2(\lambda - s))(1 - \mu^{-1}f_1(s)) =$$

$$= \mu^{-1}\Big(\Phi_+(s,\mu) - \Phi_+(s, f_1(s)\Big) +$$

$$+\mu z(1 - \mu^{-1}f_1(s))\Phi_-(s,\mu) + \frac{z(1-\mu^{-1}f_1(s))}{f_+(s,\lambda,z)} I_+\left[\frac{d(s,0,\lambda}{f_-(s,\lambda,z)}\right] +$$

$$+z(D(s,\mu,\lambda) - D(s,f_1(s),\lambda)) + \mu^{-1}(1 - z\mu f_2(\lambda - s))(1 - \mu^{-1}f_1(s))v_+(s), \quad (15)$$

where

$$D(s,\mu,\lambda) = \sum_{k=0}^{\infty} \mu^k \psi_+(s,k,\lambda).$$

It is easy to see that

$$\Phi_+(s,\mu) - \Phi_+(s,f_1(s)) = (1 - \mu^{-1}f_1(s)) \sum_{k=1}^{\infty} \mu^k \hat{\phi}_+(s,k),$$

$$D(s,\mu,\lambda) - D(s,f_1(s),\lambda) = (1 - \mu^{-1}f_1(s)) \sum_{k=1}^{\infty} \mu^k \hat{\psi}_+(s,k,\lambda).$$

And now from equation (15) we have

$$v_+(s,\mu) = I_+\left[\frac{\sum_{k=1}^{\infty}\mu^k\hat{\phi}_+(s,k)}{1 - z\mu f_2(\lambda - s)}\right] + z\mu I_+\left[\frac{I_+[\frac{d(s,0,\lambda)}{f_-(s,\lambda,z)}]}{(1 - z\mu f_2(\lambda - s))f_+(s,\lambda,z)}\right] +$$

$$+z\mu I_+\left[\frac{\sum_{k=1}^{\infty}\mu^k\hat{\psi}_+(s,k,\lambda)}{1 - z\mu f_2(\lambda - s)}\right] + v_+(s),$$

$$v_-(s,\mu) = z^{-1}\mu^{-2}I_-\left[\frac{\sum_{k=1}^{\infty}\mu^k\hat{\phi}_+(s,k)}{1 - z\mu f_2(\lambda - s)}\right] + \mu^{-1}I_-\left[\frac{I_+[\frac{d(s,0,\lambda}{f_-(s,\lambda,z)}]}{(1 - z\mu f_2(\lambda - s))f_+(s,\lambda,z)}\right] +$$

$$+\mu^{-1}I_-\left[\frac{\sum_{k=1}^{\infty}\mu^k\hat{\psi}_+(s,k,\lambda)}{1 - z\mu f_2(\lambda - s)}\right] + \frac{\Phi_-(s,\mu)}{1 - z\mu f_2(\lambda - s)}.$$

The representations stated in the theorem are inversions of these formulas on μ.

4. Representation for characteristics and limit theorems

Now we come back to system (3).

For $0 \le u < \lambda$ we obviously have

$$z^{n+1} f_2^n(\lambda - u)\left(\int_x^\infty e^{-uy} dF_1(y), -\int_x^\infty e^{-(\lambda-u)y} dF_2(y)\right) \in W_{\lambda-u}.$$

The solution of system (3) can be obtained from Theorem 1 if in it we take $e^{-uy}dF_1(y)$ and $e^{-(\lambda-u)}dF_2(y)$ instead of $dF_1(y)$ and $e^{-\lambda y}dF_2(y)$ respectively. (In formulas (5–8) we must put $f_1(s+u)$ and $f_2(\lambda-u-s)$ instead of $f_1(s)$ and $f_2(\lambda-s)$ respectively). To apply formulas (5–8) it is necessary to do some obvious calculations.

Now

$$\phi_+(s,n) = z^{n+1} f_2^n(\lambda-u)\frac{f_1(u)-f_1(s+u)}{s},$$

$$\phi_-(s,n) = z^{n+1} f_2^n(\lambda-u)\frac{f_2(\lambda-u)-f_2(\lambda-u-s))}{s},$$

$$\psi_+(s,n,\lambda) = z^{n+1} f_2^{n+1}(\lambda-u)\frac{f_1(s+u)-f_1(u)}{s},$$

and hence

$$\hat{\phi}_+(s,n) = z^{n+1} f_2^{n+1}(\lambda-u)\frac{f_1(u)-f_1(s+u)}{s(1-zf_1(s+u)f_2(\lambda-u))},$$

$$\hat{\psi}_+(s,n,\lambda) = z^{n+1} f_2^{n+1}(\lambda-u)\frac{f_1(s+u)-f_1(u)}{s(1-zf_1(s+u)f_2(\lambda-u))},$$

$$d(s,n,\lambda) = z^{n+1} f_2^n(\lambda-u)\frac{(f_1(s+u)-f_1(u))(f_2(\lambda-u)-f_2(\lambda-u-s))}{s(1-zf_1(s+u)f_2(\lambda-u))},$$

$$I_+\left[\frac{d(s,0,u,\lambda}{f_-(s+u,\lambda,z}\right] = \frac{f_+(u,\lambda,z)}{s} + \frac{f_+(s+u,\lambda,z)(1-zf_1(u)f_2(\lambda-u))}{s(1-zf_2(\lambda-u)f_1(s+u))}.$$

Now from Theorem 1 using the representations stated above, and after necessary transformations we receive such results.

Theorem 2. *Let $0 \le u < \lambda$, then for $0 \le \operatorname{Re} s < \lambda - u$ and $n \ge 1$ the following representations hold*

$$\int_0^\infty e^{-(s-\lambda)x} W_1(x,n+1)dx = z^n f_+(u,\lambda,z) I_+\left[\frac{f_2^{n-1}(\lambda-s)(f_+(s,\lambda,z)-f_+(u,\lambda,z))}{f_+(s,\lambda,z)(u-s)}\right],$$

$$\int_0^\infty e^{-(s-\lambda)x} W_1(x,1)dx = \frac{z(f_1(s)f_+(u,\lambda,z) - f_1(u)f_+(s,\lambda,z))}{f_+(s,\lambda,z)(u-s)}. \qquad (16)$$

Let the function $F_+(x,\lambda,z)$ be defined by the correlation

$$\frac{1}{f_+(s,\lambda,z)} = 1 + \int_0^\infty e^{-sx} dF_+(x,\lambda,z)$$

and denote

$$G_+(x,u,\lambda,z) = \int_x^\infty e^{-uy} dF_+(y,\lambda,z).$$

Inverting formula (16) on s we receive such results.

Theorem 3. *For any $x > 0$ the following representations hold*

$$E\Big\{e^{-\lambda\tau - u\nu} z^{N(\tau)} \mid \xi_0 = n+1, \kappa_0 = 1, \beta_0 = x\Big\} =$$

$$= z^n f_+(u,\lambda,z) e^{-(\lambda-u)x} \int_0^\infty e^{-(\lambda-u)y} G_+(x+y,u,\lambda,z) dF_2^{(n-1)\star}(y),$$

$$E\Big\{e^{-\lambda\tau - u\nu} z^{N(\tau)} \mid \xi_0 = 1, \kappa_0 = 1, \beta_0 = x\Big\} = z e^{-(\lambda-u)x} \Big(\int_x^\infty e^{-uy} dF_1(y) +$$

$$+ f_+(u,\lambda,z) \int_0^x e^{-uy} G_+(x-y,u,\lambda,z) dF_1(y)\Big),$$

where $F_2^{n\star}$ stands for an n-fold convolution of the distribution $F_2(x)$ with itself.

Particularly as $x \downarrow 0$ we get

$$E\{e^{-\lambda\tau - u\nu} z^{N(\tau)} | \xi_0 = n\} =$$

$$= z^n f_+(u,\lambda,z) \int_0^\infty e^{-(\lambda-u)v} G_+(v,u,\lambda,z) dF_2^{(n-1)\star}(v) \qquad (17)$$

and

$$E\{e^{-\lambda\tau - u\nu} z^{N(\tau)} | \xi_0 = 1\} = z(1 - f_+(u,\lambda,z)).$$

The last formula was derived in [1].

Now suppose that the following conditions hold

A: $s_- = \inf\{s < 0 : f_1(s) + f_2(s) < \infty\} < 0$.

B: for some $n \geq 0$ the distribution $F^{n\star}(.)$ has an absolutely continuous component, where

$$F(x) = \int_{-x}^\infty F_1(x+y) dF_2(y).$$

Put $a_i = -f_i'(0), \quad b_i = f_i''(0) < \infty.$

Theorem 4. *Assume conditions A and B hold and $a_1 = a_2$ then for $\lambda, \mu, u \geq 0$*

$$\lim_{n\to\infty} E\{e^{-\frac{\lambda\tau+\mu N(\tau)}{n^2}-u\nu} \mid \xi_0 = n\} = \frac{f_+(u)}{uf_+'(0)} e^{-q(\lambda,\mu)}, \tag{18}$$

where

$$q(\lambda, \mu) = a_1 \sqrt{\frac{2(a_1\lambda + \mu)}{b_1 + 2a_1^2 + b_2}}.$$

and $f_+(s)$ is the component of factorisation

$$1 - f_1(s)f_2(-s) = f_+(s)f_-(s), \qquad \text{Re } s = 0. \tag{19}$$

Remark 2. In the conditions of the theorem, the component of factorisation (19) is such that $f_\pm(0) = 0, \; f_\pm'(0) \neq 0$ and $f_\pm(s, \lambda, z) \to f_\pm(s)$ as $\lambda \downarrow 0, \; z \uparrow 1$.

Proof. Assume λ and μ are fixed. Below we will often omit these letters in the notation. If condition A holds then the equation

$$e^{-\mu n^{-2}} f_1(s)f_2(\lambda n^{-2} - s) = 1$$

has in the strip $|\text{Re } s| \leq -\varepsilon$ for some $\varepsilon > 0$ only two solutions $s_\pm(\lambda, \mu, n) = s_\pm(n)$ such that $s_-(n) < 0 < s_+(n)$ for all sufficiently large n . The following properties of these roots follow from the theorem of implicit function

$$s_\pm(n) = \pm\sqrt{\frac{2(a_1\lambda + \mu)}{b_1 + 2a_1^2 + b_2}} n^{-1} + o(n^{-1})$$

as $n \to \infty$.

Now for the components of factorisation

$$1 - e^{-\mu n^{-2}} f_1(s)f_2(\lambda n^{-2} - s) = f_+(s, n)f_-(s, n)$$

we can write the following representations

$$f_\pm(s, n) = \frac{s - s_\mp(n)}{s - a_\mp} \hat{f}_\pm(s, n) \tag{20}$$

where $\pm a_\pm > \pm s_-$ are some fixed numbers. If condition B holds, the function $\hat{f}_+(s, n)$ has the following properties

$$(\hat{f}_+(s, n))^{-1} = 1 + \int_0^\infty e^{-sx} dF_+(x, n) \tag{21}$$

and for some $\varepsilon > 0$

$$Var_{[x,\infty)]} F_+(y,n) = O(e^{-\varepsilon x}). \tag{22}$$

uniformly on n. Since

$$\begin{aligned} f_+(s,n) &= 1 + \frac{1}{f'_+(s_-(n),n)(s - s_-(n))} \\ &+ (s_-(n) - a_-) \int_0^\infty e^{-sx} \int_x^\infty e^{s_-(n)(x-y)} dF_+(y,n)dx, \end{aligned}$$

where

$$f'_+(s_-(n),n) = \frac{d}{ds} f_+(s,n)|_{s=s_-(n)},$$

then

$$dF_+(x,n) = \frac{e^{s_-(n)x}}{f'_+(s_-(n),n)} dx + Q(x,n)dx$$

and for some $\varepsilon > 0$

$$Var_{[x,\infty)]} Q(y,n) = O(e^{-\varepsilon x}). \tag{23}$$

uniformly on n.

Hence

$$G_+(x,n) = \int_x^\infty e^{-uy} dF_+(y,n) = \frac{e^{-(u-s_-(n))x}}{f'_+(s_-(n),n)(u - s_-(n))} + \int_x^\infty Q(y,n)dy$$

and formula (17) can be rewritten in the following form

$$E\{e^{-\frac{\lambda\tau + \mu N(\tau)}{n^2} - u\nu} | \xi_0 = n,\} - e^{-\mu/n^2} \frac{f_+(u,n)}{f'_+(s_-(n),n)(u - s_-(n))} (f_2^{n-1}(\lambda n^{-2} - s_-(n)) +$$

$$+ \int_0^\infty e^{-(n^{-2}\lambda - u)v} \int_v^\infty e^{-uy} Q(y,n)dy dF_2^{(n-1)\star}(v)). \tag{24}$$

The first integral in this formula, which we denote as $I(n)$, can be easily estimated from (23) and

$$|I(n)| \leq C f_2^{n-1}(\varepsilon) \to 0, \quad n \to \infty.$$

It is obvious that

$$f_2^{n-1}(n^{-2}\lambda - s_-(n)) = \left(1 - \frac{q(\lambda,\mu)}{n} + o(n^{-1})\right)^{n-1} \to e^{-q(\lambda,\mu)}.$$

And now (18) follows from (24).

Assume $a_1 > a_2$ and $\lambda, \mu, u \geq 0$ are fixed again. Now the equation

$$f_1(s) f_2(\lambda n^{-1} - s) = e^{n^{-1}\mu}$$

has for some $\varepsilon > 0$ in the strip $-\varepsilon < \text{Re } s < 0$ a single solution $s_-(n)$ and

$$s_-(n) = \frac{\lambda a_2 + \mu}{a_2 - a_1} n^{-1} + o(n^{-1})$$

as $n \to \infty$, and for function $f_+(s, n)$ the formulas (20–22) take place again. By the same arguments as above we have

Theorem 5. *Let conditions A and B hold and* $a_1 > a_2$*, then for* $\lambda, \mu, u \geq 0$

$$\lim_{n \to \infty} E\{e^{-\frac{\lambda\tau + \mu N(\tau)}{n} - u\nu} | \xi_0 = n\} = \frac{f_+(u)}{u f'_+(0)} e^{-\frac{a_2(\lambda a_2 + \mu}{a_1 - a_2}}.$$

Particularly for the first idle time we have that under the conditions of Theorems 4 and 5

$$\lim_{n \to \infty} E\{e^{-u\nu} | \xi_0 = n\} = \frac{f_+(u)}{u f'_+(0)}.$$

And hence the first idle time is asymptotically independent on $\tau, N(\tau)$ as $\xi_0 \to \infty$.

Remark 3. Condition B was introduced to simplify the proof. All statements in Section 4 are true without this condition. It is essential to have the estimate for the convergence rate in Theorems 4 and 5.

References

1. **Prabhu, N.U.** (1980). *Stochastic Storage Processes.* Springer. New York.

2. **Bratiĭchuk, N.S. and Pirdjanov B.** (1992). A new approach in studying the busy period of system GI/G/1. *Proceedings of the Sixth USSR-Japan Symposium on Probability theory and mathematical statistics.* Kiev. August 5–10. World Scientific. Singapore.

3. **Borovkov, A.A.** (1972). *Stochastic Processes the Queueing Theory.* Nauka. Moscow. (In Russian).

Exploring Stochastic Laws. pp. 91–100.
Eds. A.V. Skorokhod and Yu.V. Borovskikh

Continuity of vector-valued Gaussian Markov processes

V.V. BULDYGIN*

Department of Higher Mathematics No.1, Kiev Polytechnical Institute, Kiev, Ukraine

Received May 19, 1995

Abstract. Let $Y(t)$ be a stochastically continuous Gaussian Markov process. We will prove in this paper that almost surely a sample path of this process may have discontinuities only at the times of the jumps of the rank of the variance matrix $D_Y(t)$ of this process.

1. Notations and assumptions

Let $T \subseteq \mathrm{R} = (-\infty, \infty)$ be a closed connected set, $\mathrm{R}^m, m \geq 1$ an m-dimensional Euclidean real space of column vectors, with Euclidean norm $\|\cdot\|$, and M_m — the class of $m \times m$-dimensional real matrices

$$A = [a_{jk}] = [a_{jk}]_{j,k=1}^m$$

with Euclidean norm

$$\|A\| := (\sum_{j,k=1}^m |a_{jk}|^2)^{\frac{1}{2}}.$$

Hereinafter the sign "$\ldots^{\mathrm{T}}$" denotes the operation of the conjugation.

Now, let $C(T, \mathrm{R}^m)$ be the space of continuous R^m-valued functions $f = (f(t), t \in T)$. If $\tau \in T$, then let $C_\tau(T, \mathrm{R}^m)$ be the space of R^m-valued functions on T, which are continuous at the point τ:

$$f \in C_\tau(T, \mathrm{R}^m) \Longleftrightarrow \lim_{\varepsilon \downarrow 0} \sup_{t,s \in B_\varepsilon(\tau)} \|f(t) - f(s)\|,$$

where $B_\varepsilon(\tau) = \{t \in T : |t - \tau| < \varepsilon\}, \ \ \varepsilon > 0$. For $\tau \in T$ we shall consider the set

$$\mathcal{N}_\tau^{+}(T) = \{(t_n, \ n \geq 1) \subset T : t_n \uparrow \tau \quad \text{or} \quad t_n \downarrow \tau\}.$$

Let $Y = (Y(t), \ t \in T)$ be a *centered stochastically continuous Gaussian Markov process* on a basic probability space $(\Omega, \mathcal{F}, P)$. Since we shall consider the conditions

* Supported by grant 1(1) 3/31 from the Ukrainian State Committee of Science and Technology

of continuity of process Y, then the assumptions about centrality and stochastic continuity are completely natural. We recall, that for Gaussian processes, stochastic continuity and continuity in mean square are equivalent. Further, let

$$K_Y(t,s) := EY(t)Y^{\mathrm{T}}(s), \quad t,s \in T$$

be the *matrix covariance function* of process Y;

$$D_Y(t) := K_Y(t,t), \quad t \in T,$$

the *matrix variance function* of process Y;

$$A_Y(t) := D_Y^{\frac{1}{2}}(t), \quad t \in T,$$

the square root of symmetric positive semi-definite matrix $D(t)$ (i.e. $D_Y(t) = A_Y(t)A_Y^{\mathrm{T}}(t)$). The mean square continuity of process Y has as a consequence the continuity of all elements of the matrices $K_Y(t,s)$, $D_Y(t)$, $A_Y(t)$ by variables $t,s \in T$. Therefore, these matrix functions are continuous with respect to the Euclidean matrix norm.

For Gaussian Markov process Y the following relation holds:

$$\mathrm{R}_Y(t,s) = \mathrm{R}_Y(t,u)\mathrm{R}_Y(u,s), \quad t \le u \le s, \tag{1}$$

where $\mathrm{R}_Y(t,s)$, $t,s \in T$ is a matrix function defining the conditional expectation of process Y:

$$E(Y(t)|Y(s)) = \mathrm{R}_Y(t,s)Y(s), \quad t,s \in T.$$

If $\det D_Y(s) \neq 0$, then

$$\mathrm{R}_Y(t,s) = K_Y(t,s)D_Y^{-1}(s).$$

For all $t \in T$ we consider the vector space

$$L_t := A_Y(t)\mathrm{R}^m. \tag{2}$$

The vector space L_t is a *linear support* of the distribution of the random vector $Y(t)$, i.e. L_t is the smallest vector space by dimension for which $P\{Y(t) \in L_t\} = 1$. In the general case $0 \le \dim L_t \le m$. Denote

$$r_Y(t) := \text{rank } D(t), \quad t \in T.$$

Since $D(t)$ is a symmetric positive semi-define matrix, then (Lancaster (1969))

$$r_Y(t) = \text{rank } A_Y(t) = \dim L_t, \quad t \in T. \tag{3}$$

The function $r_Y(t), t \in T$ may have values $0, 1, \ldots, m$. Let T_c be the set of the points of continuity of the function $r_Y(t), t \in T$:

$$T_c = \{t \in T : (\exists \varepsilon = \varepsilon(t) > 0)(\forall s \in B_\varepsilon(t) \Longrightarrow r_Y(s) = r_Y(t))\}. \tag{4}$$

Then the set

$$T_d = T \backslash T_c$$

is the set of the points of the jumps of the function $r_Y(t), t \in T$.

2. Main Theorems

In the article (Buldygin and Solntsev (1988)) it was proved that almost surely a sample path of process Y may have discontinuities only at the points of change of the spaces L_t, $t \in T$. This result may be made more precise. Later on it will be proved that almost surely discontinuities may be only at the points of change of the rank of the variance matrix $D_Y(t)$ or (which is equivalent) at the points of change of a dimension of the linear supports L_t, $t \in T$, of values of process Y.

Theorem 1. *For each $\tau \in T_c$*

$$P\{Y \in C_\tau(T, \mathrm{R}^m)\} = 1.$$

Theorem 2. *The following statements are equivalent:*

(a) $P\{Y \in C(T, \mathrm{R}^m)\} = 1$;

(b) *for each* $\tau \in T_d$

$$P\{Y \in C_\tau(T, \mathrm{R}^m)\} = 1;$$

(c) *for any* $\tau \in T_d$, *and any* $(t_n, n \geq 1) \in \mathcal{N}_\tau^t(T)$

$$P\{\lim_{n \to \infty} Y(t_n) = Y(\tau)\} = 1.$$

Before formulating Theorem 3 we recall a general variant of Ito–Nisio's Theorem on the oscillation of a Gaussian process (Ito and Nisio (1968); Fernique (1975); Buldygin and Solntsev (1988)).

Let (S, ρ) be a σ-compact metric space, $(\mathfrak{X}, \|\cdot\|)$ a separable Banach space and $X = (X(s),\ s \in S)$ a separable stochastically continuous $\mathfrak{X}$-valued Gaussian random process on $(\Omega, \mathcal{F}, P)$. Consider the oscillation process

$$H_X(s) := \lim_{h \downarrow 0} \sup_{u,v \in B_h(s)} \|X(u) - X(v)\|, \quad s \in S,$$

where $B_h(s) = \{u \in S : \rho(u, s) < h\}, \quad h > 0$.

Ito–Nisio Theorem. *For the given Gaussian process X there exists a non-random function $\alpha_X(s)$, $s \in S$ $(\alpha_X(s) \in \mathrm{R} \cup \{+\infty\},\ s \in S)$ such that*

$$P\{H_X(s) = \alpha_X(s),\ \ s \in S\} = 1.$$

Consider the following set

$$\mathcal{T}_d(X) = \{s \in S : \ \alpha_X(s) \neq 0\}.$$

Ito–Nisio Theorem shows, that almost surely a sample path of process X may have discontinuities only at the points from the set $\mathcal{T}_d(X)$. Since X is a stochastically continuous process, the behaviour of its sample path at a point of discontinuity has an oscillatory character. In the general case, a description of the set $\mathcal{T}_d$ is represented by a difficult problem. But, for Gaussian Markov process Y the structure of the set $\mathcal{T}_d(Y)$ can be made more precise.

Theorem 3. *The following relation*

$$\mathcal{T}_d(Y) \subseteq T_d$$

is valid.

Remark 1. It can not be asserted, that at each point of change of the rank $r_Y(t)$ a sample path of the process Y has a discontinuity almost surely. This generally, is not correct. But, if with positive probability (therefore almost surely) a discontinuity of a sample path at some point t_0 takes place, then it is a point of change of the rank $r_Y(t)$, i.e. $t_0 \in T_d$.

Corollary 1. *Let for some $d \in \{0, 1, \ldots, m\}$ the following condition*

$$\operatorname{rank} D_Y(t) = d, \quad t \in T$$

be fulfilled. Then

$$P\{Y \in C(T, \mathrm{R}^m)\} = 1.$$

Remark 2. It is possible that $L_t \neq L_s$ if $t \neq s$. In Corollary 1 it is required only that $\dim L_t \equiv$ constant.

Corollary 2. *Let in addition Gaussian Markov process Y be a stationary process. Then*

$$P\{Y \in C(T, \mathrm{R}^m)\} = 1.$$

Let $m = 1$, i.e. let Y be a real-valued process. Then

$$D_Y(t) = E\,Y^2(t), \quad t \in T.$$

Therefore a value of the function $r_Y(t)$ may be 0 or 1, and

$$r_Y(t) = 0 \Leftrightarrow D_Y(t) = 0.$$

Thus

$$T_c = \{t \in T : D_Y(t) > 0\},$$
$$T_d = \{t \in T : D_Y(t) = 0\}.$$

By Theorem 3 almost surely a sample path of process Y may have a discontinuity only at those moments of time, for which the variance of the process is equal to zero.

Corollary 3. *Let $m = 1$, and let the following condition*

$$\inf_{t \in T} D_Y(t) > 0$$

be fulfilled. Then

$$P\{Y \in C(T, \mathrm{R})\} = 1.$$

Corollary 3 was proved earlier in (Buldygin and Solntsev (1988)). In (Buldygin and Solntsev (1988)) and in (Solntsev (1994)) different tests were proved of the continuity of Gaussian Markov processes.

3. Some preliminary results

Let $X = (X(s),\ \ s \in S)$ be a Gaussian process, which was defined in Section 2. In accordance with the notations of Section 1, let $C(S, \mathfrak{X})$ be the space of continuous $\mathfrak{X}$-valued functions on S, and let $C_v(S, \mathfrak{X})$ be the space of $\mathcal{X}$-valued functions on S, which are continuous at a point $v \in S$. Then, let $\mathcal{N}_u(S)$ be a set of all sequences $(s_n,\ n \geq 1) \subset S$, such that $\lim_{n \to \infty} \rho(u, s_n) = 0$.

The next result is a corollary to Ito–Nisio Theorem (Section 2)

Lemma 1. *The following statements are equivalent:*

(*i*) $P\{X \in C(T, \mathfrak{X})\} = 1$;

(*ii*) *for each $v \in S$*

$$P\{X \in C_v(T, \mathfrak{X})\} = 1.$$

The next result was proved in (Buldygin and Solntsev (1988)).

Lemma 2. *Take $v \in S$. The following statements are equivalent:*

(*i*) $P\{X \in C_v(S, \mathfrak{X})\} = 1$;

(*ii*) *for any $(s_n, n \geq 1) \in \mathcal{N}_v(S)$*

$$P\{\lim_{n \to \infty} X(s_n) = X(v)\} = 1.$$

Remark 3. If $S = T$, then in Lemma 2 the set $\mathcal{N}_v(S)$ may be replaced by the set $\mathcal{N}_v^+(S)$.

Lemma 3. *Take $\tau \in T_c$. Then there exists $\delta > 0$ and there exists a family of unitary linear transforms U_t, $t \in B_\delta(\tau)$ of the space R^m, such that the following statements are valid:*

(*i*) *for each* $t \in B_\delta(\tau)$

$$U_t : L_t \longleftrightarrow L_\tau;$$

(*ii*)
$$U_t \underset{t \to \tau}{\longrightarrow} I,$$
where I *is an identity mapping in* R^m.

Proof. By the condition of the lemma and relation (3), the point τ is a point of continuity of the rank $r_Y(t)$ of the matrix $A_Y(t) = [a_{jk}]$, $t \in T$ (Section 1). Thus there exists numbers $\varepsilon > 0$, $d \in \{0, 1, \ldots, m\}$ such that

$$r_Y(t) = d, \quad t \in B_\varepsilon(\tau). \tag{5}$$

If $d = 0$, then $L_t = \{0\}$, $t \in B_\varepsilon(\tau)$ and so we can put $U_t = I$, $t \in B_\varepsilon(\tau)$. If $d = m$, then $L_t = \mathrm{R}^m$, $t \in B_\varepsilon(\tau)$ and so we can also put $U_t = I$, $t \in B_\varepsilon(\tau)$.

Let $1 \le d \le m - 1$, $m \ge 2$. Then there exists a submatrix $\tilde{A}(\tau) \in M_d$ of the matrix $A_Y(\tau)$, such that $\det \tilde{A}(t) \ne 0$. Moreover, any minor of the matrix $A(\tau)$, which has dimension more then d, equals zero. We may assume, without loss of generality, that matrix $\tilde{A}(t)$ is composed of elements belonging to the first d columns and the first d rows of matrix $A_Y(\tau)$:

$$\tilde{A}(\tau) = [a_{jk}(\tau)]_{j,k=1}^{d}.$$

Since (Section 1) the elements of matrix $A(t)$, $t \in T$ are continuous functions of t, and the determinant of a matrix is a continuous function of its elements, then there exists $\delta \in (0, \varepsilon]$ such that

$$\det \tilde{A}(t) \ne 0, \quad t \in B_\delta(\tau).$$

Moreover, by (4) any minor of matrix $A(t)$, $t \in B_\delta(\tau)$, which has dimension more than d, equals zero. It follows from this that for each $t \in B_\delta(\tau)$ the columns

$$a_k(t) = (a_{1k}(t), \ldots, a_{mk}(t))^{\mathrm{T}}, \quad k = 1, \ldots, d$$

of matrix $A_Y(t)$ form a basis of the space L_t. In addition

$$\lim_{t \to \tau} \|a_k(\tau) - a_k(t)\| = 0, \quad k = 1, \ldots, d.$$

Next, by means of standard methods we may complement the vectors $a_1(t), \ldots, a_d(t)$ to a basis in the space R^m and transform it into an orthonormalized basis. After these procedures we finally obtain that for each $t \in B_\delta(\tau)$ there exists orthonormalized basis $e_1(t), \ldots, e_m(t)$ in R^m such that the vectors $e_1(t), \ldots, e_d(t)$ are composed of orthonormalized basis in the space L_t. Since these procedures are continuous with respect to elements of matrix $A_Y(t)$, $t \in B_\delta(\tau)$, then for each $t \in B_\delta(\tau)$

$$\lim_{t\to\tau} \|e_k(\tau) - e_k(t)\| = 0, \quad k = 1, \ldots, d. \tag{6}$$

Now for each $t \in B_\delta(\tau)$ we define a unitary linear transform U_t of the space R^m by means of the following relations:

$$U_t(e_k(t)) = e_k(\tau), \quad k = 1, \ldots, m.$$

It is obvious, that for these transforms statement (i) is fulfilled. Statement (ii) is a corollary to relation (5). Lemma 3 is proved.

4. Proofs of main results

Proof of Theorem 1. Take $\tau \in T_c$. According to Lemma 2 (see also Remark 3) to prove the Theorem it is sufficient to show that for each $(t_n, n \geq 1) \in \mathcal{N}_\tau^+$ the relation

$$P\{\lim_{n\to\infty} Y(t_n) = Y(\tau)\} = 1. \tag{7}$$

holds. Since τ is a point of continuity of the rank $r_Y(t),\ t \in T$ of matrix $A_Y(t),\ t \in T$ then there exists a positive integer ε and a natural number $d \in \{0, 1, \ldots, m\}$ such that $r(t) = d,\ t \in B_\varepsilon(\tau)$. If $d = 0$, then almost surely $Y(t) = 0$. In this case relation (5) holds.

Let $1 \leq d \leq m-1,\ m \geq 2$. Then by Lemma 3 there exists a positive integer δ and family $\{U_t, t \in B_\delta(\tau)\}$ of unitary linear transforms of the space R^m for which relations (i), (ii) of Lemma 3 hold. For the sake of simplicity, a matrix defining a transform U_t shall also be denoted as U_t. Then

$$Y(t) = U_t^{\mathrm{T}} Z(t),$$

$$Z(t) = U_t Y(t), \quad t \in B_\delta(\tau).$$

Since the space L_τ is the linear support of the random vector $Y(t)$, then according to (ii) of Lemma 3 for any $t \in B_\delta(\tau)$ this space is a linear support of the random vector $Z(t)$. In addition, in view of (i) of Lemma 3 the process $(Z(t),\ t \in B_\delta(\tau))$ is a centered Gaussian Markov process in the space $L = L_\tau$. After a change of coordinates the process $(Z(t),\ t \in B_\delta(\tau))$ can be regarded as a process in the space R^d, and without loss of generality we may assume that $d = m$, i.e. $L = \mathrm{R}^m$.

Next, for any $t, s \in B_\delta(\tau)$

$$K_Z(t,s) = E\, Z(t) Z^{\mathrm{T}}(s) = U_t K_Y(t,s) U_s^{\mathrm{T}},$$

$$D_Z(t) = K_Z(t,t) = U_t D_Y U_t^{\mathrm{T}}.$$

Since for each $t \in B_\delta(\tau)$ the space L is a linear support of $Z(t)$, then there exists a reciprocal matrix $D_Z^{-1}(t)$.

Take $(t_n,\ n \geq 1) \in \mathcal{N}_\tau^+(T)$. We may assume without loss of generality that $t_n \in B_h(\tau)$, $n \geq 1$, where $h \in (0, \delta)$. The value of h will be made more precise later on. The sequences $(Z_n = Z(t_n),\ n \geq 1)$ are centered Gaussian Markovian sequences in L. These sequences may be represented in the form of recurrence relations

$$Z_{n+1} = C_{n+1} Z_n + W_{n+1}, \quad n \geq 1,$$

$$Z_1 = W_1,$$

where

$$\begin{aligned} C_n &:= \mathrm{R}_Z(t_n, t_{n-1}) \\ &= K_Z(t_n, t_{n-1}) D_Z^{-1}(t_{n-1}), \quad n \geq 2, \end{aligned}$$

and $(W_n,\ n \geq 1)$ is a sequence of independent centered Gaussian random vectors in L.

Thus for any $n \geq 2$ the relation

$$Z_n = \sum_{k=1}^{n-1} (C_n \cdot \ldots \cdot C_{k+1}) W_k + W_n \tag{8}$$

holds.

For any $t, s \in B_\delta(\tau)$ we may present the covariance matrix $K_Z(t,s)$ in the form

$$K_Z(t,s) = D_Z(\tau)[I + \Delta(t,s)],$$

where

$$\Delta(t,s) = D_Z^{-1}(\tau)[K_Z(t,s) - K_Z(\tau,\tau)].$$

In view of the continuity of matrix function $K_Y(t,s)$, $t, s \in T$ (Section 1) and Lemma 3 and the theorem on the norm of a reciprocal matrix (Lancaster (1969)) for any $t, s \in (0, \delta)$ there exists the reciprocal matrix $K_Z^{-1}(t,s)$. In addition we may put numbers $h > 0$ such that the relation

$$\sup_{s,t \in B_h(\tau)} \|K_Z^{-1}(t,s)\| \, \|A_Y(t,s)\|^3 < \infty \tag{9}$$

holds. Then for each $n \geq 2$ there exists reciprocal matrix C_n^{-1} and we may represent relation (6) in the form

$$Z_n = (C_n \cdot \ldots \cdot C_2) \sum_{k=1}^{n} F_k W_k, \quad n \geq 2, \tag{10}$$

where F_k, $k \geq 1$ are some matrices. By relation (1) for any $n \geq 2$

$$\begin{aligned} C_n \cdot \ldots \cdot C_2 &= \mathrm{R}_Z(t_n, t_1) \\ &= K_Z(t_n, t_1) D^{-1}(t_1). \end{aligned}$$

Let

$$V_n = \mathrm{R}_Z^{-1}(t_n, t_1) Z_n, \quad n \geq 2,$$

then in accordance with relation (8)

$$V_n = \sum_{k=1}^{n} F_k W_k, \quad n \geq 2.$$

Since

$$E\|V_n\|^2 \leq m \left[\|K_Z^{-1}(t_n, t_1)\| \, \|A_Y(t_1)\|^3 \right]^2,$$

then by (7)

$$\sum_{k=1}^{\infty} E\|F_k W_k\|^2 = \sup_{n \geq 1} E\|V_n\|^2 < \infty.$$

Therefore the series $\sum_{k=1}^{\infty} F_k W_k$ of independent centered Gaussian random vectors almost surely converges. Since

$$\lim_{n \to \infty} \|\mathrm{R}_Z(t_n, t_1) - \mathrm{R}_Z(\tau, t_1)\| = 0,$$

then the sequence $(Z_n, n \geq 1)$ also almost surely converges. Since for each $n \geq 1$

$$Y(t_n) = U_{t_n}^{\mathrm{T}} Z_n,$$

then, in view of statement (*ii*) of Lemma 3, the sequence $(Y(t_n),\ n \geq 1)$ almost surely converges. Now, by the stochastic continuity of process Y, relation (5) is obvious. Theorem 1 is proved.

Proof of Theorem 2. The implication $(a) \Longrightarrow (c)$ is obvious. The implication $(c) \Longrightarrow (b)$ follows from Lemma 2. The implication $(b) \Longrightarrow (a)$ follows from Theorem 1 and Lemma 1. Theorem 2 is proved.

Proof of Theorem 3. Theorem 3 is a corollary to Theorem 1 and Ito–Nisio Theorem (Section 2).

Proof of Corollary 1. Since $T_d = \phi$ then Corollary 1 directly follows from Theorem 2.

Proof of Corollary 2. Corollary 2 directly follows from Corollary 1.

Proof of Corollary 3. Since $r_Y(t) = 1, \quad t \in T$, then Corollary follows from Corollary 1.

References

Buldygin, V.V. and Solntsev, S.A. (1988). Equivalence of sample and sequential continuity of Gaussian processes and continuity of Gaussian Markov processes. *Teor. Veroyatnost. i Primenen.* **33**, 672–686. (In Russian). (English translation: Theory Probab. Appl. **33**).

Fernique, X. (1974). Regularite des trajectoires des fonctions aleatoires gaussiennes. In: *Lecture Notes in Math.* **480**, Springer-Verlag, pp. 1–96.

Ito, K. and Nisio, M. (1968). On the oscillation functions of Gaussian processes. *Math. Scaud.* 22, 209–223.

Lancaster, P. (1969). *Theory of Matrices.* Academic Press, New York.

Solntsev, S.A. (1993). On sample continuity of multivariate Gaussian Markov processes. *Teor. Veroyatnost. i Primenen.* **38**, 827–841. (In Russian). (English translation: Theory Probabil. **38**).

Exploring Stochastic Laws. pp. 101
Eds. A.V. Skorokhod and Yu.V. Borovskikh

Both an n-dimensional Wiener process and its continuous one-dimensional image generate the same filtration

E.V. CHERKASHIN

National Academy of Sciences of Ukraine, International Mathematical Center

Received April 1, 1995

Abstract. An example of a continuous function which transforms an n-dimensional Wiener process onto a real line in such a way that the filtration does not change is given.

$$f(x_1,\ x_2,\ \dots\ ,x_n) = \sum_{k=1}^{n}(|x_k| + x_k I_{\{x_k \geq 0\}})^{2-2^{2(1-k)}}.$$

$$\langle f(\vec{w})\rangle_t' = 1 + 3I_{\{w_1\geq 0\}} + \textstyle\sum_{k=2}^{n}(2-2^{2(1-k)})(|w_k| + 3w_k I_{\{w_k\geq 0\}})^{2-2^{3-2k}};$$

$$\langle f(\vec{w})\rangle_t^{(2)} = \langle\langle f(\vec{w})\rangle'\rangle_{(s,t)}'$$

$$= \textstyle\sum_{k=2}^{n} \alpha_k(|w_k| + \beta w_k I_{\{w_k\geq 0\}})^{2-2^{2(1-(k-1))}}, \quad s\uparrow t;$$

$$\dots$$

$$\langle f(\vec{w})\rangle_t^{(2(n-1))} = \gamma(|w_n| + \delta w_n I_{\{w_n\geq 0\}}), \quad s\uparrow t;$$

$$\langle f(\vec{w})\rangle_t^{(2n-1)} = \gamma^2 I_{\{w_n<0\}} + (\gamma+\delta)^2 I_{\{w_n\geq 0\}}.$$

$$w_n(t) = \langle f(\vec{w})\rangle_t^{(2(n-1))}/(\lim\nolimits_{s\uparrow t}\langle f(\vec{w})\rangle_s^{(2n-1)})^{1/2},$$

$$\dots$$

$$w_1(t) = (f(\vec{w}_t) - \textstyle\sum_{k=2}^{n})/(\lim_{s\uparrow t}(\langle f(\vec{w})\rangle_s' - \sum_{k=2}^{n}))^{1/2}.$$

Exploring Stochastic Laws. pp. 103–109.
Eds. A.V. Skorokhod and Yu.V. Borovskikh

Formal stochastic calculus and admissible transformations

ANDREI A. DOROGOVTSEV

Institute of Mathematics, National Academy of Sciences of Ukraine, Tereschenkivska St. 3, 252601 Kiev-4, Ukraine

Received June 20, 1995

Abstract. In this article a process having a logarithmic derivative with respect to the formal differentiation rule is studied. It is proved that for such a process there exists nontrivial admissible transformations for its distribution. The relation with formal stochastic calculus is discussed and the correspondent examples are considered.

The aim of this paper is to prove the formalization of the following statement. If a process has a logarithmic derivative with respect to the formal differentiation rule, then there exists nontrivial admissible transformations for its distribution. This paper consists of two parts. In the first part the main result is formulated and proved, and in the second part its relation with formal stochastic calculus is discussed and the correspondent examples are considered.

1. Existence of the admissible flow. Let H be a separable real Hilbert space with inner product $(\cdot,\cdot)$ and norm $\|\cdot\|$. Let μ be a Gaussian measure on H with zero mean and correlation operator S. Suppose that $\overline{S(H)} = H$. Denote by H_0 the reproducing kernel of μ with the corresponding inner product $(\cdot,\cdot)_0$ and norm $\|\cdot\|_0$. Let H_+ be the dual of H with respect to the product $(\cdot,\cdot)_0$. Then $H_+ \subset H_0 \subset H$ with Hilbert–Schmidt embeddings. Consider a new space $E = H \oplus H_1$, where H_1 is another separable real Hilbert space. The inner product and the norm in E is denoted by $\langle\cdot,\cdot\rangle$ and $|\cdot|$ respectively. Let us construct the measure ν on E in the following way. For any Borel $A \subset E$

$$\nu(A) = \int_{\Phi^{-1}(A)} p(x)\mu(dx).$$

Here $p : H \to R$ is a nonnegative function with a bounded continuous derivative;

$$\forall x \in H: \quad \Phi(x) = x \oplus \varphi(x),$$

where function $\varphi : H \to H_1$ has two bounded continuous derivatives.

Theorem 1. *Let functions $\alpha : E \to E$ and $\beta : E \to R$ satisfy the following conditions:*

1) *the function $P \cdot Q(x) \cdot \alpha(x, \varphi(x))$, $x \in H$ maps H into H_+ and has a bounded continuous derivative. Here for every $x \in H$, $Q(x)$ is a projector in E on the subspace $L(x) = \{z = u \oplus \varphi'(x)u, \quad u \in H\}$ and P is a projector in E on the first coordinate of H;*

2) *β is continuous and bounded;*

3) *α is bounded and Lipschitz on E;*

4) *for every bounded function $f : E \to R$, which has a continuous bounded derivative,*

$$\int_E \langle f', \alpha \rangle d\nu = \int_E f \cdot \beta d\nu.$$

Then, the mapping $U_t : E \to E$, which satisfies the equation

$$\frac{dU_t(v)}{dt} = \alpha(U_t(v)), \quad U_0(v) = v, \quad v \in E,$$

are admissible for the measure ν, i.e.

$$\forall t \geq 0 : \quad \nu U_t^{-1} \ll \nu.$$

Proof. First, prove the following Lemma.

Lemma. *Let functions α, β and the probability measure ν on the Borel σ-algebra on E satisfy conditions 2)– 4).*

Then $\forall x \in \text{supp}\nu : \alpha(x) \in K_\nu$, where K_ν is a linear support of ν, i.e. the smallest closed linear subspace of E, which contains supp ν.

Proof of Lemma. Suppose, for a certain $x_0 \in \text{supp}\nu$, $\alpha(x_0) \notin K_\nu$. Since α is continuous on E, then

$$\exists \delta > 0 : \forall x \in B(x_0, \delta) : \alpha(x) \notin K_\nu.$$

Here

$$B(x_0, \delta) = \{z : |z - x_0| < \delta\}.$$

Denote by R_1 and R_2 the orthogonal projectors on K_ν and $K_\nu^\perp$ respectively. Consider function f, which is of the following kind

$$\forall y \in E : f(y) = \varphi_\delta(R_1 y) \cdot \text{arctg}\langle y, R_2\alpha(y)\rangle.$$

Here φ_δ is a function on K_ν with a continuous bounded derivative such that

$$\forall z \in K_\nu \backslash B(x_0, \delta) : \varphi_\delta(z) = 0,$$

$$\forall z \in K_\nu \cap B(x_0, \delta) : \varphi_\delta(z) > 0.$$

Function f is not differentiable in general due to function α. Consider the sequence $\{\alpha_n; n \geq 1\}$ of continuously differentiable E-valued functions on E with bounded derivatives such that

$$\alpha_n \to \alpha, \quad n \to \infty \quad (\operatorname{mod} \nu).$$

Such a sequence can be constructed due to the Hilbertian structure of E. For the function

$$f_n(y) = \varphi_\delta(\mathrm{R}_1 y) \cdot \operatorname{arctg}\langle y,\ P\alpha_n(y)\rangle$$

under fixed $n \geq 1$, due to 4) and usual approximation arguments,

$$\begin{aligned} 0 &= \int_E \langle f_n', \alpha\rangle d\nu = \\ &= \int_E \varphi_\delta(\mathrm{R}_1 y) \frac{1}{1 + \langle y, \mathrm{R}_2\alpha_n(y)\rangle^2} \cdot \langle \mathrm{R}_2\alpha_n(y), \alpha(y)\rangle \nu(dy). \end{aligned}$$

So, under $n \to \infty$

$$0 = \int_E \varphi_\delta(\mathrm{R}_1 y) \cdot \frac{1}{1 + \langle y, \mathrm{R}_2\alpha(y)\rangle^2} \cdot \langle \mathrm{R}_2\alpha(y), \alpha(y)\rangle \nu(dy).$$

With the choice of x_0 and δ this is a contradicion. So the statement of the Lemma is true.

After the change of variables we can conclude in the general situation, that

$$\forall x \oplus \varphi(x) \in \operatorname{supp} \nu : \alpha(x, \varphi(x)) \in L(x).$$

Hence, the new function

$$\tilde{\alpha}(x, \varphi(x)) = Q(x)\alpha(x, \varphi(x)), \quad x \in H$$

satisfies 4). Let us prove, that $\tilde{\alpha}$ satisfies the Lipschitz condition on the set

$$\{x \oplus \varphi(x); x \in H\}.$$

To prove this, we must find the form of $Q(x), x \in H$. For every $u \oplus \theta \in E$ denote by u_0 the element of H, such that

$$Q(x)(u \oplus \theta) = u_0 \oplus \varphi'(x)u_0.$$

Then

$$u - u_0 + \varphi'(x)^*\theta - \varphi'(x)^*\varphi'(x)u_0 = 0.$$

Hence

$$u_0 = (I + \varphi'(x)^*\varphi'(x))^{-1}(u + \varphi'(x)^*\theta),$$

where I is an identity. Operator $\varphi'(x)^*\varphi'(x)$ is nonnegative. So $(I+\varphi'(x)^*\varphi'(x))^{-1}$ exists and

$$(I+\varphi'(x)^*\varphi'(x))^{-1} = \int_0^{+\infty} e^{-t}\exp\{-\varphi'(x)^*\varphi'(x)t\}dt.$$

Now, the Lipschitz property of Q follows from this expansion and from the existence of the bounded derivative φ'. Consequently $\tilde{\alpha}$ is Lipschitz on $\{x\oplus\varphi(x); x\in H\}$. Then, the equation

$$\frac{d\tilde{U}_t}{dt} = \tilde{\alpha}(\tilde{U}_t), t\geq 0$$

$$\tilde{U}_0(x\oplus\varphi(x)) = x\oplus\varphi(x),$$

has the unique solution $\tilde{U}_t, t\geq 0$ in $\{x\oplus\varphi(x); x\in H\}$.

Now, let us consider the following change of variables

$$\forall u\oplus\theta\in E: \quad G(u\oplus\theta) = u\oplus[\theta-\varphi(u)].$$

Note, that there exists G' and G^{-1}. Also

$$G(\{x\oplus\varphi(x); x\in H\}) = H.$$

Consider the flow on H:

$$\tilde{\tilde{U}}_t = G\tilde{U}_tG^{-1}, \quad t\geq 0.$$

The derivative

$$\frac{d\tilde{\tilde{U}}_t}{dt} = \tilde{\alpha}_1(\tilde{\tilde{U}}_t, \varphi(\tilde{\tilde{U}}_t)).$$

It follows from the conditions of the Theorem and the property of differentiability with respect to the initial data, that under fixed $t\geq 0$, $\tilde{\tilde{U}}_t$ has a continuous derivative as a map from H to H. Moreover, $\tilde{\tilde{U}}_t$ satisfies all the conditions of the Theorem about admissible transformations of Gaussian measures (Ramer, 1974). Consequently, $\mu\tilde{\tilde{U}}_t^{-1} \ll \mu$. So, for every bounded function f with bounded continuous derivative,

$$\int_H f(\tilde{\tilde{U}}_t(u))\mu(du) = \int_H f(v)\cdot\frac{d\mu\tilde{\tilde{U}}_t^{-1}}{d\mu}(v)\mu(dv),$$

and

$$\frac{d\mu\tilde{\tilde{U}}_t^{-1}}{d\mu}(v) = |\det D\tilde{\tilde{U}}_t(v)|\cdot\exp(-((\tilde{\tilde{U}}_t-I)\nu, v)_0 - \frac{1}{2}\|(\tilde{\tilde{U}}_t-I)v\|_0^2).$$

Here $(\cdot,\cdot)_0$ and $\|\cdot\|_0$ are the inner product and the norm in H_0 respectively. D is the symbol of the derivative with respect to H_0. Note, that due to the condition on $\alpha, D\tilde{\tilde{U}}_t(v)-I$ is a nuclear operator in H_0 for every $v\in H$. So the formula for the density is correct. Consequently, for the function f the following integration by parts formula is valid

$$\int_H (f'(v), \tilde{\tilde{\alpha}}(v))\mu(dv) = \int_H f(v) \cdot (tr\, D\tilde{\tilde{\alpha}}(v) - (\tilde{\tilde{\alpha}}(v), v)_0)\mu(dv).$$

Denote

$$tr\, D\tilde{\tilde{\alpha}}(v) - (\tilde{\tilde{\alpha}}(v), v)_0 = \gamma(v), \quad v \in H.$$

Note, that due to the conditions of the Theorem and the preceding considerations, γ is continuous and locally bounded on H. From another side, by change of variables,

$$\int_H (f'(v), \tilde{\tilde{\alpha}}(v)) \cdot p(v)\mu(dv) = \int_H f(v)\beta(G^{-1}(v))p(v)\mu(dv).$$

Let

$$\beta(G^{-1}(v)) = \tilde{\tilde{\beta}}(v), \quad v \in H.$$

$\tilde{\tilde{\beta}}$ is continuous and locally bounded on H. Now consider $u_0 \in \{u : p(u) > 0\}$. Let O_0 be the coherent component of the set $\{u : p(u) > 0\}$ which contains u_0. Let us prove that for every $t > 0$ the solution $\tilde{\tilde{U}}_t(u_0) \in O_0$. Suppose, that this is not true. Then consider $t_0 > 0$ — the first moment when $\tilde{\tilde{U}}_t(u_0) \in \partial O_0$. Note, that

$$p(\tilde{\tilde{U}}_{t_0}(u_0)) = 0.$$

From previous considerations we have

$$\tilde{\tilde{\beta}}(U_t(u_0)) = \gamma(\tilde{\tilde{U}}_t(u_0)) - \frac{(p'(\tilde{\tilde{U}}_t(u_0)), \tilde{\tilde{\alpha}}(\tilde{\tilde{U}}_t(u_0)))}{p(\tilde{\tilde{U}}_t(u_0))}.$$

Consider a new function,

$$q(t) = p(\tilde{\tilde{U}}_t(u_0)), \quad 0 \le t \le t_0.$$

Then, due to the properties of $\tilde{\tilde{\beta}}$ and γ,

$$\exists C > 0: \quad \forall 0 \le t \le t_0: \quad |q'(t)| < C\, q(t), \quad q(t_0) = 0.$$

From the Grownuoll Lemma

$$q(t) = 0, \quad t \in [0; t_0].$$

This contradiction proves our statement. Consequently, $\tilde{\tilde{U}}_t$ is admissible for $p \cdot \mu$. It follows from the structure of $\tilde{\alpha}$, $\tilde{\tilde{\alpha}}$ and α, that U_t is admissible for ν. The Theorem is now proved.

2. Examples of formal stochastic calculus. In the articles (Dorogovtsev A. A. 1994, 1995) the following formal construction of non-Gaussian stochastic calculus was proposed. For smooth finite-dimensional functionals for the initial process $\xi(t)$; $t \in [0;1]$ the stochastic derivative D was defined with the help of the given family of derivatives $\zeta(t,s)$; $t,s \in [0;1]$ in the following way

$$D\varphi(\xi(t_1),\ldots,\xi(t_n))(s) = \sum_{i=1}^{n} \varphi_i'(\xi(t_1),\ldots,\xi(t_n)) \cdot \zeta(t_i,s).$$

For this definition to be productive some assumptions must be made. An important assumption is the existence of the formal stochastic integral i.e. the random functional I on $L_2([0;1])$, such that

$$E\int_0^1 D\,\alpha(s)\cdot\varphi(s)ds = E\,\alpha \cdot I(\varphi).$$

Here α is an arbitrary smooth finite-dimensional functional from ξ, and φ is an arbitrary function from $L_2([0;1])$. Existence of the stochastic integral means that the distribution of ξ in a certain space of trajectories satisfies the integration by parts formula.

Example 1. Let ξ be a Wiener process on $[0;1]$, $\zeta(t,s) = \chi_{[0;1]}(s)$, $t,s \in [0;1]$. Then all random variables, which are measurable with respect to ξ, can be considered as Borel functions on $C_0([0;1])$ with Wiener measure. In our case the derivative D is a usual derivative with respect to the space $H_0 = \{\varphi \in C([0;1]) : \varphi(0) = 0,\ \exists\varphi', \int_0^1 \varphi'(s)^2 ds < +\infty\}$. The existence of the stochastic integral is now guaranteed by the integration by parts formula for Gaussian measures. Note also, that the directions from H_0 are the admissible shifts of the Wiener measure.

Example 2. Let $\xi = w$ be a Wiener process as in Example 1 and $\zeta(t,s) = w(1)\cdot\chi_{[0;1]}(s)$, $t,s \in [0;1]$. From the properties of Wiener measures we can conclude now, that the stochastic integral exists and

$$\forall\varphi \in L_2([0;1]) : I(\varphi) = w(1)\cdot\int_0^1 \varphi(s)dw(s) - \int_0^1 \varphi(s)ds.$$

To apply the Theorem from Part 1 let us consider the fixed continuously differentiable function φ_0 and

$$\begin{aligned} \alpha(u) &= u(1)\cdot\varphi_0 \\ \beta(u) &= u(1)\cdot\left(u(1)\varphi_0(1) - \int_0^1 u(s)d\varphi(s)\right) - \int_0^1 \varphi(s)ds, \\ &\qquad u \in C_0([0;1]). \end{aligned}$$

Now $C_0([0;1])$ is not a Hilbert space, but due to the Wiener distribution of w the statement of the Theorem remains true. So the flow of transformations, which satisfy the equation

$$\begin{aligned} \frac{dU_t}{dt} &= U_t(1)\cdot\varphi_0, \\ U_0 &= u \end{aligned}$$

is admissible for the Wiener measure. Consequently, in this case, the formal derivative is a derivative with respect to linear transformations.

References

1. **Ramer, R.** (1974). On nonlinear transformations of Gaussian measures. *Journal of Funct. Anal.* **15**, 166–187.

2. **Dorogovtsev, A.A.** (1994). The family of Ito formulas for logarithmic processes. *Random evolutions.* Kiev, In-t of Math.

3. **Dorogovtsev, A.A.** (1995). One variant of the non-Gaussian stochastic calculus. *Journal of Appl. Math. and Stoch. Anal.* **8**.

Exploring Stochastic Laws. pp. 111–119.
Eds. A.V. Skorokhod and Yu.V. Borovskikh

Stability of periodic solutions of operator equations with perturbation coefficients

A.YA. DOROGOVTSEV*

Kiev State University, Vladimirskaya St. 64, 252001 Kiev, Ukraine

Received January 20, 1995

Abstract. We will prove the result of the continuous dependence on coefficients for the periodic in distribution solutions of difference and differential operator equations in a Banach space.

1. Introduction

This article is devoted to the problem of the stability of periodic in a distribution solutions of operator equations with respect to the perturbation of operator coefficients. In the deterministic case such stability was studied by many authors, cf (Massera and Schaffer, 1966), (Cheban, 1980), (Henry, 1981), (Maslov, 1988), (Hinrichesen and Pritchard, 1992), (Wirth and Hinrichesen, 1992). For equations, perturbed by random processes or for stochastic equations, this problem is investigated less. One recent paper is the article (Da Prado and Tudor, 1993). The statements of our article are based on the properties of periodic processes and the existence results for periodic solutions of equations in a Banach space from the book (Dorogovtsev, 1992).

Let $(B, \|\cdot\|)$ — be a complex separable Banach space, let $L(B)$ — be a space of all bounded linear operators in B with operator norm, which is also denoted by $\|\cdot\|$, and let $\sigma(A)$ — be the spectrum of $A \in L(B)$. Continuity or differentiability of B- or $L(B)$-valued functions will be treated as a continuity or a differentiability in a corresponding norm. All random elements, which will be considered, will be defined on the complete probability space (Ω, F, P). B-valued random processes are assumed to be continuous. All equalities and their correlations between the random elements are assumed to be fulfilled with probability 1. The random process satisfies the differential equation if almost all trajectories of the process satisfies this equation.

* Supported by the International Soros Science Education Program of International Renaissance Foundation, Grant ISSEP SPU 041023

The uniqueness of the solution is treated in the mean of the stochastic equivalence. Note, that the trajectories of the periodic processes in general are not periodic or bounded on R. So, the statements about periodicity in distribution of solutions do not follow directly from deterministic results.

2. Stability of periodic solutions of linear difference equations

Consider the equation

$$x(n+1) = A(n)x(n) + y(n), \quad n \in Z, \tag{1}$$

for the B-valued random process $\{x(n); n \in Z\}$, where

$$\{A(n) : n \in Z\} \subset L(B); \quad \forall\, n \in Z : \; A(n+\tau) = A(n) \tag{2}$$

with some fixed number $\tau \in N$ and $\{y(n) : n \in Z\}$ is a periodic with period τ B-valued process such that

$$E\|y(k)\| < +\infty, \quad 1 \le k \le \tau.$$

For details about periodic processes see (Dorogovtsev, 1992).

$$D := A(\tau - 1)A(\tau - 2) \cdot \ldots \cdot A(1)A(0).$$

The existence of a periodic solution of equation (1) has been given in (Dorogovtsev, 1992).

Theorem. *Equation* (1) *has a unique solution with period* τ $\{x(n) : n \in Z\}$ *with* $E\|x(k)\| < +\infty$, $k = 1, 2, \ldots, \tau$, *for every periodic process with period* τ $\{y(n) : n \in Z\}$ *with* $E\|y(k)\| < +\infty$, $k = 1, 2, \ldots, \tau$, *if and only if*

$$\sigma(D) \cap S = \emptyset \tag{3}$$

where $S := \{z \in C| \; |z| = 1\}$.

Perturbed equation (1) can be written as follows

$$x_\varepsilon(n+1) = (A(n) + B_\varepsilon(n))x_\varepsilon(n) + y(n), \quad n \in Z, \tag{4}$$

a perturbation $\{B_\varepsilon(n) : n \in Z\} \subset L(B)$, $\varepsilon > 0$ is not assumed to be periodic in n. We assume that

$$\Delta(\varepsilon) := \sup_{n \in Z} \|B_\varepsilon(n)\| < +\infty, \quad \varepsilon > 0.$$

Lemma 1. *Suppose that the sequence of operators* $\{A(n): n \in Z\}$ *satisfies conditions* (2) *and* (3). *Then, the sequence* $\{A(n): n \in Z\}$ *has discrete dichotomies.*

Proof. For discrete dichotomies see (Henry, 1981). In order to establish Lemma 1, we need to prove that equation (1) has a unique bound solution for every bound sequence $\{y(n): n \in Z\}$. The existence of the bound solution of equation (1) is shown similarly to Theorem 3 from (Dorogovtsev, 1992) and therefore the proof is omitted.

Lemma 2. *Suppose that the sequence of operators* $\{A(n): n \in Z\}$ *satisfies conditions* (2) *and* (3). *Then there exists a number* $\delta > 0$ *such that for every* $\varepsilon > 0$ *with* $\Delta(\varepsilon) < \delta$ *the sequence* $\{A(n) + B_\varepsilon(n): n \in Z\}$ *has a discrete dichotomy.*

Proof of Lemma 2 follows from result 7.6.7 (Henry, 1981).

Lemma 3. *Suppose that the sequence of operators* $\{A(n): n \in Z\}$ *satisfies conditions* (2) *and* (3) *and that the number* $\varepsilon > 0$ *such that* $\Delta(\varepsilon) < \delta$. *Then equation* (4) *has a unique* B*-valued solution* $\{x_\varepsilon(n): n \in Z\}$ *with*

$$\sup_{n\in Z} E\|x_\varepsilon(n)\| < +\infty$$

for every B*-valued process* $\{y(n): n \in Z\}$ *with*

$$\sup_{n\in Z} E\|y(n)\| < +\infty.$$

Proof. Lemma 3 is taken from (Henry, 1981) and (Dorogovtsev, 1992).

Theorem 1. *Suppose that the sequence of operators* $\{A(n): n \in Z\}$ *satisfies conditions* (2) *and* (3) *and that* $\Delta(\varepsilon) \to 0$, $\varepsilon \to 0$. *Let* $\{y(n): n \in Z\}$ *be a periodic with period* τ B*-valued process with* $E\|y(k)\| < +\infty$, $k = 1, 2, ..., \tau$. *Then for every sufficiently small* $\varepsilon > 0$ *equation* (4) *has a unique* B*-valued solution* $\{x_\varepsilon(n): n \in Z\}$ *such that*

$$\sup_{n\in Z} E\|x_\varepsilon(n)\| < +\infty.$$

Moreover,

$$\sup_{n\in Z} E\|x_\varepsilon(n) - x(n)\| \to 0, \quad \varepsilon \to 0,$$

where $\{x(n): n \in Z\}$ *is a unique periodic solution of equation* (1) *with period* τ *with* $E\|x(k)\| < +\infty$, $k = 1, 2, ..., \tau$.

Proof. Let $\{G(n, m)\}$ denote the Green's function for equation (1). By conditions

(2) and (3), we have

$$g := \sup_{n\in Z}\sum_{k\in Z}\|G(n,k)\| < +\infty$$

and

$$x(n) = \sum_{k\in Z}G(n,k)y(k-1), \quad n \in Z.$$

By Lemmas 1–3, (4) has a solution $\{x_\varepsilon(n) : n \in Z\}$ and

$$x_\varepsilon(n) = \sum_{k\in Z}G(n,k)B_\varepsilon(k-1)x_\varepsilon(k-1) + x(n), \quad n \in Z.$$

The series with random elements is convergent with probability 1 in the norm of B for every $n \in Z$. Therefore,

$$u(\varepsilon) \le g\Delta(\varepsilon)u(\varepsilon) + V,$$

where

$$u(\varepsilon) := \sup_{n\in Z}E\|x_\varepsilon(n)\|, \quad V := \sup_{1\le k\le \tau}E\|x(k)\|.$$

Let number $\varepsilon > 0$, such that for a fixed number $r \in (0,1)$ we have $g\Delta(\varepsilon) < \gamma$. For such an ε

$$\sup_{n\in Z}E\|x_\varepsilon(n) - x(n)\| \le g\Delta(\varepsilon)\frac{V}{1-\gamma}.$$

3. Stability of periodic solutions of nonlinear difference equations

Let $\{A(n) : n \in Z\}$ be operators satisfying (2) and (3) and let the function f: $Z \times B^2 \to B$ such that

$$\forall\, n \in Z \quad \forall\, \{x,y\} \subset B: \quad f(n+\tau,x,y) = f(n,x,y);$$
$$\exists\, L_1 > 0 \quad \forall\, n \in Z \quad \forall\{x,y\} \subset B:$$
$$\|f(n,x,y)\| \le L_1(1+\|x\|+\|y\|);$$
$$\exists\, L_2 > 0 \quad \forall\, n \in Z \quad \forall y \in B \quad \forall\{x',x''\} \subset B:$$
$$\|f(n,x',y) - f(n,x'',y)\| \le L_2\|x'-x''\|. \tag{5}$$

For the stability of the periodic solution of the equation

$$x(n+1) = A(n)x(n) + f(n,x(n),y(n)), \quad n \in Z, \tag{6}$$

we have the following Theorem.

Theorem 2. *Let operators $\{A(n) : n \in Z\}$ satisfy conditions (2) and (3) and let $\Delta(\varepsilon) \to 0$, $\varepsilon \to 0$, for operators $\{B_\varepsilon(n) : n \in Z\}$, $\varepsilon > 0$. Let $\{y(n) : n \in Z\}$ be* *a periodic with period τ B-valued process with $E\|y(k)\| < +\infty$, $k = 1, 2, ..., \tau$. Let $gL_2 < 1$. Then for every sufficiently small $\varepsilon > 0$ the equation*

$$x_\varepsilon(n+1) = (A(n) + B_\varepsilon(n))x_\varepsilon(n) + f(n, x_\varepsilon(n), y(n)), \quad n \in Z \tag{7}$$

has a unique solution $\{x_\varepsilon(n) : n \in Z\}$ such that

$$\sup_{n\in Z} E\|x_\varepsilon(n)\| < +\infty.$$

Moreover,

$$w(\varepsilon) := \sup_{n\in Z} E\|x_\varepsilon(n) - x(n)\| \to 0, \quad \varepsilon \to 0,$$

where $\{x(n) : n \in Z\}$ is a unique periodic solution with period τ solution of equation (6).

Before proving Theorem 2 we need to establish some auxiliary results. Let us assume that the conditions of Theorem 2 are satisfied.

Lemma 4. *For every sufficiently small $\varepsilon > 0$ equation (7) has a unique solution $\{x_\varepsilon(n) : n \in Z\}$ with*

$$\sup_{n\in Z} E\|x_\varepsilon(n)\| < +\infty.$$

Proof. Let $x^1_\varepsilon(n) := \bar{0}$, $n \in Z$, where $\bar{0}$ is a zero element in B. By Lemma 3 for all sufficiently small $\varepsilon > 0$ equation

$$x^2_\varepsilon(n+1) - (A(n) + B_\varepsilon(n))x^2_\varepsilon(n) + f(n, \bar{0}, y(n)), \quad n \in Z, \tag{8}$$

has a unique solution $\{x^2_\varepsilon(n) : n \in Z\}$ such that

$$\sup_{n\in Z} E\|x^2_\varepsilon(n)\| < +\infty; \tag{9}$$

and the processes $\{y(n) : n \in Z\}$ and $\{x^2_\varepsilon(n) : n \in Z\}$ are τ-periodic connected. Suppose for $j \geq 2$ that the process $\{x^j_\varepsilon(n) : n \in Z\}$ satisfying condition (9) is defined. Then we define for the same $\varepsilon > 0$ the process $\{x^{j+1}_\varepsilon(n) : n \in Z\}$ satisfying condition (9) as a unique solution of the following equation satisfying (9)

$$x^{j+1}_\varepsilon(n+1) = (A(n) + B_\varepsilon(n))x^{j+1}_\varepsilon(n) + f(n, x^j_\varepsilon(n), y(n)), \quad n \in Z. \tag{10}$$

Let

$$v(j, \varepsilon) := \sup_{n\in Z} E\|x^{j+1}_\varepsilon(n) - x^j_\varepsilon(n)\|, \quad j \geq 2.$$

Then we have

$$v(j, \varepsilon) \leq g\Delta(\varepsilon)v(j, \varepsilon) + gL_2 v(j-1, \varepsilon), \quad j \geq 2.$$

Therefore, for $\varepsilon > 0$ such that

$$gL_2(1 - g\Delta(\varepsilon))^{-1} < 1$$

for every $n \in Z$ the sequence $x_\varepsilon^j(n)$, $j \geq 1$ converges in the mean in B to some random element $x_\varepsilon(n)$. By (10) we have

$$x_\varepsilon(n+1) = (A(n) + B_\varepsilon(n))x_\varepsilon(n) + f(n, x_\varepsilon(n), y(n)), \quad n \in Z.$$

Process $\{x_\varepsilon(n) : n \in Z\}$ satisfying (9) uniformly in $\varepsilon > 0$ is obtained. The uniqueness of the solution can be easily proved.

Lemma 5. *Equation* (6) *has a unique periodic solution with period* τ.

Proof. By using the Theorem of existence of solution (1) Lemma 5 may be proved like Lemma 4.

Proof of Theorem 2. The existence of the unique solution of equations (6) and (7) follows directly from Lemma 4 and Lemma 5. Consider $x_\varepsilon(n) - x(n), n \in Z$. We have

$$w(\varepsilon) \leq gL_2 w(\varepsilon) + g\Delta(\varepsilon)\sup_{n\in Z} E\|x_\varepsilon(n)\|.$$

Hence the proof is completed.

4. Stability of periodic solutions of linear differential equations

Let τ be a fixed positive number.

Let function f: $R \to L(B)$ such that

$$A \in C(R, L(B)); \quad \forall t \in R : A(t+\tau) = A(t), \tag{11}$$

and let $U(\tau)$ be a monodromy operator corresponding to A. In (Dorogovtsev, 1992) it is proved that the condition

$$\sigma(U(\tau)) \cap S = \emptyset \tag{12}$$

is necessary and sufficient for the existence of a unique periodic solution with period τ $\{x(t) : t \in R\}$ such that

$$\sup_{0\leq s\leq \tau} E\|x(s)\| < +\infty, \tag{13}$$

of equation

$$x'(t) = A(t)x(t) + \xi(t), \quad t \in R \tag{14}$$

for every periodic with period τ B-valued process $\{\xi(t) : t \in R\}$ with

$$\int_0^\tau E\|\xi(s)\|ds < +\infty. \tag{15}$$

For equation (14) the following result on the continuous dependence on coefficient A will be useful.

For every $\varepsilon > 0$ let

$$B_\varepsilon \in C(R, L(B)); \quad \tilde{\Delta}(\varepsilon) := \sup_{t\in R}\|B_\varepsilon(t)\| < +\infty, \tag{16}$$

B_ε does not need to be periodic in t.

Theorem 3. *Suppose that A and B_ε satisfy conditions* (11), (12), *and* (16) *for $\varepsilon > 0$ and that $\Delta(\varepsilon) \to 0$, $\varepsilon \to 0$. Let $\{\xi(t) : t \in R\}$ be a periodic with period τ B-valued process satisfying condition* (15). *Then for every sufficiently small $\varepsilon > 0$ equation*

$$x'_\varepsilon(t) = (A(t) + B_\varepsilon(t))x_\varepsilon(t) + \xi(t), \quad t \in R, \tag{17}$$

has a unique solution $\{x_\varepsilon(t) : t \in R\}$ such that

$$\sup_{t\in R} E\|x_\varepsilon(t)\| < +\infty.$$

In addition

$$\sup_{t\in R} E\|x_\varepsilon(t) - x(t)\| \to 0, \quad \varepsilon \to 0,$$

where $\{x(t) : t \in R\}$ is a unique periodic solution with period τ of equation (14) *satisfying condition* (13).

To prove Theorem 3 we shall need the following auxiliary results. Let the conditions of Theorem 3 be satisfied.

Lemma 6. *Equation* (14) *has a unique periodic solution with period τ satisfying condition* (13).

Lemma 6 and its proof comes from (Dorogovtsev, 1992).

Lemma 7. *For every sufficiently small $\varepsilon > 0$ equation*

$$x'(t) = (A(t) + B_\varepsilon(t))x(t), \quad t \in R,$$

has an exponential dichotomy.

Lemma 7 and its proof are known, see (Massera and Schaffer, 1966), (Henry, 1981), (Cheban, 1980).

Proof of Theorem 3. Let G denote Green's function for equation (14). By condition (12)

$$\tilde{g} := \sup_{t\in R}\int_R \|G(t,s)\|ds < +\infty.$$

The existence of the unique solutions of equations (14) and (17) follows from Lemma 6 and Lemma 7.

We have

$$\sup_{t\in R} \quad E\|x_\varepsilon(t) - x(t)\| \le g\tilde{\Delta}(\varepsilon)\sup_{t\in R}E\|x_\varepsilon(t)\|,$$

$$\sup_{t\in R} \quad E\|x_\varepsilon(t)\| \le g\tilde{\Delta}(\varepsilon)\sup_{t\in R}\|x_\varepsilon(t)\| + \sup_{t\in R}\int_R \|G(t,s)\|E\|\xi(s)\|ds.$$

Theorem 3 is proved.

5. Stability of periodic solutions of nonlinear differential equations

Let the function $f : R \times B^2 \to B$ satisfy the following conditions:

$$\begin{gathered}\forall\, t \in R \quad \forall\, \{x,y\} \subset B: \quad f(t+\tau,x,y) = f(t,x,y);\\ \exists\, \tilde{L}_1 > 0 \quad \forall\, t \in R \quad \forall\, \{x,y\} \subset B:\\ \|f(t,x,y)\| \le \tilde{L}_1(1 + \|x\| + \|y\|);\\ \exists\, \tilde{L}_2 > 0 \quad \forall\, t \in R \quad \forall\, \{x',x''\} \subset B:\\ \|f(t,x',y) - f(t,x'',y)\| \le \tilde{L}_2\|x' - x''\|.\end{gathered} \tag{18}$$

In a similar way, by using Theorem 3 we can establish the following result.

Theorem 4. *Suppose that A and B_ε for $\varepsilon > 0$ satisfy conditions* (11), (12) *and* (16), *and f satisfies condition* (18). *Let $\tilde{\Delta}(\varepsilon) \to 0$, $\varepsilon \to 0$ and $\tilde{g}\tilde{L}_2 < 1$. Let $\{\xi(t) : t \in R\}$ be a periodic with period τ B-valued process satisfying condition* (15). *Then for every sufficiently small $\varepsilon > 0$ equation*

$$x'_\varepsilon(t) = (A(t) + B_\varepsilon(t))x_\varepsilon(t) + f(t, x_\varepsilon(t), \xi(t)), \quad t \in R$$

has a unique solution $\{x_\varepsilon(t) : t \in R\}$ such that

$$\sup_{t\in R}E\|x_\varepsilon(t)\| < +\infty.$$

In addition

$$\sup_{t\in R}E\|x_\varepsilon(t) - x(t)\| \to 0, \quad \varepsilon \to 0,$$

where $\{x(t) : t \in R\}$ is a unique periodic solution with period τ of equation

$$x'(t) = A(t)x(t) + f(t, x(t), \xi(t)), \quad t \in R.$$

References

Cheban, D.N. (1980). *Theory of Linear Differential Equations.* Kishinev, Schtiinca.

Da Prato, G. and Tudor, C. (1993). *Periodic and Almost Periodic Solutions for Semi-linear Stochastic Equations.* Scuola Normale Superiore. Pisa. Preprints di Matematica, No. 14.

Dorogovtsev, A.Ya. (1992). *Periodic and stationary regimes of infinite-dimensional deterministic and stochastic dynamical systems.* Vishcha Shkola. Kiev. (In Russian).

Henry, D. (1981). *Geometric Theory of Semi-linear Parabolic Equations.* Lect. Notes Math. No. 840, Springer Verlag. Berlin–Heidelberg–New York.

Hinrichsen, D. and Pritchard, A.J. (1992). *Robust stability of linear evolution operators on Banach spaces.* Report No. 269, Intitut fur Dynamische Systeme, Bremen, Germany.

Maslov, V.P. (1988). *Asymptotic Methods and Perturbation Theory.* Nauka, Moscow. (In Russian).

Massera, L.J. and Schaffer, I.I. (1966). *Linear Differential Equations and Functions Spaces.* Academic Press, New York.

Wirth, F. and Hinrichsen, D. (1992). *On stability radii of infinite dimensional time-varying discrete-time systems.* Report No. 276, Institut fur Dynamische Systeme, Bremen, Germany.

Exploring Stochastic Laws. pp. 121–128.
Eds. A.V. Skorokhod and Yu.V. Borovskikh

The number of crossings through a hyperplane by some generalized diffusion process

A.K. DURDIYEV AND N.I. PORTENKO

National Academy of Sciences of Ukraine, Institute of Mathematics, Kiev, Ukraine

Received February 15, 1995

Abstract. The number of crossings through a hyperplane by a Wiener process is investigated under the assumption that the hyperplane is a partly reflecting barrier for the process being reflected in an oblique direction that is constant over the hyperplane.

Introduction

We consider a generalized diffusion process in a d-dimensional Euclidean space $\mathbb{R}^d$ whose diffusion matrix is an identity matrix and the drift can be written in the form $A\delta_S(x)$, where A is a fixed vector in $\mathbb{R}^d$, S is a given hyperplane in $\mathbb{R}^d$ and $\delta_S(x)$ is a generalized function on $\mathbb{R}^d$ that acts on a test function φ on $\mathbb{R}^d$ by the formula

$$\int_{\mathbb{R}^d} \varphi(x)\, \delta_S(x)\, dx = \int_S \varphi(x)\, d\sigma \tag{1}$$

(the integral on the right-hand side of (1) is a surface integral).

Let us fix an orthonormal basis in $\mathbb{R}^d$ such that the hyperplane S can be given as follows $S = \{x \in \mathbb{R}^d\colon x^d = 0\}$ (x^k means the k-th coordinate of a vector $x \in \mathbb{R}^d$ in our basis, $k = 1, 2, \dots, d$). Denote by α the projection of vector A on S and by q its projection on the axis Ox^d. We assume that $\alpha^1, \dots, \alpha^{d-1}$ are arbitrary real numbers. As for q, it is assumed to be a real number in the interval $[-1, 1]$.

It was proven in Kopytko and Portenko (1983), that there exists a continuous homogeneous Markov process $(x(t), M_t, \mathbb{P}_x)$ in $\mathbb{R}^d$ (we use Dynkin's notation, see Dynkin (1963)) such that for all $\theta \in \mathbb{R}^d$ and $\varphi \in C_0(\mathbb{R}^d)$ (this is the space of all continuous compactly supported functions on $\mathbb{R}^d$) the following two conditions hold:

$$\begin{aligned}
\lim_{t\downarrow 0} \int_{\mathbb{R}^d} \varphi(x)\, \frac{1}{t}\, \mathbf{E}_x(x(t) - x(0), \theta)\, dx &= (A, \theta) \int_S \varphi(x)\, d\sigma, \\
\lim_{t\downarrow 0} \int_{\mathbb{R}^d} \varphi(x)\, \frac{1}{t}\, \mathbf{E}_x(x(t) - x(0), \theta)^2\, dx &= |\theta|^2 \int_{\mathbb{R}^d} \varphi(x)\, dx.
\end{aligned} \tag{2}$$

This process is determined by the semigroup of operators $T_t\varphi(x) = \mathbf{E}_x\varphi(x(t))$ that act on an arbitrary bounded measurable function φ on $\mathbb{R}^d$ according to the formula ($t > 0$, $x \in \mathbb{R}^d$)

$$T_t\varphi(x) = \int_{\mathbb{R}^d} g(t,x,y)\,\varphi(y)\,dy + \int_0^t d\tau \int_S G(t-\tau,x,y)\,V_1(\tau,y,\varphi)\,d\sigma_y + \\ + \int_0^t d\tau \int_S H(t-\tau,x-y)\,V_2(\tau,x,y)\,d\sigma_y, \qquad (3)$$

where $g(t,x,y) = (2\pi t)^{-d/2}\exp\{-|y-x|^2/2t\}$ is the transition probability density of a d-dimensional Wiener process, the functions $V_1(t,x,\varphi)$ and $V_2(t,x,\varphi)$ are given by the formulas

$$V_1(t,x,\varphi) = \int_{\mathbb{R}^d} \operatorname{sign} y^d \,\frac{\partial g(t,x,y)}{\partial y^d}\,\varphi(y)\,dy,$$

$$V_2(t,x,\varphi) = -\int_{\mathbb{R}^d} (q + \operatorname{sign} y^d)\,\frac{\partial g(t,x,y)}{\partial y^d}\,\varphi(y)\,dy$$

for all $t > 0$, $x \in \mathbb{R}^d$ and bounded measurable functions φ on $\mathbb{R}^d$, and the function $H(t,x)$ for $t > 0$, $x \in \mathbb{R}^d$ can be written as follows

$$H(t,x) = (2\pi t)^{-d/2} \int_0^\infty \frac{\theta + |x^d|}{t} \exp\Big\{-\frac{(\theta + |x^d|)^2 + |\bar{x} + \alpha\theta|^2}{2t}\Big\}\,d\theta$$

(here $\bar{x}$ means the projection on S of the vector x).

For given integers n and N we denote by $\nu_N^{(n)}$ the number of crossings through the hyperplane S by the sequence of random vectors $x(0)$, $x(\frac{1}{n})$, ..., $x(\frac{N}{n})$. In other words,

$$\nu_N^{(n)} = \sum_{k=1}^N \xi_k^{(n)}, \qquad (4)$$

where $\xi_k^{(n)} = 1$ if $x^d\big(\frac{k-1}{n}\big)\,x^d\big(\frac{k}{n}\big) < 0$ and $\xi_k^{(n)} = 0$ otherwise, $k = 1, 2, \ldots$. Our goal is to investigate the behavior of the distribution of the random variable $\nu_N^{(n)}$ as n and N tend to infinity simultaneously, namely, $N = [nt]$ for a fixed $t > 0$ (here $[\beta]$ denotes the integer part of a real number β).

In the case of $\alpha = 0$ this problem was solved by Portenko and Yephimenko (1987) (in fact, the problem of Portenko and Yephimenko (1987) was somewhat more complicated, namely, the vector A was assumed to have the form $A = (0, 0, \ldots, 0, q(x))$, where $q(x)$ is a given continuous function on S with the property $|q(x)| \le 1$). As we shall see below, the limit distribution of the sequence of random variables $\nu_{[nt]}^{(n)}$ does not depend on vector α.

Our approach to the problem is similar to that used by Portenko and Yephimenko (1987) and, as a matter of fact, was proposed by I. I. Gikhman (see Gikhman (1957), Gikhman (1958)) but the process under consideration in this paper is more complicated than those investigated by Portenko and Yephimenko (1987), Gikhman (1957) and Gikhman (1958).

The main result

First we note that according to the A. V. Skorokhod's lemma from the work of Skorokhod (1961), the limit distribution of variables $n^{-\frac{1}{2}}\nu_{[nt]}^{(n)}$, as $n \to \infty$, exists if and only if it exists for variables $n^{-\frac{1}{2}}\eta_{[nt]}^{(n)}$, where

$$\eta_{[nt]}^{(n)} = \sum_{k=1}^{[nt]} v_n\left(x\left(\frac{k}{n}\right)\right), \tag{5}$$

$v_n(x) = \mathbf{E}_x\xi_1^{(n)}$ for $x \in \mathbb{R}^1$ and $n = 1, 2, \ldots$. In the case of existence, both the limit distribution of (4) and of (5) coincide. If we put $D_+ = \{x \in \mathbb{R}^d \colon x^d > 0\}$, $D_- = \{x \in \mathbb{R}^d \colon x^d < 0\}$, then the function $v_n(x)$ can be written in the form $v_n(x) = v_n'(x) + v_n''(x)$, where $v_n'(x) = \mathbf{1}_{D_+}(x)\,T_{\frac{1}{n}}\mathbf{1}_{D_-}(x)$, $v_n''(x) = \mathbf{1}_{D_-}(x)\,T_{\frac{1}{n}}\mathbf{1}_{D_+}(x)$ (we denote by $\mathbf{1}_\Gamma(x)$ the indicator function of a set $\Gamma \subset \mathbb{R}^d$).

We will thus consider variables (5) instead of (4). To begin with the behavior of functions $v_n(x)$, as $n \to \infty$, should be investigated.

Lemma 1. *For an arbitrary real-valued continuous function φ on $D_+ \cup S$ with the property*

$$\sup_{\rho \geq 0} \int_{S_\rho} |\varphi(x)|\, d\sigma < \infty$$

(here $S_\rho = \{x \in \mathbb{R}^d \colon x^d = \rho\}$ for $\rho \in \mathbb{R}^1$) the following relation

$$\lim_{n\to\infty} \sqrt{n} \int_{\mathbb{R}^d} \varphi(x)\, v_n'(x)\, dx = \frac{1-q}{\sqrt{2\pi}} \int_S \varphi(x)\, d\sigma \tag{6}$$

is true and the estimate

$$\left| \int_{\mathbb{R}^d} \sqrt{n}\, \varphi(x)\, v_n'(x)\, dx \right| \leq \frac{4}{\sqrt{2\pi}} \sup_{\rho \geq 0} \int_{S_\rho} |\varphi(x)|\, d\sigma \tag{7}$$

holds.

Proof. Obviously, the integral on the left-hand side of (6) can be written in the form

$$\sqrt{n} \int_{\mathbb{R}^d} \varphi(x)\, v_n'(x)\, dx = I_1 + I_2 + I_3,$$

where

$$I_1 = \sqrt{n} \int_{D_+} \varphi(x)\, dx \int_{D_-} g\left(\frac{1}{n}, x, y\right) dy,$$

$$I_2 = \sqrt{n} \int_{D_+} \varphi(x)\, dx \int_0^{\frac{1}{n}} d\tau \int_S g\left(\frac{1}{n} - \tau, x, y\right) V_1(\tau, y, \mathbf{1}_{D_-})\, d\sigma_y,$$

$$I_3 = \sqrt{n} \int_{D_+} \varphi(x)\, dx \int_0^{\frac{1}{n}} d\tau \int_S H\left(\frac{1}{n} - \tau, x - y\right) V_2(\tau, y, \mathbf{1}_{D_-})\, d\sigma_y.$$

The behavior of I_1 and I_2, as $n \to \infty$, was investigated in Portenko and Yephimenko (1987). It was proved that $\lim_{n\to\infty}(I_1 + I_2) = 0$. Moreover, the estimates

$$|I_k| \le \frac{1}{\sqrt{2\pi}} \sup_{\rho\ge 0} \int_{S_\rho} |\varphi(x)|\, d\sigma, \qquad k = 1,2,$$

were established.

Taking into account the formulas

$$V_2(\tau, y, \mathbf{1}_{D_-}) = \frac{1-q}{\sqrt{2\pi\tau}}, \qquad \int_S H(\tau, x-y)\, d\sigma_y = \frac{1}{\sqrt{2\pi\tau}} e^{-(x^d)^2/2\tau},$$

we can write

$$I_3 = \sqrt{n} \int_{D_+} \varphi(x)\, \Phi_n(x^d)\, dx,$$

where

$$\Phi_n(\rho) = (1-q) \int_0^{\frac{1}{n}} \left(2\pi\left(\frac{1}{n} - \tau\right)\right)^{-1/2} (2\pi\tau)^{-1/2} e^{-\rho^2/2\tau}\, d\tau$$

for real ρ. Now, the quantity I_3 can be given in the following form

$$I_3 = \sqrt{n} \int_0^\infty \Phi_n(\rho)\, d\rho \int_{S_\rho} \varphi(x)\, d\sigma.$$

One can easily verify that for any continuous bounded function $\Psi(\rho)$ on $[0, +\infty)$ the relation

$$\lim_{n\to\infty} \sqrt{n} \int_0^\infty \Phi_n(\rho)\, \Psi(\rho)\, d\rho = \frac{1-q}{\sqrt{2\pi}}\, \Psi(0)$$

and the estimate

$$\left|\sqrt{n} \int_0^\infty \Phi_n(\rho)\, \Psi(\rho)\, d\rho\right| \le \frac{|1-q|}{\sqrt{2\pi}} \sup_{\rho\ge 0} |\Psi(\rho)|$$

are true. Therefore

$$\lim_{n\to\infty} I_3 = \frac{1-q}{\sqrt{2\pi}} \int_S \varphi(x)\, d\sigma$$

and I_3 have estimates similar to that that was written above for I_k, $k = 1,2$. The lemma is proved.

The following result can be proved in the same way: the relation

$$\lim_{n\to\infty} \int_{\mathbb{R}^d} v_n''(x)\, \varphi(x)\, dx = \frac{1+q}{\sqrt{2\pi}} \int_S \varphi(x)\, d\sigma \tag{8}$$

holds for every continuous function $\varphi(x)$ on $D_- \cup S$ such that it satisfies the condition

$$\sup_{\rho\le 0} \int_{S_\rho} |\varphi(x)|\, d\sigma < \infty.$$

Moreover, the estimate

$$\left|\int_{\mathbb{R}^d} v_n''(x)\, \varphi(x)\, dx\right| \le \frac{4}{\sqrt{2\pi}} \sup_{\rho\le 0} \int_{S_\rho} |\varphi(x)|\, d\sigma \tag{9}$$

is true for all $n = 1, 2, \ldots$.

Corollary 1. *Let $\varphi(x)$ be a given measurable function on $\mathbb{R}^d$ with the property*

$$\sup_{\rho \in \mathbb{R}^1} \int_{S_\rho} |\varphi(x)| \, d\sigma < \infty$$

and assume that there exist non-tangent limits $\varphi(x+)$ and $\varphi(x-)$ at every point $x \in S$ from the sides D_+ and D_- respectively. Then the relation

$$\lim_{n\to\infty} \sqrt{n} \int_{\mathbb{R}^d} v_n(x)\, \varphi(x)\, dx = \frac{1}{\sqrt{2\pi}} \int_S \Big[(1+q)\varphi(x-) + (1-q)\varphi(x+)\Big]\, d\sigma \quad (10)$$

is true and the estimate

$$\left| \int_{\mathbb{R}^d} v_n(x)\, \varphi(x)\, dx \right| \le \frac{4}{\sqrt{2\pi}} \sup_{\rho\in\mathbb{R}^1} \int_{S_\rho} |\varphi(x)|\, d\sigma \quad (11)$$

holds.

As it follows from formula (3), there exists transition probability density $G(t, x, y)$ ($t > 0$, $x \in \mathbb{R}^d$, $y \in \mathbb{R}^d$) of the process $(x(t), M_t, \mathbb{P}_x)$ and for $y \notin S$ it can be written in the following form

$$G(t,x,y) = g(t,x,y) + \operatorname{sign} y^d \int_0^t d\tau \int_S g(t-\tau, x, z)\, \frac{\partial g(\tau, z, y)}{\partial y^d}\, d\sigma_z -$$
$$- (q + \operatorname{sign} y^d) \int_0^t d\tau \int_S H(t-\tau, x-z)\, \frac{\partial g(\tau, z, y)}{\partial y^d}\, d\sigma_z . \quad (12)$$

Using the theorem on a jump of the normal derivative of a single-layer potential (see Ladyzhenskaya, Solonnikov and Ural'tseva (1967), Portenko (1990)), from (12) we obtain the following result.

Lemma 2. *For each $y \in S$ there exists non-tangent limits $G(t, x, y+)$ and $G(t, x, y-)$ from the sides D_+ and D_- respectively and these limits can be given by the formula*

$$G(t, x, y\pm) = (1 \pm q)\, H(t, x-y) \quad (13)$$

valid for $t > 0$, $x \in \mathbb{R}^d$, $y \in S$.

The following lemma contains technical results. Let us denote by $\Phi_T(K)$ the set of all real-valued measurable functions $\varphi(t, x)$ defined for $(t, x) \in [0, \infty) \times \mathbb{R}^d$ such that

$$\sup_{(t,x)\in[0,T]} |\varphi(t,x)| \le K$$

(here T and K are arbitrary positive constants).

Lemma 3. *For arbitrary $\varepsilon > 0$, $T > 0$, $K > 0$ there exists a positive number δ such that the function*

$$f_n(t,x) = \sqrt{n}\int_0^t d\tau \int_{\mathbb{R}^d} G(t-\tau,x,y)\,\varphi(\tau,y)\,v_n(y)\,dy$$

(here $t \geq 0$, $x \in \mathbb{R}^d$, $n = 1,2,\dots$) satisfies the inequality

$$|f_n(t',x') - f_n(t,x)| < \varepsilon$$

for all $x \in \mathbb{R}^d$, $x' \in \mathbb{R}^d$, $t \in [0,T]$, $t' \in [0,T]$, $\varphi \in \Phi_T(K)$ and integer n under the condition $|t-t'| + |x-x'| < \delta$.

The proof of this statement is similar to that of the prove of Lemma 2 from Portenko and Yephimenko (1987), but this proof is more complicated than that one because of the third term in (12). We only note that the proof makes use of the following inequality (see Kopytko (1995))

$$|D_t^n D_t^m H(t,x-y)| \leq C_T\, t^{-(d+2n+m)/2} \exp\Big\{-\mu\frac{|y-x|^2}{t}\Big\}$$

valid in the domain $(t,x,y) \in (0,T] \times \mathbb{R}^d \times \mathbb{R}^d$, where n and m are non-negative integers such that $2n+m \leq 2$ and C_T and μ are constants. We omit the details of the proof.

Now, we put

$$u_n(t,x,\lambda) = \mathbf{E}_x \exp\big\{i\lambda n^{-\frac{1}{2}}\eta^{(n)}_{[nt]}\big\}$$

for $t > 0$, $x \in \mathbb{R}^d$, $\lambda \in \mathbb{R}^1$ and integer n. This function satisfies the following integral equation

$$u_n(t,x,\lambda) = 1 + \\ + n\int_{[0,[nt]/n)} d\tau \int_{\mathbb{R}^d}\Big(1 - e^{-i\lambda n^{-\frac{1}{2}} v_n(y)}\Big)\, u_n(\tau,y,\lambda)\, G\Big(\frac{[nt]-[n\tau]}{n},x,y\Big)\,dy.$$

Using Lemma 3, we arrive at the conclusion that $\lim_{n\to\infty}\big[u_n(t,x,\lambda) - \tilde{u}_n(t,x,\lambda)\big] = 0$, where the function $\tilde{u}_n(t,x,\lambda)$ is defined by the relation

$$\tilde{u}_n(t,x,\lambda) = \mathbf{E}_x \exp\Big\{i\lambda n^{\frac{1}{2}}\int_0^t v_n(x(\tau))\,d\tau\Big\}$$

and satisfies the integral equation

$$\tilde{u}_n(t,x,\lambda) = 1 + i\lambda\sqrt{n}\int_0^t d\tau \int_{\mathbb{R}^d} G(t-\tau,x,y)\,\tilde{u}_n(\tau,y,\lambda)\,v_n(y)\,dy. \tag{14}$$

As it follows from Lemma 3, a sequence $n_k \to \infty$ can be chosen such that there exists $\lim_{n_k\to\infty}\tilde{u}_{n_k}(t,x,\lambda)$, which will be denoted by $u(t,x,\lambda)$. Moreover, the convergence here is uniform with respect to $x \in \mathbb{R}^d$ and locally uniform with respect to t (the parameter λ is fixed in our reasoning). Putting $n = n_k$ in (14) and letting $n_k \to \infty$, we find from Lemma 2 and Corollary 1 that the function $u(t,x,\lambda)$ satisfies the integral equation ($t \geq 0$, $x \in \mathbb{R}^d$, $\lambda \in \mathbb{R}^1$)

$$u(t,x,\lambda) = 1 + i\lambda(1-q^2)\sqrt{\frac{2}{\pi}}\int_0^t d\tau \int_S u(\tau,y,\lambda)\, H(t-\tau, x-y)\, d\sigma_y. \quad (15)$$

The unique solution of this equation can be written in the following form

$$u(t,x,\lambda) = \int_{\mathbb{R}^1} e^{i\lambda a}\, F_{t,x}(da),$$

where ($t > 0$, $x \in \mathbb{R}^d$, $a \in \mathbb{R}^1$)

$$F_{t,x}(a) = \mathbf{1}_{(0,\infty)}(a)\sqrt{\frac{2}{\pi t}}\int_0^{\frac{a}{\gamma}+x^d} \exp\left\{-\frac{b^2}{2t}\right\} db, \quad (16)$$

$\gamma = (1-q^2)\sqrt{\frac{2}{\pi}}$.

The fact of the uniqueness of the solution of equation (15) implies that $\lim_{n\to\infty} u_n(t,x,\lambda) = u(t,x,\lambda)$. We have thus already proved the following theorem.

Theorem. *For $t > 0$, $x \in \mathbb{R}^d$, $a > 0$ the relation*

$$\lim_{n\to\infty} \mathbb{P}_x\left\{n^{-\frac{1}{2}}\nu_{[nt]}^{(n)} < a\right\} = F_{t,x}(a)$$

is true, where function $F_{t,x}(a)$ is defined by formula (16).

References

Kopytko, B. I. and Portenko, N. I. (1983). Analytical methods of joining together diffusion processes. *Lecture Notes in Mathematics.* **1021,** 320–326.

Dynkin, E. B. (1963). *Markov processes.* Fizmatgiz, Moscow.

Portenko, N. I. and Yephimenko, S. V. (1987). On the number of crossings of a partly reflecting hyperplane by a multidimensional Wiener process. *Lecture Notes in Control and Information Sciences.* **96,** 194–203.

Gikhman, I. I. (1957). Some limit theorems for the number of crossings by a random function on the boundary of a given domain. *Naukovi zapiski Kiev. universitetu.* **16,** 149–164. (In Ukrainian).

Gikhman, I. I. (1958). Asymptotic distributions for the number of crossings by a random function on the boundary of a domain. *Visnyk Kiev. universitetu. Series Astr., Mathem., Mech.* **1,** 25–46. (In Ukrainian).

Ladyzhenskaya, O. A., Solonnikov, V. A. and Ural'tseva, N. N. (1967). *Linear and quasilinear equations of parabolic type.* Nauka, Moscow. (In Russian).

Portenko, N. I. (1990). *Generalized diffusion processes.* American Mathematical Society.

Skorokhod, A. V. (1961). Some limit theorems for additive functionals of sequences of independent random variables. *Ukrainian mathem. Journ.* **13**(4), 67–78.

Kopytko, B. I. (1995). *Analytical methods in the theory of diffusion processes.* Doctoral dissertation. Kiev.

Exploring Stochastic Laws. pp. 129–134.
Eds. A.V. Skorokhod and Yu.V. Borovskikh

Bounds for absolute moments of order statistics

NADEZHDA V. GRIBKOVA

Department of Applied Mathematics, Transport University, 190031, St Petersburg, Moskovsky Prospekt 9, Russia

Received February 23, 1995

Abstract. The bounds for absolute moments of members of variational series are established. Let $X_1, \ldots, X_n$ be independent identically distributed real-valued random variables and let $X_{1,n} \leq \ldots \leq X_{n,n}$ be the corresponding order statistics. The absolute moments $E|X_{i,n}|^k, k > 0$, are estimated via the absolute moment $E|X_1|^\delta, \delta > 0$, for all i such that $k\delta^{-1} \leq i \leq n - k\delta^{-1} + 1$ with order $(n^2 i^{-1}(n-1)^{-1})^{k\delta^{-1}}$ in i and n. These estimates are able to be of some use as a tool to argue in different probability limit theorems.

Let $X_1, \ldots, X_n$ be independent identically distributed real-valued random variables with common distribution function F, and let $X_{1,n} \leq \ldots \leq X_{n,n}$ denote the corresponding order statistics. Let k and δ be arbitrary positive numbers. Put $\rho = k\delta^{-1}$ and set $g(u) = u(1-u)$.

Theorem 1. *For all $n \geq 2\rho + 1$ and for all i such that $\rho \leq i \leq n - \rho + 1$ the following inequality holds*

$$E|X_{i,n}|^k < C(\rho)\left\{E|X_1|^\delta g^{-1}\left(\frac{i}{n+1}\right)\right\}^\rho, \tag{1}$$

where one can put $C(\rho) = 2\sqrt{\rho}\exp(\rho + 7/6)$.

Consequence. *For arbitrary $0 < \alpha < \beta < 1$ and for all n and i such that $\rho \leq n\alpha < i < n\beta \leq n - \rho + 1$, the following inequality holds*

$$E|X_{i,n}|^k < C(\alpha, \beta, \rho)\left\{E|X_1|^\delta\right\}^\rho,$$

where $C(\alpha, \beta, \rho)$ is constant, and depends only on α, β and ρ.

Estimate (1) was applied in Gribkova (1988, 1993) for obtaining estimates in the Central Limit Theorem for L-statistics, and it was applied in Korolyuk and Borovskikh (1993) for the analysis of the asymptotic behaviour of random permanent measures for L-statistics. Inequalities of type (1) might possibly be useful in other areas of probability theory. It should be noted that many inequalities and other auxiliary results for order statistics are included in Helmers (1982) and van Zwet (1964).

Proof of Theorem 1. To begin with let us note that by Chebyshev's inequality,

$$\left(F^{-1}(u)\right)^{\delta}(1-u) \leq E|X_1|^{\delta}$$

for $u \geq F(0)$ and

$$\left|F^{-1}(u)\right|^{\delta} u \leq E|X_1|^{\delta}$$

for $u < F(0)$ where

$$F^{-1}(u) = \inf\{x : F(x) \geq u\}.$$

We first assume that $\rho + 1 \leq i \leq n - \rho$. Then we obtain

$$\begin{aligned} E|X_{i,n}|^k &= \frac{1}{B(i,n-i+1)} \int_0^1 \left|F^{-1}(u)\right|^k u^{i-1}(1-u)^{n-i} du \\ &\leq \frac{\left\{E|X_1|^{\delta}\right\}^{\rho}}{B(i,n-i+1)} \left(\int_0^{F(0)} u^{i-\rho-1}(1-u)^{n-i} du \right. \\ &\quad \left. + \int_{F(0)}^1 u^{i-1}(1-u)^{n-\rho-i} du \right) \\ &< \left\{E|X_1|^{\delta}\right\}^{\rho} \left(\frac{B(i-\rho,n-i+1)}{B(i,n-i+1)} + \frac{B(i,n-\rho-i+1)}{B(i,n-i+1)} \right). \end{aligned} \tag{2}$$

To evaluate the ratios of beta functions on the right hand side of (2) we apply the inequalities

$$\sqrt{2\pi} x^{x+1/2} e^{-x} < \Gamma(1+x) < \sqrt{2\pi} x^{x+1/2} e^{-x+1/12x} \tag{3}$$

where $x \in R, x > 0$, which follows from Stirling's expansion. By using (3), we will prove that for all $\rho > 0$ and for all $i \geq \rho + 1$

$$B(i-\rho,n-i+1)/B(i,n-i+1) < e^{\rho+7/6}(n/i)^{\rho} \tag{4}$$

Relation (4) and the symmetry of the beta function imply that

$$B(i,n-\rho-i+1)/B(i,n-i+1) < e^{\rho+7/6}(n/(n-i+1))^{\rho} \tag{5}$$

for all $\rho > 0$ and for all $i \leq n - \rho$. Together (2) and bounds (4)–(5) yield (1) in the case $\rho + 1 \leq i \leq n - \rho$. Now we will prove (4). We have

$$B(i-\rho, n-i+1)/B(i, n-i+1) = \frac{\Gamma(i-\rho)\cdot\Gamma(n+1)}{\Gamma(i)\Gamma(n-\rho+1)}. \tag{6}$$

First suppose that $i \geq \rho+2$, so therefore $i-\rho-1 \geq 1$, and applying (3) on the right hand side of (6), we find that (6) is less than

$$e^{1/6}\left(\frac{(i-\rho-1)n}{(n-\rho)(i-1)}\right)^{1/2}\left(\frac{n}{i-1}\right)^{\rho}\left(\frac{n}{n-\rho}\right)^{n-\rho}\left(\frac{i-\rho-1}{i-1}\right)^{i-\rho-1}. \tag{7}$$

To continue let us note that

$$\left(\frac{n}{n-\rho}\right)^{n-\rho} < e^{\rho},$$

$$\left(\frac{(i-\rho-1)n}{(n-\rho)(i-1)}\right)^{1/2} < 1, \qquad \left(\frac{i-\rho-1}{i-1}\right)^{i-\rho-1} < e^{-\rho+\rho^2/(i-1)},$$

so the magnitude of (7) is less than

$$e^{1/6+\rho^2/(i-1)}\left(\frac{n}{i-1}\right)^{\rho} < e^{\rho+7/6}\left(\frac{n}{i}\right)^{\rho}. \tag{8}$$

Now we will estimate (6) for the case

$$\rho+1 \leq i \leq \rho+2. \tag{9}$$

Then we have $1 \leq i-\rho < 2, \Gamma(i-\rho) \leq 1$, and so the right hand side of (6) is not larger than

$$\Gamma(n+1)/(\Gamma(i)\Gamma(n-\rho+1)). \tag{10}$$

If ρ is an integer, then $\Gamma(n+1)/\Gamma(n-\rho+1) < n^{\rho}$ and i from (9) is $i=\rho+1$. For such i we have

$$\begin{aligned}\Gamma(n+1)/(\Gamma(i)\cdot\Gamma(n-\rho+1)) &< \frac{n^{\rho}e^{\rho}}{\sqrt{\rho}\rho^{\rho}} \\ &< e^{\rho+1}(n/i)^{\rho} \end{aligned} \tag{11}$$

and so (4) follows. And further, if ρ is not an integer, then i from (9) is $i=[\rho]+2$ (here and further $[k]$ means the whole part of positive k). Suppose at first that $[\rho]=0$. Then $i=2$ and (10) is equal to

$$\Gamma(n+1)/\Gamma(n-\rho+1) < e^{1/12}n^{\rho}\sqrt{1+\rho} < e^{1/12+\rho/2+1}(n/i)^{\rho}, \tag{12}$$

and so (4) is true. Now we will suppose that $[\rho] \geq 1$, then we have

$$\Gamma(n+1)\ /\ (\Gamma(i)\cdot\Gamma(n-\rho+1)) < e^{1/12}n^{\rho}\sqrt{1+\rho}\,\Gamma^{-1}([\rho]+2)$$

$$< e^{1/12}\sqrt{\frac{1+\rho}{1+[\rho]}}\frac{1}{\sqrt{2\pi}}\left(\frac{n}{[\rho]+1}\right)^{\rho}e^{\rho}\left(\frac{e}{[\rho]+1}\right)^{[\rho]+1-\rho}$$

$$< \frac{1}{\sqrt{2\pi}}\left(\frac{3}{2}\right)^{3/2}e^{\rho+1/12}\left(\frac{n}{[\rho]+1}\right)^{\rho}$$

$$< e^{\rho+1}(n/i)^{\rho}. \tag{13}$$

Together bounds (6)-(8) and (11)-(13) imply (4). Thus we have proved our lemma for the case $\rho+1\le i\le n-\rho$ for all $\rho>0$. It still remains to establish (1) for the extreme values of i. Now let

$$\rho\le i<\rho+1 \tag{14}$$

If $\rho\le 1$ (i.e. $k\le\delta$), then i from (14) is equal to 1. Now by Hölder's inequality we obtain

$$\begin{aligned} E|X_{1,n}|^k &= B^{-1}(1,n)\int_{-\infty}^{\infty}|x|^k(1-F(x))^{n-1}dF(x)\\ &\le B^{-\rho}(1,n)\left\{\int_{-\infty}^{\infty}|x|^{\delta}dF(x)\right\}^{\rho}\\ &= n^{\rho}\left\{E|X_1|^{\delta}\right\}^{\rho}, \end{aligned}$$

and so (1) is true. Now let $\rho>1$ (i.e. $k>\delta$). If ρ is an integer , then i from (14) is equal to ρ. Now for that i we have

$$\begin{aligned} E|X_{i,n}|^k &= B^{-1}(\rho,n-\rho+1)\int_{-\infty}^{\infty}|x|^kF^{\rho-1}(x)(1-F(x))^{n-\rho}dF(x)\\ &\le B^{-1}(\rho,n-\rho+1)\left\{E|X_1|^{\delta}\right\}^{\rho-1}\int_{-\infty}^{\infty}|x|^{\delta}(1-F(x))^{n-2\rho+1}dF(x)\\ &\le B^{-1}(\rho,n-\rho+1)\left\{E|X_1|^{\delta}\right\}^{\rho}. \end{aligned} \tag{15}$$

Further,

$$\begin{aligned} B^{-1}(\rho,n-\rho+1) &= \Gamma(n+1)/(\Gamma(\rho)\cdot\Gamma(n-\rho+1))\\ &< e^{1/12-1}\frac{1}{\sqrt{2\pi}}\sqrt{\rho-1}\frac{n^n}{(\rho-1)^{\rho}(n-\rho)^{n-\rho}}\sqrt{\frac{n}{n-\rho}} \end{aligned}$$

$$
\begin{aligned}
&< e^{1/12-1}\frac{\sqrt{\rho-1}}{\sqrt{2\pi}}\left(\frac{n}{\rho-1}\right)^{\rho} e^{\rho}\sqrt{2} \\
&< \frac{1}{\sqrt{\pi}}e^{1/12+\rho+1}\sqrt{\rho-1}\,(n/i)^{\rho}. \qquad (16)
\end{aligned}
$$

Suppose now that $\rho > 1$ and is not an integer. Then i satisfying (14) has representation $i = [\rho]+1$. For such values of i we obtain

$$
\begin{aligned}
E|X_{i,n}|^k &= B^{-1}([\rho]+1, n-[\rho]) \int_{-\infty}^{\infty} |x|^k F^{[\rho]}(x)(1-F(x))^{n-[\rho]-1} dF(x) \\
&\leq \frac{\left\{E|X_1|^\delta\right\}^{[\rho]}}{B([\rho]+1, n-[\rho])} \int_{-\infty}^{\infty} |x|^{\delta(\rho-[\rho])}(1-F(x))^{n-2[\rho]-1} dF(x). \qquad (17)
\end{aligned}
$$

By Hölder's inequality the right hand side of (17) is not larger than

$$
\begin{aligned}
&\frac{\left\{E|X_1|^\delta\right\}^{[\rho]}}{B([\rho]+1, n-[\rho])} \cdot \frac{\left\{E|X_1|^\delta\right\}^{\rho-[\rho]}}{\{B(1, n-2[\rho])\}^{\rho-[\rho]-1}} = \\
&= \left\{E|X_1|^\delta\right\}^{\rho} \frac{n!(n-2[\rho])^{\rho-[\rho]-1}}{[\rho]!\cdot(n-[\rho]-1)!} \leq \\
&\leq \left\{E|X_1|^\delta\right\}^{\rho} \frac{n^\rho}{[\rho]!}\left(\frac{n}{n-2[\rho]}\right)^{[\rho]+1-\rho}. \qquad (18)
\end{aligned}
$$

If $[\rho] = 1$, then the right hand side of (18) is equal to

$$
\left\{E|X_1|^\delta\right\}^{\rho} n^\rho \left(1+\frac{2}{n-2}\right)^{2-\rho}
$$

Further, $n-2 \geq 2$ (as $n \geq 2\rho+1 > 3$) and the last magnitude is less than

$$
\left\{E|X_1|^\delta\right\}^{\rho} n^\rho \cdot 2.
$$

So, for $i = [\rho]+1 = 2$ we have inequality

$$
\begin{aligned}
E|X_{2,n}|^k &< \left\{E|X_1|^\delta\right\}^{\rho} \cdot n^\rho \cdot 2 \\
&< 2e^\rho \left\{\frac{n^2}{2(n-1)}\right\}^{\rho},
\end{aligned}
$$

which implies (1). It still remains to consider the case $[\rho] \geq 2$. Then by (3) the right hand side of (18) is less than

$$
\left\{E|X_1|^\delta\right\}^{\rho} \frac{1}{\sqrt{2\pi}} e^{[\rho]} \left(\frac{n}{[\rho]}\right)^{\rho} \sqrt{[\rho]}(2+\frac{1}{[\rho]})^{[\rho]+1-\rho} < \left\{E|X_1|^\delta\right\}^{\rho} \sqrt{[\rho]}(n/[\rho])^\rho e^{[\rho]}.
$$

So, for $i = [\rho] + 1$ we have the bound

$$
\begin{aligned}
E|X_{i,n}|^k &< \left\{E|X_1|^\delta\right\}^\rho e^\rho \sqrt{\rho} e^{\frac{\rho-[\rho]}{[\rho]} - (\rho-[\rho])+1} \left(\frac{n}{i}\right)^\rho \\
&< e^{\rho+1}\sqrt{\rho}\left\{E|X_1|^\delta\right\}^\rho \left(g^{-1}\left(\frac{i}{n+1}\right)\right)^\rho .
\end{aligned}
$$

Thus (1) is proved for all i such that $\rho \leq i < \rho + 1$ and for all $\rho > 0$. By a standard argument, with change of signs at underlying random variables we obtain (1) for $n - \rho < i \leq n - \rho + 1$ and for all $\rho > 0$. This completes the proof of Theorem 1.

Analysing the proof of Theorem 1 we see that the following theorem is valid. Let δ_1, δ_2 and k be arbitrary positive numbers. Put $\rho_1 = k/\delta_1$ and $\rho_2 = k/\delta_2$, and denote I_A as the indicator of set A.

Theorem 2. *For all $n \geq \rho_1 + \rho_2 + 1$ and for all i such that $\rho_1 \leq i \leq n - \rho_2 + 1$ the following inequality holds*

$$
E|X_{i,n}|^k < e^{\rho_1 + \frac{1}{6}}\sqrt{\rho_1}\left[\frac{n}{i-1}E|X_1^-|^{\delta_1}\right]^{\rho_1} + e^{\rho_2 + \frac{1}{6}}\sqrt{\rho_2}\left[\frac{n}{n-i}E|X_1^+|^{\delta_2}\right]^{\rho_2},
$$

where $X_1^- = X_1 \cdot I_{(-\infty,0]}, \quad X_1^+ = X_1 \cdot I_{[0,\infty)}$.

References

Gribkova, N.V. (1988). On the rate of convergence to the normal law of finitely truncated linear combinations of order statistics. *Theory Probab. Appl.* 33, 781–784. (in Russian).

Gribkova, N.V. (1993). On the analogues of Berry–Esseen's inequality for the reduced linear combinations of order statistics. *Theory Probab. Appl.* 38, 78–84. (in Russian).

Helmers, R. (1982) *Edgeworth Expansions for Linear Combinations of Order Statistics.* Mathematical Centre Tracts 105, Amsterdam.

Korolyuk, V.S. and Borovskikh, Yu.V. (1993). *Random Permanents.* Inst. Mat. Akad. Nauk Ukrainy, Kiev. (in Russian).

van Zwet, W.R. (1964). *Convex Transformations of Random Variables.* Mathematical Centre Tracts 7, Amsterdam.

Exploring Stochastic Laws. pp. 135–146.
Eds. A.V. Skorokhod and Yu.V. Borovskikh

Fundamental identities for boundary functionals of additive sequences

D.V. GUSAK

National Academy of Sciences of Ukraine, Institute of Mathematics, Tereschenkivska St. 3, 252601 Kiev-4, Ukraine

Received January 25, 1995

Abstract. Let $\eta, \eta_1, \eta_2, \ldots$ and $\xi, \xi_1, \xi_2, \ldots$ be two independent sequences of independent identically distributed random variables; let $\eta > 0$ have a discrete distribution and

$$\nu_n = \max\{r : \sum\nolimits_{k\leq r} \eta_k \leq n\}.$$

For the additive sequence $\xi(n) = \sum_{k=0}^{\nu_n} \xi_k \quad (\xi_0 = 0)$ we establish some fundamental identities, from which the distributions of its boundary functionals (extreme values, the moment of the first achievement of the maximum, the moment and the value of the first overjump through a fixed level) can be expressed.

Important contributions to the investigation of the distributions of boundary functionals for sums of independent identically distributed random variables and homogeneous processes with independent increments were accomplished by A.A. Borovkov, V.S. Korolyuk, L. Takacs and others. The list of papers devoted to these boundary problems can be found in monographs: Borovkov A.A. (1972), Korolyuk V.S. *et al.* (1987), Bratijchuk N.S. and Gusak D.V. (1990).

A.A. Borovkov and B.A. Rogozin (1964) considered a two-dimensional random walk defined by sums:

$$\sigma_n = \sum\nolimits_{k=1}^{\infty} \eta_k \quad (\eta_k > 0,\ \nu(t) = \sup\{n : \sigma_n < t\}),$$

$$\xi(t) = \sum\nolimits_{k\leq\nu(t)} \xi_k = S_{\nu(t)},$$
$$S_n = \sum_{k\leq n} \xi_k.$$

They studied there the asymptotic behaviour of the joint distribution for

$$\{\sup\nolimits_{u\leq t} \xi(u),\ \xi(t),\ \sigma_{\nu(t)+1}\},$$

when $t \to \infty$. Gusak D.V. (1981, 1982) considered different boundary functionals for sums $\xi(t)$ (the number of addenda was defined by the number of renewals for the sequence of non-discrete independent identically distributed random variables).

The joint distribution of $\{\sup_{u \leq t} \xi(u), \xi(t)\}$ was studied by L. Takacs (1970).

In this paper we consider analogous sums with random addenda (continuously or discretely distributed), when the number of addenda is equal to the number of renewals for the sequence of integer-valued positive random variables. When ν_n has a binomial distribution, the sums S_{ν_n} form a homogeneous additive sequence, which is an analogue of a homogeneous process with independent increments.

The factorization approach to the investigation of boundary problems, which is based on stochastic relations for boundary functionals and on the results of M.G. Krein (1958) gives a possibility to prove the analogue of factorization relations (previously obtained for sums and homogeneous processes), in which the connection between the distributions of functionals for the considered additive sequence is established.

Let $\eta, \eta_1, \eta_2, \ldots; \xi, \xi_1, \xi_2, \ldots$ $(P\{\xi \neq 0\} = 1)$ be two independent sequences of independent identically distributed random variables with $\eta > 0$ geometrically distributed:

$$P\{\eta = k\} = (1-\rho)\rho^{k-1} \qquad (k \geq 1, \quad 0 < \rho < 1). \tag{1}$$

Then the sequence of sums

$$\xi(n) = \sum\nolimits_{k=0}^{\nu_n} \xi_k \quad (n \geq 0, \xi_0 = 0),$$

$$\nu_n = \max\{r : \sum\nolimits_{k \leq r} \eta_k \leq n\} \tag{2}$$

we shall call a homogeneous additive sequence.

It is easy to prove that the number of renewals ν_n for $\{\eta_k\}_{k \geq 1}$ has a binomial distribution

$$P\{\nu_n = k\} = C_n^k (1-\rho)^k \rho^{n-k} \quad (0 \leq k \leq n).$$

If ξ is continuously distributed ($\varphi(\alpha) = Ee^{i\alpha\xi}$), then $\xi(n)$ is also continuously distributed and has characteristic function (ch.f.)

$$\varphi_n(\alpha) = Ee^{i\alpha\xi(n)} = \varphi_1^n(\alpha),$$

$$\varphi_1(\alpha) = (1-\rho)\varphi(\alpha) + \rho.$$

If ξ are integer-valued random variables ($P\{\xi = 0\} = 0$, $p(u) = Eu^{\xi}$), then $\xi(n)$ has a generating function (g.f.)

$$P_n(u) = Eu^{\xi(n)} = P_1^n(u),$$

$$P_1(u) = (1-\rho)p(u) + \rho.$$

If we note that ε_t is a random variable with the geometrical distribution

$$P\{\varepsilon_t = k\} = (1-t)t^k (k \geq 0)$$

then correspondingly for continuous and discrete cases

$$Ee^{i\alpha\xi(\varepsilon_t)} = \frac{1-t}{1-t\varphi_1(\alpha)}, \tag{3}$$

$$Eu^{\xi(\varepsilon_t)} = \frac{1-t}{1-tp_1(u)}. \tag{4}$$

Let $\eta > 0$ have an arbitrary discrete distribution. Then ν_n is not binomialy distributed and its distribution is expressed by distributions of sums

$$\sigma_n = \sum\nolimits_{k\leq n} \eta_k \; (P\{\eta = k\} = g_k, G(n) = P\{\eta \leq n\}):$$

$$P\{\nu_n = 0\} = P\{\eta > n\} = \bar{G}(n),$$

$$g(t) = Et^\eta,$$

$$P\{\nu_n = k\} = \sum\nolimits_{r=k}^{n} P\{\sigma_k = r\}\bar{G}(n-r)(1 \leq k \leq n).$$

The sequence of sums

$$\xi(n) = \sum\nolimits_{k=0}^{\nu_n} \xi_k \qquad (\xi(0) = \xi_0 = 0, \quad n \geq 0)$$

is called an additive sequence. It is not homogeneous without assumption (1), ξ_k are called jumps of the sequence $\xi(n)$.

For the ch.f. $\varphi_n(\alpha) = Ee^{i\alpha\xi(n)}$ from the stochastic relation

$$\xi(n) \doteq \begin{cases} 0, & \eta > n, \\ \xi + \xi(n-\eta), & \eta \leq n, \end{cases}$$

(the equality $\xi_1 \doteq \xi_2$ means the identity of distributions for ξ_1 and ξ_2), the following equation is derived

$$\varphi_n(\alpha) = P\{\eta > n\} + \varphi(\alpha)\sum\nolimits_{k=1}^{n} \varphi_{n-k}(\alpha) g_k.$$

From here after a generating transformation we find out that for continuously distributed jumps

$$\varphi(t,\alpha) = Ee^{i\alpha\xi(\varepsilon_t)} = \frac{1-g(t)}{1-\varphi(\alpha)g(t)} \quad (0 < t < 1). \tag{5}$$

Similarly for the case of discretely distributed jumps it is proved that

$$p(t,u) = Eu^{\xi(\varepsilon_t)} = \frac{1-g(t)}{1-p(u)g(t)} \quad (0 < t < 1). \tag{6}$$

If $g(t) = \frac{(1-\rho)t}{1-\rho t}$, then after substituting it into (5) and (6) we obtain relations (3) and (4) for the homogeneous case.

If $P\{\eta = 1\} = 1$, then we obtain the ordinary sequence of independent identically distributed random variables

$$\xi(n) = S_n = \sum\nolimits_{k \leq n} \xi_n,$$

$$Eu^{\xi(n)} = (p(u))^n.$$

In the case, when $P\{\xi = 1\} = 1$, then

$$\xi(n) = \nu_n, \quad Eu^{\xi(n)} = [(1-\rho)u + \rho]^n.$$

We now consider the case of integer valued jumps.

We introduce the notations of basic boundary functionals, for which we establish fundamental identities, namely: for extreme values of $\xi(n)$, for the moment and the value of the first over jump through the fixed integer-valued level:

$$\begin{aligned}
\xi^+(m) &= \max\nolimits_{r \leq m} \xi(r); \\
\xi^-(m) &= \min\nolimits_{r \leq m} \xi(r); \\
\tau^+(n) &= \max\{r : \xi(r) > n\}, \\
\gamma^+(n) &= \xi(\tau^+(n)) - n; \\
\bar{\tau}^+(n) &= \max\{r : \xi(r) \geq n\}, \\
\bar{\gamma}^+(n) &= \xi(\bar{\tau}^+(n)) - n; \\
T_+(n) &= \inf\{k : \xi(k) = \xi^+(n)\}.
\end{aligned}$$

The first identity is established for the distributions of the extreme and with the help of it the connection between the generating functions

$$P(t,u) = Eu^{\xi(\varepsilon_t)}, \quad P_\pm(t,u) = Eu^{\xi^\pm(\varepsilon_t)} \tag{7}$$

is expressed.

The second identity is proved for the joint distribution of $\{\tau^+(n), \gamma^+(n)\}$ and with the help of it it is established that the generating transformation of this distribution can be expressed in terms of $P_+(t,u)$.

Both these identities are corresponding analogues of identities (Borovkov A.A. (1970)) for the sums of independent identically distributed random variables and homogeneous processes and for the processes defined by Gusak D.V. (1981, 1982).

Let us denote the denominator in (6)

$$k(t,u) = 1 - p(u)g(t), \quad g(t) = Et^\eta;$$

$$p(u) = \sum\nolimits_{r\neq 0} u^r p_r = Eu^{\xi}(p_r = P\{\xi = r\}).$$

Later for the g.f. of a sequence $\{q_r\}_{r=-\infty}^{\infty}$

$$q(u) = \sum\nolimits_r u^r q_r, \quad |u| = 1, \quad \sum\nolimits_r |q_r| < \infty,$$

we will use the notation of projective operations

$$q_+^0(u) = [q(u)]_+^0 = \sum\nolimits_{r\geq 0} u^r q_r \quad (|u| \leq 1)$$

$$q_-(u) = [q(u)]_- = \sum\nolimits_{r<0} u^r q_r \quad (|u| \geq 1)$$

$$q_+^0(u) + q_-(u) = q(u) \quad (|u| = 1)$$

and Φ_c is the class of g.f. $\{q(u)\}$ with $q_0 = c \neq 0$.

$$\Phi_+^0 = \{q_+^0(u)\}; \quad \Phi_- = \{q_-(u)\}.$$

Lemma 1. The function $k_*(t,u) = \dfrac{k(t,u)}{k(t,1)} \in \Phi_c$ allows the factorization

$$k_*(t,u) = k_*^+(t,u)k_*^-(t,u), \quad |u| = 1, \quad (0 < t < 1) \tag{8}$$

where $k_*^{\pm}(t,y) \in \Phi_{\pm}^0$ are analytic for $|u| < 1$ $(|u| > 1)$ and

$$k_*^{\pm}(t,u) = \exp\{\sum\nolimits_{n=1}^{\infty} \frac{1}{n} E[t^{\sigma_n}(1 - u^{S_n}), \ \pm S_n > 0]\}. \tag{9}$$

Proof of Lemma 1 follows from the relations

$$\begin{aligned} k(t,u) &= \exp\{ln(1 - p(u)g(t))\} \\ &= \exp\{-\sum\nolimits_{n=1}^{\infty} \frac{1}{n}(p(u)g(t))^n\} \\ &= k_+(t,u)k_0(t)k - (t,u)(|u| = 1) \end{aligned}$$

where

$$k_0(t) = \exp\{-\sum\nolimits_{n>1} \frac{1}{n} E[t^{\sigma_n}, S_n = 0]\},$$

$$k_{\pm}(t,u) = \exp\{-\sum\nolimits_{n\geq 1} \frac{1}{n} E[u^{S_n} t^{\sigma_n}, \pm S_n > 0]\}.$$

Theorem 1. *The g.f. $p(t,u)$ allows the factorization*

$$P(t,u) = P_+(t,u)P_-(t,u) \quad (|u| = 1) \tag{10}$$

$$P_{\pm}(t,u) = Eu^{\xi^{\pm}(e_t)} = [k_*^{\pm}(t,u)]^{-1}(|u|^{\pm 1} < 1). \tag{11}$$

Proof. From the stochastic relation

$$\tau^+(n) \doteq \begin{cases} \eta, & \xi > n, \\ \eta + \tau^+(n-\xi), & \xi \le n, \end{cases}$$

it is easy to derive the equation for the g.f.

$$B(n,u) = E[u^{\tau^+(n)},\ \tau^+(n) < \infty],$$

$$B(n,u) = \bar{G}(n)g(u) + \sum\nolimits_r P\{\tau^+(n) = r\}B(n-r,u) \quad (n > 0). \tag{12}$$

With the help of generating transformations equation (12) is transformed to the following one

$$P_+(t,u)(1 - p(u)g(t)) = g_-^0(t,u), \quad g_-^0 \in \Phi_-^0.$$

By adjusting the results of M.G. Krein (1958) to probabilistic boundary problems and using the factorization (8), we find the solution of the last equation

$$P_+(t,u) = \frac{k_+(t,1)}{k_+(t,u)} = [k_*^+(t,u)]^{-1} \in \Phi_+^0.$$

Similarly the equation for $P_-(t,u)$ is derived and its solution has an analogous form

$$P_-(t,u) = [k_*^-(t,u)]^{-1} \in \Phi_-^0.$$

Taking into account (11) and the decomposition (8), we can easily obtain relation (10), which is called the first factorization identity.

In the homogeneous case (see (1)), $g(t) = \dfrac{(1-\rho)t}{1-\rho t}$ the factorization for

$$k(t,u) = \exp\{-\sum\nolimits_{n\ge 1}\frac{1}{n}(tP_1(u))^n\}$$

is simplified ($P_1(u) = Eu^{\xi_k^1}$) and it is reduced to the factorization problem for sums $S_n^1 = \sum_{k\le n}\xi_k^1 (n \ge 1)$. Then the components of factorization (9) have the form

$$k_*^\pm(t,u) = \exp\{\sum\nolimits_{n\ge 1}\frac{t^n}{n}E[(1-u^{S_n^1}), \pm S_n^1 > 0]\} \tag{13}$$

and the components $P_\pm(t,u)$ of the factorization (10) are expressed by means of (13), where the dependence $k_*^\pm(t,u)$ on the distribution η is reduced to the dependence on one parameter ρ, which has a probabilistic sense

$$\xi_k^1 = \begin{cases} \xi_k & \text{with probability } 1-\rho, \\ 0 & \text{with probability } \rho. \end{cases}$$

To consider the second factorization identity we must remark that for integer-valued random walks and processes we need to consider two versions. The first version is established for the pair $\{\bar\tau^+(n),\bar\gamma^+(n)\}$ and the second version — for $\{\tau^+(n),\gamma^+(n)\}$.

If the condition

$$P\{\xi \geq m\} > 0 \text{ for } m \geq 1, \quad E|\xi(1)| < \infty, \tag{14}$$

is accomplished then for $\bar\tau^+(n+m)$ $(n \geq 1, m \geq 1)$ on the set $\{\omega : \bar\tau_{n+m} < \infty\}$ the following stochastic relation is true

$$\bar\tau^+ \doteq \begin{cases} \bar\tau(n), & \bar\gamma^+(n) \geq m, \\ \tau^+(n) + \tau^+(m - \bar\gamma^+(n)), & \bar\gamma^+(n) < m. \end{cases} \tag{15}$$

Taking into account, that

$$E[u^{\bar\tau^+(n)}, \bar\tau^+(n) < \infty] = P\{\xi^+(\varepsilon_u) \geq n\}$$

on the basis of (15) we derive the equation

$$\begin{aligned} P\{\xi^+(\varepsilon_u) \geq m+n\} &= E[u^{\bar\tau^+(n)}, \tau^+(n) < \infty] - \\ &- E[u^{\bar\tau^+(n)}, \bar\gamma^+(n) < m, \bar\tau^+(n) < \infty] + \\ &+ E[u^{\tau^+(n)+\bar\tau^+(m-\gamma^+(n))}, \quad \gamma^+(n) < m]. \end{aligned} \tag{16}$$

Theorem 2. *Under condition* (14) *the joint distribution of* $\{\bar\tau^+(n),\bar\gamma^+(n)\}$ *is defined by the relations*

$$\begin{aligned} &E[u^{\bar\tau^+(n)} t^{\bar\gamma^+(n)}, \bar\tau^+(n) < \infty] = \\ &= E[t^{\xi^+(\varepsilon_u)-n}, \xi^+(\varepsilon_u) \geq n](Et^{\xi^+(\varepsilon_u)})^{-1}, \end{aligned} \tag{17}$$

$$\begin{aligned} &E[u^{\bar\tau^+(\varepsilon_v)} t^{\bar\gamma^+(\varepsilon_v)}, \bar\tau^+(\varepsilon_v) < \infty] = \\ &= \frac{1-v}{t-v}[tEt^{\xi^+(\varepsilon_u)} - vEv^{\xi^+(\varepsilon_u)}](Et^{\xi^+(\varepsilon_u)})^{-1}; \end{aligned} \tag{18}$$

$$E[u^{\bar\tau^+(\varepsilon_v)}, \bar\tau^+(\varepsilon_v) < \infty] = 1 - vEv^{\xi^+(\varepsilon_u)}. \tag{19}$$

Proof. For the simplification of the following transformation we omit the condition $\{\tau^+(.) < \infty\}$. Then equation (16) can be rewritten in the form

$$\begin{aligned} P\{n \leq \xi^+(\varepsilon_u) < n+m\} &= E[u^{\bar\tau^+(n)}, \bar\gamma^+(n) < m] - \\ &- E[u^{\bar\tau^+(n)+\tau^+(n-\bar\gamma^+(n))}, \bar\gamma^+(n) < m]. \end{aligned}$$

From here it follows that

$$P\{0 \le \xi^+(\varepsilon_u) - n < m\} =$$

$$= \sum\nolimits_{r<m} E[u^{\bar\tau^+(n)}, \bar\gamma^+(n) = r] P\{\xi^+(\varepsilon_u) < m - r\}$$

and after generating transformation with respect to m we find that

$$E[t^{\xi^+(\varepsilon_u)-n}, \xi^+(\varepsilon_u) \ge n] = E[u^{\bar\tau^+(n)} t^{\bar\gamma^+(n)}] E t^{\xi^+(\varepsilon_u)}.$$

Therefore formula (17) is true.

By generating transformation with respect to n the last relation reduces to the equation

$$\frac{1}{t-v}[tEt^{\xi^+(\varepsilon_u)} - vEv^{\xi^+(\varepsilon_u)}] = \frac{1}{1-v} Et^{\xi^+(\varepsilon_v)} E[u^{\bar\tau^+(\varepsilon_v)} t^{\bar\gamma^+(\varepsilon_v)}],$$

from which formula (18) follows. For $t = 1$ we obtain the relation (19).

If $P\{\xi_k = 1\} = p_1 > 0$, $P\{\xi_k \le 1\} = 1$, then the sequence $\xi(n)$ is called semi-continuous. In this case $\bar\gamma^+(n) \equiv 0$, relation (15) has a meaning only for $m = 1$ and the generating functions for $\bar\tau^+(n)$ and for $\xi^+(\varepsilon_u)$ can be presented in terms of probabilities

$$P^0_+(u) = P\{\xi^+(\varepsilon_u) = 0\},$$

$$\bar P^0_+(u) = 1 - P_+(u).$$

If $E\xi_1 < 0$, then

$$\lim\nolimits_{u\to 1} P^0_+(u) = P^0_+(1) = P\{\xi^+ = 0\} < 1.$$

If $E\xi_1 \ge 0$, then

$$\lim\nolimits_{u\to 1} \bar P^0_+(u) = P\{\xi^+ > 0\} = 1.$$

Corollary 1. *For the semi-continuous case the following relations are true*

$$E[u^{\bar\tau^+(\varepsilon_v)}, \bar\tau^+(\varepsilon_v) < \infty] = \frac{1-v}{1 - v\bar P^0_+(u)}, \tag{20}$$

$$E[v^{\xi^+(\varepsilon_u)}] = \frac{P^0_+(u)}{1 - v\bar P^0_+(u)}.$$

After substituting (20) into (18) we obtain the relation

$$Eu^{\bar\tau^+(\varepsilon_v)} t^{\bar\gamma^+(\varepsilon_v)} = \frac{1-v}{1 - v\bar P^0_+(u)}.$$

This means that $\bar\gamma^+(\varepsilon_v) = 0$ with probability 1.

For $\{\tau^+(n), \gamma^+(n)\}$ a similar assertion can be proved.

Theorem 3. *Under condition* (14)

$$E[u^{\tau^+(n)}t^{\gamma^+(n)}, \tau^+(n) < \infty] = E[t^{\xi^+(\varepsilon_u)-n}, \xi^+(\varepsilon_u) > n](Et^{\xi^+(\varepsilon_u)})^{-1}, \tag{21}$$

$$E[u^{\tau^+(\varepsilon_v)}t^{\gamma^+(\varepsilon_v)}, \tau^+(\varepsilon_v) < \infty] = \frac{t(1-v)}{t-v}\times$$
$$\times[Et^{\xi^+(\varepsilon_u)} - Ev^{\xi^+(\varepsilon_u)}](Et^{\xi^+(\varepsilon_u)})^{-1}. \tag{22}$$

Proof is based on the stochastic representation

$$\tau^+(n+m) \doteq \begin{cases} \tau^+(n), & \gamma^+(n) > m, \\ \tau^+(n) + \tau^+(m - \gamma^+(n)), & \gamma^+(n) \le m. \end{cases}$$

After substituting (20) into (22) for the semi-continuous case we have the relation

$$E[u^{\tau^+(\varepsilon_v)}t^{\gamma^+(\varepsilon_v)}, \tau^+(\varepsilon_v) < \infty] = tE[u^{\tau^+(\varepsilon_v)}, \tau^+(\varepsilon_v) < \infty],$$

which means that $\gamma^+(\varepsilon_v) = 1$ with probability 1.

According to the terminology of Borovkov A.A. (1970) we shall call relation (22) the *second factorization identity.* Really for discretely distributed jumps ξ_k we have two versions: relation (18) for $\{\tau(\cdot), \gamma(\cdot)\}$ and (22) for $\{\tau(\cdot), \gamma(\cdot)\}$. If ξ_k are continuously distributed, then $\bar{\gamma}(\cdot) = \gamma(\cdot)$.

For the homogeneous case according to (3) and (4), $\xi(n) = S_n^1 = \sum_{k\le n}\xi_k^1$. Correspondingly the components of the first factorization identity $P_\pm(t,u)$ are expressed in terms of $k_*^\perp(t,u)$ (see (13)), and respectively the joint distributions of $\{\bar{\tau}^+(\cdot), \bar{\gamma}^+(\cdot)\}$ and $\{\tau^+(\cdot), \gamma^+\}$ are expressed in terms of $p_+(t,u)$. Thus, in the homogeneous case we obtain previously known results for ordinary random walks (see Borovkov A.A., 1970).

As for the two-dimensional scheme of independent random variables which is described by sums: $S_n = \sum_{k\le n}\xi_k$, $S'_n = \sum_{k\le n}\eta_k$, we are restricted by the consideration of pairs:

$$\{\xi(n) = S_{\nu_n};\ \eta(n) = S'_{\nu_n}\},$$

$$\{\xi(n);\ \eta_*(n) = S'_{\nu_n+1}\}.$$

Remark that $\eta(n)$ defines the value of sum S' up to the crossing of the level n; $\eta_*(n)$ — the values of sum S' after this crossing.

On the base of the stochastic representation of $\xi(n)$, $\eta(n)$, $\eta_*(n)$, particularly

$$\eta(n) \doteq \begin{cases} 0, & \eta > n, \\ \eta + \eta(n-\eta), & \eta \le n, \end{cases}$$

$$\eta_*(n) \doteq \begin{cases} 0, & \eta > n, \\ \eta + \eta_*(n-\eta), & \eta \le n, \end{cases}$$

the following assertion is established.

Theorem 4. *The g.f. for $\{\xi(\varepsilon_t), \eta(\varepsilon_t)\}$ and $\{\xi(\varepsilon_t), \eta_*(\varepsilon_t)\}$ are defined by the relations*

$$Eu^{\xi(\varepsilon_t)}v^{\eta(\varepsilon_t)} = \frac{1-g(t)}{1-g(vt)p(u)}, \tag{23}$$

$$Eu^{\xi(\varepsilon_t)}v^{\eta_*(\varepsilon_t)} = \frac{g(v)-g(vt)}{1-g(vt)p(u)};$$

in the homogeneous case

$$Eu^{\xi(\varepsilon_t)}v^{\eta(\varepsilon_t)} = \frac{1-t}{1-t\rho}\frac{1-vt\rho}{1-vtp_1(u)},$$

$$Eu^{\xi(\varepsilon_t)}v^{\eta_*(\varepsilon_t)} = \frac{v(1-\rho)(1-t)}{(1-\rho v)(1-vtp_1(u))}.$$

For the investigation of the joint distribution of triples:

$$\{\xi^+(n), \xi(n); \eta(n)\},$$
$$\{\xi^+(n), \xi(n); \eta_*(n)\}$$

it is necessary to derive equations for their g.f. These equation have a common kernel function

$$k(t,u,v) = 1 - g(tv)p(u).$$

To factorize it with respect to u we introduce

$$k_*(t,u,v) = \frac{1-g(tv)p(u)}{1-g(tv)}$$

and then we can use factorization (8) with factors

$$k_*^{\pm}(t,u,v) = k_*^{\pm}(tv,u),$$

represented in (9).

The following assertion shows that the required g.f. for these triples can be represented in terms of $p_{\pm}(tv,u)$.

Theorem 5. *Under condition $E|\xi(1)| < \infty$*

$$Eu^{\bar{\xi}(\varepsilon_t)}z^{\xi^+(\varepsilon_t)}v^{\eta(\varepsilon_t)} = Ev^{\eta(\varepsilon_t)}Eu^{\bar{\xi}(\varepsilon_{vt})}Ez^{\xi^+(\varepsilon_{vt})}. \tag{24}$$

Proof is based on the equation for ($x < 0,\ r > 0$):

$$A_n(r,x,v) = E[v^{\eta(n)}, \xi(n) < x + r, \xi^+(n) < r],$$

which is derived from the stochastic relations for $\{\eta(n),\ \xi(n),\ \tau^+(r)\}$.

On the basis of the stochastic relations

$$\xi(n) \doteq \gamma^+(0) + \xi(n - \tau^+(0)),$$
$$\xi^+(n) \doteq \gamma^+(0) + \xi^+(n - \tau^+(0)),$$

and

$$T_+(n) = \tau^+(0) + T_+(n - \tau^+(0))$$

it is easy to prove the following assertion.

Theorem 6. *The joint g.f. for $[T(\varepsilon_t), \xi(\varepsilon_t), \bar{\xi}(\varepsilon_t)]$ is defined by the relation*

$$\begin{aligned} T_t(u,v,w) &= E[u^{T(\varepsilon_t)} v^{\xi(\varepsilon_t)} w^{\bar{\xi}(\varepsilon_t)}] \\ &= Ev^{\xi^+(\varepsilon_{tu})} E(vw)^{\bar{\xi}(\varepsilon_t)} \cdot \frac{P\{\xi^+(\varepsilon_t) = 0\}}{P\{\xi^+(\varepsilon_{tu}) = 0\}}, \end{aligned} \tag{25}$$

$$Eu^{T(\varepsilon_t)} v^{\xi^+(\varepsilon_t)} = \frac{P\{\xi^+(\varepsilon_t) = 0\}}{P\{\xi^+(\varepsilon_{tu}) = 0\}} Ev^{\xi^+(\varepsilon_{tu})}. \tag{26}$$

Under the condition:

$$E|\xi(1)| < \infty, \quad E\xi < 0, \tag{27}$$

the non-degenerate distribution of

$$\xi^+ = \sup_{n<\infty} \xi(n)$$

exists and

$$\xi^+ = S^+ = \sup_{n<\infty} S_n.$$

Under condition (27) *the limit distribution for $\{T(\infty), \xi^+\}$ is defined by the limit g.f.*

$$Eu^{T(\infty)} v^{\xi +} = \frac{P\{\xi^+ = 0\}}{P\{\xi^+(\varepsilon_u) = 0\}} Ev^{\xi^+}(\varepsilon_u). \tag{28}$$

It should be mentioned that the property of monotonicity of functionals $T(t)$, $\xi^+(t)$ provides the existence of their limit g.f.. Therefore we can use the limit passage $t \to 1$ in (26), after which we obtain (28).

Under condition (27) the following limit relation follows from (22)

$$\frac{1}{1-v}E\left[u^{\tau^+(\varepsilon_v)}t^{\gamma^+(\varepsilon_v)}\tau^+(\varepsilon_v)<\infty\right]=$$

$$=\frac{1}{1-v}E\left[t^{\gamma^+(\varepsilon_v)}(1-v^{\xi^+(\varepsilon_u)})\right]\overset{v\to 1}{\to}E\left[\xi^+(\varepsilon_u)t^{\gamma^+(\infty)}\right]=$$

$$=\frac{t}{t-1}\left[Et^{\xi^+(\varepsilon_u)}-1\right]\left(Et^{\xi^+(\varepsilon_u)}\right)^{-1}.$$

Under $u\to 1$ we obtain the relation

$$E\left[\xi^+t^{\gamma^+(\infty)}\right]=\frac{t}{t-1}\left(Et^{\xi^+}-1\right)\left(Et^{\xi^+}\right)^{-1}.$$

References

Borovkov, A.A. (1972). *Random processes in queueing theory.* Moscow. Nauka. (In Russian).

Borovkov, A.A. and Rogozin, B.A. (1964). Boundary problems for some two-dimensional random walks. *Theor. Probab. Appl.* **9**(3), 401–430.

Bratijchuk, N.S. and Gusak, D.V. (1990). *Boundary problems for processes with independent increments.* Kiev. Naukova Dumka. (In Russian).

Gusak, D.V. (1981). Bondary functionals for sums of random number of summands. *Analytical Methods for the Investigation in Probability Theory.* Kiev. Institute of mathematics of Ukr. Ac. Sci. pp. 20–35. (In Russian).

Gusak, D.V. (1982). Factorization identities for sums with a random number of summands. *Applied Problems of Probability Theory.* Kiev. Institute of mathematics of Ukr. Ac. Sci. pp. 25–44. (In Russian).

Korolyuk V.S., Bratijchuk N.S. and Pirdjanov, B. (1987). *Boundary problems for random walks.* Ashkhabad. Ylym. (In Russian).

Krein, M.G. (1958). Integral equations on a semi-axis with the kernel depending on difference of arguments. *Uspekhi Mat. Nauk.,* **13**(5), 3–120. (In Russian).

Takacs L. (1970). On the distribution of the supremum for stochastic processes. *Ann. Inst. H. Poincare.* **B6**(3), 237–247.

Exploring Stochastic Laws. pp. 147–152.
Eds. A.V. Skorokhod and Yu.V. Borovskikh

A note on stochastic variational equations

TAKEYUKI HIDA

Department of Mathematics, Meijo University, Nagoya 468, Japan

Received June 5, 1995

Abstract. As a mathematical model of random phenomena in a fluctuating environment we propose a random field $X(C)$ which is a generalized white noise functional of a particular type depending on a manifold C. It is expressed as the so-called Hitsuda–Skorohod integral and its variational equation can be discussed within the framework of white noise analysis.

Introduction

The purpose of this paper is to propose a variational calculus for a certain class of random fields which are simplified mathematical models of random phenomena in a fluctuating environment.

To describe propagation of these phenomena extended in a space as time continues, the parameter set is taken to be space-time, namely a multi dimensional space, say R^d or a domain in R^d.

For the fluctuation we may take a basic mathematical object: it is an independent random variable at distinct points in space-time, elementary and hopefully Gaussian in distribution. Such a system does exist in probability theory, that is *white noise*. It is really a system of *idealized elementary random variables* (these words were born after a conversation with John R. Klauder).

Our theory starts therefore with functionals of white noise describing the phenomena. They are expressed in terms of the Hitsuda–Skorohod integrals and shall be dealt within the framework of white noise analysis (see Section 2).

Since the functionals in question stand for propagation or development of the phenomena, they may be expressed as $X(C)$, C being a manifold, and changes as C deforms. It is therefore necessary to specify the class of C, as well as the way of the deformation of C. This can be done consistently in Section 3.

Having reviewed the notions of the creation and annihilation operators, the Hitsuda–Skorohod integral is defined to have $X(C)$ in Section 2. It should be noted that the "non-anticipating property" of the integrand is meaningless in the present case.

With a suitable choice of integrand we can see a particular form of the variation $\delta X(C)$, where infinitesimal deformation δC is specified. We can finally get a characteristic property of the variation as the main Theorem. An example of Langevin type illustrates how $X(C)$ shows the actual phenomena.

1. Background

As a system of idealized elementary random variables, we introduce white noise with parameter set R^d. Take the basic nuclear space E to be the Schwartz space $\mathcal{S}(R^d)$ and to have a Gel'fand triple

$$E \subset L^2(R^d) \subset E^*, \tag{1.1}$$

E^* being the space of tempered distributions.

White noise is defined to be a probability measure space (E^*, μ) such that the characteristic functional $C(\xi)$, $\xi \in E$, of μ is given by

$$\begin{aligned} C(\xi) &= \exp\left[-\frac{1}{2}\|\xi\|^2\right], \quad \|\cdot\| \text{ being } L^2(R^d)\text{-norm}, \\ &= \int_{E^*} \exp[i\langle x, \xi\rangle] d\mu(x), \end{aligned} \tag{1.2}$$

where $\langle\cdot,\cdot\rangle$ is the canonical bilinear form connecting E and E^*.

The measure μ is a standard Gaussian, and with this measure each $x(t)$, $x \in E^*$, may be thought of as a mathematical expression of fluctuation at a point $t \in R^d$.

The complex Hilbert space $L^2(E^*, \mu)$ is denoted simply by (L^2). We have a Fock space:

$$(L^2) = \bigoplus H_n. \tag{1.3}$$

Let A be an operator

$$A = -\Delta + |u|^2 + 1, \quad \Delta: \text{ Laplacian}, \quad u \in R^d,$$

acting on $L^2(R^d)$. By using the second quantization $\Gamma(A)$ of A, Hilbertian norms are introduced

$$\|\varphi\|_p = \|\Gamma(A^p)\varphi\|, \quad \varphi \in D(\Gamma(A^p)), \ p \in \mathbf{Z}. \tag{1.4}$$

With the help of these countable Hilbertian norms, we are given a σ-Hilbert nuclear space $(\mathcal{S})$ and further given another Gel'fand triple

$$(\mathcal{S}) \subset (L^2) \subset (\mathcal{S})^*. \tag{1.5}$$

A member of $(\mathcal{S})^*$ is called a *generalized white noise functional.*

We then introduce one of the basic tools of white noise analysis, namely the S-transform:

$$S : \varphi \longrightarrow (S\varphi)(\xi) = \int \varphi(x+\xi)d\mu(x), \quad \varphi \in (\mathcal{S})^*.$$

$(S\varphi)(\xi)$ is often called the U-functional associated with φ.

Theorem 1. (Potthoff–Streit). *A functional $F(\xi)$ is a U-functional associated with some $\varphi \in (\mathcal{S})^*$ if and only if*

1) *for any fixed ξ and η, $F(\lambda) = F(\xi + \lambda\eta)$, λ real, extends to an entire function $F(z)$, $z \in C$;*

2) *$f(z)$ is of exponential type of order 2.*

See [3] for details.

We now remember a formal meaning of $x(t)$, the value of x evaluated at t. It is, as it were, one of the variables of white noise functional $\varphi(x) = \varphi(x(t), t \in R^d)$. Such an understanding leads us to define the differential operator ∂_t :

$$\partial_t \varphi(x) = S^{-1} \frac{\delta}{\delta\xi(t)} (S\varphi)(\xi),$$

where $\frac{\delta}{\delta\xi(t)}$ is the Fréchet derivative.

∂_t acts as an annihilation operator on $(\mathcal{S})$, while its adjoint, denoted by ∂_t^*, acts on $(\mathcal{S})^*$ as a creation operator. The S-transform of $\partial_t^* \varphi$ is expressed as $\xi(t)(S\varphi)(\xi)$. The commutation relations are

$$[\partial_t, \partial_s] = [\partial_t^*, \partial_s^*] = 0, \tag{1.6}$$

$$[\partial_t, \partial_s^*] = \delta(t-s), \quad t, s \in R^d. \tag{1.7}$$

For more details see [3].

2. Hitsuda–Skorohod integral

We are now ready to define the Hitsuda–Skorohod integral in some particular cases in a slightly modified manner, so as to fit into our purposes.

Let D be a domain or surface in R^d, and let $F(u, x)$, $u \in \overline{D}$, be a family of $(\mathcal{S})^*$-functionals such that $S(F(u, x))(\xi)$ is continuous in u. Take $f \in E$. Then we can define

$$X = \int_D F(u, x) \partial_u^* f(u)\, du, \tag{2.1}$$

where du is the volume element or surface element dependent on D.

Proposition 1. *The integral* (2.1) *is well-defined and* X *is a member of* $(\mathcal{S})^*$.

Proof. By assumption, the S-transform of the integrand is $\xi(t)f(u)S(F(u,x))(\xi)$, since $F(u,x)$ itself is expressed in terms of the products of ∂_u^*'s. Applying the conditions in Theorem 1 and using integration we can see that the result still satisfies the conditions of Theorem 1.

Our interest is focused on the case when the fluctuation comes from the domain (C) enclosed by a surface C. Thus, the integrand of (2.1) is specified to be

$$F(C,u,x)\partial_u^* f(u). \tag{2.2}$$

Again, whilst examining the actual problems, causality is taken into account. It means that if $\mathbf{B}(C)$ denotes the σ-field with respect to which all the $\langle x, \xi\rangle$ with $\text{supp}(\xi) \subset (C)$ is measurable, then $F(C,u,x)$ is $\mathbf{B}(C)$-measurable for any $u \in (C)$.

Now the $X(C)$ is considered as a mathematical model, indeed a white noise model of random phenomena that we are going to discuss.

3. Variational equation for $X(C)$

To make the description simpler we assume $d = 2$ in the following.

Propagation of $X(C)$ is made by deformation of C. We therefore have to introduce an admissible class of C's. Set

$$\mathbf{C} = \{C : C \text{ is a contour diffeomorphic to } S^1\}.$$

We often take a sub-class and still use the same symbol. It is topologized (in a sense) in such a way that the deformation from C to $C + \delta C$ is measured by the normal distance $\delta n(s)$ at $s \in C$. The C^∞- topology can be introduced to $\{\delta n(s)\}$.

Another concept to be introduced is how to deform C. Let G be a topological group consisting of certain diffeomorphisms of $R^2 \cup \{\infty\}$. We usually assume that the two dimensional conformal group (shifts, isotropic dilation, $SO(2)$ and special conformal transformations) is a subgroup of G.

The above two concepts are in the relation

$$\mathbf{C} \text{ is invariant under } G. \tag{3.1}$$

Coming back to the integrand (2.2), the random field $X(C)$, $C \in \mathbf{C}$ is now given by

$$X(C) = \int_{(C)} F(C,u,x)\partial_u^* f(u)du. \tag{3.2}$$

The assumptions for F are as follows.

1) The S-transform of F, denoted by $U(C,u,\xi)$ is continuous in u.

2) The derivative $\frac{\delta}{\delta C}U(C,u,\xi)(s)$, $s \in C$ exists and is a U-functional.

3) $F(C,u,x)$ is $\mathbf{B}(C)$-measurable.

With these assumptions, we can easily prove

Proposition 2. *Let $X(C)$ be given by (3.2). Then, we have*

$$\delta X(C) = \int_{(C)} \delta F(C,u,x)\partial_u^* f(u)\, du + \int_C F(C,s,x)\partial_s^* f(s)\delta n(s) ds. \tag{3.3}$$

By assumption, δF can be expressed in the form

$$\delta F(C,u,x) = \int_C F'(C,u,x)(s)\delta n(s)\, ds, \tag{3.4}$$

$F'(C,u,x)(s)$ being $\frac{\delta F(C,u,x)}{\delta C}(s)$.

Hence (3.3) is expressible as a Volterra form

$$Y = \int_C G(C,u,x,s)\delta n(s)\, ds, \tag{3.5}$$

where G is $\mathbf{B}(C)$-measurable.

An interesting question is, "Given an expresion of the form (3.5), is it a variation of some random field indexed by C?" To answer this question we need to prove a Theorem. Set

$$T_n(C) = \begin{cases} \frac{1}{n!}\partial_{t_1}\cdots\partial_{t_n}, & t_1,\ldots,t_n \in (C), \quad n \geq 1, \\ 0, & \text{otherwise.} \end{cases}$$

Theorem 2. *Let integral (3.5) be given. If*

$$\sum_n \int\cdots\int \partial_{t_1}^*\cdots\partial_{t_n}^*\langle T_n(C)G, 1\rangle_\mu\, dt^n$$

is defined in $(\mathcal{S})^*$ and is a derivative in C of some $\mathbf{B}(C)$-measurable functional $F(C,u,x)$, then Y is a variation of a random field $X(C)$ expressed in the form (3.2).

Remark. The operator ∂_t stands for a sudden random action, intuitively speaking, so it is reasonable to use it in a characterization of $\delta X(C)$ as in Theorem 2.

Proof of the Theorem. Operators $T_n(C)$ guarantee the Fock space expansion within the class of $\mathbf{B}(C)$-measurable functionals, although the domain is limited, while the operator annihilates $F(C, s, x)\partial_s^* f(s)$ for $s \in C$. Thus, it is possible to separate the integral of G in (3.5) into two terms as in (3.3) and eventually $F\hat{\otimes}f$ is unique. With these notes the Theorem is proved.

Example. Variational equation of Langevin type ([4]). Set

$$X(C) = X_0 \int_{(C)} e^{-\rho(C,u)g(u)} \partial_u^* f(u)\, du, \quad (C) \supset (C_0)$$

where ρ denotes the distance and $X_0 = X(C_0)$. Then

$$\delta X(C) = -X(C) \int_C g(s)\delta n(s) ds + X_0 \int_C f(s) \partial_s^* \delta n(s) ds.$$

Acknowledgement. The author is grateful to Professor K. Saitô for his help on the final version of this paper.

References

1. **Lévy, P.** (1951). *Problèmes concret d'analyse fonctionnalle.* Gauthier–Villars.

2. **Hida, T.** (1980). *Brownian motion.* Springer–Verlag. Russian translation (1986). Nauka.

3. **Hida, T., Kuo, H.-H., Potthoff, J and Streit, L.** (1993). *White Noise, An Infinite Dimensional Calculus.* Kluwer Academic Publishers.

4. **Hida, T.** (1995). *White noise analysis, An overview and some future directions.* Working paper MS-1, IIAS (International Inst. for Advanced Studies) Kizu-cho, Kyoto, 619-02, Japan.

Exploring Stochastic Laws. pp. 153–165.
Eds. A.V. Skorokhod and Yu.V. Borovskikh

Summability method in some ergodic type theorems

CVETAN JARDAS[1] AND NIKOLA SARAPA[2]

[1] *Department of Mathematics, University of Rijeka, 51000 Rijeka, Omladinska 14, Croatia*
[2] *Department of Mathematics, University of Zagreb, 41000 Zagreb, Bijenicka 30, Croatia*

Received March 20, 1995

Abstract. In this paper we generalize some well known classical ergodic type theorems (ergodic theorems due to Yoshida (see [12], p.213), Birkhoff's pointwise ergodic Theorem (see [2]) and others). Instead of the usual Cesàro $(c, 1)$ summability of a sequence of iterates $(T^n,\ n \in \mathbf{N})$, where T is a linear operator with some additional properties, we consider the problem of the summability of this sequence by strongly regular matrices. As a consequence of our results we obtain a generalization of Kolmogorov's strong law of large numbers for independent identically distributed random variables.

Introduction and preliminaries

For the sake of completeness, following [4], [8] and [5] chapter 8.4 we recall some definitions and known results from summability theory and from classical ergodic theory.

Let $A = [a_{nj}]$ $(n, j \in \mathbf{N})$ be a complex infinite matrix. We say that A is *regular* or a Toeplitz matrix if it transforms convergent sequences into other convergent sequences and if it preserves limits. In other words A is a regular matrix if

$$A\text{-}\lim(x) = \underset{n\to\infty}{A\text{-}\lim}\, x_n = \lim_{n\to\infty} \sum_{j=1}^{\infty} a_{nj} x_j = \lim_{n\to\infty} x_n,$$

for every convergent sequence $x = (x_n,\ n \in \mathbf{N})$ of complex numbers. By the well known Silverman–Toeplitz Theorem we have that a matrix A is regular if and only if the following three conditions are satisfied:

$$M = \sup_{n\in\mathbf{N}} \sum_{j=1}^{\infty} |a_{nj}| < \infty \tag{A}$$

$$\lim_{n\to\infty} a_{nj} = 0, \quad \text{for every} \quad j \in \mathbf{N}. \tag{B}$$

$$\lim_{n\to\infty}\sum_{j=1}^{\infty} a_{nj} = 1. \tag{C}$$

Definition 1. A matrix A is called *strongly regular* if it is regular and if in addition it satisfies the following condition

$$\lim_{n\to\infty}\sum_{j=1}^{\infty} |a_{nj} - a_{n,j+1}| = 0. \tag{D}$$

As an example of a stochastic strongly regular matrix we have Cesàro $(c, 1)$ summability matrix $C = [a_{nj}]$ defined by

$$a_{nj} = \begin{cases} \frac{1}{n}, & 1 \le j \le n \\ 0, & j > n, \end{cases} \qquad n, j \in \mathbf{N}.$$

Example 1.

(i) Let $A = [a_{nj}]$ $(n, j \in \mathbf{N})$ be defined by

$$a_{nj} = \frac{\binom{r+j-1}{j}}{\binom{n+r}{n}}, \quad r > 0, \quad 0 \le j \le n,$$
$$a_{nj} = 0, \qquad j > n.$$

It is easy to prove that A is a strongly regular matrix.

(ii) Let $(\varepsilon_n, n \in \mathbf{N})$ be a sequence of positive real numbers such that

$$\lim_{n\to\infty} \varepsilon_n = 0.$$

Let $A = [a_{nj}]$ $(n, j \in \mathbf{N})$ be defined by

$$a_{nj} = (e^{\varepsilon_n} - 1)e^{-j\varepsilon_n}, \quad n, j \in \mathbf{N}.$$

It can be proved (see [11]) that A is a stochastic strongly regular matrix.

Let X be a vector space and let $T : X \longrightarrow X$ be a linear operator. By $R(T)$ we denote the range of T and by $N(T)$ we denote the null-subspace of T. We denote by I the identity operator on X. If X is a topological space and if K is a subset of X, we denote by $\overline{K}$ the closure of K.

Let $(X, \mathcal{A}, \mu)$ be a measure space and let T be a linear operator defined on $L^p = L^p(X, \mathcal{A}, \mu)$, where $1 \le p < \infty$. The operator T is said to be:

(a) *a contraction*, if $\|Tf\|_p \le \|f\|_p$ for all $f \in L^p$;

(b) *positive*, if $Tf \ge 0$ whenever $f \ge 0$.

Suppose now that $(X, \mathcal{A}, \mu)$ is a measure space and let $T : X \longrightarrow X$ be a measurable map. In this case T is said to be *a transformation*. T is called *measure preserving* if $\mu \circ T^{-1} = \mu$ or equivalently if $\mu(T^{-1}(A)) = \mu(A)$ for all $A \in \mathcal{A}$.

A set $A \in \mathcal{A}$ is called *T-invariant* if $T^{-1}(A) = A$. The collection $\mathcal{A}_T$ of all T-invariant sets in $\mathcal{A}$ is a σ-algebra on X. The transformation T is called *ergodic*, if for every $A \in \mathcal{A}_T$ we have either $\mu(A) = 0$ or $\mu(A^c) = 0$.

For $1 \leq p \leq \infty$ the symbol $\|T\|_p$ will be used for the norm of T when it is considered as a bounded linear operator in L^p.

Example 2. Let $(A, \mathcal{A}, P)$ be a probability space and let $X^{\mathbf{N}}$ be the space of all sequences $(x_n, n \in \mathbf{N})$ of elements $x_n \in X$. Then on $X^{\mathbf{N}}$ there is a product σ-algebra and a product probability $P^{\mathbf{N}}$ of copies of P in the well known sense. The **shift** transformation T is defined by $T((x_n, n \in \mathbf{N})) = (x_{n+1}, n \in \mathbf{N})$. Then T is a measurable, measure preserving transformation and it can be proved (see 8.4 in [5]) that T is ergodic.

Theorem (Birkhoff's pointwise ergodic Theorem, see 8.4 in [5]). *Let* $(X, \mathcal{A}, \mu)$ *be a σ-finite measure space, and let T be a measure-preserving transformation of X. Then for every $f \in L^1$ there exists $f^* \in L^1(X, \mathcal{A}_T, \mu)$ satisfying $\|f^*\|_1 \leq \|f\|_1$ such that*

$$\lim_{n\to\infty} \frac{1}{n} \sum_{j=0}^{n-1} f \circ T^j - f^*, \quad \mu - \text{a.e.}$$

If T is ergodic, then f^* equals (a.e.) some constant c.

Results

Strongly regular matrices play an important role in ergodic theory. We shall see that the well known ergodic type theorems hold true if we consider the summability of the sequence of iterates of an operator by strongly regular matrices instead of the usual Cesàro $(c, 1)$ summability.

1⁰. We first generalize the ergodic type theorems due to Yoshida (see [12], Theorem 1 and Theorem 2, p. 213).

Theorem 1. *Let X be a sequentially complete, weakly and sequentially compact locally convex vector space with topology generated by family ε of seminorms and let $T : X \longrightarrow X$ be a continuous linear operator such that the family of operators $\{T^n, n \in \mathbf{N}\}$ is equicontinuous in the sense that for every seminorm $q \in \varepsilon$, there exists a continuous seminorm q' on X such that*

$$\sup_{n\in\mathbf{N}} q(T^n x) \le q'(x), \quad \text{for every} \quad x \in X. \tag{1}$$

Further, let $A = [a_{nj}]$ $(n, j \in \mathbf{N})$ *be a strongly regular matrix and*

$$T_n = \sum_{j=1}^{\infty} a_{nj} T^j (n \in \mathbf{N}).$$

Then, $T_n (n \in \mathbf{N})$ *are well defined on* X *and for every* $x \in X$ *there exists* $\lim_{n\to\infty} T_n x$. *If we put* $T_0 x = \lim_{n\to\infty} T_n x (x \in X)$ *then* T_0 *is a continuous linear operator not dependent on* A*, such that*

$$T_0 = T_0^2 = T\,T_0 = T_0\,T, \tag{2}$$

$$R(T_0) = N(I-T), \tag{3}$$

$$N(T_0) = \overline{R(I-T)} = R(I-T_0). \tag{4}$$

Moreover, we have a decomposition into a direct sum:

$$X = \overline{R(I-T)} \oplus N(I-T). \tag{5}$$

Proof. By using (1), for $q \in \varepsilon$ and every $n \in \mathbf{N}$ we get

$$\begin{aligned} q(\sum_{j=s}^{t} a_{nj} T^j x) &\le \sum_{j=s}^{t} |a_{nj}| q(T^j x) \\ &\le q'(x) \sum_{j=s}^{t} |a_{nj}| \longrightarrow 0, \quad \text{as} \quad t, s \longrightarrow \infty. \end{aligned}$$

Since X is a sequentially complete space, it follows that T_n are defined on X. It is obvious that the operators $T_n (n \in \mathbf{N})$ are linear and continuous. For $n \in \mathbf{N}$ we have

$$T_n(I-T) = a_{n1} T_1 + \sum_{j=1}^{\infty} (a_{n,j+1} - a_{nj}) T^{j+1}.$$

If $w \in R(I-T)$ then there exists $x \in X$ such that $w = (I-T)x$, so by using (1) for every $q \in \varepsilon$ we get

$$q(T_n w) = q(T_n(I-T)x) \le q'(x)(|a_{n1}| + \sum_{j=1}^{\infty} |a_{n,j+1} - a_{nj}|).$$

Since A is a strongly regular matrix we conclude

$$\lim_{n\to\infty} T_n w = 0, \quad \text{for every} \quad w \in R(I-T). \tag{6}$$

Let us prove that we have

$$\overline{R(I-T)} \subset \{x \in X : \lim_{n\to\infty} T_n x = 0\}. \tag{7}$$

Suppose that $z \in \overline{R(I-T)}$. Since q' is a continuous seminorm on X, then for every $\varepsilon > 0$ there exists $w \in R(I-T)$ such that $q'(z-w) < \frac{\varepsilon}{M}$. Using (1) we get

$$q(T_n(z-w)) \leq q'(z-w)\sum_{j=1}^{\infty} |a_{nj}| < \varepsilon.$$

It follows from here that

$$\begin{aligned} q(T_n z) &\leq q(T_n w) + q(T_n(z-w)) \\ &\leq q(T_n w) + \varepsilon, \quad \text{for every} \quad q \in \varepsilon, \end{aligned}$$

so by (6) we conclude that (7) holds true.

In fact (7) (see [8], Theorem 3) is an equality.

Suppose now, that $x \in X$ is arbitrary. Since X is weakly and sequentially compact, it follows that there exists a subsequence $(T_{n_k}x, k \in \mathbf{N}) \subset (T_n x, n \in \mathbf{N})$ such that the weak limit $w - \lim_{k\to\infty} T_{n_k}x = x_0$ exists. It follows from (7) that

$$\lim_{n\to\infty} T_n(x - T\,x) = 0. \tag{8}$$

Since T and T_n commute, the operators T_{n_k} and T satisfy the conditions of Cohen's lemma (see [3]) and therefore $T\,x_0 = x_0$. From $T^j x_0 = x_0$ $(j \in \mathbf{N})$ it follows that $T^j x = x_0 + T^j(x - x_0)$ so we have

$$T_n x = (\sum_{j=1}^{\infty} a_{nj})x_0 + T_n(x - x_0), \quad n \in \mathbf{N}. \tag{9}$$

By the continuity of T we have

$$\begin{aligned} x - T_{n_k}x &= (1 - \sum_{j=1}^{\infty} a_{n_k}j)x + \\ &\quad + \sum_{j=1}^{\infty} a_{n_{kj}}(I-T)(I + T + T^2 + \ldots + T^{j-1})x \\ &= (1 - \sum_{j=1}^{\infty} a_{n_{kj}})x + (I-T)z_{n_k}, \quad k \in \mathbf{N}, \end{aligned} \tag{10}$$

where $z_{n_k} = \sum_{j=1}^{\infty} a_{n_{kj}}(I + T + T^2 + \ldots + T^{j-1})x$. Taking the weak limit in (10) (as $k \longrightarrow \infty$) we get $x - x_0 = w - \lim_{k\to\infty}(I-T)z_{n_k}$. Since $(I-T)z_{n_k} \in R(I-T)$ then $x - x_0$ is an element of the weak closure of $R(I-T)$. But in locally convex vector spaces the weak closure of a convex subset is equal to the original closure of this subset (see [10], Theorem 3. 12 p. 64) so we conclude that $x - x_0 \in \overline{R(I-T)}$. Now, by (7) we have $\lim_{n\to\infty} T_n(x-x_0)=0$, so by (9) we obtain $\lim_{n\to\infty} T_n x = x_0$. Put $x_0 = T_0 x = \lim_{n\to\infty} T_n x = \lim_{n\to\infty} \sum_{j=1}^{\infty} a_{nj} T^j x, x \in X$. Linearity of T_0 is obvious. By use of (1), for $q \in \varepsilon$ we get

$$q(T_n x) \;\leq\; \sum_{j=1}^{\infty} |a_{nj}| q(T^j x) \leq M\, q'(x), \tag{11}$$

for every $x \in X$ and every $n \in \mathbf{N}$.

Therefore the family of operators $\{T_n, n \in \mathbf{N}\}$ is equicontinuous in the sense of (1). It follows easily from here that T_0 is a continuous operator. Now, from $T x_0 = x_0$ and $x_0 = T_0 x$ we get

$$T\, x_0 = T\, T_0 x = x_0 = T_0 x, \quad \text{for every} \quad x \in X$$

and therefore $T\, T_0 = T_0 \implies T^j T_0 = T_0$ for every $j \in \mathbf{N}$. Now we have $T_n T_0 = (\sum_{j=1}^{\infty} a_{nj}) T_0$, for every $n \in \mathbf{N}$, which implies $T_0^2 = T_0$. On the other hand we have

$$T_n - T_n T = a_{n1} T + \sum_{j=1}^{\infty} (a_{n,j+1} - a_{nj}) T^{j+1}.$$

By use of (1) and the fact that A is a strongly regular matrix we obtain $T_0 = T_0 T$, so (2) holds true.

Let us prove that the limiting operator T_0 does not depend on strongly regular matrix A. For this purpose let $B = [b_{nj}]$ $(n, j \in \mathbf{N})$ be another strongly regular matrix and let $W_n = \sum_{j=1}^{\infty} b_{nj} T^j$ $(n \in \mathbf{N})$. From the previous discussion it follows that there exists the continuous linear operator T_1 on X such that

$$T_1 x = \lim_{n\to\infty} W_n x, \quad x \in X, \tag{12}$$

and

$$T_1 = T_1^2 = T\, T_1 = T_1 T. \tag{13}$$

It follows from (12) that for every $j \in \mathbf{N}$ we have

$$T_1 = T_1^2 = T^j T_1 = T_1 T^j. \tag{14}$$

Multiplying (13) by a_{nj} and summing over j we get

$$(\sum_{j=1}^{\infty} a_{nj}) T_1 = T_n T_1 = T_1 T_n, \quad n \in \mathbf{N}. \tag{15}$$

By letting $n \longrightarrow \infty$ we obtain $T_1 = T_0 T_1 = T_1 T_0$. In the same way we get $T_0 = T_1 T_0 = T_0 T_1$ and therefore $T_0 = T_1$.

Relations (3), (4) and (5) follow now from the Corollary in [12], p. 214.

Corollary 1. *Let X be a Banach space. If the strong limit $T_0 = \lim_{n\to\infty} T_n$ exists, it is a projection on X on the subspace $\{x \in X : Tx = x\}$ of all fixed points of T and its complementary projection has $\overline{R(I-T)}$ as its range. Moreover projection T_0 does not depend on strongly regular matrix A.*

2^0. Now we prove two mean ergodic theorems for contractions in the spaces $L^p(1 \le p < \infty)$.

Theorem 2. *Let T be a positive contraction in L^p, where $1 < p < \infty$, and let $A = [a_{nj}]$ $(n, j \in \mathbf{N})$ be a strongly regular matrix. Then there exists a positive contraction P in L^p satisfying*

$$P = P^2 = TP = PT, \tag{16}$$

such that

$$\sum_{j=1}^{\infty} a_{nj} T^j f \longrightarrow Pf \quad \text{in} \quad L^p, \tag{17}$$

for every $f \in L^p$, as $n \longrightarrow \infty$. Moreover, contraction P does not depend on strongly regular matrix A.

Proof. We can use the results of Theorem 1 for the case of the Banach space L^p. Since T is a contraction, condition (1) is satisfied. From the proof of Theorem 1 it follows that for the proof of (16) it is sufficient to prove that the set

$$K = \{\sum_{j=1}^{\infty} a_{nj} T^j f : n \in \mathbf{N}\} \tag{18}$$

is weakly and sequentially compact for every $f \in L^p$. By [6], IV.8.4, p. 289, this is equivalent to the boundedness of K. We have

$$\|\sum_{j=1}^{\infty} a_{nj} T^j f\|_p \le M \|f\|_p, \quad n \in \mathbf{N},$$

and therefore (16) holds true. Relation (15) follows from (2). Moreover it follows from Theorem 1 that P is a bounded linear operator not dependent on strongly regular matrix A. Therefore we get the same limit P if we take the usual Cesàro $(c, 1)$ summability of the sequence of iterates $(T^n, n \in \mathbf{N})$. Thus we have

$$\frac{1}{n} \sum_{j=1}^{n} T^j f \longrightarrow Pf \quad \text{in} \quad L^p, \tag{19}$$

for every $f \in L^p$, as $n \longrightarrow \infty$. Now, by Theorem 2. 1. 1 in [7], p. 19, we conclude that if T is a positive contraction, then the same is true for P. Notice also that since

P is a contraction and also an idempotent (relation (15)) we conclude that $\|P\| = 1$ or $P = 0$.

Remark 1. Suppose that operator T in Theorem 2 is not a contraction, but the sequence of iterates $(T^n, n \in \mathbf{N})$ is uniformly bounded, that is $M_1 = \sup_{n\in\mathbf{N}} \|T^n\| < \infty$ (in this case we say that operator T is power bounded). Then Theorem 2 remains true (this follows from Theorem 1). The limiting operator P is bounded (not necessarily a contraction) and we have $\|P\| \le M\, M_1$.

Now we prove the mean ergodic theorem for the space L^1.

Theorem 3. *Let T be a contraction in L^1 and suppose that there exists a strictly positive $q \in L^1$ such that*

$$|f| \le g \Longrightarrow |Tf| \le g, \quad f \in L^1. \tag{20}$$

Further, let $A = [a_{nj}]$ $(n, j \in \mathbf{N})$ be a strongly regular matrix. Then there exists a contraction P in L^1 satisfying (15) *such that*

$$\sum_{j=1}^{\infty} a_{nj} T^j f \longrightarrow P f \quad \text{in} \quad L^1, \tag{21}$$

for every $f \in L^1$, as $n \longrightarrow \infty$. Moreover, contraction P does not depend on strongly regular matrix A.

Proof. As in the proof of Theorem 2, for proving (20) it is sufficient to prove that the set K in (17) is weakly and sequentially compact for every $f \in L^1$. By [6], IV.8.9, p. 292, this is equivalent to the boundedness of K and the condition that for each decreasing sequence $(E_k, k \in \mathbf{N}) \subset \mathcal{A}$ with void intersection we have

$$\lim_{k\to\infty} \int_{E_k} h\, d\mu = 0, \quad \text{uniformly for} \quad h \in K.$$

The proof that K is bounded is the same as in Theorem 2. Therefore, for the proof of (20) it is sufficient to prove that for every $f \in L^1$ we have

$$\lim_{k\to\infty} \int_{E_k} (\sum_{j=1}^{\infty} a_{nj} T^j f)\, d\mu = 0, \quad \text{uniformly in} \quad n \in \mathbf{N}. \tag{22}$$

We have

$$|\int_{E_k} (\sum_{j=1}^{\infty} a_{nj} T^j f)\, d\mu| \le \sum_{j=1}^{\infty} |a_{nj}| \int_{E_k} |T^j f|\, d\mu. \tag{23}$$

Since function $g \in L^1$ is strictly positive, then for every $f \in L^1$ and any $\varepsilon > 0$ there is a constant $c > 0$ and a splitting

$$f = f_c + \tilde{f}_c$$

with $|f_c| \leq cg$ and $\|\tilde{f}_c\|_1 < \varepsilon$ (see [1], Theorem 2.4.13, p. 88). By use of (19) and the fact that T is a contraction we obtain

$$\begin{aligned}\int_{E_k} |T^j f| d\mu &\leq c\int_{E_k} g\, d\mu + \int_X |T^j \tilde{f}_c| d\mu \\ &\leq c\int_{E_k} g\, d\mu + \varepsilon. \end{aligned} \tag{24}$$

(21) follows from (22), (23) and (A) since $\lim_{k\to\infty} \int_{E_k} g\, d\mu = 0$. Therefore (20) holds true. It follows from Theorem 1 that P is a bounded linear operator in L^1 not dependent on strongly regular matrix A. Therefore we get the same limit P if we take for A the usual Cesàro $(c, 1)$ summability matrix. By a consequence of Theorem 1.1 in [9] (p. 72 and 73) we then conclude that P is a contraction. In the same way as in Theorem 2 we conclude that $\|P\| = 1$ or $P = 0$.

Remark 2. Similarly as in Remark 1 we conclude that Theorem 3 remains true if we suppose that operator T is not a contraction, but T is power bounded.

3⁰. Now we generalize the well known classical pointwise ergodic theorem due to G.D. Birkhoff (see [2]).

We first need the following lemma.

Lemma 1. *Let $(X, \mathcal{A}, \mu)$ be a σ-finite measure space and let T be a linear operator in L^1 with $\|T\|_\infty \leq 1$ and $\|T\|_1 \leq 1$. If $A = [a_{nj}]$ $(n, j \in \mathbf{N})$ is a stochastic strongly regular matrix, then for $1 \leq p < \infty$ and $f \in L^p$ we have*

$$\sup_{n\in\mathbf{N}} |\sum_{j=1}^{\infty} a_{nj}(T^j f)(x)| < \infty, \quad \mu-\text{a.e.}$$

Proof. Put $T_n = \sum_{j=1}^{\infty} a_{nj} T^j (n \in \mathbf{N})$. Since space L^1 is complete and for $f \in L^1$

$$\|\sum_{j=k}^{r} a_{nj} T^j f\|_1 \leq \|f\|_1 \sum_{j=k}^{r} a_{nj},$$

it follows from the fact that A is a stochastic matrix that, $T_n (n \in \mathbf{N})$ are defined on L^1. Further $\|T_n\|_1 \leq 1 (n \in \mathbf{N})$. Let g be a bounded function in L^p. Since $\|T\|_\infty \leq 1$ we have

$$\begin{aligned}\|T_n g\|_\infty &\leq \sum_{j=1}^{\infty} a_{nj} \|g\|_\infty \\ &= \|g\|_\infty. \end{aligned}$$

Thus in proving the lemma we may assume as in Lemma VIII.6.5 in [6] (p. 675) that f is in L^1, $f \geq 0$ and that T is a positive operator. Now, the proof of the lemma is similar to the proof of Lemma VIII.6.5 in [6].

Theorem 4. (Pointwise ergodic theorem). *Let $(X, \mathcal{A}, \mu)$ be a σ-finite measure space and let T be a linear operator in L^1 with $\|T\|_\infty \leq 1$ and $\|T\|_1 \leq 1$. If $A = [a_{nj}]$ $(n \in \mathbf{N})$ is a stochastic strongly regular matrix, then for every p with $1 \leq p < \infty$ and every function $f \in L^p$, the limit*

$$\lim_{n\to\infty} \sum_{j=1}^{\infty} a_{nj}(T^j f)(x), \tag{25}$$

exists for almost all $x \in X$ and the limiting function does not depend (μ-a.e.) on matrix A.

Proof. Since $\|T\|_\infty \leq 1$ and $\|T\|_1 \leq 1$ it follows from the corollary of the Riesz convexity theorem that $\|T\|_p \leq 1$ for all $p, 1 \leq p \leq \infty$ (see [6], VI.10.12, p. 526). For p with $1 < p < \infty$ the space L^p is reflexive ([6], IV.8.2, p. 288). We have

$$\begin{aligned} \|T_n f\|_p &= \|\sum_{j=1}^{\infty} a_{nj} T^j f\|_p \\ &\leq \|f\|_p, \end{aligned}$$

since A is a stochastic matrix. Therefore the set $\{T_n f : n \in \mathbf{N}\}$ is bounded in L^p for every $f \in L^p$ and by II.3.28 in [6] (p. 68) it follows that this set is weakly and sequentially compact. Since A is a strongly regular matrix we conclude from Theorem 1 and Corollary 1 that the sequence of operators $(T_n, n \in \mathbf{N})$ converges strongly to the bounded linear operator $T_0 : L^p \longrightarrow L^p$, not dependent on A, which is a projection on the subspace $\{f \in L^p : Tf = f\}$ of all fixed points of T. From Corollary 1 it follows that vectors

$$h = f^* + (I - T)g,$$

with $f^*, g \in L^p, Tf^* = f^*$ and g bounded are dense in L^p. Moreover the vector f^* is uniquely determined by h. For such a vector h we have

$$T_n h = f^* + a_{n1} T g + \sum_{j=1}^{\infty} (a_{n,j+1} - a_{nj}) T^{j+1} g,$$

and therefore

$$\|T_n h - f^*\|_\infty \leq (a_{n1} + \sum_{j=1}^{\infty} |a_{n,j+1} - a_{nj}|) \|g\|_\infty.$$

Since A is a strongly regular matrix it follows that

$$\|T_n h - f^*\|_\infty \longrightarrow 0 \qquad \text{as } n \longrightarrow \infty,$$

and therefore $T_n h \longrightarrow f^*$, uniformly μ-a.e. as $n \longrightarrow \infty$, for every h in dense set in L^p. By Lemma 1 $\sup_{n\in\mathbf{N}} |(T_n f)(x)| < \infty$ almost everywhere for every $f \in L^p$.

Thus, by Theorem IV.11.2 in [6], p. 332, the sequence $\sum_{j=1}^{\infty} a_{nj}T^j f$ converges μ-a.e. for every $f \in L^p$.

For $f^* \in N(I-T)$ we have $T f^* = f^*$ and therefore

$$T^j f^* = f^*(j \in \mathbf{N}) \Longrightarrow T_n f^* = (\sum_{j=1}^{\infty} a_{nj}) f^* = f^*(n \in \mathbf{N})$$

and since $T_n f^* \longrightarrow T_0 f^*$ in L^p (as $n \longrightarrow \infty$) we conclude that

$$T_0 f^* = f^*.$$

On the other hand by using relation $T_0 = T_0 T$ (see (2)) we get

$$T_n h = T_n f^* + T_n(g - T g) \overset{\text{in } L^p}{\longrightarrow} T_0 f^* + T_0 g - T_0 T g = T_0 f^* = f^*.$$

We conclude from this that the limiting function in (24) does not depend (μ-a.e.) on stochastic strongly regular matrix A.

Since L^p is dense in L^1 we may apply Lemma 1 and Theorem IV.11.2 in [6] to see that the sequence $\sum_{j=1}^{\infty} a_{nj}T^j f$ converges μ-a.e. for every $f \in L^1$.

Corollary 2. *Let T be any measure-preserving transformation of a σ-finite measure space $(X, \mathcal{A}, \mu)$ and let $A = [a_{nj}]$ $(n, j \in \mathbf{N})$ be a stochastic strongly regular matrix. Then for every $f \in L^1$ there exists $f^* \in L^1(X, \mathcal{A}, \mu)$ such that*

$$\lim_{n\to\infty} \sum_{j=1}^{\infty} a_{nj} f \circ T^{j-1} = f^*, \quad \mu - \text{a.e.} \tag{26}$$

Moreover the function f^* does not depend (μ-a.e.) on matrix A and we have $\|f^*\|_1 \le \|f\|_1$. If T is ergodic, then f^* equals (μ a.e.) some constant c.

Proof. Put $U f := f \circ T, f \in L^1$. Then $U : L^1 \longrightarrow L^1$ is a positive linear contraction on L^1, since by the image measure theorem we have

$$\begin{aligned} \|U f\|_1 &= \int_X |f \circ T| d\mu = \int_X |f| d(\mu \circ T^{-1}) = \\ &= \int_X |f| d\mu = \|f\|_1, f \in L^1. \end{aligned}$$

We have $U^j f = f \circ T^j, j \in \mathbf{N}, f \in L^1$. Since T is a measure-preserving transformation we also obtain $\|U\|_\infty \le 1$. Therefore we can apply Theorem 4, and show that the limit

$$\lim_{n\to\infty} \sum_{j=1}^{\infty} a_{nj}(U^{j-1} f)(x) = \lim_{n\to\infty} \sum_{j=1}^{\infty} a_{nj} f(T^{j-1}x) = f^*(x),$$

exists for almost all $x \in X$ and the limiting function f^* does not depend (μ-a.e.) on matrix A. Therefore we get the same limit f^* if we take the usual Cesàro $(c, 1)$

summability of the sequence $(f \circ T^{n-1}, n \in \mathbf{N})$. The remaining statements follow now from Birkhoff's pointwise ergodic Theorem (see p. 5).

We now generalize the well known Kolmogorov's strong law of large numbers for independent identically distributed random variables.

Theorem 5. *Let $(X_n, n \in \mathbf{N})$ be a sequence of independent identically distributed random variables with finite mean and let $A = [a_{nj}]$ $(n, j \in \mathbf{N})$ be a stochastic strongly regular matrix. Then we have*

$$\text{(a.s.)} \lim_{n\to\infty} \sum_{j=1}^{\infty} a_{nj} X_j = E X_1. \tag{27}$$

Proof. Let $(\Omega, \mathcal{F}, P)$ be the probability space on which variables X_n are defined. Consider the function $Y : \Omega \longrightarrow \mathbf{R}^{\mathbf{N}}$, defined by $Y(\omega) = (X_n(\omega), n \in \mathbf{N}), \omega \in \Omega$. Then by independence the image measure $P \circ Y^{-1}$ equals $Q^{\mathbf{N}}$, where $Q = P_{X_1}$ is the probability law on X_1. So we may assume $\Omega = \mathbf{R}^{\mathbf{N}}, X_n$ are the coordinates and $P = Q^{\mathbf{N}}$. By Example 2 the shift transformation T in $\Omega = \mathbf{R}^{\mathbf{N}}$ is ergodic. If we take for f in (25) the first projection, we conclude that

$$\text{(a.s.)} \lim_{n\to\infty} \sum_{j=1}^{\infty} a_{nj} X_j \tag{28}$$

exists and that this limit does not depend on stochastic strongly regular matrix A. Therefore we get the same limit if we take the usual Cesàro $(c, 1)$ (a.s.) summability of the sequence $(X_n,\ n \in \mathbf{N})$. Now (26) follows from Kolmogorov's strong law of large numbers.

References

Ash, R.B. (1972). *Real Analysis and Probability.* Academic Press, New York.

Birkhoff, G.D. (1931). Proof of the ergodic theorem. *Proc. Nat. Acad. Sci. U.S.A.* 17, 656–660.

Cohen, L.W. (1940). On the mean ergodic theorem. *Ann. of Math.* **3**(41), 505–509.

Cooke, R.G. (1950). *Infinite matrices and sequence spaces.* MacMillan, London.

Dudley, R.M. (1989). *Real Analysis and Probability.* Wadsworth, Inc., Belmont, California.

Dunford, N. and Schwartz, J. (1958). *Linear Operators. Part I.* Intersc. Publish. Inc., New York.

Garsia, A.M. (1970). *Topics in Almost Everywhere Convergence. Lectures in Advanced Mathematics 4.* Markham Publishing Company.

Jardas, C. and Sarapa, N. (1994). On some generalizations of classical ergodic type theorems. *Grazer Mathematische Berichte.* **323**, 51–62.

Krengel, U. (1985). *Ergodic Theorems.* Walter de Gruyter and Co., Berlin.

Rudin, W. (1978). *Functional Analysis.* Tata McGraw-Hill, TMH edition.

Sarapa, N. (1977). On the (c, α) summability of probability distributions defined on compact semigroups. *Glasnik Mat.* **12**(32), 359–372.

Yoshida, K. (1978). *Functional Analysis.* Springer-Verlag, Berlin, Heidelberg, New York.

Exploring Stochastic Laws. pp. 167–174.
Eds. A V Skorokhod and Yu.V Borovskikh

On the Laplace transform of finite-dimensional distribution functions of semi-continuous random processes with reflecting and delaying screens

TAHIR A. KHANIEV[1] AND HALIM ÖZDEMIR[2]

[1] *The State University of Baku, Faculty of Mechanics and Mathematics, Z Khalilov St. 23, 370145, Baku, Azerbaijan*

[2] *Karadoniz Technical University, Faculty of Arts and Sciences, Department of Mathematics, 61080, Trabzon, Turkey*

Received February 1, 1995

Abstract. In this paper, we have constructed semi-continuous random process $X(t)$ with reflecting and delaying screens which plays an important role in stock control theory Then, by using the probability characteristics of renewal processes $\{T_n\}$ and $\{Y_n\}$, the Laplace transform of one dimensional distribution functions of $X(t)$ is obtained

1. Introduction

Stochastic processes called semi-continuous random processes play an important role in the solution of most of the problems in stock control theory. Processes of this type have been widely studied (see, for example, Smith (1961) and Spitzer (1969)). But sometimes some problems occur in stock control theory, such that in order to get an adequate solution we have to consider some processes which are more complex than semi-continuous random processes. In order to characterize the quantity of stock in a storehouse it is necessary to use stochastic processes called semi-continuous random processes with two screens in which one of the screens is reflecting and the other one is delaying. This type of problem may occur, for example, in the control of military stocks. Note that semi-continuous random processes with two screens, namely reflecting and delaying, have not been studied enough.

In the following the process mentioned above is mathematically constructed and the Laplace transform of one dimensional distribution functions of it is obtained.

2. Construction of process $X(t)$

In order to solve the stock control problem considered above let us construct the following stochastic process $X(t)$. Suppose

$$\{(\xi_i, \eta_i)\}, \quad i = 1, 2, \ldots$$

is a sequence of identically and independently distributed pairs of random variables, defined in a probability space $(\Omega, \mathcal{F}, P)$ such that ξ_1 and η_1 are positively valued with the distribution functions $\Phi(t) = P\{\xi_1 < t\}$ and $F(x) = P\{\eta_1 < x\}$, $t > 0$, $x > 0$, respectively.

Before stating the corresponding process let us construct the following renewal processes:

$$\begin{aligned} T_0 &= Y_0 = 0, \\ T_n &= \sum_{i=1}^{n} \xi_i, \\ Y_n &= \sum_{i=1}^{n} \eta_i, \qquad n = 1, 2, \ldots . \end{aligned}$$

The main probability characteristic of these renewal processes are well-known (see Ginlar (1975), Feller (1966), Smith (1961), Spitzer (1969), etc.).

By using the renewal processes $\{T_n\}$ and $\{Y_n\}$ let us construct the following random walks:

$$\begin{aligned} E_{2n} &= T_n - Y_n, \\ E_{2n-1} &= T_n - Y_{n-1}, \qquad n = 1, 2 \ldots . \end{aligned}$$

Moreover, let us introduce the following sequence of random variables with screens:

$$\begin{aligned} \chi_0 &= z \in [0, \beta], \\ \chi_n &= |\min\{\beta, \chi_{n-1} + \xi_n\} - \eta_n|, \qquad n = 1, 2, \ldots . \end{aligned}$$

Now we construct process $X(t)$ which has a reflecting screen below and a delaying screen above and which represents the level of stock at time t:

$$X(t) = \min\{\beta, t + \chi_{n-1} - T_{n-1}\}, \qquad \text{if} \quad t \in [T_{n-1}, T_n), \quad n = 1, 2, \ldots .$$

This process is called a semi-continuous random process with reflecting and delaying screens.

The main aim of this study is to express the Laplace transform of one dimensional distribution functions of $X(t)$ by the probability characteristics of renewal processes $\{T_n\}$ and $\{Y_n\}$.

3. Main result

In order to formulate the main result of this paper let us state the following notations and probability characteristics of renewal processes $\{T_n\}$ and $\{Y_n\}$:

$$a_n(t,z,x) = P\{z + E_k \in [0,\beta],\ k = 1,\dots,2n,\ Y_n \ge t + z - x,\ T_n \le t < T_{n+1}\},\quad n \ge 0, x \in [0,\beta],$$

$$c_n(s,z,v) = P\{z + E_i \in [0,\beta],\ i = 1,\dots,2n-1,\ z + E_{2n} < v,\ T_k < s\},\quad n \ge 1, v < 0,$$

$$A(t,z,x) = \sum_{n=0}^{\infty} a_n(t,z,x), \qquad C(s,z,v) = \sum_{n=1}^{\infty} c_n(s,z,v).$$

For any function $\psi(t,z)$, $z \in [0,\beta]$, denote by $\bar{\psi}(t,\beta)$ the following integral,

$$\bar{\psi}(t,\beta) = \int_0^{2\beta} \psi(t,|\beta - u|)\, dF(u).$$

The Laplace and Laplace–Stieltjes transforms of any function $M(t,z)$ with respect to t will be denoted by $\tilde{M}(p,z)$ and $M^*(p,z)$ respectively, i.e.,

$$\tilde{M}(p,z) = \int_0^{\infty} e^{-pt} M(t,z)\, dt,$$

and

$$M^*(p,z) = \int_0^{\infty} e^{-pz}\, d_t M(t,z).$$

Moreover, for any two functions $M_1(t,z)$ and $M_2(t,z)$, by $M_1(t,z) * M_2(t,z)$ let us denote the following composition,

$$M_1(t,z) * M_2(t,z) = \int_0^t M_2(t-s,z)\, d_s M_1(s,z).$$

Finally, by $P(t,z,x)$ we denote the conditional one-dimensional distribution function of $X(t)$:

$$P(t,z,x) = P_z\{X(t) < x\}.$$

Now we can formulate the main result of this study as the following.

Theorem 1. *If $\{(\xi_i,\eta_i),\ i = 1,2,\dots\}$ is the initial sequence of random pairs mentioned above then, in terms of the probability characteristics of renewal processes $\{T_n\}$ and $\{Y_n\}$, the Laplace transform of the one dimensional distribution function of $X(t)$, has the following form*

$$\tilde{P}(\Lambda,z,x) = \frac{1}{\Lambda}\frac{\bar{\tilde{R}}(\Lambda,\beta,x)}{\bar{\tilde{R}}(\Lambda,\beta,+\infty)} + \frac{\tilde{R}(\Lambda,z,x)\bar{\tilde{R}}(\Lambda,\beta,+\infty) - \tilde{R}(\Lambda,z,+\infty)\bar{\tilde{R}}(\Lambda,\beta,x)}{\bar{\tilde{R}}(\Lambda,\beta,+\infty)},$$

where

$$\tilde{R}(\Lambda,z,x)=\tilde{A}(\Lambda,z,x)+\sum_{l=1}^{\infty}\int_{-\beta}^{0}\cdots(l)\cdots\int_{-\beta}^{0}\prod_{i=1}^{l}C^{*}(\Lambda,|v_{i-1}|,dv_i)\tilde{A}(\Lambda,|v_l|,x).$$

Proof. Denote by $\hat{\gamma}_1$ the moment of time that the level goes down below the delaying screen for the first time. According to the total probability formula we can write

$$P(t,z,x)=P_z\{X(t)<x\}=P_z\{\hat{\gamma}_1\geq t;\ X(t)<x\}+P_z\{\hat{\gamma}_1<t;\ X(t)<x\}. \quad (1)$$

For simplicity, denote by $R(t,z,x)$ the first term in (1). For the second term in (1) we have

$$\begin{aligned}P_z\{\hat{\gamma}_1<t;\ X(t)<x\}&=\int_{s=0}^{t}\int_{v=0}^{2\beta}P_z\{\hat{\gamma}_1\in ds\}P\{\eta_1\in dv\}P_{(\beta-v)\vee 0}\{X(t-s)<x\}\\&=\int_0^t H(ds,z)\int_0^{2\beta}P(t-s,|\beta-v|,x)\,dF(v)\\&=\int_0^t H(ds,z)\Big[\int_0^{2\beta}P(t-s,|\beta-v|,x)\,dF(v)\Big]\\&=\int_0^{2\beta}dF(v)\int_0^t P(t-s,|\beta-v|,x)\,H(ds,z)\\&=\int_0^{\infty}dF(v)\,H(t,z)*P(t,|\beta-v|,x),\end{aligned} \quad (2)$$

where

$$H(ds,z)=P_z\{\hat{\gamma}_1\in ds\}.$$

Substituting expression (2) into formula (1) we get

$$P(t,z,x)=R(t,z,x)+\int_0^{2\beta}dF(v)\,H(t,z)*P(t,|\beta-v|,x). \quad (3)$$

Applying the Laplace transform to formula (3) with respect to t, we have

$$\begin{aligned}\tilde{P}(\Lambda,z,x)&=\tilde{R}(\Lambda,z,x)+\int_0^{2\beta}dF(v)\,H^{*}(\Lambda,z)\tilde{P}(\Lambda,|\beta-v|,x)\\&=\tilde{R}(\Lambda,z,x)+H^{*}(\Lambda,z)\int_0^{2\beta}P(\Lambda,|\beta-v|,x)\,dF(v)\\&=\tilde{R}(\Lambda,z,x)+H^{*}(\Lambda,z)\bar{\tilde{P}}(\Lambda,\beta,x).\end{aligned} \quad (4)$$

In equality (4), substituting $|\beta-z|$ instead of z and integrating in the sense of Lebesgue–Stieltjes with respect to the second argument, we have

$$\bar{\tilde{P}}(\Lambda,\beta,x)=\bar{\tilde{R}}(\Lambda,\beta,x)+\bar{H}^{*}(\Lambda,\beta)\bar{\tilde{P}}(\Lambda,\beta,x). \quad (5)$$

Hence,

$$\bar{\tilde{P}}(\Lambda,\beta,x)[1-\bar{H}^{*}(\Lambda,\beta)]=\bar{\tilde{R}}(\Lambda,\beta,x) \quad (6)$$

or

$$\tilde{\tilde{P}}(\Lambda,\beta,x)=\frac{\tilde{\tilde{R}}(\Lambda,\beta,x)}{1-\bar{H}^*(\Lambda,\beta)}. \tag{7}$$

Substituting expression (7) into formula (4) we get

$$\tilde{P}(\Lambda,z,x)=\tilde{R}(\Lambda,z,x)+\frac{H^*(\Lambda,z)\tilde{\tilde{R}}(\Lambda,\beta,x)}{1-\bar{H}^*(\Lambda,\beta)},\quad \text{for } z\in[0,\beta]. \tag{8}$$

So, finally we have obtained the following formula for the Laplace transform of distribution functions of random process $X(t)$:

$$\tilde{P}(\Lambda,z,x)=\tilde{R}(\Lambda,z,x)+H^*(\Lambda,z)\tilde{\tilde{R}}(\Lambda,\beta,x)[1-\bar{H}^*(\Lambda,\beta)]^{-1}. \tag{9}$$

Taking into account the formula

$$H^*(\Lambda,z)=1-\Lambda\tilde{R}(\Lambda,z,+\infty),$$

we can write

$$\tilde{P}(\Lambda,z,x)=\frac{1}{\Lambda}\frac{\tilde{\tilde{R}}(\Lambda,\beta,x)}{\tilde{\tilde{R}}(\Lambda,\beta,+\infty)}+\frac{\tilde{R}(\Lambda,z,x)\tilde{\tilde{R}}(\Lambda,\beta,+\infty)-\tilde{R}(\Lambda,z,+\infty)\tilde{\tilde{R}}(\Lambda,\beta,x)}{\tilde{\tilde{R}}(\Lambda,\beta,+\infty)}. \tag{10}$$

Now, let us express function $R(t,z,x)$ by the probability characteristics of renewal processes $\{T_n\}$ and $\{Y_n\}$. Firstly, let us denote by $V_0(t)$ the number of times in which the process $X(t)$ reflects from the reflecting screen. Applying the total probability formula we can write

$$R(t,z,x)=\sum_{l=0}^{\infty}P_z\{V_0(t)=l;\ \hat{\gamma}_1\geq t;\ X(t)<x\}. \tag{11}$$

For simplicity, denote $r_l(t,z,x)$ by

$$r_l(t,z,x)=P_z\{V_0(t)=l;\ \hat{\gamma}_1\geq t;\ X(t)<x\},\qquad l=0,1,2,\dots .$$

In order to develop a general formula of $r_l(t,z,x)$ in terms of the probability characteristics of the renewal processes $\{T_n\}$ and $\{Y_n\}$ let us compute $r_0(t,z,x)$ and $r_1(t,z,x)$ successively.

$$\begin{aligned}
r_0(t,x,z)&=P_z\{V_0(t)=0;\ \hat{\gamma}_1\geq t;\ X(t)<x\}\\
&=\sum_{n=0}^{\infty}P_z\{V_0(t)=0;\ \hat{\gamma}_1\geq t;\ X(t)<x;\ T_n\leq t<T_{n+1}\}\\
&=\sum_{n=0}^{\infty}P\{z+T_i-Y_{i-1}\in[0,\beta],\ z+T_i-Y_i\in[0,\beta],\ i=1,2,\dots,n,\\
&\qquad\qquad z+t-Y_n\in[0,x),\ T_n\leq t<T_{n+1}\}\\
&=\sum_{n=0}^{\infty}P\{z+E_{2i-1}\in[0,\beta],\ z+E_{2i}\in[0,\beta],\ i=1,2,\dots,n,\\
&\qquad\qquad z+t-Y_n\in[0,x),\ T_n\leq t<T_{n+1}\}\\
&=\sum_{n=0}^{\infty}P\{z+E_k\in[0,\beta],\ k=1,2,\dots,2n,\ Y_n>t+z-x,\ T_n\leq t<T_{n+1}\}.
\end{aligned}$$

Thus

$$
\begin{aligned}
r_0(t,z,x) &= \sum_{n=0}^{\infty} a_n(t,z,x) = A(t,z,x). \qquad (12)\\
r_1(t,z,x) &= P_z\{V_0(t)=1;\ \hat{\gamma}_1 \ge t;\ X(t)<x\}\\
&= \sum_{n=0}^{\infty} P_z\{V_0(t)=1;\ \hat{\gamma}_1 \ge t;\ X(t)<x;\ T_n \le t < T_{n+1}\}\\
&= \sum_{n=0}^{\infty}\sum_{k=1}^{n}\int_0^t P_z\{V_0(t)=1;\ \hat{\gamma}_1 \ge t;\ X(t)<x;\ T_n \le t < T_{n+1};\ T_k \in ds\}\\
&= \sum_{n=1}^{\infty}\sum_{k=1}^{n}\int_0^t\int_{-\beta}^{0}[c_k(ds,z,dv)][a_{n-k}(t-s,|v|,x)]\\
&\quad + \sum_{n=0}^{\infty}\int_0^t\int_{-\infty}^{-\beta} c_n(ds,z,dv)\bar{\Phi}(t-s)\varepsilon(x-\beta)\\
&= \int_0^t\int_{-\beta}^{0}\Big[\sum_{k=1}^{\infty} c_k(ds,z,dv)\Big]\Big[\sum_{l=0}^{\infty} a_l(t-s,|v|,x)\Big]\\
&\quad + \int_0^t\int_{-\infty}^{-\beta}\Big[\sum_{n=1}^{\infty} c_n(ds,z,dv)\Big]\bar{\Phi}(t-s)\varepsilon(x-\beta)\\
&= \int_0^t\int_{-\beta}^{0} C(ds,z,dv)A(t-s,|v|,x)\\
&\quad + \int_0^t\int_{-\infty}^{-\beta} C(ds,z,dv)\bar{\Phi}(t-s)\varepsilon(x-\beta)\\
&= \int_0^t\int_{-\beta}^{0} C(ds,z,dv)A(t-s,|v|,x) + \int_0^t C(ds,z,-\beta)\bar{\Phi}(t-s)\varepsilon(x-\beta)\\
&= \int_{-\beta}^{0} C(t,z,dv) * A(t,|v|,x) + C(t,z,-\beta) * \bar{\Phi}(t)\varepsilon(x-\beta),
\end{aligned}
$$

where

$$\bar{\Phi}(t) = 1 - \Phi(t)$$

and $\varepsilon(x)$ is the Heaviside function.

Thus

$$r_1(t,z,x) = \int_{-\beta}^{0} C(t,z,dv) * A(t,|v|,x) + C(t,z,-\beta) * \bar{\Phi}(t)\varepsilon(x-\beta). \qquad (13)$$

In our case, the probability $P(t,z,x)$ can easily be found if $x \notin [0,\beta]$ since

$$P(t,z,x) = \begin{cases} 0, & \text{if } x < 0,\\ 1, & \text{if } x > \beta. \end{cases}$$

Therefore, we have to calculate the probability $P(t,z,x)$ for $x \in [0,\beta]$. Taking into account the definition of the Heaviside function, from (13) we get

$$r_1(t,z,x) = \int_{-\beta}^{0} C(t,z,dv) * A(t,|v|,x). \qquad (14)$$

In a similar way it is possible to show that

$$r_2(t,z,x) = \int_{-\beta}^{0}\int_{-\beta}^{0} C(t,z,dv_1) * C(t,|v_1|,dv_2) * A(t,|v_2|,x), \tag{15}$$

for $x \in [0,\beta]$.

Applying the Laplace transform for expressions (12), (14), and (15) we have

$$\tilde{r}_0(\Lambda,z,x) = \tilde{A}(\Lambda,z,x)$$
$$\tilde{r}_1(\Lambda,z,x) = \int_{-\beta}^{0} C^*(\Lambda,z,dv)\tilde{A}(\Lambda,|v|,x)$$

and

$$\tilde{r}_2(\Lambda,z,x) = \int_{-\beta}^{0}\int_{-\beta}^{0} C^*(\Lambda,z,dv_1)C^*(\Lambda,|v_1|,dv_2)\tilde{A}(\Lambda,|v_2|,x).$$

By using mathematical induction, it is possible to prove that

$$\begin{aligned}\tilde{r}_l(\Lambda,z,x) &= \int_{-\beta}^{0}\cdots(l)\cdots\int_{-\beta}^{0} C^*(\Lambda,z,dv_1)C^*(\lambda,|v_1|,dv_2)C^*(\Lambda,|v_2|,dv_3)\\ &\qquad\cdots C^*(\Lambda,|v_{l-1}|,dv_l)\tilde{A}(\Lambda,|v_l|,x)\\ &= \int_{-\beta}^{0}\cdots(l)\cdots\int_{-\beta}^{0}\prod_{i=1}^{l} C^*(\Lambda,|v_{i-1}|,dv_i)\tilde{A}(\Lambda,|v_l|,x),\end{aligned} \tag{16}$$

where $|v_0| = z \in [0,\beta]$, $l = 2,3,\ldots$.

Applying the Laplace transform for (11) and substituting the expressions

$$\tilde{r}_l(\Lambda,z,x), \quad l = 0,1,2,\ldots$$

in it, we have

$$\tilde{R}(\Lambda,z,x) = \tilde{A}(\Lambda,z,x) + \sum_{l=1}^{\infty}\int_{-\beta}^{0}\cdots(l)\cdots\int_{-\beta}^{0}\prod_{i=1}^{l} C^*(\Lambda,|v_{i-1}|,dv_i)\tilde{A}(\Lambda,|v_l|,x).$$

Hence the theorem is proved.

Acknowledgement: We wish to express our thanks to Academician A.V. Skorokhod for the formulation of the general problems in connection with the study of semi-continuous random processes with two screens, his support and valuable advice.

References

Ginlar, E. (1975). *Introduction to stochastic processes.* Prentice-Hall, Inc., Englewood Cliffs, New Jersey.

Feller, W. (1966). *An introduction to probability theory and its applications, vol. II.* New York: John Wiley and Sons, Inc.

Nasirova, T.H. (1984). *Process of semi-Markov random walks.* Baku, ELM.

Prachbu, N. (1969). *Methods of queuing process and stock control.* Moscow.

Senturia, L., Puri Prem, S.A. (1973). A semi-Markov storage model. *Adv. Probab.* **5**(2), 362–378.

Smith, W.L. (1961). Renewal theory and its applications. *Mathematics, Moscow,* **5**(3).

Spitzer, F. (1969). *Principles of random walk.* Moscow, Mir. (In Russian).

Exploring Stochastic Laws. pp. 175–181.
Eds. A.V. Skorokhod and Yu.V. Borovskikh

Large deviation probabilities for U-statistics based on samples from finite populations

P.N. KOKIC[1] AND N.C. WEBER[2]

[1] *A.B.A.R.E. GPO Box 1563 Canberra, A.C.T. 2601 Australia*

[2] *School of Mathematics and Statistics, F07 University of Sydney N.W.S. 2006 Australia*

Received March 10, 1995

Abstract. Let U_{Nn} be a U-statistic of order 2 based on a sample of n observations drawn without replacement from a finite population of size N. In this paper sufficient conditions are given to establish a Cramér-type large deviation theorem for a suitably standardized version of U_{Nn}. This extends the result of Vandemaele (1982) to the finite population case.

1. Introduction

A central limit theorem for U-statistics based on samples drawn from finite populations was established by Nandi and Sen (1963) as part of a study of U-statistics based on dependent samples. More recently Zhao and Chen (1987) established a Berry–Esseen bound for such statistics and Kokic and Weber (1990) obtained an Edgeworth expansion for these U-statistics. In this paper we will consider a class of symmetric statistics that includes the U-statistics of order 2 studied by Nandi and Sen and establish a Cramér-type large deviation probability result for standardized statistics. A review of large deviation results for classical U-statistics based on independent and identically distributed (i.i.d.) random variables can be found in Chapter 8 of the monograph by Korolyuk and Borovskikh (1994).

Let $W = \{w_{ij}\}$ be an $N \times N$ array of real numbers satisfying $w_{ij} = w_{ji}$ and $w_{ii} = 0$. Let $(R_1, \ldots, R_N)$ be a vector selected at random from the set of $N!$ permutations of $(1, 2, \ldots, N)$. If $1 < n < N$, we will be interested in the statistics

$$U_{Nn} = \sum_{i=2}^{n} \sum_{j=1}^{i-1} W_{ij},$$

where $W_{ij} = w_{R_i R_j}$. Note that if we have a finite population $\mathcal{A} = \{a_1, a_2, \ldots, a_N\}$, a symmetric, real valued $h: \mathcal{A}^2 \to \mathbb{R}$ and we set $w_{ij} = h(a_i, a_j)$ $i \neq j$, $w_{ii} = 0$ then

$\binom{n}{2}^{-1}U_{Nn}$ is the U-statistic with kernel h based on the sample consisting of the n terms $a_{R_1}, \ldots, a_{R_n}$ drawn without replacement from $\mathcal{A}$.

Robinson (1977) obtained a general large deviation result for the mean of a sample drawn from a finite population. We will use this result and techniques similar to those used by Malevich and Abdalimov (1979) and Vandemaele (1982) to obtain a Cramér-type large deviation result for U_{Nn}. We will set up our notation and state the main result in Section 2. In Section 3 we will provide some technical lemmas and give the proof of the theorem.

2. Main Theorem

Assume $\sum_{i=1}^{N}\sum_{j=1}^{N} w_{ij} = 0$ and $\sum_{i=2}^{N}\sum_{j=1}^{i-1} w_{ij}^2 = 1$. Let $p = n/N$, $q = 1-p$, $w_i = (\sum_{j=1}^{N} w_{ij})$, $i = 1, \ldots, N$ and define

$$\nu_N^2 = p^3 q \sum_{i=1}^{N} w_i^2 \quad \text{and} \quad b_N = \max_{1\le j\le N} |w_j| \Big(\sum_{i=1}^{N} w_i^2\Big)^{-1/2}.$$

Note that ν_N^2 is closely related to the variance of U_n and to avoid trivial cases we assume $\nu_N^2 > 0$. Further let

$$v_{ij} = w_{ij} - \hat{w}_i - \hat{w}_j, \qquad i \neq j,$$

where $\hat{w}_i = (N-2)^{-1} w_i$, $i = 1, \ldots, N$, and write

$$U_{Nn} = (n-1)S_{Nn} + V_{Nn}, \tag{1}$$

where $V_{Nn} = \sum_{i=2}^{n}\sum_{j=1}^{i-1} v_{R_iR_j}$ and $S_{Nn} = \sum_{i=1}^{n} \hat{w}_{R_i}$.

We can now state our main result.

Theorem 1. *Suppose there exist constants K and $\gamma \geq 0$ such that for all N and $t \geq 2$,*

$$N^t\Big(N^{-1}\sum_{j=1}^{N} w_j^2\Big)^{-t/2}\binom{N}{2}^{-1}\sum_{i=2}^{N}\sum_{j=1}^{i-1}|v_{ij}|^t \leq K^t t^{\gamma t}. \tag{2}$$

Then uniformly for z satisfying $1 \leq z = o(((pq)^{-\frac{1}{2}}b_N)^{-(3+2\gamma)^{-1}})$,

$$P(\nu_N^{-1}U_{Nn} \geq z) \sim 1 - \Phi(z)$$

if $(pq)^{-\frac{1}{2}}b_N \to 0$ as $N \to \infty$.

Note that the moment condition at (2) will be satisfied with $\gamma = 0$ for bounded variables: $|v_{ij}| < c$ such that $\sum_{j=1}^{N} w_j^2 \geq c^{-1}N^3$ for some $c > 0$. In this special case it is also worth noting that $b_N = O(N^{-\frac{1}{2}})$.

Since for $N \geq 3$,

$$\sum_{i \neq j}^{N} \sum^{N} |v_{ij}|^t \leq c_1^t \sum_{i \neq j}^{N} \sum^{N} |w_{ij}|^t,$$

for some constant c_1, condition (2) is implied by

$$N^t (N^{-1} \sum_{j=1}^{N} w_j^2)^{-t/2} \binom{N}{2}^{-1} \sum_{i=2}^{N} \sum_{j=1}^{i-1} |w_{ij}|^t \leq K^t t^{\gamma t}.$$

Also note that since $b_N \geq N^{-\frac{1}{2}}$, $(pq)^{-\frac{1}{2}} b_N \to 0$ implies that $Npq \to \infty$ as $n \to \infty$. In the case $b_N = O(N^{-\frac{1}{2}})$ and $\gamma = 0$ the natural range $1 \leq z = o((Npq)^{1/6})$ is achieved, which corresponds to the result of Vandemaele (1982).

3. Proof of Theorem 1

Prior to establishing the theorem we obtain several useful lemmas.

Lemma 1. *Let U_{Nn} be based on an $N \times N$ array as above and further assume $w_i = 0$, $i = 1, \ldots, N$. If t is an even integer satisfying $2 < t < 2N(p \wedge q)$, then*

$$E|U_{Nn}|^t \leq (12e\,t)^t (Npq)^t \nu_{Nt},$$

where $\nu_{Nt} = \binom{N}{2}^{-1} \sum_{i=2}^{N} \sum_{j=1}^{i-1} |w_{ij}|^t$.

Proof. The proof is similar to that of Lemma 2 in Kokic and Weber (1991). By restricting the range of values for t the bound given in Lemma 6.1 of Bickel (1974) can be used instead of the result of Dharmadhikari et al. (1968) to obtain the tighter moment bound.

Lemma 2. *If Condition (2) holds for all even integers t, where $2 \leq t \leq 2Npq$, then*

$$E|\nu_N^{-1} V_{Nn}|^t \leq L^t (Npq)^{-t/2} t^{t(1+\gamma)},$$

for some constant L independent of N, n and t.

Proof. Using Lemma 1 with $c = 12e$,

$$\begin{aligned}
E|\nu_N^{-1} V_{Nn}|^t &\leq (ct)^t (Npq)^t (p^3 q \sum_{j=1}^{N} w_j^2)^{-t/2} \binom{N}{2}^{-1} \sum_{i=2}^{N} \sum_{j=1}^{i-1} |v_{ij}|^t, \\
&\leq (ct)^t (Npq)^{-t/2} N^t (N^{-1} \sum_{j=1}^{N} w_j^2)^{-t/2} \binom{N}{2}^{-1} \sum_{i=2}^{N} \sum_{j=1}^{i-1} |v_{ij}|^t, \\
&\leq (ct)^t (Npq)^{-t/2} K^t t^{\gamma t}, \qquad \text{by (2)}, \\
&\leq (Kc)^t (Npq)^{-t/2} t^{t(1+\gamma)}.
\end{aligned}$$

Let λ_N be as defined in Robinson (1977, p. 919), where the subscript N is used instead of n to represent population size. Then λ_N is a power series with real coefficients and no constant term.

Lemma 3. *If $\{a_N\}$ is a sequence of real numbers converging to zero as $N \to \infty$, then*

$$\lambda_N(a_N) = O(a_N) \quad \text{as } N \to \infty.$$

Proof. Now $\lambda_N(s) = \sum_{k=1}^{\infty} \tilde{\lambda}_{Nk} s^k$. According to Robinson (1977), $\lambda_N(s)$ is dominated by the power series $\sum_{k=1}^{\infty} |\tilde{\lambda}_k| s^k$, say, which is convergent in a neighbourhood of 0. Suppose this neighbourhood contains the interval $[-\epsilon, \epsilon]$, where $\epsilon > 0$. For all $N > N_0$, $|a_N| \le \epsilon$, and so

$$\begin{aligned} |a_N^{-1}\lambda_N(a_N)| &\le \left|a_N^{-1}\sum_{k=1}^{\infty} |\tilde{\lambda}_k| a_N^k\right| \\ &\le \sum_{k=1}^{\infty} |\tilde{\lambda}_k|\,|a_N|^{k-1} \\ &\le \epsilon^{-1}\sum_{k=1}^{\infty} |\tilde{\lambda}_k|\epsilon^k < \infty. \end{aligned}$$

Proof of Theorem 1. Assume, without loss of generality, that $n \ge 2$. Let $0 < \alpha \le 1/3$ and assume that $1 \le z = o(((pq)^{-\frac{1}{2}} b_N)^{-\alpha})$, where $(pq)^{-\frac{1}{2}} b_N \to 0$ as $N \to \infty$. We will show that the best choice of α that can be made is $\alpha = (3+2\gamma)^{-1}$.

Notice that for any $\epsilon < 0$,

$$\begin{aligned} &P(\nu_N^{-1}(n-1)S_{Nn} \ge z - \epsilon) + P(|\nu_N^{-1}V_{Nn}| > \epsilon) \ge P(\nu_N^{-1}U_{Nn} \ge z) \ge \\ &\qquad \ge P(\nu_N^{-1}(n-1)S_{Nn} \ge z + \epsilon) - P(|\nu_N^{-1}V_{Nn}| > \epsilon). \end{aligned} \tag{3}$$

We will set $\epsilon = ((pq)^{-\frac{1}{2}} b_N)^{\alpha}$ in this proof. The theorem is established by showing that the first terms in the lower and upper bounds in the expression above may be approximated by $1 - \Phi(z)$, while the remainder is negligible.

Note that

$$P(\nu_N^{-1}(n-1)S_{Nn} \ge z - \epsilon) = P\left(\left(pq\sum_{j=1}^{N} w_j^2\right)^{-\frac{1}{2}} \sum_{j=1}^{n} W_j \ge y\right),$$

where $W_j = w_{R_j}$ and $y = (N-2)N^{-1}(z-\epsilon)$. To approximate the term on the right of this expression we apply Theorem 1 in Robinson (1977).

Since $\epsilon = o(1)$ as $N \to \infty$, for all sufficiently large N, $y \ge \frac{1}{2}$. Furthermore, as $y \le 2z$, $pq \le \frac{1}{4}$ and $z = o(((pq)^{-\frac{1}{2}} b_N)^{-\alpha})$,

$$\begin{aligned} y b_N &\le z(pq)^{-1/2} b_N \to 0, \\ y^3 b_N &\le 4z(pq)^{-3/2} b_N^3 \to \infty \qquad \text{as } N \to \infty, \end{aligned}$$

and also $b_N \to 0$ as $N \to \infty$. Hence by Lemma 3, $y^2\lambda_N(yb_N) = O(y^3 b_N) \to 0$ as $N \to \infty$, and so by Theorem 1 in Robinson (1977), which holds for $y \geq \frac{1}{2}$,

$$P(\nu_N^{-1}(n-1)S_{Nn} \geq z - \epsilon) - 1 - \Phi(y) \qquad \text{as } N \to \infty. \tag{4}$$

Next we show that $1 - \Phi(y) \sim 1 - \Phi(z)$ as $N \to \infty$. Let $0 < \delta < 1$ be arbitrary. By Feller (1968, p. 175), for $z > \delta^{-1}$,

$$z^{-1}\phi(z) \geq 1 - \Phi(z) \geq z^{-1}\phi(z)(1-\delta). \tag{5}$$

Now choose N_δ so that for all $N \geq N_\delta$

$$|z - y| \leq |z - y|\{1 - \Phi(\delta^{-1} + \delta)\}^{-1} < \delta, \tag{6}$$

$$z\epsilon < \delta \quad \text{and} \quad \epsilon^2 < \delta. \tag{7}$$

By (6) when $z > \delta^{-1} + \delta$ we must have $y > \delta^{-1}$ and so from (5),

$$\begin{aligned}
\frac{1-\Phi(y)}{1-\Phi(z)} &\geq (1-\delta)zy^{-1}\exp\{-\tfrac{1}{2}(y^2 - z^2)\} = \\
&= (1-\delta)\{1 - (z-y)/z\}^{-1}\exp\Big[\frac{1}{2}\Big\{z^2\Big(1 - \Big(\frac{n}{n-1}\cdot\frac{N-2}{N}\Big)^2\Big) \\
&\quad + 2\Big(\frac{n}{n-1}\cdot\frac{N-2}{N}\Big)^2 z\epsilon - \Big(\frac{n}{n-1}\cdot\frac{N-2}{N}\Big)^2\epsilon^2\Big\}\Big] \\
&\geq (1-\delta)(1 + |z-y|)^{-1}\exp\Big[-\frac{1}{2}\Big\{z^2\Big(\Big(\frac{n}{n-1}\cdot\frac{N-2}{N}\Big)^2 - 1\Big) + 4\epsilon^2\Big\}\Big], \\
&\qquad \text{as } n \geq 2 \text{ and } z \geq 1, \\
&\geq (1-\delta)(1+\delta)^{-1}\exp(-6\delta),
\end{aligned} \tag{8}$$

using (7) and noting $\epsilon \geq n^{-\frac{1}{2}}$.

Likewise, for $z > \delta^{-1} + \delta$ and $N \geq N_\delta$

$$\frac{1-\Phi(y)}{1-\Phi(z)} \leq (1-\delta)^{-2}\exp(8\delta). \tag{9}$$

When $z \leq \delta^{-1} + \delta$, we may deduce from Taylor's theorem and (6) that for $N \geq N_\delta$

$$\begin{aligned}
\left|\frac{1-\Phi(y)}{1-\Phi(z)}\right| &= \left|\frac{(y-z)\phi(\theta)}{1-\Phi(z)}\right| \\
&\leq (2\pi)^{-\frac{1}{2}}|y - z|\{1 - \Phi(\delta^{-1} + \delta)\}^{-1} \\
&\leq \delta/\sqrt{2\pi},
\end{aligned} \tag{10}$$

where θ is on the line segment joining z and y. By (4), (8), (9) and (10) we see that

$$P(\nu_N^{-1}(n-1)S_{Nn} \geq z - \epsilon) \sim 1 - \Phi(z) \qquad \text{as} \quad N \to \infty.$$

Similarly,

$$P(\nu_N^{-1}(n-1)S_{Nn} \geq z + \epsilon) \sim 1 - \Phi(z) \qquad \text{as} \quad N \to \infty.$$

Therefore, to complete the proof it suffices to show that

$$P(|\nu_N^{-1}V_{Nn}| \geq \epsilon) = o(1-\Phi(z)) \qquad \text{as} \quad N \to \infty \tag{11}$$

for any $\alpha \leq (3+2\gamma)^{-1}$. By Lemma 2, for any even integer $k \in [2, 2Npq]$,

$$\begin{aligned} &P(|\nu_N^{-1}V_{Nn}| \geq (b_N(pq)^{-\frac{1}{2}})^{\alpha}) \\ &\quad \leq (b_N(pq)^{-\frac{1}{2}})^{-\alpha k} L^k (Npq)^{-k/2} k^{k(1+\gamma)} \\ &\quad \leq L^k((pq)^{-\frac{1}{2}} b_N)^{k(1-\alpha)} k^{k(1+\gamma)}, \qquad \text{as} \quad b_N \geq N^{-\frac{1}{2}}. \end{aligned} \tag{12}$$

Now let k be the largest even integer less then or equal to

$$(e^{\frac{1}{2}}L)^{-(1+\gamma)^{-1}}(b_N(pq)^{-\frac{1}{2}})^{-(1-\alpha)/(1+\gamma)}.$$

As $b_N^{-1} \leq N^{\frac{1}{2}}$ and $b_N(pq)^{-\frac{1}{2}} \to 0$ as $N \to 0$, for all sufficiently large N, $k \in [2, 2Npq]$. Therefore by (12), for all sufficiently large N,

$$\begin{aligned} P(|\nu_N^{-1}V_{Nn}| \geq \epsilon) &\leq L^k((pq)^{-\frac{1}{2}}b_N)^{k(1-\alpha)}((pq)^{-\frac{1}{2}}b_N)^{-k(1-\alpha)}e^{\frac{1}{2}}L)^{-k} \\ &\leq \exp\{1 - c((pq)^{-\frac{1}{2}}b_N)^{-(1-\alpha)/(1+\gamma)}\} \end{aligned} \tag{13}$$

for $c = \frac{1}{2}(e^{\frac{1}{2}}L)^{-(1+\gamma)^{-1}} > 0$.

Choose δ as before except that we restrict its range so that $0 < \delta^2 < 2c$. From (7) we know that $b_N(pq)^{-\frac{1}{2}} \leq 1$, and so by (13), if $z \geq \delta^{-1}$ and $N \geq N_\delta$

$$\begin{aligned} &\exp\{-c((pq)^{-\frac{1}{2}}b_N)^{-(1-\alpha)/(1+\gamma)}\}/(1-\Phi(z)) \\ &\quad \leq (1-\delta)^{-1}\sqrt{2\pi}\, z \exp\Big\{\frac{1}{2}z^2 - c((pq)^{-\frac{1}{2}}b_N)^{-(1-\alpha)/(1+\gamma)}\Big\} \\ &\quad \leq (1-\delta)^{-1}\sqrt{2\pi}\, \delta(b_N(pq)^{-\frac{1}{2}})^{-\alpha} \exp\Big\{\frac{1}{2}\delta^2((pq)^{-\frac{1}{2}}b_N)^{-2\alpha} \\ &\qquad\qquad - c((pq)^{-\frac{1}{2}}b_N)^{-(1-\alpha)/(1+\gamma)}\Big\}, \qquad \text{by (7)}, \\ &\quad \leq (1-\delta)^{-1}\sqrt{2\pi}\, \delta(b_N(pq)^{-\frac{1}{2}})^{-\alpha} \exp\{-c_1((pq)^{-\frac{1}{2}}b_N)^{-2\alpha}\}, \end{aligned} \tag{14}$$

provided $2\alpha \leq (1-\alpha)/(1+\gamma)$, where $c_1 = c - \frac{1}{2}\delta^2 > 0$. If $z < \delta^{-1}$ then

$$\begin{aligned} &\exp\Big\{-c((pq)^{-\frac{1}{2}}b_N)^{-(1-\alpha)/(1+\gamma)}\Big\}/(1-\Phi(z)) \leq \\ &\leq \exp\Big\{-c((pq)^{-\frac{1}{2}}b_N)^{-(1-\alpha)/(1+\gamma)}\Big\}/(1-\Phi(\delta^{-1})). \end{aligned} \tag{15}$$

The result at (11) and hence the Theorem follows on applying the estimates at (14) and (15) to (13), and by noting that $2\alpha \leq (1-\alpha)/(1+\gamma)$ if and only if $\alpha \leq (3+2\gamma)^{-1}$.

References

Bickel, P.J. (1974). Edgeworth expansions in nonparametric statistics. *Ann. Statist.* **2**, 1–20.

Dharmadhikari, S.W., Fabian, V. and Jogdeo, K. (1968). *Bounds on the moments of martingales.* Ann. Math. Statist. **39**, 1719–1723.

Feller, W. (1968). *An Introduction to Probability Theory and its Applications, Vol. 1*, (3-rd edition), Wiley, New York.

Kokic, P.N. and Weber, N.C. (1990). An Edgeworth expansion for U-statistics based on samples from finite populations. *Ann. Probab.* **18**, 390–404.

Kokic, P.N. and Weber, N.C. (1991). Rates of strong convergence for U-statistics in finite populations. *J. Austr. Math. Soc. (A)* **50**, 468–480.

Korolyuk, V.S. and Borovskich, Y.V. (1994). *Theory of U-statistics.* Kluwer Academic Publishers, Dordrecht.

Malevich, T.L. and Abdalimov, B. (1979). Large deviation probabilities for U-statistics. *Theory of Probab. and Appl.* **24**, 215–219.

Nandi, H.K. and Sen, P.K. (1963). Unbiased estimation of the parameters of a finite population. *Calcutta Statist. Assoc. Bull.* **12**, 124–148.

Robinson, J. (1977). Large deviation probabilities for samples from a finite population. *Ann. Probab.* **5**, 913–925.

Vandemaele, M. (1982). On large deviation probabilities for U-statistics. *Theory of Probab. and Appl.* **27**, 614.

Zhao, L. and Chen, X. (1987). Berry–Esseen bounds for finite population U-statistics. *Sci. Sinica Ser. A* **30**, 113–127.

Exploring Stochastic Laws. pp. 183–199.
Eds. A.V. Skorokhod and Yu.V. Borovskikh

Spatial quantiles and L-estimators

V.I. KOLTCHINSKII

Department of Mathematics and Statistics, The University of New Mexico, Albuquerque, NM 87131-1141

Received May 10, 1995

Abstract. The spatial quantile function defined recently by Dudley and Koltchinskii (1992) and by Chaudhuri (1993) is used to introduce L-parameters and L-estimators in multivariate case. The asymptotic normality of L-estimators is shown to hold under certain smoothness of the weight function of the estimator. It is also shown that such L-estimators can be used to estimate the center of symmetric distributions.

1. Introduction

Let $(X_1, \dots, X_n)$ be a sample from unknown distribution P in $\mathbb{R}^1$, and let $(X_{[1]}, X_{[2]}, \dots, X_{[n]})$ be its increasing ordering. L-estimators are defined as linear combinations of order statistics $X_{[1]} \le X_{[2]} \le \dots \le X_{[n]}$ with certain weights $m_{n,j}$:

$$L_n := \sum_1^n m_{n,j} X_{[j]}. \tag{1.1}$$

The weights can often be and are chosen as

$$m_{n,j} := \int_{(j-1)/n}^{j/n} m(t)\, dt, \qquad j = 1, \dots, n, \tag{1.2}$$

where m is an integrable function on $[0,1]$. Taking the weight function $m(t)$ close enough to 0 for the values of t close to the endpoints of the interval $[0,1]$ one could control the influence of the extreme values of the sample, and thus to increase robustness of the estimator L_n (see Huber (1981)).

A trivial extension of (1.1) is a class of statistics of the form

$$L_n := \sum_1^n m_{n,j}\, h(X_{[j]}) \tag{1.3}$$

with a transformation function $h: \mathbb{R}^1 \mapsto \mathbb{R}^1$. They are also often called L-estimators.

* Research partially supported by the Alexander von Humboldt Foundation

Asymptotics of L-estimators has been investigated by many authors (see Shorack (1972), Shorack and Wellner (1986), Stigler (1974), Boos (1979), Serfling (1980), Mason (1981), Norvaisha (1994) among others).

There is no common extension of this class of estimators to the multivariate case. The main problem is that there is no really natural ordering of vectors in the space $\mathbb{R}^d$ for $d > 1$. But there is another way to handle L-estimators. One can represent these statistics as $L_n = L(P_n)$, where P_n is the empirical measure based on the sample $(X_1, \ldots, X_n)$, and the functional $L(P)$ (an L-parameter of P) is defined as

$$L(P) := \int_0^1 h(F_P^{-1}(t))\, m(t)\, dt,$$

F_P being the distribution function and F_P^{-1} being the quantile function of P. Thus if one has a reasonable extension of the distribution and quantile functions to multivariate case, then it could be possible to get an extension of L-parameters and L-estimators without dealing with ordering of multivariate data. Such an extension has been provided recently by Dudley and Koltchinskii (1992) and Chaudhuri (1993) (see also Koltchinskii (1994a, b, c)). Chaudhuri (1993) was, in fact, the one who suggested to use this spatial quantile function in order to define L-estimators in multivariate case. We are going to discuss this extension in the next section, and then (in sections 3, 4) to define a spatial version of L-estimators and to study their asymptotics.

In what follows $\langle\cdot,\cdot\rangle$ denotes an inner product in $\mathbb{R}^d$, and $|\cdot|$ denotes the norm $|\cdot| := \langle\cdot,\cdot\rangle^{1/2}$. Given two vectors $u, v \in \mathbb{R}^d$, $u \otimes v$ denotes the operator in $\mathbb{R}^d$ $x \mapsto u\langle v, x\rangle$. Given an orthonormal basis in $\mathbb{R}^d$ (with respect to the inner product $\langle\cdot,\cdot\rangle$), the matrix of the operator $u \otimes v$ is going to be $(u_i v_j)_{i,j=1}^d$, where $(u_1, \ldots, u_d)$, $(v_1, \ldots, v_d)$ are coordinate representations of vectors u, v, respectively, in this basis.

Let $B(s;\delta)$ denote the open ball and $\bar{B}(s;\delta)$ denote the closed ball with the center $s \in \mathbb{R}^d$ and with the radius $\delta > 0$. Let $B_d := B(0;1)$.

We denote $l^\infty(\mathbb{R}^d)$ the space of all uniformly bounded functions from $\mathbb{R}^d$ into $\mathbb{R}^d$ with the norm

$$\|Y\|_\infty := \sup\{|Y(t)| : t \in \mathbb{R}^d\}.$$

2. Spatial quantiles and their properties

Let us consider a kernel

$$f(s,x) := 0.5\,(|s-x| - |x| + s), \qquad s \in \mathbb{R}^1,\ x \in \mathbb{R}^1. \tag{2.1}$$

Note that $f(s,x)$ is a convex function with respect to s for any $x \in \mathbb{R}^1$. Therefore a function $f_P : \mathbb{R}^1 \mapsto \mathbb{R}^1$, defined by

$$f_P(s) := \int_{\mathbb{R}^1} f(s,x)\, P(dx), \qquad s \in \mathbb{R}^1,$$

is convex. The distribution function F_P of P is a subgradient of f_P and the quantile function F_P^{-1} is a subgradient of the Young–Fenchel conjugate (or the Legendre transformation) of f_P:

$$f_P^*(t) := \sup_{s\in\mathbb{R}^1}[\langle s,t\rangle - f_P(s)].$$

It could be shown that for any $t \in (0,1)$ a t-th quantile of P is a minimal point of the functional

$$f_{P,t}(s) := f_P(s) - \langle s,t\rangle, \qquad s \in \mathbb{R}^1.$$

Let now P be a Borel probability on $\mathbb{R}^d$. An extension of quantiles to multivariate case suggested by Dudley and Koltchinskii (1992) and Chaudhuri (1993) is based on replacing the kernel (2.1) by the following one:

$$f(s;x) := |s-x| - |x|, \qquad x \in \mathbb{R}^d.$$

A function

$$f_P(s) := \int_{\mathbb{R}^d}(|s-x|-|x|)\,P(dx), \qquad s \in \mathbb{R}^d$$

is convex and a map $F_P: \mathbb{R}^d \mapsto \mathbb{R}^d$, defined by

$$F_P(s) := \int_{\{x\neq s\}} \frac{s-x}{|s-x|}\,P(dx), \qquad s \in \mathbb{R}^d,$$

is a subgradient of f_P.

2.1. Main properties of the map F_P.

(i) The map F_P is monotone, i.e.,

$$\Big\langle s_1 - s_2, F_P(s_1) - F_P(s_2)\Big\rangle \geq 0, \qquad s_1, s_2 \in \mathbb{R}^d.$$

(ii) Let A_P be the set of all atoms of P, and let $p(s) := P(\{s\})$, $s \in A_P$. The map F_P is continuous in $\mathbb{R}^d \setminus A_P$. For $s \in A_P$ and for all $v \in \mathbb{R}^d$ with $|v| = 1$

$$F_P(s+\varepsilon v) \to F_P(s) + p(s)v, \qquad \varepsilon \to 0.$$

(iii) For all $v \in \mathbb{R}^d$ with $|v| = 1$

$$F_P(\lambda v) \to v, \qquad \lambda \to \infty.$$

Moreover, we have

$$F_P(s) = \frac{s}{|s|} + o(1), \qquad s \to \infty.$$

Let us assume now that P is not concentrated in a straight line. Under this condition the function f_P is going to be strictly convex and we have the following additional properties of the map F_P.

(iv) F_P is a strictly monotone map, i.e.,

$$\Big\langle s_1 - s_2, F_P(s_1) - F_P(s_2)\Big\rangle > 0, \qquad s_1, s_2 \in \mathbb{R}^d,\ s_1 \neq s_2.$$

(v) Let

$$U_P := B_d \setminus \bigcup_{s\in A_P} \bar{B}(F_P(s); p(s)).$$

All the balls $\bar{B}(F_P(s); p(s))$, $s \in A_P$ are mutually disjoint. F_P is a one-to-one map from $\mathbb{R}^d$ onto $U_P \cup F_P(A_P)$, and, moreover, it is a homeomorphism of $\mathbb{R}^d \setminus A_P$ and U_P. For a nonatomic measure P the map F_P is a homeomorphism of $\mathbb{R}^d$ and B_d.

The map F_P^{-1} (defined quite naturally on the set U_P) could be extended to the whole ball B_d by assuming that for $t \in B_d \setminus U_P$ $F_P^{-1}(t)$ is the unique $s \in A_P$ such that $t \in \bar{B}(F_P(s); p(s))$.

(vi) For any $t \in B_d$ $F_P^{-1}(t)$ is a minimal point of the functional

$$f_{P,t}(s) := f_P(s) - \langle s, t\rangle, \qquad s \in \mathbb{R}^d.$$

In particular, $F_P^{-1}(0)$ is the so called L_1- or Haldane's spatial median of P (defined as a minimal point of the functional f_P on $\mathbb{R}^d$).

(vii) The map F_P completely characterizes the distribution P, i.e., $F_P = F_Q$ implies $P = Q$.

(viii) If P_θ denotes a θ-shift of P, $\theta \in \mathbb{R}^d$, then

$$\begin{aligned} F_{P_\theta}(s) &= F_P(s-\theta), \qquad & s \in \mathbb{R}^d, \\ F_{P_\theta}^{-1}(t) &= F_P^{-1}(t) + \theta, \qquad & t \in B_d. \end{aligned}$$

(ix) If O is an orthogonal transformation of $\mathbb{R}^d$, then

$$\begin{aligned} F_{P\circ O^{-1}}(s) &= OF_P(O^{-1}s), \qquad & s \in \mathbb{R}^d, \\ F_{P\circ O^{-1}}^{-1}(t) &= OF_P^{-1}(O^{-1}t), \qquad & t \in B_d. \end{aligned}$$

(x) If P is spherically symmetric with the center $a \in \mathbb{R}^d$, then there exists a continuous distribution function ψ_P on $(0, +\infty)$ such that $\psi_P(0+) = P(\{0\})$ and

$$F_P(s) = \begin{cases} \psi_P(|s-a|)\dfrac{s-a}{|s-a|} & \text{for } s \neq a, \\ 0 & \text{for } s = a, \end{cases}$$

$$F_P^{-1}(t) = \begin{cases} a + \psi_P^{-1}(|t|)\dfrac{t}{|t|} & \text{for } t \neq 0, \\ a & \text{for } t = 0. \end{cases}$$

It's easy to show that

$$\psi_P(\lambda) = \int_{\mathbb{R}^d} \frac{\lambda - x_1}{((\lambda - x_1)^2 + x_2^2 + \cdots + x_d^2)^{1/2}} P_{-a}(dx), \qquad \lambda > 0, \tag{2.2}$$

(recall that P_{-a} is the $-a$-shift of P, so the center of symmetry of P_{-a} is 0).

(xi) If P is symmetric with the center $a \in \mathbb{R}^d$ (in the sense that the random vectors $X - a$ and $a - X$ are identically distributed, given X has the distribution P), then

$$\begin{aligned} F_P(s) &= -F_P(2a - s), \qquad s \in \mathbb{R}^d, \\ F_P^{-1}(-t) &= 2a - F_P^{-1}(t), \qquad t \in B_d. \end{aligned}$$

(xii) Suppose that F_P is continuously differentiable in a neighborhood of the point $F_P^{-1}(t)$ with the nondegenerate derivative $F_P' \circ F_P^{-1}(t)$. Then the influence curve (see Huber (1981)) of the functional $P \mapsto F_P^{-1}(t)$ is given by

$$IC(x; P; F_P^{-1}(t)) = \begin{cases} -\mathrm{inv}(F_P' \circ F_P^{-1}(t))(F_P - F_{\delta_x}) \circ F_P^{-1}(t) & \text{for } t \neq F_P(x), \\ 0 & \text{for } t = F_P(x). \end{cases}$$

We denote here and in what follows $\mathrm{inv}(A)$ the inverse of a nondegenerate linear operator A in $\mathbb{R}^d$.

In what follows we assume that $(X_1, \ldots, X_n)$ is a sample from the distribution P on $\mathbb{R}^d$, and P_n is the empirical measure based on this sample:

$$P_n := n^{-1} \sum_1^n \delta_{X_k},$$

δ_x being the unit point mass at x.

(xiii) For any Borel nonatomic probability P on $\mathbb{R}^d$ the sequence of empirical processes

$$n^{1/2}(F_{P_n}(s) - F_P(s)), \qquad s \in \mathbb{R}^d$$

converges weakly in the space $l^\infty(\mathbb{R}^d)$ to the vector valued continuous Gaussian process $\xi_P(s)$, $s \in \mathbb{R}^d$ with the mean 0 and the covariance[1]

$$E\xi_P(s_1) \otimes \xi_P(s_2) = \int_{\{x \neq s_1, x \neq s_2\}} \frac{s_1 - x}{|s_1 - x|} \otimes \frac{s_2 - x}{|s_2 - x|} P(dx) - F_P(s_1) \otimes F_P(s_2).$$

2.2. Remarks. Dudley and Koltchinskii (1992) and Chaudhuri (1993) suggested to use F_P^{-1} as an extension of the quantile function to multivariate case. They studied the properties of such quantiles (described briefly above) and investigated

[1] We understand weak convergence in the space $l^\infty(\mathbb{R}^d)$ in the sense of Hoffmann-Jorgensen, i.e. we assume continuity of the limit process and for all bounded $\|\cdot\|_\infty$-continuous functionals Ψ from $l^\infty(\mathbb{R}^d)$ to $\mathbb{R}^1$ we have

$$E^*\Psi(n^{1/2}(F_{P_n} - F_P)) \to E\Psi(\xi_P), \qquad n \to \infty,$$

where E^* denotes the outer expectation.

asymptotic behavior of the empirical quantile function $F_{P_n}^{-1}$. Koltchinskii (1994a, b) obtained Bahadur–Kiefer type asymptotic representations for such quantiles.

Koltchinskii (1994c) further extended this approach considering inverses of a map

$$F_P(s) := \int_{\{x\neq s\}} \Phi(|s-x|)\,\frac{s-x}{|s-x|}\,P(dx), \qquad s\in\mathbb{R}^d$$

with a nonnegative increasing function Φ on $(0,+\infty)$. Then $F_P^{-1}(0)$ is an M-parameter of the distribution P, defined by the minimization of the convex functional

$$\int_{\mathbb{R}^d} (\varphi(|s-x|) - \varphi(|s_0-x|))\,P(dx).$$

Here φ is a convex nondecreasing function on $[0,+\infty)$ with the derivative Φ in $(0,+\infty)$, and $s_0\in\mathbb{R}^d$ is a fixed point.

3. L-estimators: definition and asymptotic properties

In one dimensional case an L-parameter of a nonatomic distribution P could be written as

$$L(P) = \int_{[0,1]} h(F_P^{-1}(t))\,m(t)\,(P\circ F_P^{-1})\,(dt),$$

but in this case $P\circ F_P^{-1}$ is, in fact, the uniform distribution on $[0,1]$. In the case $d>1$ our extension of the quantile function does not possess this remarkable property. But let us assume that P is nonatomic and it is not concentrated in a straight line. Then we could still define an L-parameter as

$$L(P) := L_{h,m}(P) := \int_{B_d} h(F_P^{-1}(t))\,m(t)\,(P\circ F_P^{-1})\,(dt) = \int_{\mathbb{R}^d} h(s)\,m(F_P(s))\,P(ds),$$

where $h\colon \mathbb{R}^d\mapsto\mathbb{R}^l$, $l\geq 1$ and $m\colon B_d\mapsto\mathbb{R}^1$ are Borel functions, such that

$$\int_{\mathbb{R}^d} |h(s)|\,|m(F_P(s))|\,P(ds) < +\infty.$$

The corresponding L-estimator is defined as

$$L_n := L(P_n) := L_{h,m}(P_n) := \int_{\mathbb{R}^d} h(s)\,m(F_{P_n}(s))\,P_n(ds).$$

Suppose there exists a Borel function $m'\colon B_d\mapsto\mathbb{R}^d$ such that

$$\int_{\mathbb{R}^d} |h(s)|\,r_m(F_P(s);\delta)\,P(ds)\to 0, \qquad \delta\to 0, \tag{3.1}$$

where

$$r_m(t;\delta) := \delta^{-1}\sup\Big\{|m(t+h)-m(t)-\langle m'(t),h\rangle| : |h|\leq\delta,\ t+h\in B_d\Big\}, \qquad t\in B_d.$$

Let us also introduce the following conditions

$$\int_{\mathbb{R}^d} |h(s)|^2\,|m'(F_P(s))|^2\,P(ds) < +\infty. \tag{3.2}$$

and

$$\int_{\mathbb{R}^d} |h(s)|^2 m^2(F_P(s))\, P(ds) < +\infty. \tag{3.3}$$

3.1. Theorem. *Suppose that conditions (3.1)–(3.3) hold. Then the sequence $n^{1/2}(L_n - L(P))$ is asymptotically normal with mean 0 and the covariance*

$$\int_{\mathbb{R}^d} f \otimes f\, dP - \int_{\mathbb{R}^d} f\, dP \otimes \int_{\mathbb{R}^d} f\, dP$$

where $f := f_{P,h,m}$ is a map from $\mathbb{R}^d$ into $\mathbb{R}^l$, defined by

$$f(s) := f_{P,h,m}(s) := h(s)\, m(F_P(s)) + \\ + \int_{\mathbb{R}^d\setminus\{s\}} h(x) \left\langle m'(F_P(x)), \frac{x-s}{|x-s|} \right\rangle P(dx), \qquad s \in \mathbb{R}^d.$$

Moreover, the following asymptotic representation holds

$$n^{1/2}(L_n - L(P)) = n^{1/2} \int_{\mathbb{R}^d} f(s)\,(P_n - P)(ds) + o_p(1), \qquad n \to \infty. \tag{3.4}$$

Proof. First observe that the following identity holds

$$\begin{aligned} n^{1/2}(L_n - L(P)) &= n^{1/2} \int_{\mathbb{R}^d} h(s)\, m(F_P(s))\,(P_n - P)(ds) + \\ &+ n^{1/2} \int_{\mathbb{R}^d} h(s) \left[m(F_{P_n}(s)) - m(F_P(s))\right] P(ds) + \\ &+ n^{1/2} \int_{\mathbb{R}^d} h(s) \left[m(F_{P_n}(s)) - m(F_P(s))\right] (P_n - P)(ds). \end{aligned} \tag{3.5}$$

Next note that

$$\begin{aligned} & n^{1/2} \int_{\mathbb{R}^d} h(s) \left[m(F_{P_n}(s)) - m(F_P(s))\right] P(ds) = \\ &= n^{1/2} \int_{\mathbb{R}^d} h(s) \Big[m(F_{P_n}(s)) - m(F_P(s)) - \\ &\qquad - \langle m'(F_P(s)), F_{P_n}(s) - F_P(s)\rangle\Big] P(ds) + \\ &\quad + n^{1/2} \int_{\mathbb{R}^d} h(s)\, \langle m'(F_P(s)), F_{P_n}(s) - F_P(s)\rangle\, P(ds) = \\ &= n^{1/2} \int_{\mathbb{R}^d} h(s) \Big[m(F_{P_n}(s)) - m(F_P(s)) - \\ &\qquad - \langle m'(F_P(s)), F_{P_n}(s) - F_P(s)\rangle\Big] P(ds) + \\ &\quad + n^{1/2} \int_{\mathbb{R}^d} \int_{\mathbb{R}^d\setminus\{s\}} h(x) \left\langle m'(F_P(x)), \frac{x-s}{|x-s|} \right\rangle P(dx)\,(P_n - P)(ds). \end{aligned} \tag{3.6}$$

By the definition of the remainder r_m we get

$$\begin{aligned} & n^{1/2} \left| \int_{\mathbb{R}^d} h(s) \left[m(F_{P_n}(s)) - m(F_P(s)) - \langle m'(F_P(s)), F_{P_n}(s) - F_P(s)\rangle\right] P(ds) \right| \le \\ &\le n^{1/2} \|F_{P_n} - F_P\|_\infty \int_{\mathbb{R}^d} |h(s)|\, r_m\Big(F_P(s); \|F_{P_n} - F_P\|_\infty\Big)\, P(ds). \end{aligned} \tag{3.7}$$

By condition (3.1) and property (xiii) of Section 2 we have

$$\int_{\mathbb{R}^d} |h(s)|\, r_m\Big(F_P(s); \|F_{P_n} - F_P\|_\infty\Big)\, P(ds) = o_p(1), \qquad n \to \infty. \tag{3.8}$$

It follows from (3.7) and (3.8) that

$$\begin{aligned} n^{1/2} \Big| \int_{\mathbb{R}^d} h(s) \Big[& m(F_{P_n}(s)) - m(F_P(s)) - \\ & - \langle m'(F_P(s)), F_{P_n}(s) - F_P(s)\rangle \Big] P(ds) \Big| = o_p(1), \qquad n \to \infty. \end{aligned} \tag{3.9}$$

Now let us take a look at the last term of (3.5). We have

$$\begin{aligned} & n^{1/2} \int_{\mathbb{R}^d} h(s) \Big[m(F_{P_n}(s)) - m(F_P(s)) \Big] (P_n - P)(ds) = \\ & = n^{1/2} \int_{\mathbb{R}^d} h(s) \Big\langle m'(F_P(s)), F_{P_n}(s) - F_P(s) \Big\rangle (P_n - P)(ds) + \\ & \quad + n^{1/2} \int_{\mathbb{R}^d} h(s) \Big[m(F_{P_n}(s)) - m(F_P(s)) - \\ & \qquad - \Big\langle m'(F_P(s)), F_{P_n}(s) - F_P(s) \Big\rangle \Big] (P_n - P)(ds). \end{aligned} \tag{3.10}$$

The first term in the right-hand side of (3.10) has the following representation

$$\begin{aligned} & n^{1/2} \int_{\mathbb{R}^d} h(s) \Big\langle m'(F_P(s)), F_{P_n}(s) - F_P(s) \Big\rangle (P_n - P)(ds) = \\ & = n^{1/2} \int_{\mathbb{R}^d} h(s) \Big[n^{-1} \sum_{j: X_j \neq s} \Big\langle m'(F_P(s)), \frac{s - X_j}{|s - X_j|} \Big\rangle - \\ & \qquad - \langle m'(F_P(s)), F_P(s) \rangle \Big] (P_n - P)(ds) = \\ & = n^{-3/2} \sum_{i \neq j} \bar{U}(X_i, X_j), \end{aligned} \tag{3.11}$$

where

$$\begin{aligned} \bar{U}(x,y) \;:=\; & U(x,y) - \int_{\mathbb{R}^d} U(x,z)\, P(dz) - \\ & - \int_{\mathbb{R}^d} U(z,x)\, P(dx) + \int_{\mathbb{R}^d \times \mathbb{R}^d} U(z_1, z_2)\, P(dz_1)\, P(dz_2), \end{aligned}$$

and

$$U(x,y) := \begin{cases} h(x) \Big\langle m'(F_P(x)), \dfrac{x-y}{|x-y|} \Big\rangle & \text{for } x \neq y, \\ 0 & \text{for } x = y. \end{cases}$$

Under condition (3.2) we have (by the properties of U-statistics, see, for example, Serfling (1980))

$$\sum_{i \neq j} \bar{U}(X_i, X_j) = O_p(n), \qquad n \to \infty. \tag{3.12}$$

It follows from (3.11) and (3.12) that

$$n^{1/2} \int_{\mathbb{R}^d} h(s) \Big\langle m'(F_P(s)), F_{P_n}(s) - F_P(s) \Big\rangle\, d(P_n - P) = O_p(n^{-1/2}), \qquad n \to \infty. \tag{3.13}$$

Now we are going to estimate the last term of representation (3.10). By the definition of the remainder r_m

$$\left|m(F_{P_n}(s)) - m(F_P(s)) - \left\langle m'(F_P(s)), F_{P_n}(s) - F_P(s)\right\rangle\right| \leq$$
$$\leq \|F_{P_n} - F_P\|_\infty r_m\left(F_P(s); \|F_{P_n} - F_P\|_\infty\right),$$

and we have

$$\Bigg| n^{1/2} \int_{\mathbb{R}^d} h(s) \left[m(F_{P_n}(s)) - m(F_P(s)) - \right.$$
$$\left. - \left\langle m'(F_P(s)), F_{P_n}(s) - F_P(s)\right\rangle\right] (P_n - P)(ds)\Bigg| \leq$$
$$\leq n^{1/2} \int_{\mathbb{R}^d} |h(s)| \left|m(F_{P_n}(s)) - m(F_P(s)) - \right.$$
$$\left. - \left\langle m'(F_P(s)), F_{P_n}(s) - F_P(s)\right\rangle\right| P_n(ds) +$$
$$+ n^{1/2} \int_{\mathbb{R}^d} |h(s)| \left|m(F_{P_n}(s)) - m(F_P(s)) - \right.$$
$$\left. - \left\langle m'(F_P(s)), F_{P_n}(s) - F_P(s)\right\rangle\right| P(ds) \leq$$
$$\leq n^{1/2} \|F_{P_n} - F_P\|_\infty \int_{\mathbb{R}^d} |h(s)| r_m\left(F_P(s); \|F_{P_n} - F_P\|_\infty\right) P_n(ds) +$$
$$+ n^{1/2} \|F_{P_n} - F_P\|_\infty \int_{\mathbb{R}^d} |h(s)| r_m\left(F_P(s); \|F_{P_n} - F_P\|_\infty\right) P(ds). \quad (3.14)$$

It follows from (3.1) and property (xiii) of Section 2 that

$$n^{1/2}\|F_{P_n} - F_P\|_\infty \int_{\mathbb{R}^d} |h(s)| r_m(F_P(s); \|F_{P_n} - F_P\|_\infty) P(ds) = o_p(1), \qquad n \to \infty.$$

If $\|F_{P_n} - F_P\|_\infty \leq \delta$ (which holds for any positive δ and large n with the probability close to 1), then

$$\int_{\mathbb{R}^d} |h(s)| r_m(F_P(s); \|F_{P_n} - F_P\|_\infty) P_n(ds) \leq \int_{\mathbb{R}^d} |h(s)| r_m(F_P(s); \delta) P_n(ds) \leq$$
$$\leq \int_{\mathbb{R}^d} |h(s)| r_m(F_P(s); \delta) (P_n - P)(ds) + \int_{\mathbb{R}^d} |h(s)| r_m(F_P(s); \delta) P(ds).$$

The first term tends to 0 by the Law of Large Numbers (for a fixed $\delta > 0$). The second one could be made arbitrarily small by choosing δ small enough (see condition (3.1)). Thus

$$\int_{\mathbb{R}^d} |h(s)| r_m(F_P(s); \|F_{P_n} - F_P\|_\infty) P_n(ds) = o_p(1), \qquad n \to \infty,$$

and by (3.14) we have

$$n^{1/2} \int_{\mathbb{R}^d} h(s) \left[m(F_{P_n}(s)) - m(F_P(s)) - \right.$$
$$\left. - \langle m'(F_P(s)), F_{P_n}(s) - F_P(s)\rangle\right] (P_n - P)(ds) = o_p(1). \qquad (3.15)$$

It follows from relationships (3.10), (3.13) and (3.15) that

$$n^{1/2}\int_{\mathbb{R}^d} h(s)\,[m(F_{P_n}(s)) - m(F_P(s))]\,(P_n - P)(ds) = o_p(1), \qquad n \to \infty. \tag{3.16}$$

Finally, (3.5), (3.6), (3.9) and (3.16) imply representation (3.4), which, in turn, along with the Central Limit Theorem imply asymptotic normality of $n^{1/2}(L_n - L(P))$ (under conditions (3.1)–(3.3)).

4. Some versions and examples of L-estimators

It's reasonable to normalize L-parameters and L-estimators, i.e., to consider instead of $L(P)$

$$\tilde{L}(P) := \tilde{L}_{h,m}(P) := \frac{L_{h,m}(P)}{L_{1,m}(P)} = \frac{\int_{\mathbb{R}^d} h(s)\,m(F_P(s))\,P(ds)}{\int_{\mathbb{R}^d} m(F_P(s))\,P(ds)}$$

(we assume that $L_{1,m}(P) \neq 0$ and

$$\int_{\mathbb{R}^d}(|h(s)| + 1)\,|m(F_P(s))|\,P(ds) < +\infty),$$

and instead of L_n

$$\tilde{L}_n := \tilde{L}(P_n) := \tilde{L}_{h,m}(P_n) = \frac{L_{h,m}(P_n)}{L_{1,m}(P_n)} = \frac{\int_{\mathbb{R}^d} h(s)\,m(F_{P_n}(s))\,P_n(ds)}{\int_{\mathbb{R}^d} m(F_{P_n}(s))\,P_n(ds)}.$$

Using Theorem 3.1 it's easy to get asymptotic normality of $\tilde{L}_n$. Namely, the following corollary holds.

4.1. Corollary. *Suppose that*

$$\int_{\mathbb{R}^d}(|h(s)| + 1)\,r_m(F_P(s);\delta)\,P(ds) \to 0, \qquad \delta \to, \tag{4.1}$$

$$\int_{\mathbb{R}^d}(|h(s)| + 1)^2\,|m'(F_P(s))|^2\,P(ds) < +\infty, \tag{4.2}$$

and

$$\int_{\mathbb{R}^d}(|h(s)| + 1)^2\,m^2(F_P(s))\,P(ds) < +\infty. \tag{4.3}$$

Suppose also that $L_{1,m}(P) \neq 0$. Then the sequence $n^{1/2}(\tilde{L}_n - \tilde{L}(P))$ is asymptotically normal with the mean 0 and the covariance

$$\int_{\mathbb{R}^d} \tilde{f} \otimes \tilde{f}\,dP - \int_{\mathbb{R}^d} \tilde{f}\,dP \otimes \int_{\mathbb{R}^d} \tilde{f}\,dP,$$

where $\tilde{f} := \tilde{f}_{P,h,m}$ is a map from $\mathbb{R}^d$ into $\mathbb{R}^l$, defined by

$$\tilde{f}(s) := \frac{f_{P,h,m}(s)}{L_{1,m}(P)} - \frac{L_{h,m}(P) f_{P,1,m}(s)}{L^2_{1,m}(P)}.$$

Moreover, the following asymptotic representation holds

$$n^{1/2}(\tilde{L}_n - \tilde{L}(P)) = n^{1/2} \int_{\mathbb{R}^d} \tilde{f}(s)\,(P_n - P)(ds) + o_p(1), \qquad n \to \infty. \tag{4.4}$$

Proof. Indeed, we have the following representation

$$\begin{aligned} n^{1/2}(\tilde{L}_n - \tilde{L}(P)) &= \\ &= \frac{L_{1,m}(P)}{L_{1,m}(P_n)} \Big[\frac{n^{1/2}\,(L_{h,m}(P_n) - L_{h,m}(P))}{L_{1,m}(P)} - \\ &\qquad - \frac{L_{h,m}(P)\, n^{1/2}\,(L_{1,m}(P_n) - L_{1,m}(P))}{L_{1,m}^2(P)} \Big]. \end{aligned} \tag{4.5}$$

Under conditions (4.1)-(4.3) we have (by Theorem 3.1) asymptotic normality of the sequence $n^{1/2}\,(L_{1,m}(P_n) - L_{1,m}(P))$, which under the condition $L_{1,m}(P) \neq 0$ implies that

$$\frac{L_{1,m}(P)}{L_{1,m}(P_n)} = 1 + o_p(1), \qquad n \to \infty.$$

Next representation (3.4) holds (under conditions (4.1)-(4.3) and $L_{1,m}(P) \neq 0$) for the sequence $n^{1/2}\,(L_{1,m}(P_n) - L_{1,m}(P))$ as well as for the sequence $n^{1/2}\,(L_{h,m}(P_n) - L_{h,m}(P))$. Using this fact and representation (4.5) we get (4.4), and asymptotic normality follows immediately.

In what follows we assume that $h(s) \equiv s$. Then $\tilde{L}(P)$ and $\tilde{L}_n$ possess equivariance properties (which $L(P)$ and L_n do not have). If P_θ denotes the θ-shift of P we would have (see property (ix), Section 2)

$$\begin{aligned} \tilde{L}(P_\theta) &= \frac{\int_{\mathbb{R}^d} s\, m(F_{P_\theta}(s))\, P_\theta(ds)}{\int_{\mathbb{R}^d} m(F_{P_\theta}(s))\, P_\theta(ds)} = \frac{\int_{\mathbb{R}^d} s\, m(F_P(s-\theta))\, P_\theta(ds)}{\int_{\mathbb{R}^d} m(F_P(s-\theta))\, P_\theta(ds)} = \\ &= \frac{\int_{\mathbb{R}^d} (s+\theta)\, m(F_P(s))\, P(ds)}{\int_{\mathbb{R}^d} m(F_P(s))\, P(ds)} \\ &= \theta + \frac{\int_{\mathbb{R}^d} s\, m(F_P(s))\, P(ds)}{\int_{\mathbb{R}^d} m(F_P(s))\, P(ds)} = \theta + \tilde{L}(P). \end{aligned}$$

4.2. Corollary. *Suppose P is symmetric with the center $a \in \mathbb{R}^d$ (in the sense that $X - a$ and $a - X$ are identically distributed, given X has the distribution P). Let m be a Borel function from B_d into $\mathbb{R}^1$, satisfying the conditions $m(t) = m(-t)$, $t \in B_d$ and*

$$\int_{\mathbb{R}^d} (|s| + 1)^2\, m^2(F_P(s))\, P(ds) < +\infty. \tag{4.6}$$

Suppose there exists a Borel function m' from B_d into $\mathbb{R}^d$ such that

$$\int_{\mathbb{R}^d} (|s|+1)^2 \, |m'(F_P(s))|^2 \, P(ds) < +\infty \tag{4.7}$$

and

$$\int_{\mathbb{R}^d} (|s|+1) \, r_m(F_P(s);\delta) \, P(ds) \to 0, \qquad \delta \to 0. \tag{4.8}$$

Finally, assume that $L_{1,m}(P) \neq 0$. Then the sequence $n^{1/2}(\tilde{L}_n - a)$ is asymptotically normal with the mean 0 and the covariance

$$\int_{\mathbb{R}^d} \tilde{f} \otimes \tilde{f} \, dP - \int_{\mathbb{R}^d} \tilde{f} \, dP \otimes \int_{\mathbb{R}^d} \tilde{f} \, dP,$$

where $\tilde{f} := \tilde{f}_{P,h,m}$.

Proof. It follows immediately from Corollary 4.1 since for a symmetric with respect to the center a distribution P and for a symmetric with respect to 0 function m we have $\tilde{L}(P) = a$.

Let $d \geq 2$ and let p be a locally bounded nonnegative function on $[0, +\infty)$, such that

$$\int_0^{+\infty} \rho^{d-1} \, p(\rho) \, d\rho = \frac{1}{c_d},$$

where c_d denotes the surface area of the unit sphere S^{d-1} in the space $\mathbb{R}^d$.

Now we assume that P is absolutely continuous with a density $p(|\cdot - a|)$, $a \in \mathbb{R}^d$. It means that it is spherically symmetric about the center a. Given a random vector X with the distribution P, let π denote the distribution of the random variable $|X - a|$. Then π is absolutely continuous with the density

$$\frac{d\pi}{d\rho} = c_d \, \rho^{d-1} \, p(\rho), \qquad \rho > 0.$$

4.3. Corollary. *Let μ be a Borel function from $[0,1]$ into $\mathbb{R}^1$, such that*

$$\mu(u) = \mu(0) + \int_0^u \mu'(v) \, dv, \qquad u \in [0,1],$$

where μ' is a Borel function on $[0,1]$, satisfying the condition[2]

$$\int_0^{+\infty} (\rho+1) \, \omega_{\mu'}(\psi_P(\rho);\delta) \, \pi(d\rho) \to 0, \qquad \delta \to 0. \tag{4.9}$$

[2] We denote

$$\omega_{\mu'}(u;\delta) = \sup\{|\mu'(v) - \mu'(u)| : v \in [0,1], \ |v-u| \leq \delta\}, \qquad u \in [0,1], \ \delta > 0.$$

Suppose also that

$$\int_0^{+\infty} (\rho+1)^2 \, (\mu \circ \psi_P)^2(\rho) \, \pi(d\rho) < +\infty, \tag{4.10}$$

and

$$\int_0^{+\infty} (\rho+1)^2 \, (\mu' \circ \psi_P)^2(\rho) \, \pi(d\rho) < +\infty. \tag{4.11}$$

Define $m(t) := \mu(|t|)$. *Assume that*

$$\int_0^{+\infty} (\mu \circ \psi_P)(\rho) \, \pi(d\rho) \neq 0. \tag{4.12}$$

Then the sequence $n^{1/2}(\tilde{L}_n - a)$ *is asymptotically normal with the mean* 0 *and the covariance*

$$\int_{\mathbb{R}^d} \tilde{f} \otimes \tilde{f} \, dP - \int_{\mathbb{R}^d} \tilde{f} \, dP \otimes \int_{\mathbb{R}^d} \tilde{f} \, dP,$$

where $\tilde{f} := \tilde{f}_{P,h,m}$.

Proof. Since $m(t) = \mu(|t|)$, $t \in B_d$, it's easy to show (using property (x), Section 2) that for any orthogonal transformation O of $\mathbb{R}^d$ we have

$$\tilde{L}(P \circ O^{-1}) = O\tilde{L}(P).$$

Since P is spherically symmetric with the center $a \in \mathbb{R}^d$, we have $\tilde{L}(P) = a$. Note that P is also symmetric about the center a in the sense of Corollary 4.2, so we can apply this corollary. Conditions (4.6) and (4.7) follow immediately from (4.10) and (4.11), respectively. Indeed, we can use representation (2.1) to get $|F_P(s)| = \psi_P(|s-a|)$, $s \in \mathbb{R}^d$. Obviously, convergence of the integral (4.6) is equivalent to

$$\int_{\mathbb{R}^d} (|s-a|+1)^2 \, m^2(F_P(s)) \, P(ds) < +\infty.$$

But the last integral is equal to

$$\int_{\mathbb{R}^d} (|s-a|+1)^2 \, \mu^2(\psi_P(|s-a|)) \, P(ds) = \int_0^{+\infty} (\rho+1)^2 \, (\mu \circ \psi_P)^2(\rho) \, \pi(d\rho) < +\infty,$$

which is finite by (4.10). Condition (4.7) could be proved similarly. It's also easy to check that condition (4.12) implies $L_{1,m}(P) \neq 0$.

Now we have to prove that (4.8) also holds. We have

$$m(t+h) - m(t) = \mu(|t+h|) - \mu(|t|) = \int_{|t|}^{|t+h|} \mu'(v) \, dv.$$

Thus

$$m(t+h) - m(t) - \mu'(|t|)\,(|t+h| - |t|) = \int_{|t|}^{|t+h|} [\mu'(v) - \mu'(|t|)] \, dv$$

and we got

$$\Big| m(t+h) - m(t) - \mu'(|t|)\,(|t+h| - |t|) \Big| \leq \int_{|t|-|h|}^{|t|+|h|} \Big| \mu'(v) - \mu'(|t|) \Big| \, dv.$$

(We assumed here for simplicity that $|t| \pm |h| \in [0,1]$. In the general case the last inequality should be slightly modified). Therefore

$$\delta^{-1} \sup\{|m(t+h) - m(t) - \mu'(|t|)(|t+h| - |t|)| : |h| \leq \delta,\ t+h \in B_d\} \leq \omega_{\mu'}(|t|;\delta). \tag{4.13}$$

Take $\delta > 0$. We are going to use the following identity

$$|t+h| - |t| = \int_0^1 \left\langle \frac{t+\lambda h}{|t+\lambda h|}, h \right\rangle d\lambda,$$

which holds for any t with $|t| > \delta$ and for all h with $|h| \leq \delta$. (In this case the function $\lambda \mapsto |t + \lambda h|$ is continuously differentiable in the interval $(0,1)$ with the derivative $\left\langle \frac{t+\lambda h}{|t+\lambda h|}, h \right\rangle$). It follows that

$$\left| |t+h| - |t| - \left\langle \frac{t}{|t|}, h \right\rangle \right| \leq \int_0^1 \left| \left\langle \frac{t+\lambda h}{|t+\lambda h|} - \frac{t}{|t|}, h \right\rangle \right| d\lambda \leq$$

$$\leq \sup_{\lambda \in [0,1]} \left| \frac{t+\lambda h}{|t+\lambda h|} - \frac{t}{|t|} \right| |h| \leq \frac{2\delta^2}{|t|}.$$

(We use an elementary inequality

$$\left| \frac{s}{|s|} - \frac{t}{|t|} \right| \leq \frac{2|s-t|}{|s| \vee |t|} \tag{4.14}$$

at this point). Thus for $|t| > \delta$ and $|h| \leq \delta$ we have

$$\left| \mu'(|t|)(|t+h| - |t|) - \mu'(|t|) \left\langle \frac{t}{|t|}, h \right\rangle \right| \leq 2\,|\mu'(|t|)| \frac{\delta^2}{|t|}. \tag{4.15}$$

It follows from (4.13) and (4.15) that for $|t| > \delta$

$$r_m(t;\delta) \leq \omega_{\mu'}(|t|;\delta) + 2\,|\mu'(|t|)| \frac{\delta}{|t|}. \tag{4.16}$$

In the case $|t| \leq \delta$, $|h| \leq \delta$ we have (assuming $\frac{t}{|t|} = 0$ for $t = 0$)

$$\left| |t+h| - |t| - \left\langle \frac{t}{|t|}, h \right\rangle \right| \leq 2\,|h| \leq 2\delta,$$

so that

$$r_m(t;\delta) \leq \omega_{\mu'}(|t|;\delta) + 2\,|\mu'(|t|)|. \tag{4.17}$$

Denote

$$\hat{r}_m(u;\delta) := \sup\{r_m(t;\delta) : |t| = u\}, \qquad u \in [0,1].$$

Using representation (2.1) and inequalities (4.16) and (4.17), we can get

$$\int_{\mathbb{R}^d}(|s-a|+1)\,r_m(F_P(s);\delta)\,P(ds) =$$

$$= \int_{\mathbb{R}^d}(|s-a|+1)\,r_m\Big(\psi_P(|s-a|)\frac{s-a}{|s-a|};\delta\Big)\,P(ds) \leq$$

$$\leq \int_0^{+\infty}(\rho+1)\,\hat{r}_m(\psi_P(\rho);\delta)\,\pi(d\rho) \leq \int_0^{+\infty}(\rho+1)\omega_{\mu'}(\psi_P(\rho);\delta)\,\pi(d\rho) +$$

$$+ 2\delta\int_{\{\psi_P(\rho)>\delta\}}(\rho+1)\,\frac{(\mu'\circ\psi_P)(\rho)}{\psi_P(\rho)}\,\pi(d\rho) +$$

$$+ 2\int_{\{\psi_P(\rho)\leq\delta\}}(\rho+1)\,(\mu'\circ\psi_P)(\rho)\,\pi(d\rho). \tag{4.18}$$

The first term in the right-hand side of inequality (4.18) tends to 0 when $\delta\to 0$ due to condition (4.9).

We show now that there exists a constant $c>0$ such that for small $\rho>0$

$$\psi_P(\rho) \geq c\rho. \tag{4.19}$$

Indeed, it's easy to check (using representation (2.2)) that ψ_P is a continuously differentiable function with the derivative

$$\psi_P'(\rho) = \int_{\mathbb{R}^d}\frac{x_2^2+\cdots+x_d^2}{((\rho-x_1)^2+x_2^2+\cdots+x_d^2)^{3/2}}\,P_{-a}(dx), \qquad \rho>0. \tag{4.20}$$

To show this consider a function

$$\gamma(\rho;x_1,\ldots,x_n) = \frac{\rho-x_1}{((\rho-x_1)^2+x_2^2+\cdots+x_d^2)^{1/2}}$$

(assuming that $\gamma=0$ for $\rho=x_1, x_2=\cdots=x_d=0$). Then

$$\psi_P(\rho) = \int_{\mathbb{R}^d}\gamma(\rho;x_1,\ldots,x_n)\,P_{-a}(dx), \qquad \rho>0.$$

We have P_{-a} a.e.

$$\frac{d}{d\rho}\gamma(\rho;x_1,\ldots,x_n) = \gamma'(\rho;x_1,\ldots,x_n) = \frac{x_2^2+\cdots+x_d^2}{((\rho-x_1)^2+x_2^2+\cdots+x_d^2)^{3/2}}.$$

It's easy to get (say, using inequality (4.14)) for $\delta\neq 0$

$$\left|\frac{\gamma(\rho+\delta;x_1,\ldots,x_n)-\gamma(\rho;x_1,\ldots,x_n)}{\delta}\right| \leq \frac{2}{((\rho-x_1)^2+x_2^2+\cdots+x_d^2)^{1/2}}.$$

The right-hand side of the last inequality is an integrable function with respect to P_{-a} (recall at this point that P has a locally bounded density and $d\geq 2$). Thus, by dominated convergence, we have

$$\lim_{\delta\to 0}\frac{\psi_P(\rho+\delta)-\psi_P(\rho)}{\delta} =$$

$$= \lim_{\delta\to 0}\int_{\mathbb{R}^d}\frac{\gamma(\rho+\delta;x_1,\ldots,x_n)-\gamma(\rho;x_1,\ldots,x_n)}{\delta}\,P_{-a}(dx) =$$

$$= \int_{\mathbb{R}^d}\lim_{\delta\to 0}\frac{\gamma(\rho+\delta;x_1,\ldots,x_n)-\gamma(\rho;x_1,\ldots,x_n)}{\delta}\,P_{-a}(dx) =$$

$$= \int_{\mathbb{R}^d}\gamma'(\rho;x_1,\ldots,x_n)\,P_{-a}(dx),$$

and representation (4.20) follows. It implies (again by dominated convergence) that $\psi_P'(\rho)$ is continuous in $\rho > 0$ and

$$\psi_P'(0+) = \int_{\mathbb{R}^d} \frac{x_2^2 + \cdots + x_d^2}{(x_1^2 + x_2^2 + \cdots + x_d^2)^{3/2}} P_{-a}(dx).$$

Since P_{-a} is absolutely continuous (so, it is not concentrated in a line $x_2 = x_3 = \cdots = x_d = 0$), we can conclude that $\psi_P'(\rho) > 0$, $\rho > 0$, and, moreover, $\psi_P'(0+) > 0$. Therefore, there exists $c > 0$ such that $\psi_P'(\rho) > c > 0$ for all small enough $\rho > 0$. Since $\psi_P(0+) = 0$ (P is nonatomic) (4.19) follows.

The inequality (4.19) along with condition (4.11) imply that

$$\int_0^{+\infty} (\rho + 1) \frac{|\mu' \circ \psi_P|(\rho)}{\psi_P(\rho)} \pi(d\rho) < +\infty.$$

Therefore the second term in the right-hand side of (4.18) also tends to 0 when $\delta \to 0$. Finally, by Hölder's inequality

$$\int_{\{\psi_P(\rho) \leq \delta\}} (\rho + 1)\, |\mu' \circ \psi_P|(\rho)\, \pi(d\rho) \leq$$

$$\leq \Big(\int_0^{+\infty} (\rho + 1)^2\, |\mu' \circ \psi_P|^2(\rho)\, \pi(d\rho)\Big)^{1/2} (\pi(\{\psi_P(\rho) \leq \delta\}))^{1/2}.$$

Now condition (4.11) and the fact that

$$\{\rho \colon \psi_P(\rho) \leq \delta\} \subset [0, \delta/c]$$

(which follows from (4.19)) imply that the third term in the right-hand side of (4.18) tends to 0 when $\delta \to 0$ (recall that P is nonatomic, and so is π). Hence, (4.18) implies condition (4.7), and Corollary 4.2 implies the rest.

Suppose now that μ' is a function from $L_\infty([0,1])$ and it is a.e. continuous. Then each of the following conditions

(i) $\int_0^{+\infty} \rho^2\, \pi(d\rho) < +\infty$,

(ii) μ is supported in an interval $[0, c]$ for some $c \in (0,1)$ (with no moment assumptions on π)

implies conditions (4.9)–(4.11).

In particular, one could take $0 < \alpha < \beta < 1$ and define

$$\mu(u) := \begin{cases} 1 & \text{for } u \in [0, \alpha) \\ \dfrac{\beta - u}{\beta - \alpha} & \text{for } u \in [\alpha, \beta) \\ 0 & \text{for } u \in [\beta, 1]. \end{cases}$$

Then, clearly, (ii) holds. This example of L-estimator could be considered as an extension of linearly trimmed mean to multivariate case.

References

Boos, D. D. (1979). A differential for L-statistics. *Ann. Statist.* **7**, 955–959.

Chaudhuri, P. (1993). On a geometric notion of quantiles for multivariate data. *JASA*. (In press).

Dudley, R. M. and Koltchinskii, V. (1992). *The spatial quantiles.* Preprint.

Huber, P. J. (1981). *Robust Statistics.* John Wiley and Sons, New York.

Koltchinskii, V. (1994a). Bahadur–Kiefer approximation for spatial quantiles. In: *Probability in Banach Spaces.* J. Hoffmann–Jorgensen, J. Kuelbs and M. Marcus (Eds.), 394–408. Birkhäuser, Boston.

Koltchinskii, V. (1994b). Nonlinear transformations of empirical processes: functional inverses and Bahadur–Kiefer representations. In: *Probability Theory and Mathematical Statistics*, B. Grigelionis, J.Kubilius et al. (Eds.), 1994, 423–445. VSP-TEV, The Netherlands.

Koltchinskii, V. (1994c). *M-estimation and spatial quantiles.* Preprint.

Mason, D. M. (1981). Asymptotic normality of linear combinations of order statistics with a smooth score function. *Ann. Statist.* **9**, 899–904.

Norvaisha, R. (1994). *The central limit theorem for L-statistics.* Preprint.

Serfling, R. J. (1980). *Approximation Theorems of Mathematical Statistics.* John Wiley and Sons, New York.

Shorack, G. R. (1972). Functions of order statistics. *Ann. Math. Stat.* **43**, 412–427.

Shorack, G. R. and Wellner, J. A. (1986). *Empirical Processes with Applications to Statistics.* John Wiley and Sons, New York.

Stigler, S. M. (1974). Linear functions of order statistics with smooth weight functions. *Ann. Statist.* **2**, 676–693.

Exploring Stochastic Laws. pp. 201–218.
Eds. A.V. Skorokhod and Yu.V. Borovskikh

Asymptotic inference for nearly nonstationary complex-valued AR(1) processes

J. KORMOS[1,*], T. VAN DER MEER[2], G. PAP[3,*] AND M. VAN ZUIJLEN[4,**]

[1] *Institute of Mathematics and Informatics, Lajos Kossuth University, Debrecen, Hungary*
[2] *Department of Mathematics, University of Nijmegen, Nijmegen, The Netherlands*
[3] *Institute of Mathematics and Informatics, Lajos Kossuth University, Debrecen, Hungary*
[4] *Department of Mathematics, University of Nijmegen, Nijmegen, The Netherlands*

Received March 20, 1995

Abstract. In this paper nearly nonstationary complex-valued AR(1) processes (in other words, models with characteristic root near the unit circle) are studied. Our main aim is to clarify the relation between the discrete and continuous time models. First we give some convergence results related to the random step functions defined by the help of the discrete models. As a consequence, we obtain weak convergence of the least-squares estimator of the parameter in the discrete models to the maximum likelihood estimator of the parameter of the connected continuous time model. The above results will be used to solve similar problems for nearly nonstationary (real-valued) AR(p) processes in a forthcoming paper.

1. Introduction

In the first order autoregressive AR(1) model the observations X_k are generated according to the scheme

$$\begin{cases} X_k = \beta X_{k-1} + \varepsilon_k, & k = 1, 2, \dots \\ X_0 = 0, \end{cases} \tag{1}$$

where ε_k is the (unobservable) random disturbance (noise) at time k, and β is an unknown parameter. Let us assume that (ε_k) is a square-integrable martingale difference sequence satisfying the conditional form of the Lindeberg condition (see, e.g. Chan and Wei (1987), Jeganathan (1991)). We mention that the results are

* Part of this work has been carried out while these authors were visiting the Department of Mathematics, University of Nijmegen
** Part of this work has been carried out while this author was visiting the Institute of Mathematics and Informatics, University of Debrecen, Hungary and was supported by Grant ERB-CIPA-CT-92-1512 of the European Economic Community.

similar if (ε_k) is a sequence satisfying some mixing condition (see, e.g. Phillips (1987)).

The least-squares estimator (LSE) of β is

$$\hat{\beta}_n = \frac{\sum_{k=1}^n X_k X_{k-1}}{\sum_{k=1}^n X_{k-1}^2}. \tag{2}$$

When $|\beta| < 1$ it is well known (see, for example, Mann and Wald (1943) and Anderson (1959)) that the standardized LSE is asymptotically normal:

$$\left(\sum_{k=1}^n X_{k-1}^2\right)^{1/2} (\hat{\beta}_n - \beta) \xrightarrow{\mathcal{D}} \mathcal{N}(0,1), \tag{3}$$

where $\xrightarrow{\mathcal{D}}$ denotes convergence in distribution. Remark, that with another standardization

$$\sqrt{n}\,(\hat{\beta}_n - \beta) \xrightarrow{\mathcal{D}} \mathcal{N}(0, 1-\beta^2). \tag{4}$$

However, when $\beta = 1$, the limit distribution of the properly standardized sequence of the least-squares estimators is non-normal, and it was shown by White (1958) (see also Rao (1978)) that

$$n(\hat{\beta}_n - 1) \xrightarrow{\mathcal{D}} \frac{\int_0^1 W(t)\,dW(t)}{\int_0^1 W^2(t)\,dt}, \tag{5}$$

where $W(t)$, $t \geq 0$ is a standard Wiener process.

The result (5) led to the study of the following so-called nearly nonstationary AR(1) model:

$$\begin{cases} X_{n,k} = \beta_n X_{n,k-1} + \varepsilon_{n,k}, & k = 1, 2, \dots, n \\ X_{n,0} = 0, \end{cases} \tag{6}$$

where $\beta_n = 1 + \gamma_n/n$, $\gamma_n \to \gamma$ (see, for example, Chan and Wei (1987), Phillips (1987), Jeganathan (1991), Dzhaparidze *et al.* (1992)). (The above model is called also near integrated and is applied often in economic theory; see Phillips (1987).) It was shown that

$$\left(\sum_{k=1}^n X_{n,k-1}^2\right)^{1/2} (\hat{\beta}_n - \beta_n) \xrightarrow{\mathcal{D}} \frac{\int_0^1 Y(t)\,dW(t)}{\left(\int_0^1 Y^2(t)\,dt\right)^{1/2}}, \tag{7}$$

where $Y(t)$, $t \in [0,1]$ is an Ornstein–Uhlenbeck process defined as the solution of the stochastic differential equation

$$dY(t) = \gamma Y(t)\,dt + dW(t), \qquad Y(0) = 0. \tag{8}$$

By another standardization

$$n\,(\hat{\beta}_n - \beta_n) \xrightarrow{\mathcal{D}} \frac{\int_0^1 Y(t)\,dW(t)}{\int_0^1 Y^2(t)\,dt}. \tag{9}$$

In this paper nearly nonstationary complex-valued AR(1) processes are studied. We suppose that for every $n = 1, 2, \ldots$ the observations are generated according to the scheme

$$\begin{cases} X_{nk} = \beta_n X_{n,k-1} + \varepsilon_{nk}, & k = 1, 2, \ldots, n \\ X_{n0} = 0, \end{cases} \tag{10}$$

where (ε_{nk}) is an array of complex random variables and $\beta_n = e^{\gamma_n/n + i\theta}$, where (γ_n) is a sequence of unknown complex numbers such that $\gamma_n \to \gamma$, $\gamma \in \mathbb{C}$ is unknown, and $\theta \in (-\pi, \pi]$ is known. The real-valued cases, where $\beta_n = e^{\gamma_n/n}$ or $\beta_n = -e^{\gamma_n/n}$ are contained in Jeganathan (1991).

Our main aim in this paper is to clarify the relation between discrete and continuous time models. This relationship was pointed out by Kormos (1989, 1990) and by Kormos and Piterbarg (1991) in connection with hypothesis testing of nonstationarity for a (real-valued) Gaussian autoregressive time series.

We shall examine the limit behaviour of the sequence of random step functions

$$X_{n,[nt]}, \quad t \in [0,1], \quad n = 1, 2, \ldots,$$

and we give a necessary and sufficient condition on the innovations (ε_{nk}) for convergence (in the Skorokhod space $D([0,1] \to \mathbb{C})$) of the *rotated* sequence

$$\frac{1}{\sqrt{n}} e^{-i[nt]\theta} X_{n,[nt]}, \quad t \in [0,1], \quad n = 1, 2, \ldots$$

to a complex-valued Ornstein–Uhlenbeck process $Y(t)$, $t \in [0,1]$, which depends only on the parameter γ (and does not depend on the parameter θ). We remark that the unrotated sequence does not converge. The Ornstein–Uhlenbeck process $Y(t)$, $t \in [0,1]$ can be considered as a continuous time AR(1) process (see Arató (1982)).

As a consequence, we obtain a sufficient condition for weak convergence of the least-squares estimators (LSE) $\hat{\gamma}_n$ in the discrete models to the maximum likelihood estimator (MLE) $\hat{\gamma}$ of the parameter of the limit process $Y(t)$, $t \in [0,1]$.

Moreover, we establish weak convergence of the sequence of quadratic forms

$$\sum_{k=1}^{n} |X_{nk} - \beta_n X_{n,k-1}|^2 - \sum_{k=1}^{n} |X_{nk} - e^{i\theta} X_{n,k-1}|^2,$$

(which are connected with the LSE), to the log-likelihood function of the process $Y(t)$, $t \in [0,1]$. The above results will be used to solve similar problems for nearly nonstationary (real-valued) AR(p) processes in a forthcoming paper.

In the last section we generalize the results to the case where the sequence of the partial sum processes

$$M_n(t) = \frac{1}{\sqrt{n}} \sum_{k=1}^{[nt]} e^{-ik\theta} \varepsilon_{nk}, \quad t \in [0,1], \qquad n = 1, 2, \ldots$$

of the rotated noises converges to an arbitrary complex-valued Wiener process (which then governs the corresponding complex-valued Ornstein–Uhlenbeck process).

2. Convergence results

For every $n = 1, 2, \dots$ consider the complex-valued AR(1) model

$$\begin{cases} X_{nk} = \beta_n X_{n,k-1} + \varepsilon_{nk}, & k = 1, 2, \dots, n \\ X_{n0} = 0, \end{cases}$$

where (ε_{nk}) is an array of complex random variables and $\beta_n = e^{\gamma_n/n + i\theta}$, where (γ_n) is a sequence of complex numbers such that $\gamma_n \to \gamma$ and $\theta \in (-\pi, \pi]$.

The *rotated* observations

$$Z_{nk} = e^{-ik\theta} X_{nk}, \quad k = 0, 1, \dots, n; \quad n = 1, 2, \dots$$

form again a nearly nonstationary complex AR(1) model

$$\begin{cases} Z_{nk} = e^{\gamma_n/n} Z_{n,k-1} + \zeta_{nk}, & k = 1, 2, \dots, n \\ Z_{n0} = 0, \end{cases} \tag{11}$$

where $(\zeta_{nk}) = (e^{-ik\theta}\varepsilon_{nk})$ is the rotated array of the random disturbances.

The random step functions

$$\begin{aligned} Y_n(t) &= \frac{1}{\sqrt{n}} Z_{n,[nt]}, \quad t \in [0,1] \\ M_n(t) &= \frac{1}{\sqrt{n}} \sum_{k=1}^{[nt]} \zeta_{nk}, \quad t \in [0,1] \end{aligned}$$

can be considered as random elements in the complex Skorokhod space $D([0,1] \to \mathbb{C})$. We shall use the space $\mathbb{C}^2$ endowed with the norm $|z| = (|z_1|^2 + |z_2|^2)^{1/2}$, $z = (z_1, z_2) \in \mathbb{C}^2$. We denote by $C([0,1] \to \mathbb{R}^2)$ the space of $\mathbb{R}^2$-valued continuous functions endowed with the supremum norm. The supremum norm and the Skorokhod metric on the spaces $D([0,1] \to \mathbb{C})$, $D([0,1] \to \mathbb{C}^2)$ and $D([0,1] \to \mathbb{R}^2)$ will be denoted by $\|\cdot\|_\infty$ and ρ, respectively. We shall need the following simple lemma, which is based on the continuous mapping theorem and the Skorokhod-construction.

Lemma 1. *Let* $\Phi, \Phi_n : D([0,1] \to \mathbb{C}) \to D([0,1] \to \mathbb{C}^2)$, $n = 1, 2, \dots$ *be measurable mappings such that* $\|\Phi_n(x_n) - \Phi(x)\|_\infty \to 0$ *for all* $x, x_n \in D([0,1] \to \mathbb{C})$ *with* $\|x_n - x\|_\infty \to 0$. *Let* Z, Z_n, $n = 1, 2, \dots$ *be stochastic processes with values in* $D([0,1] \to \mathbb{C})$ *such that* $Z_n \xrightarrow{\mathcal{D}} Z$ *in* $D([0,1] \to \mathbb{C})$ *and almost all trajectories of* Z *are continuous. Then* $\Phi_n(Z_n) \xrightarrow{\mathcal{D}} \Phi(Z)$ *in* $D([0,1] \to \mathbb{C}^2)$.

Proof. Due to the Skorokhod-construction we can find processes $\widetilde{Z}_n$ and a process $\widetilde{Z}$, such that $\widetilde{Z}_n \stackrel{\mathcal{D}}{=} Z_n$, $\widetilde{Z} \stackrel{\mathcal{D}}{=} Z$ and

$$\rho(\widetilde{Z}_n, \widetilde{Z}) \to 0 \quad \text{a.s.}$$

Using the fact that $\widetilde{Z}$ has continuous trajectories a.s., we conclude that

$$\|\widetilde{Z}_n - \widetilde{Z}\|_\infty \to 0 \quad \text{a.s.}$$

Thus we have

$$\|\Phi_n(\widetilde{Z}_n) - \Phi(\widetilde{Z})\|_\infty \to 0 \quad \text{a.s.}$$

and hence

$$\Phi_n(\widetilde{Z}_n) \xrightarrow{\mathcal{D}} \Phi(\widetilde{Z}) \quad \text{in} \quad D([0,1] \to \mathbb{C}^2).$$

The last relation implies the desired result.

A *complex Wiener process* is a process $W(t) = W_1(t) + iW_2(t)$, $t \in [0,1]$, where $(W_1(t), W_2(t))$, $t \in [0,1]$, is a — not necessarily standard — Wiener process in $\mathbb{R}^2$.

Consider the complex-valued Ornstein–Uhlenbeck process defined as the solution of the stochastic differential equation

$$dY(t) = \gamma Y(t)\,dt + dW(t), \qquad Y(0) = 0. \tag{12}$$

Our first theorem gives a necessary and sufficient condition for the weak convergence $Y_n \xrightarrow{\mathcal{D}} Y$ in $D([0,1] \to \mathbb{C})$.

Theorem 1. *The following statements are equivalent:*

(i) $M_n \xrightarrow{\mathcal{D}} W$ *in* $D([0,1] \to \mathbb{C})$,

(ii) $Y_n \xrightarrow{\mathcal{D}} Y$ *in* $D([0,1] \to \mathbb{C})$,

(iii) $(M_n, Y_n) \xrightarrow{\mathcal{D}} (W, Y)$ *in* $D([0,1] \to \mathbb{C}^2)$.

Remark. The statement (i) is in other words: the functional central limit theorem holds for the rotated triangular array $(e^{-ik\theta}\varepsilon_{nk})_{k=1,\ldots,n;n\geq 1}$.

Proof. (i)$\Longrightarrow$(iii). It is known that the process $Y(t)$, $t \in [0,1]$, can be expressed as follows.

$$Y(t) = \int_0^t e^{\gamma(t-s)}\,dW(s), \quad t \in [0,1].$$

The main idea of the proof is getting rid of Itô's integral using partial integration, which gives

$$Y(t) = W(t) + \gamma \int_0^t e^{\gamma(t-s)} W(s)\,ds, \quad t \in [0,1].$$

A similar formula holds for the random step functions $Y_n(t)$, $t \in [0,1]$:

$$Y_n(t) = e^{\gamma_n([nt]/n-t)} M_n(t) + \gamma_n \int_0^t e^{\gamma_n([nt]/n-s)} M_n(s)\,ds, \tag{13}$$

since (11) implies

$$Y_n(t) = \frac{1}{\sqrt{n}} \sum\nolimits_{k=1}^{[nt]} e^{\gamma_n([nt]-k)/n} \zeta_{nk},$$

and the discrete analogue of the partial integration yields

$$\begin{aligned}
\gamma_n \int_0^{[nt]/n} e^{-\gamma_n s} M_n(s) ds &= \sum\nolimits_{j=1}^{[nt]} M_n\left(\frac{j-1}{n}\right)\left(e^{-\gamma_n(j-1)/n} - e^{-\gamma_n j/n}\right) \\
&= \frac{1}{\sqrt{n}} \sum\nolimits_{j=1}^{[nt]} \sum\nolimits_{k=1}^{j} \zeta_{nk}\left(e^{-\gamma_n(j-1)/n} - e^{-\gamma_n j/n}\right) \\
&= \frac{1}{\sqrt{n}} \sum\nolimits_{k=1}^{[nt]-1} \sum\nolimits_{j=k+1}^{[nt]} \zeta_{nk}\left(e^{-\gamma_n(j-1)/n} - e^{-\gamma_n j/n}\right) \\
&= \frac{1}{\sqrt{n}} \sum\nolimits_{k=1}^{[nt]-1} \zeta_{nk}\left(e^{-\gamma_n k/n} - e^{-\gamma_n [nt]/n}\right) \\
&= \frac{1}{\sqrt{n}} \sum\nolimits_{k=1}^{[nt]} \zeta_{nk}\left(e^{-\gamma_n k/n} - e^{-\gamma_n [nt]/n}\right) \\
&= e^{-\gamma_n[nt]/n}(Y_n(t) - M_n(t)),
\end{aligned}$$

thus

$$\begin{aligned}
\gamma_n \int_0^t e^{-\gamma_n s} M_n(s)\, ds &= e^{-\gamma_n[nt]/n}(Y_n(t) - M_n(t)) + (e^{-\gamma_n[nt]/n} - e^{-\gamma_n t})M_n(t) \\
&= e^{-\gamma_n[nt]/n} Y_n(t) - e^{-\gamma_n t} M_n(t),
\end{aligned}$$

which gives the equation (13).

Hence the processes (W, Y) and (M_n, Y_n) can be expressed as

$$(W, Y) = \Phi(W), \qquad (M_n, Y_n) = \Phi_n(M_n), \quad n = 1, 2, \ldots,$$

where the measurable mappings $\Phi, \Phi_n : D([0,1] \to \mathbb{C}) \to D([0,1] \to \mathbb{C}^2)$, $n = 1, 2, \ldots$ are defined as follows

$$\begin{aligned}
\Phi(x)(t) &= \left(x(t), x(t) + \gamma \int_0^t e^{\gamma(t-s)} x(s) ds\right), \\
\Phi_n(x)(t) &= \left(x(t), e^{-\gamma_n([nt]/n-t)} x(t) + \gamma_n \int_0^t e^{\gamma_n([nt]/n-s)} x(s) ds\right).
\end{aligned}$$

Applying Lemma 1 we obtain $(M_n, Y_n) \xrightarrow{\mathcal{D}} (W, Y)$ in $D([0,1] \to \mathbb{C}^2)$.

(iii)$\Longrightarrow$(i) and (iii)$\Longrightarrow$(ii) are trivial.

(ii)$\Longrightarrow$(iii). Since the process $Y(t)$, $t \in [0,1]$, is a solution of the stochastic differential equation (12) we have

$$Y(t) = \gamma \int_0^t Y(s)\, ds + W(t),$$

thus

$$W(t) = Y(t) - \gamma \int_0^t Y(s)\, ds.$$

A similar formula holds for the partial sum process $M_n(t),\ t\in[0,1]$:

$$\begin{aligned} M_n(t) &= \frac{1}{\sqrt{n}}\sum_{k=1}^{[nt]}\zeta_{nk} = \frac{1}{\sqrt{n}}\sum_{k=1}^{[nt]}(Z_{nk} - e^{\gamma_n/n}Z_{n,k-1}) \\ &= \frac{1}{\sqrt{n}}Z_{n,[nt]} - \frac{1}{\sqrt{n}}(e^{\gamma_n/n}-1)\sum_{k=1}^{[nt]-1} Z_{nk} \\ &= Y_n(t) - n(e^{\gamma_n/n}-1)\int_0^{[nt]/n} Y_n(s)\,ds \\ &= \left(1+(t-[nt]/n)n(e^{\gamma_n/n}-1)\right)Y_n(t) - n(e^{\gamma_n/n}-1)\int_0^t Y_n(s)\,ds. \end{aligned}$$

Hence the processes (W,Y) and (M_n,Y_n) can be expressed as

$$(W,Y)=\Psi(Y), \qquad (M_n,Y_n)=\Psi_n(Y_n), \quad n=1,2,\ldots,$$

where the measurable mappings $\Psi,\Psi_n : D([0,1]\to\mathbb{C})\to D([0,1]\to\mathbb{C}^2)$, $n=1,2,\ldots$ are defined as follows

$$\begin{aligned} \Psi(x)(t) &= \left(x(t)-\gamma\int_0^t x(s)ds,\, x(t)\right), \\ \Psi_n(x)(t) &= \left(\left(1+(t-[nt]/n)n(e^{\gamma_n/n}-1)\right)x(t) - n(e^{\gamma_n/n}-1)\int_0^t x(s)\,ds,\, x(t)\right). \end{aligned}$$

Applying Lemma 1 we obtain $(M_n,Y_n)\xrightarrow{\mathcal{D}}(W,Y)$ in $D([0,1]\to\mathbb{C}^2)$.

We remark that the same method can be used in case of an arbitrary continuous semimartingale instead of a Wiener process (see also in Dzhaparidze *et al.* (1992)).

3. Least-squares estimators

The least-squares estimator of the parameter γ_n of the model (10) can be obtained minimizing the sum of squares

$$\sum_{k=1}^n |X_{nk}-e^{\gamma_n/n+i\theta}X_{n,k-1}|^2 = \sum_{k=1}^n |Z_{nk}-e^{\gamma_n/n}Z_{n,k-1}|^2.$$

First we find the least-squares estimator of the parameter $\lambda_n+i\omega_n=e^{\gamma_n/n}$, $\lambda_n,\omega_n\in\mathbb{R}$. The above sum of squares can be written as follows:

$$\sum_{k=1}^n |Z_{nk}|^2 - 2\lambda_n\Re\left(\sum_{k=1}^n \overline{Z}_{n,k-1}Z_{nk}\right) - 2\omega_n\Im\left(\sum_{k=1}^n \overline{Z}_{n,k-1}Z_{nk}\right) + (\lambda_n^2+\omega_n^2)\sum_{k=1}^n |Z_{n,k-1}|^2,$$

where $\Re(z)$ and $\Im(z)$ denotes the real and the imaginary part of $z\in\mathbb{C}$, respectively. We get the least-squares estimators

$$\hat\lambda_n = \frac{\Re\left(\sum_{k=1}^n \overline{Z}_{n,k-1}Z_{nk}\right)}{\sum_{k=1}^n |Z_{n,k-1}|^2}, \qquad \hat\omega_n = \frac{\Im\left(\sum_{k=1}^n \overline{Z}_{n,k-1}Z_{nk}\right)}{\sum_{k=1}^n |Z_{n,k-1}|^2}.$$

Finally, we obtain

$$\hat{\gamma}_n = n \log \frac{\sum_{k=1}^n \overline{Z}_{n,k-1} Z_{nk}}{\sum_{k=1}^n |Z_{n,k-1}|^2}.$$

Let $W(t) = W_1(t) + iW_2(t)$, $t \in [0,1]$, be a complex Wiener process, where $(W_1(t), W_2(t))$, $t \in [0,1]$, is a — not necessarily standard — Wiener process in $\mathbb{R}^2$ with $\mathbb{E}W_1(t) = \mathbb{E}W_1(t) = 0$ and

$$\mathbb{E}\begin{pmatrix} W_1^2(t) & W_1(t)W_2(t) \\ W_1(t)W_2(t) & W_2^2(t) \end{pmatrix} = t\Sigma.$$

Consider the complex-valued Ornstein–Uhlenbeck process defined as the solution of the stochastic differential equation

$$dY(t) = \gamma Y(t)\,dt + dW(t), \qquad Y(0) = 0. \tag{14}$$

We can identify the process $Y(t)$, $t \in [0,1]$, with the process

$$(Y_1(t), Y_2(t)) = (\Re(Y(t)), \Im(Y(t))), \qquad t \in [0,1],$$

with values in $\mathbb{R}^2$. We have

$$\begin{pmatrix} dY_1(t) \\ dY_2(t) \end{pmatrix} = \begin{pmatrix} \lambda & -\omega \\ \omega & \lambda \end{pmatrix} \begin{pmatrix} Y_1(t)dt \\ Y_2(t)dt \end{pmatrix} + \begin{pmatrix} dW_1(t) \\ dW_2(t) \end{pmatrix}, \qquad \begin{pmatrix} Y_1(0) \\ Y_2(0) \end{pmatrix} = \begin{pmatrix} 0 \\ 0 \end{pmatrix}$$

where $\lambda = \Re(\gamma)$, $\omega = \Im(\gamma)$. Hence $(Y_1(t), Y_2(t))$, $t \in [0,1]$, is an Ornstein–Uhlenbeck process in $\mathbb{R}^2$.

Suppose now that $W(t)$, $t \in [0,1]$, is a *complex standard Wiener process*, that is, $(\sqrt{2}W_1(t), \sqrt{2}W_2(t))$, $t \in [0,1]$, is a standard Wiener process in $\mathbb{R}^2$. Then $\mathbb{E}|W(t)|^2 = t$ and $\Sigma = \frac{1}{2}I$, where I is the 2×2 unit matrix.

Consider the measures $\mathbb{P}_W$ and $\mathbb{P}_Y$ on $C([0,1] \to \mathbb{R}^2)$ generated by the processes (W_1, W_2) and (Y_1, Y_2), respectively. It is known that $\mathbb{P}_Y$ is absolutely continuous with respect to the measure $\mathbb{P}_W$ and the Radon–Nikodym derivative has the form

$$\frac{d\mathbb{P}_Y}{d\mathbb{P}_W}(Y) = \exp\left\{-\int_0^1 |AY(t)|^2\,dt + 2\int_0^1 \langle AY(t), dY(t)\rangle\right\},$$

where

$$A = \begin{pmatrix} \lambda & -\omega \\ \omega & \lambda \end{pmatrix},$$

(see Arató (1982)). It can be rewritten as follows

$$\begin{aligned}\frac{d\mathbb{P}_Y}{d\mathbb{P}_W}(Y) &= \exp\Big\{-(\lambda^2+\omega^2)\int_0^1 (Y_1^2(t)+Y_2^2(t))\,dt + 2\lambda\int_0^1 Y_1(t)\,dY_1(t) \\ &\qquad +2\lambda\int_0^1 Y_2(t)\,dY_2(t) + 2\omega\int_0^1 (Y_1(t)\,dY_2(t) - Y_2(t)\,dY_1(t))\Big\} \\ &= \exp\left\{-|\gamma|^2\int_0^1 |Y(t)|^2\,dt + 2\Re\left(\int_0^1 \overline{\gamma Y(t)}\,dY(t)\right)\right\}.\end{aligned}$$

Consequently the maximum likelihood estimators of the parameters λ and ω are

$$\hat{\lambda} = \frac{\Re\left(\int_0^1 \overline{Y(t)}\,dY(t)\right)}{\int_0^1 |Y(t)|^2\,dt}, \qquad \hat{\omega} = \frac{\Im\left(\int_0^1 \overline{Y(t)}\,dY(t)\right)}{\int_0^1 |Y(t)|^2\,dt},$$

hence the maximum likelihood estimator of γ is

$$\hat{\gamma} = \frac{\int_0^1 \overline{Y(t)}\,dY(t)}{\int_0^1 |Y(t)|^2\,dt}.$$

Consider the quadratic form

$$L_n(X_{n1},\dots,X_{nn}) = \sum_{k=1}^n |X_{nk} - e^{\gamma_n/n+i\theta}X_{n,k-1}|^2 - \sum_{k=1}^n |X_{nk} - e^{i\theta}X_{n,k-1}|^2,$$

connected with the LSE $\hat{\gamma}_n$, and the log-likelihood function of the process Y

$$L(Y) = -\log\frac{d\mathbb{P}_Y}{d\mathbb{P}_W}(Y).$$

The next theorem gives a sufficient condition for the convergences $\hat{\gamma}_n \xrightarrow{\mathcal{D}} \hat{\gamma}$ and $L_n(X_{n1},\dots,X_{nn}) \xrightarrow{\mathcal{D}} L(Y)$.

We shall denote by χ_B the indicator function of the set B.

Theorem 2. *Let $(\varepsilon_{nk})_{k=1,\dots,n;n\geq 1}$ be a triangular array of complex square integrable martingale differences with respect to the filtrations $(\mathcal{F}_{nk})_{k=0,1,\dots,n;n\geq 1}$ such that for all $t\in[0,1]$*

$$\frac{1}{n}\sum_{k=1}^{[nt]}\mathbb{E}\left\{\left(\begin{array}{cc}(\Re(\zeta_{nk}))^2 & \Re(\zeta_{nk})\Im(\zeta_{nk}) \\ \Re(\zeta_{nk})\Im(\zeta_{nk}) & (\Im(\zeta_{nk}))^2\end{array}\right)\Bigg|\mathcal{F}_{n,k-1}\right\} \xrightarrow{\mathbb{P}} \left(\begin{array}{cc}\frac{t}{2} & 0 \\ 0 & \frac{t}{2}\end{array}\right), \quad \text{as } n\to\infty,$$

where $\zeta_{nk} = e^{-ik\theta}\varepsilon_{nk}$ and

$$\forall\alpha>0 \qquad \frac{1}{n}\sum_{k=1}^{[nt]}\mathbb{E}(|\varepsilon_{nk}|^2\chi_{\{|\varepsilon_{nk}|>\alpha\sqrt{n}\}}|\mathcal{F}_{n,k-1}) \xrightarrow{\mathbb{P}} 0, \quad \text{as } n\to\infty.$$

Then

$$\hat{\gamma}_n \xrightarrow{\mathcal{D}} \hat{\gamma}$$

and

$$L_n(X_{n1},\dots,X_{nn}) \xrightarrow{\mathcal{D}} L(Y).$$

Proof. Using a version of the functional central limit theorem on the space $D([0,1]\to\mathbb{R}^2)$ (Theorem 7.11 in Liptser and Shiryayev (1989)) we obtain $M_n \xrightarrow{\mathcal{D}} W$ in $D([0,1]\to\mathbb{C})$. Applying Theorem 1 we can conclude $(M_n,Y_n)\xrightarrow{\mathcal{D}}(W,Y)$ in $D([0,1]\to\mathbb{C}^2)$. This implies convergence of the stochastic integrals

$$\left(\int_0^1 |Y_n(t)|^2\,dt, \int_0^1 \overline{Y_n(t)}\,dM_n(t)\right) \xrightarrow{\mathcal{D}} \left(\int_0^1 |Y(t)|^2\,dt, \int_0^1 \overline{Y(t)}\,dW(t)\right)$$

(see Proposition 6 in Jeganathan (1991)). Using continuous mapping theorem (c.f. Theorem 5.5 in Billingsley (1968)) we obtain

$$\begin{aligned}
\hat{\gamma}_n &= n\log\frac{\sum_{k=1}^n \overline{Z}_{n,k}(e^{\gamma_n/n}Z_{n,k-1}+\zeta_{nk})}{\sum_{k=1}^n |Z_{n,k-1}|^2} \\
&= n\log\left(e^{\gamma_n/n}+\frac{\int_0^1\overline{Y_n(t)}\,dM_n(t)}{n\int_0^1|Y_n(t)|^2\,dt}\right) \\
&= \log\left[1+\frac{1}{n}\left(n(e^{\gamma_n/n}-1)+\frac{\int_0^1\overline{Y_n(t)}\,dM_n(t)}{\int_0^1|Y_n(t)|^2\,dt}\right)\right]^n \\
&\xrightarrow{\mathcal{D}} \gamma+\frac{\int_0^1\overline{Y(t)}\,dW(t)}{\int_0^1|Y(t)|^2\,dt}=\frac{\int_0^1\overline{Y(t)}\,dY(t)}{\int_0^1|Y(t)|^2\,dt}=\hat{\gamma}
\end{aligned}$$

and

$$\begin{aligned}
&L_n(X_{n1},\dots,X_{nn}) = \\
&= \sum_{k=1}^n |Z_{nk}-e^{\gamma_n/n}Z_{n,k-1}|^2-\sum_{k=1}^n|Z_{nk}-Z_{n,k-1}|^2 \\
&= (|e^{\gamma_n/n}|^2-1)\sum_{k=1}^n|Z_{n,k-1}|^2-2\Re\left(\overline{(e^{\gamma_n/n}-1)Z_{n,k-1}}Z_{nk}\right) \\
&= -n^2|e^{\gamma_n/n}-1|^2\int_0^1|Y_n(t)|^2\,dt-2n\Re\left(\int_0^1\overline{(e^{\gamma_n/n}-1)Y_n(t)}\,dM_n(t)\right) \\
&\xrightarrow{\mathcal{D}} -|\gamma|^2\int_0^1|Y(t)|^2\,dt-2\Re\left(\int_0^1\overline{\gamma Y(t)}\,dW(t)\right) \\
&= |\gamma|^2\int_0^1|Y(t)|^2\,dt-2\Re\left(\int_0^1\overline{\gamma Y(t)}\,dY(t)\right)=L(Y).
\end{aligned}$$

We give an important corollary of Theorem 2, where the variables (ε_{nk}) are real-valued.

Corollary 1. *Let* $\theta \neq 0$, $\theta \neq \pi$ *and let* $(\varepsilon_{nk})_{k=1,\dots,n;n\geq 1}$ *be a triangular array of real square integrable martingale differences with respect to the filtrations* $(\mathcal{F}_{nk})_{k=0,1,\dots,n;n\geq 1}$ *such that for all* $t \in [0,1]$

$$\mathbb{E}(\varepsilon_{nk}^2 | \mathcal{F}_{n,k-1}) = 1 \qquad a.s.$$

and

$$\forall \alpha > 0 \qquad \frac{1}{n}\sum_{k=1}^{[nt]} \mathbb{E}(\varepsilon_{nk}^2 \chi_{\{|\varepsilon_{nk}|>\alpha\sqrt{n}\}} | \mathcal{F}_{n,k-1}) \xrightarrow{\mathbb{P}} 0, \quad \text{as } n \to \infty.$$

Then

$$\hat{\gamma}_n \xrightarrow{\mathcal{D}} \hat{\gamma}$$

and

$$L_n(X_{n1}, \dots, X_{nn}) \xrightarrow{\mathcal{D}} L(Y).$$

Proof. Using the fact that for $\theta \neq 0$, $\theta \neq \pi$

$$\lim_{n\to\infty} \frac{1}{n}\sum_{k=1}^{n} \cos^2 k\theta = \frac{1}{2}, \qquad \lim_{n\to\infty} \frac{1}{n}\sum_{k=1}^{n} \sin^2 k\theta = \frac{1}{2}, \qquad \lim_{n\to\infty} \frac{1}{n}\sum_{k=1}^{n} \sin k\theta \cos k\theta = 0,$$

we conclude that the conditions of Theorem 2 are fulfilled.

We remark that the cases $\theta = 0$ and $\theta = \pi$ can be also handled in the way as we investigated the complex-valued case.

4. Generalization

Now we generalize the results of Section 3 to the case where $W(t) = W_1(t) + iW_2(t)$, $t \in [0,1]$, is a — not necessarily standard — complex Wiener process. Then the Radon–nikodym derivative has the form

$$\frac{d\mathbb{P}_Y}{d\mathbb{P}_W}(Y) = \exp\left\{-\frac{1}{2}\int_0^1 \langle \Sigma^{-1}AY(t), AY(t)\rangle\, dt + \int_0^1 \langle \Sigma^{-1}AY(t), dY(t)\rangle\right\},$$

where Σ^{-1} denotes the pseudo inverse (or generalized inverse) of the matrix Σ (see Arató (1982)).

The matrix A can be decomposed as

$$A = \lambda \begin{pmatrix} 1 & 0 \\ 0 & 1 \end{pmatrix} + \omega \begin{pmatrix} 0 & -1 \\ 1 & 0 \end{pmatrix} = \lambda I + \omega \Omega, \tag{15}$$

hence the log-likelihood function of the process Y

$$L(Y) = -\log \frac{d\mathbb{P}_Y}{d\mathbb{P}_W}(Y)$$

can be expressed as

$$\begin{aligned} L(Y) &= \frac{1}{2}\lambda^2 \int_0^1 \langle \Sigma^{-1}Y, Y\rangle\, dt + \lambda\omega \int_0^1 \langle \Sigma^{-1}\Omega Y, Y\rangle\, dt + \frac{1}{2}\omega^2 \int_0^1 \langle \Sigma^{-1}\Omega Y, \Omega Y\rangle\, dt \\ &\quad -\lambda \int_0^1 \langle \Sigma^{-1}Y, dY\rangle - \omega \int_0^1 \langle \Sigma^{-1}\Omega Y, dY\rangle. \end{aligned}$$

The maximum likelihood estimators of λ and ω are the solutions of the system of equations

$$\begin{aligned} \lambda \int_0^1 \langle \Sigma^{-1}Y, Y\rangle\, dt + \omega \int_0^1 \langle \Sigma^{-1}\Omega Y, Y\rangle\, dt - \int_0^1 \langle \Sigma^{-1}Y, dY\rangle &= 0 \\ \lambda \int_0^1 \langle \Sigma^{-1}\Omega Y, Y\rangle\, dt + \omega \int_0^1 \langle \Sigma^{-1}\Omega Y, \Omega Y\rangle\, dt - \int_0^1 \langle \Sigma^{-1}\Omega Y, dY\rangle &= 0 \end{aligned}$$

hence

$$\begin{pmatrix} \hat\lambda \\ \hat\omega \end{pmatrix} = S^{-1} \begin{pmatrix} \int_0^1 \langle \Sigma^{-1}Y, dY\rangle \\ \int_0^1 \langle \Sigma^{-1}\Omega Y, dY\rangle \end{pmatrix},$$

where

$$S = \begin{pmatrix} \int_0^1 \langle \Sigma^{-1}Y, Y\rangle\, dt & \int_0^1 \langle \Sigma^{-1}\Omega Y, Y\rangle\, dt \\ \int_0^1 \langle \Sigma^{-1}\Omega Y, Y\rangle\, dt & \int_0^1 \langle \Sigma^{-1}\Omega Y, \Omega Y\rangle\, dt \end{pmatrix}.$$

(The inverse matrix S^{-1} exists with probability 1 in the usual sense.)

Let us consider now the sum of squares

$$\sum_{k=1}^{n} |\Sigma^{-1/2}(Z_{nk} - e^{\gamma_n/n} Z_{n,k-1})|^2,$$

where we identify a complex number $z \in \mathbb{C}$ with the column vector $(\Re(z), \Im(z))$. It can be also written in the form

$$\sum_{k=1}^{n} \langle \Sigma^{-1}(Z_{nk} - e^{\gamma_n/n} Z_{n,k-1}), Z_{nk} - e^{\gamma_n/n} Z_{n,k-1}\rangle.$$

Using again the decomposition (15) we can derive the least squares estimators for the parameter $\lambda_n + i\omega_n = e^{\gamma_n/n}$, $\lambda_n, \omega_n \in \mathbb{R}$:

$$\begin{pmatrix} \hat\lambda_n \\ \hat\omega_n \end{pmatrix} = S_n^{-1} \begin{pmatrix} \sum_{k=1}^{n} \langle \Sigma^{-1} Z_{n,k-1}, Z_{n,k}\rangle \\ \sum_{k=1}^{n} \langle \Sigma^{-1}\Omega Z_{n,k-1}, Z_{n,k}\rangle \end{pmatrix},$$

where

$$S_n = \begin{pmatrix} \sum_{k=1}^n \langle \Sigma^{-1} Z_{n,k-1}, Z_{n,k-1}\rangle & \sum_{k=1}^n \langle \Sigma^{-1}\Omega Z_{n,k-1}, Z_{n,k-1}\rangle \\ \sum_{k=1}^n \langle \Sigma^{-1}\Omega Z_{n,k-1}, Z_{n,k-1}\rangle & \sum_{k=1}^n \langle \Sigma^{-1}\Omega Z_{n,k-1}, \Omega Z_{n,k-1}\rangle \end{pmatrix}.$$

The least squares estimator of γ_n is $\hat{\gamma}_n = n \log(\hat{\lambda}_n + i\hat{\omega}_n)$.

Consider the quadratic form

$$L_n(X_{n1}, \ldots, X_{nn}) = \frac{1}{2}\sum_{k=1}^n |\Sigma^{-1/2}(Z_{nk} - e^{\gamma_n/n} Z_{n,k-1})|^2 - \frac{1}{2}\sum_{k=1}^n |\Sigma^{-1/2}(Z_{nk} - Z_{n,k-1})|^2,$$

connected with the LSE $\hat{\gamma}_n$.

Theorem 3. *Let $(\varepsilon_{nk})_{k=1,\ldots,n:n\geq 1}$ be a triangular array of complex square integrable martingale differences with respect to the filtrations $(\mathcal{F}_{nk})_{k=0,1,\ldots,n;n\geq 1}$ such that for all $t \in [0,1]$*

$$\frac{1}{n}\sum_{k=1}^{[nt]} \mathbb{E}\left\{ \begin{pmatrix} (\Re(\zeta_{nk}))^2 & \Re(\zeta_{nk})\Im(\zeta_{nk}) \\ \Re(\zeta_{nk})\Im(\zeta_{nk}) & (\Im(\zeta_{nk}))^2 \end{pmatrix} \middle| \mathcal{F}_{n,k-1} \right\} \xrightarrow{\mathbb{P}} t\Sigma, \quad \text{as } n \to \infty,$$

where $\zeta_{nk} = e^{-ik\theta}\varepsilon_{nk}$ and

$$\forall \alpha > 0 \qquad \frac{1}{n}\sum_{k=1}^{[nt]} \mathbb{E}(|\varepsilon_{nk}|^2 \chi_{\{|\varepsilon_{nk}|>\alpha\sqrt{n}\}} | \mathcal{F}_{n,k-1}) \xrightarrow{\mathbb{P}} 0, \quad \text{as } n \to \infty.$$

Then

$$\hat{\gamma}_n \xrightarrow{\mathcal{D}} \hat{\gamma}$$

and

$$L_n(X_{n1}, \ldots, X_{nn}) \xrightarrow{\mathcal{D}} L(Y).$$

Proof. Using the same arguments as in the proof of Theorem 2 we obtain the weak convergence of the joint distribution of the stochastic integrals

$$\int_0^1 \langle \Sigma^{-1} Y_n, Y_n\rangle\, dt, \quad \int_0^1 \langle \Sigma^{-1}\Omega Y_n, Y_n\rangle\, dt,$$

$$\int_0^1 \langle \Sigma^{-1}\Omega Y_n, \Omega Y_n\rangle\, dt, \quad \int_0^1 \langle \Sigma^{-1} Y_n, A_n Y_n\rangle\, dt,$$

$$\int_0^1 \langle \Sigma^{-1}\Omega Y_n, A_n Y_n\rangle\, dt, \quad \int_0^1 \langle \Sigma^{-1} Y_n, dM_n\rangle, \quad \int_0^1 \langle \Sigma^{-1}\Omega Y_n, dM_n\rangle$$

to the joint distribution of

$$\int_0^1 \langle \Sigma^{-1} Y, Y\rangle\, dt, \quad \int_0^1 \langle \Sigma^{-1}\Omega Y, Y\rangle\, dt, \quad \int_0^1 \langle \Sigma^{-1}\Omega Y, \Omega Y\rangle\, dt, \quad \int_0^1 \langle \Sigma^{-1} Y, AY\rangle\, dt,$$

$$\int_0^1 \langle \Sigma^{-1}\Omega Y, AY\rangle\, dt, \quad \int_0^1 \langle \Sigma^{-1} Y, dW\rangle, \quad \int_0^1 \langle \Sigma^{-1}\Omega Y, dW\rangle,$$

where

$$A_n = \begin{pmatrix} \Re(\gamma_n) & -\Im(\gamma_n) \\ \Im(\gamma_n) & \Re(\gamma_n) \end{pmatrix}.$$

We have

$$n\begin{pmatrix} \hat\lambda_n - 1 \\ \hat\omega_n \end{pmatrix} = nS_n^{-1}\begin{pmatrix} \sum_{k=1}^n \langle \Sigma^{-1} Z_{n,k-1}, Z_{n,k} - Z_{n,k-1}\rangle \\ \sum_{k=1}^n \langle \Sigma^{-1}\Omega Z_{n,k-1}, Z_{n,k} - Z_{n,k-1}\rangle \end{pmatrix},$$

where

$$\frac{1}{n^2}S_n = \begin{pmatrix} \int_0^1 \langle \Sigma^{-1} Y_n, Y_n\rangle\, dt & \int_0^1 \langle \Sigma^{-1}\Omega Y_n, Y_n\rangle\, dt \\ \int_0^1 \langle \Sigma^{-1}\Omega Y_n, Y_n\rangle\, dt & \int_0^1 \langle \Sigma^{-1}\Omega Y_n, \Omega Y_n\rangle\, dt \end{pmatrix}$$

and using $Z_{n,k} - Z_{n,k-1} = n^{-1}A_n Z_{n,k-1} + \zeta_{nk}$ we have

$$\frac{1}{n}\begin{pmatrix} \sum_{k=1}^n \langle \Sigma^{-1} Z_{n,k-1}, Z_{n,k} - Z_{n,k-1}\rangle \\ \sum_{k=1}^n \langle \Sigma^{-1}\Omega Z_{n,k-1}, Z_{n,k} - Z_{n,k-1}\rangle \end{pmatrix}$$

$$= \begin{pmatrix} \int_0^1 \langle \Sigma^{-1} Y_n, A_n Y_n\rangle\, dt \\ \int_0^1 \langle \Sigma^{-1}\Omega Y_n, A_n Y_n\rangle\, dt \end{pmatrix} + \begin{pmatrix} \int_0^1 \langle \Sigma^{-1} Y_n, dM_n\rangle \\ \int_0^1 \langle \Sigma^{-1}\Omega Y_n, dM_n\rangle \end{pmatrix}.$$

Applying the continuous mapping theorem we conclude

$$n\begin{pmatrix} \hat\lambda_n - 1 \\ \hat\omega_n \end{pmatrix} \xrightarrow{\mathcal{D}} S^{-1}\left(\begin{pmatrix} \int_0^1 \langle \Sigma^{-1} Y, AY\, dt\rangle \\ \int_0^1 \langle \Sigma^{-1}\Omega Y, AY\, dt\rangle \end{pmatrix} + \begin{pmatrix} \int_0^1 \langle \Sigma^{-1} Y, dW\rangle \\ \int_0^1 \langle \Sigma^{-1}\Omega Y, dW\rangle \end{pmatrix}\right)$$

$$= S^{-1}\begin{pmatrix} \int_0^1 \langle \Sigma^{-1} Y, dY\rangle \\ \int_0^1 \langle \Sigma^{-1}\Omega Y, dY\rangle \end{pmatrix} = \hat\gamma$$

which implies

$$\hat\gamma_n = n\log(\hat\lambda_n + i\hat\omega_n) = \log\left(1 + \frac{n\left((\hat\lambda_n - 1) + i\hat\omega_n\right)}{n}\right)^n \xrightarrow{\mathcal{D}} \hat\gamma.$$

Moreover

$$L_n(X_{n1}, \ldots, X_{nn}) =$$
$$= \frac{1}{2}\sum_{k=1}^n |\Sigma^{-1/2}(Z_{nk} - e^{\gamma_n/n} Z_{n,k-})|^2 - \frac{1}{2}\sum_{k=1}^n |\Sigma^{-1/2}(Z_{nk} - Z_{n,k-1})|^2$$

$$
\begin{aligned}
&= -\frac{1}{2n^2}\sum_{k=1}^{n}\langle \Sigma^{-1}A_nZ_{n,k-1}, A_nZ_{n,k-}\rangle - \frac{1}{n}\sum_{k=1}^{n}\langle \Sigma^{-1}A_nZ_{n,k-1}, \zeta_{n,k}\rangle \\
&= -\frac{1}{2}\int_0^1 \langle \Sigma^{-1}A_nY_n(t), A_nY_n(t)\rangle\, dt - \frac{1}{n}\int_0^1 \langle \Sigma^{-1}A_nY_n(t), dM_n(t)\rangle \\
&\xrightarrow{\mathcal{D}} -\frac{1}{2}\int_0^1 \langle \Sigma^{-1}AY(t), AY(t)\rangle\, dt - \frac{1}{n}\int_0^1 \langle \Sigma^{-1}AY(t), dW(t)\rangle \\
&= \frac{1}{2}\int_0^1 \langle \Sigma^{-1}AY(t), AY(t)\rangle\, dt - \frac{1}{n}\int_0^1 \langle \Sigma^{-1}AY(t), dY(t)\rangle = L(Y).
\end{aligned}
$$

Finally we give a generalization of Corollary 1. Consider the nearly nonstationary complex-valued AR(1) processes generated according to the scheme

$$
\begin{cases} X_{nk} = \beta_n X_{n,k-1} + \kappa_n \eta_{nk}, & k = 1,2,\dots,n \\ X_{n0} = 0, \end{cases} \tag{16}
$$

where (η_{nk}) is an array of real random variables, $\beta_n = e^{\gamma_n/n+i\theta}$, (γ_n) and (κ_n) are sequences of unknown complex numbers such that $\gamma_n \to \gamma$, $\gamma \in \mathbb{C}$ is unknown, $\kappa_n \to \kappa$, $\kappa \in \mathbb{C}$ is known and $\theta \in (-\pi, \pi]$ is known.

Minimizing the sum of squares

$$
\sum_{k=1}^{n} |X_{nk} - e^{\gamma_n/n+i\theta}X_{n,k-1}|^2 = \sum_{k=1}^{n} |Z_{nk} - e^{\gamma_n/n}Z_{n,k-1}|^2
$$

we obtain the least squares estimator

$$
\hat{\gamma}_n = n \log \frac{\sum_{k=1}^{n} \overline{Z}_{n,k-1}Z_{nk}}{\sum_{k=1}^{n} |Z_{n,k-1}|^2}.
$$

Let $W(t) = W_1(t) + iW_2(t)$, $t \in [0,1]$, be the complex Wiener process, where $(W_1(t), W_2(t))$, $t \in [0,1]$, is a Wiener process in $\mathbb{R}^2$ with $\mathbb{E}W_1(t) = \mathbb{E}W_1(t) = 0$ and

$$
\mathbb{E}\begin{pmatrix} W_1^2(t) & W_1(t)W_2(t) \\ W_1(t)W_2(t) & W_2^2(t) \end{pmatrix} = \frac{t}{2}|\kappa|^2 I.
$$

Consider the complex-valued Ornstein–Uhlenbeck process defined as the solution of the stochastic differential equation

$$
dY(t) = \gamma Y(t)\, dt + dW(t), \qquad Y(0) = 0. \tag{17}
$$

The maximum likelihood estimator of γ is

$$
\hat{\gamma} = \frac{\int_0^1 \overline{Y(t)}\, dY(t)}{\int_0^1 |Y(t)|^2\, dt}.
$$

Consider the quadratic form

$$
L_n(X_{n1},\dots,X_{nn}) = \frac{1}{|\kappa|^2}\sum_{k=1}^{n} |X_{nk} - e^{\gamma_n/n+i\theta}X_{n,k}|^2 - \frac{1}{|\kappa|^2}\sum_{k=1}^{n} |X_{nk} - e^{i\theta}X_{n,k-1}|^2,
$$

connected with the LSE $\hat{\gamma}_n$, and the log-likelihood function of the process Y

$$L(Y) = -\log \frac{d\mathbb{P}_Y}{d\mathbb{P}_W}(Y).$$

Corollary 2. *Let* $\theta \neq 0$, $\theta \neq \pi$ *and let* $(\eta_{nk})_{k=1,\ldots,n;n\geq 1}$ *be a triangular array of real square integrable martingale differences with respect to the filtrations* $(\mathcal{F}_{nk})_{k=0,1,\ldots,n;n\geq 1}$ *such that for all* $t \in [0,1]$

$$\mathbb{E}(\eta_{nk}^2 | \mathcal{F}_{n,k-1}) = 1 \qquad a.s.$$

and

$$\forall \alpha > 0 \qquad \frac{1}{n}\sum_{k=1}^{[nt]} \mathbb{E}(\eta_{nk}^2 \chi_{\{|\eta_{nk}|>\alpha\sqrt{n}\}} | \mathcal{F}_{n,k-1}) \xrightarrow{\mathbb{P}} 0, \quad \text{as } n \to \infty.$$

Then

$$\hat{\gamma}_n \xrightarrow{\mathcal{D}} \hat{\gamma}$$

and

$$L_n(X_{n1}, \ldots, X_{nn}) \xrightarrow{\mathcal{D}} L(Y).$$

Proof. We apply Theorem 3 for $\varepsilon_{nk} = \kappa_n \eta_{nk}$. Let $\mu_n = \Re(\kappa_n)$, $\mu = \Re(\kappa)$, $\nu_n = \Im(\kappa_n)$ and $\nu = \Im(\kappa)$. Identifying again a complex number $z \in \mathbb{C}$ with the column vector $(\Re(z), \Im(z))'$ we obtain

$$\begin{aligned}
\zeta_{nk} &= e^{-ik\theta}\kappa_n\eta_{nk} = \begin{pmatrix} \cos\theta & -\sin\theta \\ \sin\theta & \cos\theta \end{pmatrix}^{-k} \begin{pmatrix} \mu_n & -\nu_n \\ \nu_n & \mu_n \end{pmatrix} \begin{pmatrix} \eta_{nk} \\ 0 \end{pmatrix} \\
&= \begin{pmatrix} \mu_n & -\nu_n \\ \nu_n & \mu_n \end{pmatrix} \begin{pmatrix} \cos\theta & -\sin\theta \\ \sin\theta & \cos\theta \end{pmatrix}^{-k} \begin{pmatrix} \eta_{nk} \\ 0 \end{pmatrix}
\end{aligned}$$

and

$$\frac{1}{n}\sum_{k=1}^{n} \mathbb{E}\left\{\zeta_{nk}\zeta_{nk}' | \mathcal{F}_{n,k-1}\right\} =$$

$$= \begin{pmatrix} \mu_n & -\nu_n \\ \nu_n & \mu_n \end{pmatrix} \frac{1}{n}\sum_{k=1}^{n} \begin{pmatrix} \cos^2 k\theta & -\sin k\theta\cos k\theta \\ \sin k\theta \cos k\theta & \sin^2 k\theta \end{pmatrix} \begin{pmatrix} \mu_n & \nu_n \\ -\nu_n & \mu_n \end{pmatrix}$$

$$\to \frac{1}{2}\begin{pmatrix} \mu & -\nu \\ \nu & \mu \end{pmatrix}\begin{pmatrix} \mu & \nu \\ -\nu & \mu \end{pmatrix} = \frac{1}{2}|\kappa|^2 I,$$

hence we conclude that the conditions of Theorem 3 are fulfilled.

Acknowledgment

The authors are grateful to M. Arató and K. Dzhaparidze for the valuable discussions.

References

Anderson, T.W. (1959). On asymptotic distributions of estimates of parameters of stochastic difference equations. *Ann. Math. Statist.* 30, 676–687.

Arató, M. (1982). *Linear stochastic systems with constant coefficients. A statistical approach.* Lecture Notes in Control and Inf., vol. 45, Springer, Berlin, 1982; in Russian - Nauka, Moscow, 1989.

Billingsley, P. (1968). *Convergence of probability measures.* John Wiley & Sons, New York.

Chan, N.H. and Wei, C.Z. (1987). Asymtotic inference for nearly nonstationary AR(1) processes. *Ann. Statist.* 15, 1050–1063.

Dzhaparidze, K., Kormos, J., van der Meer, T., van Zuijlen, M. (1992). Parameter estimation for nearly nonstationary AR(1) processes. Report 9218, Department of Mathematics, Catholic University Nijmegen, The Netherlands.

Kormos, J. (1989). Hypothesis testing for nearly nonstationary AR(1) processes. *Alkalmazott Mat. Lapok* 14, 27–35. (In Hungarian).

Kormos, J. (1990). Hypothesis testing for nearly nonstationary autogressive models. *Comput. Math. Appl.* 19, 75–82.

Kormos, J. and Piterbarg, V.I. (1991). A nonstationarity criterion for a Gaussian autoregressive time series under a close alternative. *Problemy Peredachi Informatsii* 27, 70–75; translation in *Problems Inform. Transmission* 27, 334–338. (In Russian).

Jeganathan, P. (1991). On the asymptotic behaviour of least-squares estimators in AR time series with roots near the unit circle. *Econometric Theory* 7, 269–306.

Liptser, R.Sh. and Shiryayev, A.N. (1989). *Theory of martingales.* Kluwer, Dordrecht.

Mann, H.B. and Wald, A. (1943). On the statistical treatment of linear stochastic difference equations. *Econometrica* 11, 173–220.

Phillips, P.C.B. (1987). Towards a unified asymptotic theory for autoregression. *Biometrika* 74, 535–547.

Rao, M.M. (1978). Asymptotic distribution of an estimator of the boundary parameter of an unstable process. *Ann. Statist.* 6, 185–190.

White, J.S. (1958). The limiting distribution of the serial correlation coefficient in the explosive case. *Ann. Math. Statist.* 29, 1188–1197.

Exploring Stochastic Laws. pp. 219–232.
Eds. A.V. Skorokhod and Yu.V. Borovskikh

Average and stability of dynamical systems with rapid stochastic switchings

VLADIMIR S. KOROLYUK*

National Academy of Sciences of Ukraine, Institute of Mathematics

Received January 25, 1995

Abstract. For dynamical system

$$dU^{\varepsilon}(t)/dt = C(U^{\varepsilon}(t), \kappa(t/\varepsilon))$$

with velocity $C(u,x)$ switched by stochastic process $\kappa(t)$ in a compact state space, the sufficient conditions for stability for sufficiently small ε are established under the condition of exponential stability of the averaged system

$$dU(t)/dt = C(U(t)).$$

Two cases of switching processes are considered: a) $\kappa(t)$, an ergodic Markov process; b) $\kappa(t)$, a stationary and uniformly mixing process.

1. Introduction

1.1. Modern asymptotic analysis of stochastic systems includes two types of problems: investigation of stochastic systems provided that the time parameter increases to infinity, $t \to \infty$; analysis of stochastic systems in a series scheme with a small series parameter $\varepsilon \to 0$. The first type of problem leads to the stability theory founded by Liapunov, A.M. (1950) for deterministic systems, and by Kushner, H.J. (1967), Khasminskii, R.Z. (1963), Gikhman, I.I. (1964) among others, for stochastic systems. The second type of problem leads to the average theory created by Bogolyubov, N.N. and Mitropolskii, Yu.A. (1974). Averaging methods were extended to stochastic systems by Gikhman, I.I. (1964), Khasminskii, R.Z. (1963), Skorokhod A.V. (1987) and many others.

At the same time, problems of the mixed type are also of great interest: to analyse the stochastic system in a series scheme when two parameters are changing, $t \to \infty$

* This paper was prepared when the author worked at the Universitat Autònoma de Barcelona, supported by a DGICYT grant

and $\varepsilon \to 0$. Provided that the average principle may be applied, a natural problem arises: to investigate the stability of stochastic systems provided that the averaged system is stable. Bogolyubov, N.N. was the first who considered this problem in his work with Mitropolskii, Yu.A. for deterministic dynamical systems (see also the work of Mitropolskii, Yu.A. and Samoilenko, A.M. (1987)). Stability problems for stochastic systems in a series scheme were investigated by Blankenship, G. and Papanicolaou, G.C. in [2] (see also [4, 10, 11]).

The main object under consideration in this paper is the dynamical system

$$dU^\varepsilon(t)/dt = C(U^\varepsilon(t), \kappa(t/\varepsilon)) \tag{1.1}$$

with velocity $C(u,x)$, $u \in R^d$, $x \in X$, switched by a stochastic process $\kappa(t)$, $t \geq 0$, whose state space X is compact. We associate with stochastic system (1.1) the averaged system

$$dU(t)/dt = C(U(t)), \tag{1.2}$$

where the average velocity $C(u)$ should be chosen in such a way that the weak convergence holds

$$w \cdot \lim_{\varepsilon\to 0} U^\varepsilon(t) = U(t). \tag{1.3}$$

The main goal of this paper is to establish the stability of stochastic system (1.1) for a sufficiently small ε assuming the exponential stability of averaged system (1.2). Two variants of switching processes are considered: $\kappa(t)$, a stationary and uniformly mixing process. The basic approach consists of a perturbed Liapunov function of averaged system (1.2). In both cases, the Liapunov operator (derivative along a path, see e.g. [8]) is constructed using the generators of semigroups related to the two component stochastic process $U^\varepsilon(t)$, $\kappa(t/\varepsilon)$, $t \geq 0$. In our objective of clarifying the construction of the Liapunov function in an admissible form, we deliberately weaken the assumptions we use further. At the same time, it is well-known [8] that the stability conditions for linear stochastic systems provide those for nonlinear systems of (1.1) type with a velocity $C(u)$ smooth enough in u.

Preliminary notions and notations. Let $\varphi(u)$, $u \in R^d$ be a real valued function of $u = (u_1, \dots u_d)$. The gradient $\nabla\varphi(u) := (\partial\varphi(u)/\partial u_k,\ 1 \leq k \leq d)$ is a vector-function. The scalar product of two vectors in R^d is conveniently interpreted as the product of two rectangular matrices of orders $d \times 1$ and $1 \times d$, respectively. For $\varphi = (\varphi_k, 1 \leq k \leq d)$ and $\psi = (\psi_k, 1 \leq k \leq d)$, we let

$$\varphi^*\psi := \sum_1^d \varphi_k\psi_k.$$

At the same time, observe that

$$\varphi\psi^* := [\varphi_k\psi_r, 1 \leq k, r \leq d]$$

is a $d \times d$ square matrix. The sign $*$ marks the transposed matrix. For example, the product of two vectors φ and ψ and a matirx C has the following form

$$\varphi^* C \psi = \sum_1^d \varphi_k C_{kr} \psi_r$$

and is merely a scalar.

1.2. To construct a Liapunov operator associated with dynamical system (1.2), the following argument is used. Let $V(u)$, $u \in R^d$, be a real valued sufficiently smooth function. Consider small increments of this function along a path of system (1.2)

$$\begin{aligned} \Delta_h V(u(t)) &:= V(u(t+h)) - V(u(t)) \\ &= V\left(u(t) + \int_t^{t+h} C(u(s))ds\right) \\ &= h \nabla V^*(u(t)) C(u(t)) + o(h). \end{aligned}$$

Then

$$\lim_{h \to 0} h^{-1} \Delta_h V(u(t)) = \nabla V^*(u(t)) C(u(t)).$$

The right-hand side of this formula is called a Liapunov operator associated with dynamical system (1.2) and is denoted by

$$\overset{\circ}{V}(u) := \nabla V^*(u) C(u). \tag{1.4}$$

Note that the right-hand side of (1.4) determines the generator G of semigroup G_t related to dynamical system (1.2) which acts in the following form

$$G_t V(u) := V(u(t)), \quad u(0) = u. \tag{1.5}$$

Therefore this Liapunov operator can be represented in the form

$$\overset{\circ}{V}(u) = GV(u) = \nabla V^*(u) C(u). \tag{1.6}$$

It is well-known [8] that a Liapunov operator, along a path of a Markov process can be represented in the form

$$\overset{\circ}{V}(u) = QV(u),$$

where Q is a generator of a semigroup associated with the Markov process.

In both cases, for deterministic dynamical systems and for Markov processes, Liapunov operators can be defined in terms of the generators of the associated semigroups. As expected for dynamical system (1.1) with a stationary switching process, the Liapunov operator can also be represented by means of the generator of the associated semigroup which corresponds to the two-component stochastic process $U^\varepsilon(t)$, $\kappa(t/\varepsilon)$, $t \geq 0$. Such an approach was introduced in [15] using the semigroup of conditioned shifts [13] (see also [16, 1]).

2. Ergodic Markov switchings

2.1. Assume that $\kappa(t)$, $t \geq 0$, is an ergodic Markov process in a compact metric space (X, Ω). Let B_X be the Banach space of real valued bounded functions and let Q be the generator of the associated semigroup, whose action in B_X is given by the following formula

$$Q\varphi(x) := q(x) \int_X P(x, dy)[\varphi(y) - \varphi(x)].$$

$\kappa(t)$, $t \geq 0$, thus defined, is a jump Markov process with intensities $q(x)$ of the sojourn times in states $x \in X$, and stochastic kernel $P(x, dy)$ responsible for the transition probabilities of the imbedded Markov chain. Denote by $\pi(dx)$ the stationary distribution of the Markov process $\kappa(t)$, $t \geq 0$, which satisfies

$$\int_B \pi(dx)q(x) = \int_X \pi(dx)q(x)P(x, B), \quad B \in X.$$

Introduce projector Π in B_X acting in the following form

$$\Pi\varphi(x) := \int_X \pi(dx)\varphi(x)I =: \hat{\varphi}I,$$

where, by definition,

$$\hat{\varphi} := \int_X \pi(dx)\varphi(x), \quad I := 1(x) \equiv 1.$$

It is well-known [12] that there exists a kernel $R_0(x, dy)$ which defines a bounded operator R_0 on B_X given by

$$R_0\psi(x) := \int_X R_0(x, dy)\psi(y).$$

This operator has the following property

$$R_0 Q\varphi(x) = QR_0\varphi(x) = \varphi(x)$$

for all $\varphi(x) \in B_X$ which satisfies $\Pi\varphi(x) = 0$. At the same time,

$$R_0\Pi\varphi(x) = \Pi R_0\varphi(x) = 0.$$

Moreover, each solution $\varphi(x)$ of the equation

$$Q\varphi(x) = \psi(x), \quad \Pi\psi(x) = 0 \tag{2.1}$$

satisfying the condition $\Pi\varphi(x) = 0$ can be represented by the formula

$$R_0\psi(x) = \int_X R_0(x, dy)\psi(y). \tag{2.2}$$

2.2. A stochastic system with rapid Markov switchings is defined as a solution of the linear differential equation

$$dU^{\varepsilon}(t)/dt = C(\kappa(t/\varepsilon))U^{\varepsilon}(t) \tag{2.3}$$

with velocity matrix $C(x) = [C_{kr}(x);\ 1 \le k, r \le d]$.

The associated averaged deterministic system is given by the solution of the differential equation

$$dU(t)/dt = CU(t), \quad U(0) = u \tag{2.4}$$

with the averaged velocity matrix $C = [C_{kr};\ 1 \le k, r \le d]$ defined as follows

$$C_{kr} = \int_X \pi(dx)C_{kr}(x), \quad 1 \le k, r \le d,$$

where $\pi(dx)$ is the stationary distribution of Markov process $\kappa(t)$, $t \ge 0$. This means that the averaged velocity matrix can be represented by the formula

$$C = \Pi C(x). \tag{2.5}$$

We keep in mind that projector Π acts on the elements of matrix $C(x)$.

2.3. Theorem.

A (Averaging). *Given the assumptions of Sections 2.1 and 2.2, solution $U^{\varepsilon}(t)$ of equation (2.3) with Markov ergodic switchings weakly converges as $\varepsilon \to 0$ to the solution $U(t)$ of the averaged equation (2.4),*

$$w \cdot \lim_{\varepsilon \to 0} U^{\varepsilon}(t) = U(t). \tag{2.6}$$

S (Stability). *Assume that $V(u)$ is a twice continuously differentiable Liapunov function for the averaged system (2.4) and its Liapunov operator satisfies the inequality*

$$\breve{V}(u) \le -aV(u), \quad a \ge 0. \tag{2.7}$$

Let the velocities of stochastic system (2.3) be bounded,

$$\max_{1 \le k, r \le d} \max_{x \in X} |C_{kr}(x)| \le C < \infty. \tag{2.8}$$

Then, for each $\varepsilon \in [0, \varepsilon_0]$ sufficiently small, stochastic systems $U^{\varepsilon}(t)$ are asymptotically stable,

$$P\{\lim_{t \to \infty} |U^{\varepsilon}(t)| = 0\} = 1. \tag{2.9}$$

3. Stationary and uniformly mixing switchings

3.1. Let $\kappa(t)$, $t \geq 0$, be a stationary process in a compact metric space X with sample paths in $D[0,\infty]$.

Denote by $F_t := \sigma\{\kappa(s), s \leq t\}$ and $F^t := \sigma\{\kappa(s),\ s \geq t\}$ the filtrations generated by the sample paths of $\kappa(s)$, prior and posterior to the time t, respectively. Let E_t and E^t be the correspondent conditional expectations with respect to filtrations F_t and F^t. The uniform mixing function is defined by the formula

$$\phi(t) := \sup_{t\geq 0} \sup_{A\in F_t, B\in F^{t+\tau}} |P(B/A) - P(B)|. \tag{3.1}$$

The main assumption in this section is

$$\int_0^\infty e^{c\tau}\phi(\tau)d\tau = \phi_c < \infty \tag{3.2}$$

for some $c > 0$. Let $g(t)$ be a bounded real valued F_t-measurable random function with zero mean:

$$|g(t)| \leq b, \quad E_0 g(t) = 0.$$

By the moment inequality (see [5, section 7.2]),

$$|E_t g(t+\tau)| \leq 2\phi(\tau)b. \tag{3.3}$$

Upon integrating we obtain the bound

$$\left|\int_0^\infty E_t g(t+\tau)d\tau\right| \leq 2\phi_0 b. \tag{3.4}$$

3.2. Theorem.

A (Averaging). *Given that the stationary and uniformly mixing process $\kappa(t)$ satisfies condition* (3.2), *the stochastic system $U^\varepsilon(t)$ defined by the solution of equation* (2.3) *weakly converges as $\varepsilon \to 0$ to the solution of the averaged system* (2.4),

$$w \cdot \lim_{\varepsilon\to 0} U^\varepsilon(t) = U(t), \tag{3.5}$$

whose averaged velocity matrix is

$$C = E_0 C(\kappa(t)). \tag{3.6}$$

S (Stability). *Let the Liapunov function $V(u)$ of the averaged system* (2.4) *satisfy the assumptions of Theorem* 2.2 *and assume the velocities of stochastic system* (2.3) *are bounded* (*see* (2.8)). *Then, for sufficiently small ε, stochastic systems $U^\varepsilon(t)$ are exponentially stable in mean square,*

$$E_0\|U^\varepsilon(t)\|^2 \leq A\|U\|^2 \exp[-bt], \quad b > 0. \tag{3.7}$$

4. Proofs of theorems

4.1. Markov switchings. The solution of stochastic system (2.3) with Markov ergodic switchings coupled with the switching Markov process form a two-component Markov process $U^\varepsilon(t)$, $\kappa(t/\varepsilon)$, $t \geq 0$, whose generator is the following

$$L^\varepsilon \varphi(u,x) = \varepsilon^{-1} Q\varphi(u,x) + G(x)\varphi(u,x), \tag{4.1}$$

where Q is a generator of the Markov process $\kappa(t)$, and $G(x)$ are generators of the semigroups associated with the deterministic systems

$$dU(t;x)/dt = C(x)U(t;x), \quad x \in X. \tag{4.2}$$

Operators $G(x)$ act in the following form

$$G(x)\varphi(u,x) := \nabla_u \varphi^*(u,x)C(x)u.$$

Recall that

$$\nabla_u \varphi(u,x) := (\partial\varphi(u,x)/\partial u_k,\ 1 \leq k \leq d)$$

is the vector-gradient of $\varphi(u,x)$ in u. For the sake of brevity, we shall write in the sequel

$$U_t^\varepsilon := U^\varepsilon(t), \quad \kappa_t^\varepsilon := \kappa(t/\varepsilon).$$

The main idea of the proof of Theorem 2.2 is to choose a test function for the generator (4.1) in the form of the sum

$$V^\varepsilon(u,x) = V(u) + \varepsilon\varphi(u,x) \tag{4.3}$$

of a test function $V(u)$ and a perturbing function $\varphi(u,x)$ whose dependence on x will be described later.

Following the martingale approach from [19], we introduce the martingale

$$\mu_t^\varepsilon = V^\varepsilon(U_t^\varepsilon, \kappa_t^\varepsilon) - V(u,x) - \int_0^t L^\varepsilon V^\varepsilon(U_s^\varepsilon, \kappa_s^\varepsilon)ds \tag{4.4}$$

with respect to filtration F_t^ε. We calculate the integrand in (4.4) by means of (4.1),

$$L^\varepsilon V^\varepsilon(u,x) = \varepsilon^{-1}QV(u) + [Q\varphi(u,x) + G(x)V(u)] + \varepsilon G(x)\varphi(u,x). \tag{4.5}$$

The conservative property of the ergodic Markov process with a generator implies that $QV(u) = 0$. Now, according to the properties of the solutions of equation (2.1) mentioned in Section 2.1, we are going to choose the perturbing function $\varphi(u,x)$. A well-defined equation for the perturbing function is constructed in the form of

$$Q\varphi(u,x) + G(x)V(u) = GV(u). \tag{4.6}$$

By definition (2.5) of the averaged operator G, the solvability condition for (4.6) holds

$$\Pi\left[G - G(x)\right] = 0. \tag{4.7}$$

The solution of (4.6), given that

$$\Pi\varphi(u, x) = 0$$

can be written in the form

$$\varphi(u, x) = R_0\left[G - G(x)\right]V(u) \tag{4.8}$$

with the bounded potential operator R_0 (see (2.2)). Since the perturbing function $\varphi(u, x)$ is a solution of equation (4.6), the integrand in (4.4) can be transformed to the following form

$$L^\varepsilon V^\varepsilon(u, x) = GV(u) + \varepsilon G(x)\varphi(u, x).$$

One can obtain the following asymptotic representation of martingale (4.4)

$$\mu_t^\varepsilon = V(U_t^\varepsilon) - V(u) - \int_0^t GV(U_s^\varepsilon)ds +$$

$$+\varepsilon\int_0^t G(x)\varphi(U_s^\varepsilon, \kappa_s^\varepsilon)ds + \varepsilon\left[\varphi(U_t^\varepsilon, \kappa_t^\varepsilon) - \varphi(u, x)\right], \tag{4.9}$$

which is the starting point for the proof of both parts of Theorem 2.2.

Averaging. Rewrite (4.9) in the form

$$\mu_t^\varepsilon = V(U_t^\varepsilon) - V(u) - \int_0^t GV(U_s^\varepsilon)ds + \eta_t^\varepsilon, \tag{4.10}$$

where

$$\eta_t^\varepsilon = \int_0^t G(x)\varphi(U_s^\varepsilon, \kappa_s^\varepsilon)ds + \left[\varphi(U_t^\varepsilon, \kappa_t^\varepsilon) - \varphi(u, x)\right].$$

For the test function

$$V(u) \in C^2(R^d),$$

the remainder term in (4.10) has the estimate

$$\sup_{0\le t\le T} E\left|\eta_t^\varepsilon\right| \le A < \infty.$$

The square characteristic of martingale (4.4) has the form [20]:

$$<\mu^\varepsilon>_t = \int_0^t \left[L^\varepsilon (V^\varepsilon)^2 - 2V^\varepsilon L^\varepsilon V^\varepsilon\right]ds.$$

A simple calculation based on (4.7) and (4.8) gives

$$< \mu^\varepsilon >_t = \varepsilon \int_0^t Q\varphi^2(U_s^\varepsilon, \kappa_s^\varepsilon) ds.$$

Thus,

$$< \mu^\varepsilon >_t \Rightarrow 0 \quad \text{as} \quad \varepsilon \to 0.$$

By the limit theorem for stochastic processes in the series scheme [5, Section 4.8], the weak convergence (2.6) holds. The limit process $U(t)$ satisfies the following equation

$$V(U(t)) - V(u) = \int_0^t GV(U(s)) ds.$$

This is in an equivalent form to (2.4).

Stability. In view of (2.7) and (2.8) and the boundedness of the potential operator R_0, the integrand

$$h(u, x) := G(x)\varphi(u, x)$$

can be estimated as follows

$$|h(u, x)| \le a_1 V(u), \quad |\varphi(u, x)| \le a_2 V(u). \tag{4.11}$$

Next, choosing the test function of the averaged system (2.4) to satisfy assumption (2.7) of Theorem 2.2, we derive from (4.9) the basic inequality

$$0 \le V(U_t^\varepsilon) \le a_3 V(u) + a_4 \mu_t^\varepsilon \tag{4.12}$$

in view of (4.11). This means that the process $V(U_t^\varepsilon)$ is a supermartingale which satisfies the following estimate

$$P\left\{\sup_{t \ge 0} V(U_t^\varepsilon) > \delta\right\} \le AV(u)/\delta.$$

The properties of the Liapunov function for the averaged system (2.4) lead to statement (2.8) of Theorem 2.2.

4.2. Stationary and uniformly mixing processes. Consider the two-component stochastic process $U_t^\varepsilon, \kappa_t^\varepsilon$, $t \ge 0$ under the assumptions of Theorem 3.2. As was already mentioned in Section 1.3, our main objective is to calculate the Liapunov operator on a test function along a sample path using the semigroup of conditioned shifts associated with this two-component process. Choose a test function for stochastic system (2.3) in the form of the sum

$$V^\varepsilon(u, t) = V(u) + \varepsilon\varphi(u, t)$$

of a test function $V(u) \in C^2(R^d)$ and a perturbing random function $\varphi(u,t)$ which is F_t^ε-measurable and has the representation

$$\varphi(u,t) = \int_0^\infty E_t g(u, t+s) ds, \tag{4.13}$$

with a bounded random function $g(u,t)$ which is also F_t^ε-measurable ($F_t^\varepsilon := F_{t/\varepsilon}$) and has mean zero,

$$E_0 g(u,t) = 0.$$

As we mentioned in Section 3.1, given condition (3.2), the integral (4.13) exists and may be estimated as follows

$$\left| \int_0^\infty E_t g(u, t+s) ds \right| \leq 2\phi_0 b,$$

with some constant $b \geq |g(u,t)|$.

According to the scheme discussed in Section 1.3, the Liapunov operator of the averaged system (2.4) has the form

$$\overset{\circ}{V}(u) = \nabla V^*(u) C u. \tag{4.14}$$

Let us calculate the Liapunov operator on the perturbing function $\varphi(u,t)$. First, consider the increments

$$E_t^\varepsilon \left[\varphi(U_{t+\Delta}^\varepsilon, t+\Delta) - \varphi(U_t^\varepsilon, t) \right] =$$

$$= E_t^\varepsilon \left[\varphi(U_{t+\Delta}^\varepsilon, t+\Delta) - \varphi(U_t^\varepsilon, t+\Delta) \right] +$$

$$+ E_t^\varepsilon \left[\varphi(U_t^\varepsilon, t+\Delta) - \varphi(U_t^\varepsilon, t) \right]. \tag{4.15}$$

The first term on the right-hand side of (4.15) is the increment along a sample path of system (2.3),

$$E_t^\varepsilon \left[\varphi(U_{t+\Delta}^\varepsilon, t+\Delta) - \varphi(U_t^\varepsilon, t+\Delta) \right] =$$

$$= \int_t^{t+\Delta} E_t \nabla_u \varphi^*(U_s^\varepsilon, t+\Delta) C(\kappa_s^\varepsilon) U_s^\varepsilon ds.$$

The second term on the right hand side of (4.15) is represented according to (4.13)

$$E_t^\varepsilon \left[\varphi(U_t^\varepsilon, t+\Delta) - \varphi(U_t^\varepsilon, t) \right] = - \int_t^{t+\Delta} E_t^\varepsilon g(u, s) ds.$$

Thus, the Liapunov operator has the following form on a perturbing function $\varphi(u,t)$

$$\overset{\circ}{\varphi}(u,t) = \nabla_u \varphi^*(u,t) C(\kappa_t^\varepsilon) u - g(u,t). \tag{4.16}$$

Now, we obtain the Liapunov operator on the test function $V^\varepsilon(u,t)$

$$\overset{\circ}{V}{}^\varepsilon(u,t) = \nabla V^*(u) C(\kappa_t^\varepsilon) u - \varepsilon g(u,t) +$$

$$+ \varepsilon \nabla_u \varphi^*(u,t) C(\kappa_t^\varepsilon) u. \tag{4.17}$$

The choice of function $g(u,t)$ is the next essential step of the proof. Put

$$\varepsilon g(u,t) := \nabla V^*(u) C_0(\kappa_t^\varepsilon) u \tag{4.18}$$

with $C_0(x) := C(x) - C$. Then Liapunov operator (4.17) is transformed to the following form

$$\overset{\circ}{V}{}^\varepsilon(u,t) = \overset{\circ}{V}(u) + \varepsilon \nabla_u \varphi^*(u,t) C(\kappa_t^\varepsilon) u, \tag{4.19}$$

where by definition $\overset{\circ}{V}(u) = \nabla V^*(u) C u$.

In view of (4.13) and (4.18), the perturbing function $\varphi(u,t)$ has the representation

$$\varphi(u,t) = \varepsilon^{-1} \int_0^\infty E_t^\varepsilon \nabla V^*(u) C_0(\kappa(t/\varepsilon + s/c)) u ds =$$

$$= \int_0^\infty E_t^\varepsilon \nabla V^*(u) C_0(\kappa(t/\varepsilon + s)) u ds,$$

and the gradient is written as follows

$$\nabla_u \varphi(u,t) = \int_0^\infty E_t^\varepsilon \left[\nabla^2 V^*(u) C_0(\kappa(t/\varepsilon + s) + C_0(\kappa(t/\varepsilon + s) \nabla V(u) \right] ds.$$

This is why the second term in (4.19) is represented as

$$h(u,t) =$$

$$= \int_0^\infty E_t^\varepsilon \left[u^* C_0^*(\kappa(t/\varepsilon + s)) \nabla^2 V(u) + \nabla V^*(u) C_0(\kappa(t/\varepsilon + s)) \right] ds C(\kappa_t^\varepsilon) u. \tag{4.20}$$

We need these routine calculations to complete the proof of Theorem 3.2. By the properties of conditional expectation, the basic formula is derived (see (4.19), (4.20))

$$E_{t_0}^\varepsilon V(U_t^\varepsilon) - V(U_{t_0}^\varepsilon) =$$

$$= \int_{t_0}^t \overset{\circ}{V}(U_s^\varepsilon) ds + \varepsilon \int_{t_0}^t h(U_s^\varepsilon, s) ds + \varepsilon \left[\varphi(U_t^\varepsilon, t) - \varphi(U_{t_0}^\varepsilon, t_0) \right], \quad t < t_0. \tag{4.21}$$

Averaging. By choosing the test function $V(u) \in C^2(R^d)$, the weak convergence (2.6) is obtained in view of the limit theorem for stochastic processes [5, Section 4.8]. The limit process $U(t)$ is determined by the evolution equation

$$V(U(t)) - V(u) = \int_0^t GV(U(s))ds$$

with the generator

$$GV(u) := \nabla V^*(u)Cu.$$

This is an equivalent form of the averaged system (2.4).

Stability. Using the evident equality

$$e^{ct}V(U_t) - V(u) = \int_0^t e^{cs}\left[\overset{\circ}{V}(U_s) + cV(U_s)\right] ds$$

the basic formula (4.21) can be rewritten as follows ($t_0 = 0$)

$$E_0^\varepsilon V(U_t^\varepsilon)e^{ct} = \int_0^t \left[\overset{\circ}{V}(U_s^\varepsilon) + cV(U_s^\varepsilon)\right] ds + \\ + \varepsilon E_0^\varepsilon \left[\int_0^t h_c(U_s^\varepsilon, s)ds + \varepsilon\varphi_c(U_t^\varepsilon, t)\right], \tag{4.22}$$

where

$$\varphi_c(u,t) = \int_0^t e^{cs}E_t g(u, t+s)ds, \quad h_c(u,t) := \nabla_u \varphi_c^*(u,t)C(\kappa_t^\varepsilon)u.$$

It is well-known [8] that the Liapunov function for the averaged system (2.3) can be chosen in a bilinear form

$$V(u) = u^*Vu$$

with some positive definite matrix V. In such a way,

$$\nabla V(u) = 2Vu, \qquad \nabla^2 V(u) = 2V,$$

with the following estimates holding

$$b_1 \|u\|^2 \le V(u) \le b_2 \|u\|^2. \tag{4.23}$$

The remainder term in (4.20) is now represented as follows

$$h(u,t) = \int_0^\infty E_t^\varepsilon \left[u^*C_0^*(\kappa(t/\varepsilon + s))V + u^*VC^*(\kappa(t/\varepsilon + s))\right] dsC(\kappa_t^\varepsilon)u.$$

Given conditions (2.8) and (3.2), the following estimates hold

$$|h_c(u,t)| \le b_3 V(u), \quad |\varphi_c(u,t)| \le b_4 V(u). \tag{4.24}$$

Apply the main assumption (2.7) of Theorem 2.3 together with estimates (4.23) and (4.24) to obtain the following inequality from the basic formula (4.22),

$$E_0 \|U_t^\varepsilon\|^2 \leq b \|u\|^2 e^{-ct}$$

which holds for some constant a and for all $\varepsilon \leq \varepsilon_0$, with ε_0 sufficiently small, so that

$$c + \varepsilon_0 b_3 < a.$$

References

1. **Benderskii, M.M. and Pastur, L.A.** (1973). Asymptotic behaviour of the solutions of a second order equation with random coefficients. *Teorija Funkcii & Funk. Analis.* **22,** 3–14. Khar'kov. (In Russian).

2. **Blankenship, G. and Papanicolaou, G.C.** (1978). Stability and control of stochastic systems with wide-band noise disturbances. *SIAM J. Appl. Math.* 34, 437–476.

3. **Bogolyubov, N.N. and Mitropolskii, Yu. A.** (1974). *Asymptotical Methods in the Theory of nonlinear oscillations.* Nauka, Moscow. (In Russian).

4. **Tsarkov, E.F.** (1992). Average and stability of linear differential equations with small diffusion coefficients. In: *Proceed.* VI *USSR–Japan Symposium. "Prob. Theory Math. Stat."* A.N. Shiryaev *et al.* (Eds). WSP, Singapore, pp. 390–396.

5. **Ethier, S.N. and Kurtz, T.G.** (1986). *Markov Processes: Characterization and Convergence.* Wiley, New York.

6. **Gihman, I.I.** (1964). *Differential equations with random coefficients.* Akad. Nauk Ukraine, Kiev, **MR 31**(6037).

7. **Gihman, I.I.** (1966). On stability of the solutions of stochastic differential equations. In: *Limit theorems and statistical decisions.* Tashkent: Institute of Mathematics, pp.14–45. (In Russian).

8. **Khasminskii, R.Z.** (1966). Stochastic Stability of Differential Equations (English transl. (1980). SJTHOFF, The Netherlands).

9. **Khasminskii, R.Z.** (1963). The averaging principle for parabolic and elliptic differential equations and Markov processes with small diffusion. *Theory Probab. Appl.* 8–21. **MR 28**(4253).

10. **Korolyuk, V.S.** (1991). *Averaging and stability of dynamical systems with rapid Markov switchings.* Preprint S-901 87, Univ. of Umeå, Sweden.

11. **Korolyuk, V.S.** (1991). Stability of autonomous dynamical systems with rapid Markov switchings. *Ukrainian Math. Journ.*, **33**(2), 1176–1181. (In Russian).

12. **Korolyuk, V.S. and Turbin, A.F.** (1978). *Mathematical foundations of the phase lumping of large systems.* Naukova Dumka, Kiev (English transl. (1993). Kluwer Academic Publishers).

13. **Kurtz, T.G.** (1975). Semigroups of conditioned shifts and approximation of Markov processes. *Ann. Probab.* **3**, 618–642.

14. **Kushner, H.J.** (1967). *Stochastic Stability and Control.* Academic Press, New York.

15. **Kushner, H.J.** (1979). Jump-diffusion approximation for ordinary differential equations with wide-band random right-hand sides. *SIAM J. Control Optim.* **17**, 729–744.

16. **Kushner, H.J.** (1984). *Approximation and Weak Convergence Methods for Random Processes.* MIT Press. Cambridge, Massachusetts.

17. **Liapunov, A.M.** (1950). General problem on stability of the motion. Moscow. Gostechizdat. (In Russian).

18. **Mitropolskii, Yu.A. and Samoilenko, A.M.** (1987). *Mathematical Problems of Nonlinear Mechanics.* Kiev. Institute of Math. Ukr. Ac. Sci. (In Russian).

19. **Papanicolaou, G.C., Stroock, D. and Varadhan, S.R.S.** (1977). *Martingale approach to some limit theorems.* Stat. Mech. Dynam. Syst. Duke Univ. Math. Series, 3. Durham (NC).

20. **Skorokhod, A.V.** (1987). *Asymptotic Methods in the Theory of Stochastic Differential Equations.* Naukova Dumka, Kiev.

Exploring Stochastic Laws. pp. 233–242.
Eds. A.V. Skorokhod and Yu.V. Borovskikh

Uniform exponential bounds for the pointwise availability of a repairable system

IGOR N. KOVALENKO[1,*] AND ALESSANDRO BIROLINI[2,*]

[1] *Cybernetics Institute of the Nat. Acad. of Sci. of Ukraine and STORM Research Centre University of North London 166-220 Holloway Road London N7 8DB, United Kingdom*

[2] *Reliability Laboratory of the Federal Institute of Technology Zürich*

Received May 12, 1995

Abstract. Lower and upper exponential bounds are derived for the deviation from the steady state value of the pointwise availability $P_0(t)$ of a repairable system with an up-state at time 0. It is assumed that failure free times are exponentially distributed with a parameter λ, whereas repair times have a general distribution with a finite mean $1/\mu$. Provided $\rho = \lambda/\mu$ is sufficiently small and the repair time distribution satisfies some rather weak additional conditions, inequalities

$$P_0(t) \geq \frac{1}{1+\rho} - c_L \frac{\rho}{1+\rho} e^{-(\lambda+\mu)t}, \quad t \geq 0$$

and

$$P_0(t) \leq \frac{1}{1+\rho} + c_U \frac{\rho}{1+\rho} e^{-(\lambda+\mu)t}, \quad t \geq 0$$

are established.

1. Problem statement

Consider a repairable system described by an alternating renewal process with exponentially distributed up-phases durations X_i with parameter λ and arbitrarily distributed down-phases durations Y_i. Denote by μ^{-1} the mean down-time, by $B(t)$ the distribution function and by $\beta(s)$ the Laplace–Stieltjes transform of the normed random variables μY_i, and set $\rho = \lambda/\mu$. In the exponential case, i.e. $\overline{B}(t) = \mathbf{P}\{\mu Y_i \geq t\} = e^{-t}$, we have

$$P_0(t) = \frac{1}{1+\rho} + \frac{\rho}{1+\rho} e^{-(\lambda+\mu)t}, \quad t \geq 0$$

* This paper has been elaborated in the STORM Research Centre at the University of North London (under the sponsorship of the British Council in Ukraine) and at the Federal Institute of Technology, Zürich

for probability $P_0(t)$ to be in the up state at time t. Evidently, this equation can be presented in the form

$$P_0(\mu\tau) = \frac{1}{1+\rho} + \frac{\rho}{1+\rho} e^{-(1+\rho)\tau}, \quad \tau \geq 0.$$

One can interpret τ as a dimensionless variable

$$\mu t = t/EY_i.$$

Though, in some cases (the Erlangian repair time distribution, for instance) the $P_0(t)$ is an oscillating function, and $P_0(t) < \frac{1}{1+\rho}$ for some t. Thus it is important to obtain a lower bound for $P_0(t)$. The purpose of this paper is to derive the following lower and upper bounds for $P_0(t)$:

$$P_0(t) \geq \frac{1}{1+\rho} - c_L \frac{\rho}{1+\rho} e^{-(\lambda+\mu)t}, \quad t \geq 0 \tag{1}$$

and

$$P_0(t) \leq \frac{1}{1+\rho} + c_U \frac{\rho}{1+\rho} e^{-(\lambda+\mu)t}, \quad t \geq 0 \tag{2}$$

for some constants c_L and c_U, if $B(t)$ satisfies some appropriate additional conditions. In particular, in some cases the constants c_L and c_U can be chosen to equal one, so that exact reference to the case of exponentially distributed repair times is possible.

Extensive literature on this problem exists. Most, widely used performance parameters of two-state systems have been collaborated carefully by many authors. See Gnedenko, Belyaev and Soloviev [1] for a classical theory. A variety of performance parameters of two-state systems have been presented by Baxter [2]. Methods for numerical analysis of two-state systems have been developed in a series of papers by McConalogue and others. We shall mention here the following papers. Cleroux and McConalogue [3] have suggested spline algorithm for computing convolution integrals of the form

$$(G * H)(t) = \int_0^t G(t-x)dH(x)$$

based on the cubic spline approximation. For a FORTRAN program see McConalogue [4]. Attention has been paid to practical examples concerning distributions of a special interest for reliability theory: gamma, lognormal, inverse normal, Weibull distributions. As some of them have singularities for some regions of parameters, McConalogue [5] has solved the problem of elaborating special numerical algorithms which are well suited to the case. A tabulation of renewal functions for reliability - important distributions was created by Baxter, McConalogue, Schener and Blischke [6]. Other important publications which must be mentioned, are: Stoyan and Daley [7], Min Xie [8] and Stone and Wainger [9].

Some approaches to establishing bounds have also been suggested in Birolini, Kochubinsky and Kovalenko [10]. In a recent work by Atkinson [11] some coupling type bounds for a renewal process have been derived.

We should note that the uniformity of a bound in t requires overcoming some mathematical difficulties. Indeed, bounds (1) and (2) with $t \geq t_0$ instead of $t \geq 0$ can be established in a much easier way.

2. Main relations

Set

$$P_0(\mu t) = \frac{1}{1+\rho} + \frac{\rho}{1+\rho}\Delta(t). \tag{3}$$

(We use hereinafter the variable t instead of τ in cases where there cannot be any confusion.) For deriving bounds for $\Delta(t)$ the following notations are introduced

$$b_1(t) = \overline{B}(t),$$

$$\beta_1(s) = (1-\beta(s))/s,$$

$$b_2(t) = \int_t^\infty b_1(x)dx,$$

$$\beta_2(s) = (1-\beta_1(s))/s,$$

so that $\beta_i(s)$ are Laplace transforms of the functions $b_i(t)$, $i = 1, 2$.

Then let $b_n(t)$ be defined recursively by the relation

$$b_n(t) = \int_0^t b_{n-1}(\tau)b_1(t-\tau)d\tau \tag{4}$$

for $n \geq 3$. Thus, the Laplace transform of $b_n(t)$ equals $\beta_2(s)\beta_1^{n-2}(s)$ for $n \geq 2$. Let also $B^{*(n)}(t)$ be the n-fold convolution of the distribution $B(t)$.

The following relations will play an important role in the analysis given here:

$$P_0(\mu t) = e^{-\rho t} + \sum_{n=1}^\infty \frac{\rho^n}{n!}\int_0^t e^{-\rho(t-\tau)}(t-\tau)^n dB^{*(n)}(\tau) \tag{5}$$

and

$$\Delta(t) = \sum_{n=0}^\infty (-\rho)^n b_{n+2}(t), \quad \rho < 1. \tag{6}$$

Equation (5) is well-known; it is obtained by a renewal argument. Equation (5) implies the expression $1/(s(1+\rho\beta_1))$ for the Laplace transform of function $P_0(\mu t)$. Subtracting $1/(s(1+\rho))$ from this function one obtains the expression $\rho\beta_2(s)/((1+\rho)(1+\rho\beta_1(s)))$ for the Laplace transform of the function $P_0(\mu t) - 1/(1+\rho)$. (6) can then be obtained by the inversion formula.

Each integral expression from the RHS of (5) can be written as

$$\int_0^\infty e^{-\rho(t-\tau)}(t-\tau)_+^n dB^{*(n)}(\tau)$$

where $(\ldots)_+$ denotes the positive part of the expression in the brackets.

Note that the function

$$e^{-\rho(t-\tau)}(t-\tau)_+^n$$

is non-increasing in τ at least for $0 \le \tau \le 1/\rho$, hence the monotonicity principle can be applied to it in the following form.

Statement 1. *If* $B(t) \le 1 - e^{-t}$ *for* $0 \le t \le t_0$, *then*

$$\Delta(t) \le e^{-(1+\rho)t}, \quad 0 \le t \le t_0. \tag{7}$$

For the proof of (7) it is sufficient to note that

(i) $B^{*n}(t) \le e^{-t} \sum_{k=n}^{\infty} t^k/k!$ for each $n = 1, 2, \ldots,\ 0 \le t \le t_0$, by induction;

(ii) $\int f dB \le \int f dC$ as soon as f is a non-increasing function and non-decreasing functions $B(t)$, $C(t)$ satisfy the inequality $B(t) \le C(t)$ for $0 \le t \le t_0$, and

(iii) the expressions

$$\int_0^\infty e^{-s(t-\tau)}(t-\tau)_+^n(\tau^{n-1}e^{-\tau}/(n-1)!)d\tau$$

are associated to the exponentially distributed repair time for which $\Delta(t)$ equals $e^{-(1+\rho)t}$.

Statement 2. *The inequality*

$$b_{n+2}(t) \le \beta_1^{n+1}(-z)e^{-zt}, \quad n \ge 0, \quad t \ge 0 \tag{8}$$

holds for every positive z for which the Laplace transform $\beta(-z)$ is finite.

To prove (8) it is sufficient to note that

$$\begin{aligned} b_2(t) &= \int_t^\infty b_1(x)dx \\ &\le e^{-zt}\beta_1(-z) \end{aligned}$$

and

$$\begin{aligned} e^{zt}b_{n+2}(t) &= \int_0^t e^{z(t-\tau)}\, b_{n+1}(t-\tau)\, e^{z\tau}\, b_1(\tau)\, d\tau \\ &\le \sup_t\,(e^{zt}\, b_{n+1}(t))\, \beta_1(-z), \end{aligned}$$

hence induction can be applied.

3. Lower bounds

Theorem 1.

(i) *Inequality (1) holds for*

$$c_L = \rho\,\beta_1^2(-1-\rho)/(1-\rho^2\,\beta_1^2(-1-\rho)) \tag{9}$$

if $\rho\,\beta_1(-1-\rho) < 1$.

(ii) *If* $\beta_1(-z)$ *is finite for some* $z > 1$ *then* c_L *is arbitrarily small for a sufficiently small positive* ρ.

The Theorem easily follows from equation (6) and inequality (8), taking into account the non-negativity of $b_n(t)$.

Example 1. Consider the gamma-distributed repair time so that

$$dB(t) = \frac{1}{\Gamma(m)} m^m\, t^{m-1}\, e^{-t}\, dt, \;\; t > 0 \tag{10}$$

where m is a positive constant (not necessarily an integer). We have

$$\beta_1(s) = \frac{1}{s}\left(\left(1+\frac{s}{m}\right)^{-m} - 1\right)$$

for $s > -m$, thus the inequality (1) can be stated as soon as m is larger than 1 and ρ is sufficiently small. It can easily be checked that $\beta_1(s)$ decreases in m for a negative s. Therefore, if $m \geq 2.5$ then

$$\beta_1(-1-\rho) \leq \frac{1}{1+\rho}\left(\left(1-\frac{1+\rho}{2.5}\right)^{-2.5} - 1\right).$$

Consider, for example, the case valid in most practical situations: $0 \leq \rho \leq \leq 0.1$. Then $\beta_1(-1-\rho) \leq 2.96471$ and (9) leads to the result $c_L < 0.96365$; in particular, $c_L < 1$. For $\rho \leq 0.05$ the inequality $c_L < 1$ holds for every $m \geq 1.6$. If $\rho \leq 0.01$ then the condition $m \geq 1.16$ is sufficient for the same inequality.

Example 2. Consider the Weibull distribution for repair time so that

$$\overline{B}(t) = \exp\{-(ct)^\alpha\}, \;\; c - \Gamma(1/\alpha)/\alpha. \tag{11}$$

A numerical computation becomes necessary. So, for $\alpha = 2$ it follows that

$$\beta_1(s) = \int_0^\infty \exp\{-(\sqrt{\pi}\,t/2)^2 - st\}\, dt$$

and a numerical computation shows that $\beta_1(-1-\rho) < 2.4$ for $0 \leq \rho \leq 0.1$. Hence, by (9) one obtains $c_L < 0.612$ for the stated range of ρ.

The above approach leads to a useful bound for all $t \geq 0$, though it is not very good for small t. Equation (5) implies the simplest lower bound

$$P_0(\mu t) \geq e^{-\rho t}, \quad t \geq 0 \tag{12}$$

and the more accurate bound

$$P_0(\mu t) \geq e^{-\rho t} + \rho \int_0^t e^{-\rho(t-\tau)}(t-\tau)\,dB(\tau). \tag{13}$$

Inequality (13) can be transformed to

$$P_0(\mu t) \geq e^{-\rho t}\left(1 + \rho \int_0^t (t-\tau)\,dB(\tau)\right) = e^{-\rho t}\left(1 + \rho t - \rho \int_0^t \overline{B}(\tau)\,d\tau\right),$$

and since

$$\int_0^\infty \overline{B}(\tau)d\tau = 1,$$

it follows that

$$P_0(\mu t) \geq e^{-\rho t}(1 - \rho + \rho t), \tag{14}$$

which can be used for $t \geq 1$ unless the lower bound (1) is better. Figure 1 shows the positions of the three bounds given by (1), (12) and (14), respectively. The lower bound as a whole can be given by

$$P_0(\mu t) \geq e^{-\rho t}(1 + \rho(t-1)_+) \bigvee \left(\frac{1}{1+\rho} - \frac{\rho}{1+\rho} e^{-(1+\rho)t}\right), \quad t \geq 0 \tag{15}$$

where $\bigvee$ is the sign of the maximum. Note that $P_0(\mu t) \geq 1/(1+\rho)$ for $0 \leq t \leq 1-\rho/2$ so that $t' \geq 1 - \rho/2$ and $t'' > 2$ for $0 \leq \rho \leq 0.1$.

The sense of t' and t'' is clear from Figure 1.

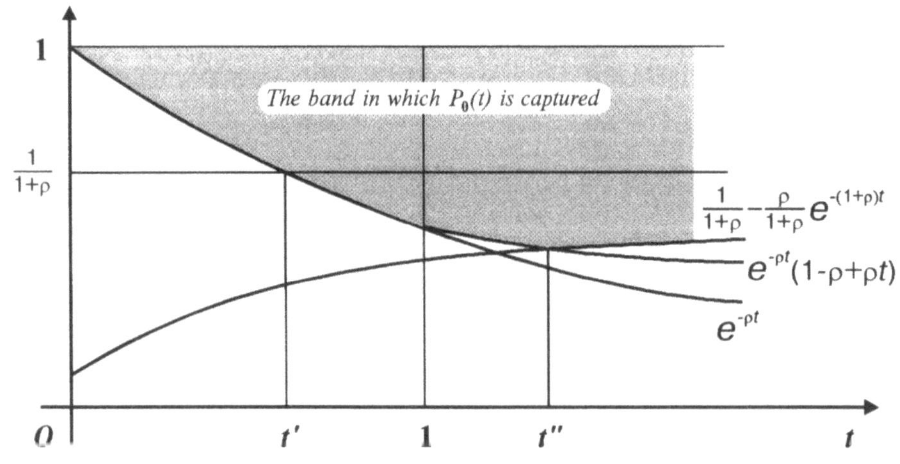

Figure 1.

4. Upper bounds

Upper bounds for $P_0(t)$ are of less importance than lower bounds, although they are of interest, at least for the completeness of mathematical results. Equation (6) implies the inequality

$$\Delta(t) \leq \sum_{n=0}^{\infty} \rho^{2n} b_{2n+2}(t), \quad t \geq 0. \tag{16}$$

The investigation of the conditions sufficient for the inequality

$$\Delta(t) \leq e^{-(1+\rho)t}, \quad t \geq 0 \tag{17}$$

leads to the following Theorem.

Theorem 2. 1. *If*

(*i*) *for a positive* t_0

$$B(t) \leq 1 - e^{-t}, \quad 0 \leq t \leq t_0, \tag{18}$$

(*ii*)

$$b_2(t) < e^{-t}, \quad t > 0 \tag{19}$$

and

(*iii*) *for some* $z_0 > 1$

$$\beta(-z_0) < \infty, \tag{20}$$

then inequality (2) *with* $c_U = 1$ *or, equivalently,* (17) *is satisfied for sufficiently small positive* ρ.

2. *If* $b_2(t) > e^{-t}$ *for some positive* t, *then inequality* (17) *cannot hold for sufficiently small positive* ρ.

Proof. In assumption (20), inequality (8) implies the bound

$$b_{n+2}(t) \leq \left(\beta_1^{n+1}(-z_0)\, e^{-(z_0-1-\rho)t}\right) e^{-(1+\rho)t} \tag{21}$$

as soon as $\rho < z_0 - 1$. Hence from (16) it follows that

$$\Delta(t) \leq \frac{\beta_1(-z_0)\, e^{-(z_0-1-\rho)t}}{1 - \rho^2 \beta_1^2(-z_0)}\, e^{-(1+\rho)t} \tag{22}$$

for sufficiently small positive ρ. Obviously the ratio in the RHS is less than one for $t > T$ where T is a sufficiently large number.

Statement 1 implies the validity of the desired inequality $\Delta(t) \leq e^{-(1+\rho)t}$ for $0 \leq t \leq t_0$.

For each t, $t_0 < t < T$, inequalities (8), (16) and (19) imply

$$\Delta(t) - e^{-(1+\rho)t} \leq A + B \tag{23}$$

where

$$A = b_2(t) - e^{-(1+\rho)t} \tag{24}$$

and

$$B = \rho^2 \beta_1^3(-1-\rho) e^{-(1+\rho)t}/(1-\rho^2\beta_1^2(-1-\rho)) \tag{25}$$

for sufficiently small positive ρ. Since $b_2(t)$ is a continuous function, (19) implies

$$A \leq -\delta e^{-(1+\rho)t} \tag{26}$$

for sufficiently small positive ρ, fixed $\delta > 0$ and $t_0 \leq t \leq T$. Finally, from (25) the inequality

$$B \leq \delta e^{-(1+\rho)t}, \quad t_0 \leq t \leq T \tag{27}$$

follows for sufficiently small positive ρ.

The second statement of Theorem 2 follows from the inequality

$$\Delta(t) \geq b_2(t) - (\rho b_3(t) + \rho^3 b_5(t) + \ldots), \tag{28}$$

implying the bound

$$\liminf_{\rho \to 0} \Delta(t) \geq b_2(t), \tag{29}$$

since $b_n(t) \leq 1$ for $n \geq 2$, from (8) with $z = 0$.

Example 3. Consider the Erlangian case, i.e. (10) with an integer $m \geq 3$. It can be checked that $B(t) \leq 1 - e^{-t}$ at least for $0 \leq t \leq 1 + 1/(3\sqrt{m})$. Hence $\Delta(t) \leq e^{-(1+\rho)t}$ for $0 \leq t \leq 1 + 1/(3\sqrt{m})$.

In the following, some inequalities are established for the case $\rho \leq 0.1$. From

$$\beta(s) = \left(1 + \frac{s}{m}\right)^{-m}$$

it follows that

$$\beta_1(s) = \frac{1}{s}\left(\left(1+\frac{s}{m}\right)^{-m} - 1\right) = \frac{1}{m}\sum_{k=1}^{m}\left(1+\frac{s}{m}\right)^{-k},$$

and thus

$$b_1(t) = e^{-mt}\sum_{k=0}^{m-1}\frac{(mt)^k}{k!}. \tag{30}$$

Similarly

$$b_2(t) = e^{-mt}\sum_{k=0}^{m-1}\left(1-\frac{k}{m}\right)\frac{(mt)^k}{k!}. \tag{31}$$

Since $\left(1 - \frac{k}{m}\right) m^k \leq (m-1)^k$, (31) implies

$$b_2(t) \leq \left(\frac{m-1}{m-1.1}\right)^{m-1}\left(e^{-(m-1.1)t}\sum_{k=0}^{m-1}\frac{(m-1.1)^k t^k}{k!}\right) e^{-1.1t}. \tag{32}$$

The function in the brackets decreases in t; its supremum in the region $\{0 \leq t \leq 1 + 1/(3\sqrt{m});\ m = 3, 4, \ldots\}$ is less than 0.671. Therefore,

$$b_2(t) < 0.671 e^{-1.1t}, \quad t \geq 1 + 1/(3\sqrt{m}). \tag{33}$$

Using (16) and (8) with $z = 1 + \rho$ for $0 \leq \rho \leq 0.1$, it follows then that

$$e^{(1+\rho)t}(\Delta(t) - b_2(t)) \leq \rho^2 \beta_1^3(-1-\rho)/(1 - \rho^2 \beta_1^2(-1-\rho)) \leq$$

$$\leq 0.01\, \beta_1^3(-1.1)/(1 - 0.01\, \beta_1^2(-1.1)) \leq 0.205 \tag{34}$$

Inequalities (33) and (34) imply (2) with $c_U = 0.876 < 1$.

Note. An explicit formula

$$\Delta(t) = e^{-(2+\rho/2)t}(\cos \sqrt{2\rho - \rho^2/4}\, t +$$

$$+ ((1 - \rho/2)/\sqrt{2\rho - \rho^2/4}) \sin \sqrt{2\rho - \rho^2/4}\, t) \tag{35}$$

is known for the case $m = 2$. Equation (35) implies the inequality $|\Delta(t)| \leq e^{-(2+\rho/2)t}$ for $m = 2$, $0 \leq \rho \leq 2$. Therefore, considering Theorems 1 and 2 and Example 1, the following statement holds.

Statement 3. *The inequality*

$$\left| P_0(\mu t) - \frac{1}{1+\rho} \right| \leq \frac{\rho}{1+\rho} e^{-(1+\rho)t} \tag{36}$$

is satisfied for every Erlangian repair time distribution at least for $0 \leq \rho \leq 0.1$.

5. Conclusions

1. Lower bound (1) with $c_L = 1$ is valid for many cases typical for practical applications. As it is shown in Theorem 1, if the abscissa of convergence of the Laplace transform $\beta(s)$ is less than -1, then (1) is valid for sufficiently small positive ρ. In Example 1 it is shown that for a gamma distribution with the shape parameter $\alpha \geq 2.5$ inequality (1) is valid as soon as $\rho \leq 0.1$. In Example 2 this inequality is proved for the Weibull distribution with the shape parameter $\alpha = 2$. Lower bound (1) can be improved for sufficiently small t; the bound as a whole is given by inequality (15).

2. Upper bound (2) with $c_U = 1$ requires some appropriate additional assumptions. Sufficient conditions are given by Theorem 2. In particular, the mentioned inequality is satisfied for Erlangian distributions as soon as $\rho \leq 0.1$.

Acknowledgements

Some numerical computations were carried out by Federico Bonzanigo, Anatoly Kochubinsky and Nikolai Kuznetsov, helping us to derive our theoretic results. Many thanks to all of them.

References

1. **Gnedenko, B.V., Belyaev, Yu.K. and Solovyov, A.D.** (1969). *Mathematical Methods of Reliability Theory.* New York: Academic Press.

2. **Baxter, L.A.** (1981). Availability measures for a two-state system. *J. Appl. Probab.* 18, 227–235.

3. **Cleroux, R. and McConalogue, D.J.** (1976). A numerical algorithm for recursively defined convolution integrals involving distribution functions. *Management. Sci.* 22, 1138–1146.

4. **McConalogue, D.J.** (1978). Convolution integrals involving probability distribution functions. *The Computer Journal* **21** 3, Algorithm 102, 270–272.

5. **McConalogue, D.J.** (1981). Numerical treatment of convolution integrals distribution functions with densities having singularities at the origin. *Communications in Statistics, Ser.B,* 10, 265–280.

6. **Baxter, L.A., McConalogue, D.J., Schener, E.M., and Blischke, W.R.** (1982). On the tabulation of the renewal function. *Technometrics* **24**(2).

7. **Stoyan, D. and Daley, D.J.** (1983). *Comparison Methods for Queues and Other Stochastic Models.* J. Wiley & Sons, Chichester.

8. **Min Xie** (1989). Some results on renewal equations. *Commun. Statist.- Theory Meth.* **18**(3), 1159–1171.

9. **Stone, C.J. and Wainger, S.** (1967). One-sided error estimates in renewal theory. *J. d'Analyse Math.* 20, 325–352.

10. **Birolini, A., Kochubinsky, A.I. and Kovalenko, I.N.** (1993). Some bounds and numerical results on renewal and alternating renewal processes. *Doklady Akademii Nauk Ukrainy* 10, 25–31.

11. **Atkinson, J.B.** (1995). The transient M/G/1/0 queue: some bounds & approximations for light traffic with applications to reliability. *School of Mathematical Sciences. University of North London,* 1–23.

Exploring Stochastic Laws. pp. 243–250.
Eds. A.V. Skorokhod and Yu.V. Borovskikh

Simple proof of Hida distribution characterization theorem

IZUMI KUBO[1] AND HUI-HSIUNG KUO[2*]

[1] *Department of Mathematics, Faculty of Science, Hiroshima University, Higashi-Hiroshima 724, Japan*

[2] *Department of Mathematics, Louisiana State University, Baton Rouge, LA 70803, USA*

Received January 21, 1995

Abstract. The Potthoff–Streit Theorem gives a characterization of Hida distributions in terms of their S-transforms. A simple proof of this characterization Theorem is given. Norm estimates and related problems are discussed.

1. The space of Hida distributions

Let E be a real separable Hilbert space with norm $|\cdot|_0$. Suppose $\{|\cdot|_p; p = 1, 2, \ldots\}$ is a sequence of densely defined inner product norms on E satisfying the conditions:

(1) $|\cdot|_0 \le \rho|\cdot|_1 \le \cdots \le \rho^p|\cdot|_p \le \cdots$ for some $0 < \rho < 1$.

(2) Let $\mathcal{E}_p \equiv \{\xi \in E; |\xi|_p < \infty\}$. For any $p \ge 0$, there exists $q > p$ such that the inclusion map $i_{q,p} : \mathcal{E}_q \hookrightarrow \mathcal{E}_p$ is a Hilbert–Schmidt operator and $\|i_{q,p}\|_{HS} < 1$.

Let $\mathcal{E}$ be the projective limit of $\{\mathcal{E}_p; p \ge 0\}$. Let μ be the standard Gaussian measure on the dual space $\mathcal{E}'$ of $\mathcal{E}$, i.e., its characteristic function is given by $e^{-\frac{1}{2}|\xi|_0^2}, \xi \in \mathcal{E}$. We use (L^2) to denote the complex Hilbert space $L^2(\mathcal{E}', \mu)$. By the Wiener-Itô decomposition Theorem, each $\varphi \in (L^2)$ can be expressed uniquely as a sum of multiple Wiener integrals, i.e., $\varphi = \sum_{n=0}^{\infty} I_n(f_n), f_n \in E_c^{\widehat{\otimes} n}$ (Here and below the sub index c denotes the complexification of a real vector space). Moreover, the (L^2)-norm $\|\varphi\|_0$ of φ is given by

$$\|\varphi\|_0 = \left(\sum_{n=0}^{\infty} n!|f_n|_0^2\right)^{1/2},$$

where $|\cdot|_0$ denotes the norm on $E_c^{\widehat{\otimes} n}$ induced by the norm $|\cdot|_0$ on E.

---

* Research supported by U.S. Army Research Office grant DAAH04-94-G-0249

For each positive integer p and for $\varphi = \sum_{n=0}^{\infty} I_n(f_n) \in (L^2)$, define

$$\|\varphi\|_p = \left(\sum_{n=0}^{\infty} n!|f_n|_p^2\right)^{1/2},$$

where $|\cdot|_p$ denotes the norm on $\mathcal{E}_{p,c}^{\widehat{\otimes} n}$ induced by the norm $|\cdot|_p$ on $\mathcal{E}_p$. Let

$$(\mathcal{E}_p) \equiv \{\varphi \in (L^2);\ \|\varphi\|_p < \infty\}.$$

Let $(\mathcal{E})$ be the projective limit of $\{(\mathcal{E}_p); p \geq 0\}$. The dual space $(\mathcal{E})^*$ of $(\mathcal{E})$ is referred as *the space of Hida distributions.* This construction of Hida distributions is due to Kubo and Takenaka [4]. Note that if $\varphi = \sum_{n=0}^{\infty} I_n(\theta_n) \in (\mathcal{E})$, then $\theta_n \in \mathcal{E}_c^{\widehat{\otimes} n}$ for all $n \geq 1$.

Let $\langle\cdot,\cdot\rangle$ denote the bilinear pairing of $(\mathcal{E}_c')^{\widehat{\otimes} n}$ and $\mathcal{E}_c^{\widehat{\otimes} n}$. For $\xi \in \mathcal{E}_c$, the *renormalization* $:e^{\langle\cdot,\xi\rangle}:$ of $e^{\langle\cdot,\xi\rangle}$ is defined by

$$:e^{\langle\cdot,\xi\rangle}: = \sum_{n=0}^{\infty} \frac{1}{n!} I_n(\xi^{\otimes n}). \tag{1}$$

Then $\|:e^{\langle\cdot,\xi\rangle}:\|_p = e^{\frac{1}{2}|\xi|_p^2}$ for all $p \geq 0$. Hence $:e^{\langle\cdot,\xi\rangle}: \in (\mathcal{E})$ for any $\xi \in \mathcal{E}_c$. The *S-transform* $S\Phi$ of a Hida distribution $\Phi \in (\mathcal{E})^*$ is defined to be the following function on $\mathcal{E}_c$:

$$S\Phi(\xi) = \langle\langle \Phi, :e^{\langle\cdot,\xi\rangle}: \rangle\rangle, \quad \xi \in \mathcal{E}_c,$$

where $\langle\langle\cdot,\cdot\rangle\rangle$ denotes the bilinear pairing of $(\mathcal{E})^*$ and $(\mathcal{E})$. The S-transform is injective, i.e., if $S\Phi = S\Psi$, then $\Phi = \Psi$.

2. The characterization Theorem

A characterization Theorem of Hida distributions in terms of their S-transforms has been proved by Potthoff and Streit [8] (see also [1, 7]) for the space $(\mathcal{S})^*$ arising from the case $E = L^2(\mathbb{R})$ and $|h|_p = \left|(-D_u^2 + u^2 + 1)^p h\right|_0$. Their characterization Theorem is also true for the space $(\mathcal{E})^*$ of Hida distributions. In this paper we present a very simple proof of this Theorem. In fact, our proof produces also a norm estimate which has been obtained in [2] for the special case $(\mathcal{S})^*$.

Let $\Phi \in (\mathcal{E})^*$ and $F = S\Phi$. First observe that for any ξ and η in $\mathcal{E}_c$, the function $F(z\xi + \eta)$, $z \in \mathbb{C}$, is an entire function on $\mathbb{C}$. Next, note that there exists some $p \geq 0$ such that Φ belongs to the dual space $(\mathcal{E}_p)^*$ of $(\mathcal{E}_p)$. Hence for any $\xi \in \mathcal{E}_c$,

$$\begin{aligned} |F(\xi)| &= |\langle\langle \Phi, :e^{\langle\cdot,\xi\rangle}: \rangle\rangle| \\ &\leq \|\Phi\|_{-p} \|:e^{\langle\cdot,\xi\rangle}:\|_p \\ &= \|\Phi\|_{-p} e^{\frac{1}{2}|\xi|_p^2}, \end{aligned}$$

where $\|\cdot\|_{-p}$ denote the norm on $(\mathcal{E}_p)^*$.

Theorem 1. *Let $\Phi \in (\mathcal{E})^*$. Then its S-transform $F = S\Phi$ satisfies the conditions:*

(a) *For any ξ and η in $\mathcal{E}_c$, the function $F(z\xi + \eta)$, $z \in \mathbb{C}$ is an entire function on the set $\mathbb{C}$.*

(b) *There exist nonnegative constants K, a, and p such that*

$$|F(\xi)| \leq K e^{a|\xi|_p^2}, \quad \forall \xi \in \mathcal{E}_c.$$

Conversely, suppose a function F defined on $\mathcal{E}_c$ satisfies the above two conditions. Then there exists a unique $\Phi \in (\mathcal{E})^*$ such that $F = S\Phi$. Moreover, for any number q such that $2ae^2\|i_{q,p}\|_{HS}^2 < 1$, the following inequality holds:

$$\|\Phi\|_{-q} \leq K\left(1 - 2ae^2\|i_{q,p}\|_{HS}^2\right)^{-1/2},$$

where $i_{q,p}$ is the inclusion map from $\mathcal{E}_q$ into $\mathcal{E}_p$.

Proof. The first assertion has already been shown above. The uniqueness part of the second assertion is obvious since the S-transform is injective. To prove the existence of Φ, note that condition (a) is equivalent to the fact that for any $n \geq 1$ and any $\xi_1, \ldots, \xi_n \in \mathcal{E}_c$, the function $F(z_1\xi_1 + \cdots + z_n\xi_n)$ is holomorphic on $\mathbb{C}^n$.

For each $n \geq 1$, define an n-linear functional G_n on $\mathcal{E}_c \times \cdots \times \mathcal{E}_c$ by

$$G_n(\xi_1, \ldots, \xi_n) = \frac{1}{n!}\frac{\partial}{\partial z_1}\cdots\frac{\partial}{\partial z_n}F(z_1\xi_1 + \cdots + z_n\xi_n)\Big|_{z_1=\cdots=z_n=0}. \tag{2}$$

By the Cauchy formula, we have

$$G_n(\xi_1, \ldots, \xi_n) = \frac{1}{n!}\frac{1}{(2\pi i)^n}\int_{|z_1|=r_1}\cdots\int_{|z_n|=r_n}\frac{F(z_1\xi_1 + \cdots + z_n\xi_n)}{z_1^2\cdots z_n^2}\,dz_1\cdots dz_n.$$

Suppose $|\xi_1|_p = \cdots = |\xi_n|_p = 1$. Then by condition (b)

$$\begin{aligned} |G_n(\xi_1, \ldots, \xi_n)| &\leq \frac{1}{n!}\frac{1}{(2\pi)^n}\int_{|z_1|=r_1}\cdots\int_{|z_n|=r_n}\frac{|F(z_1\xi_1 + \cdots + z_n\xi_n)|}{|z_1|^2\cdots|z_n|^2}|dz_1\cdots dz_n| \\ &\leq \frac{1}{n!}\frac{1}{(2\pi)^n}\int_0^{2\pi}\cdots\int_0^{2\pi}\frac{Ke^{a(r_1+\cdots+r_n)^2}}{r_1^2\cdots r_n^2}r_1\cdots r_n\,d\theta_1\cdots d\theta_n \\ &= K\frac{1}{n!}\frac{e^{a(r_1+\cdots+r_n)^2}}{r_1\cdots r_n}. \end{aligned}$$

We can minimize the last quantity by choosing $r_1 = \cdots = r_n = \frac{1}{\sqrt{2an}}$. Hence

$$|G_n(\xi_1, \ldots, \xi_n)| \leq K\frac{1}{n!}(2aen)^{n/2}.$$

This implies that for any $\xi_1,\dots,\xi_n \in \mathcal{E}_c$, we have

$$|G_n(\xi_1,\dots,\xi_n)| \le K\frac{1}{n!}(2aen)^{n/2}|\xi_1|_p\cdots|\xi_n|_p.$$

Hence G_n is continuous on $\mathcal{E}_c\times\cdots\times\mathcal{E}_c$. By the nuclear Theorem (for $n=2$, see [9, p.525], the general case follows by induction), there exists $J_n \in (\mathcal{E}_c')^{\widehat{\otimes} n}$ such that

$$\langle J_n, \xi_1\otimes\cdots\otimes\xi_n\rangle = G_n(\xi_1,\dots,\xi_n), \quad \xi_1,\dots,\xi_n\in\mathcal{E}.$$

Alternatively, we can define J_n by

$$J_n = \sum_{j_1,\dots,j_n} G_n(\zeta_{j_1},\dots,\zeta_{j_n})\zeta'_{j_1}\otimes\cdots\otimes\zeta'_{j_n},$$

where $\{\zeta_j; j=1,2,\dots\}\subset\mathcal{E}$ is an orthonormal basis for $\mathcal{E}_q$ and $\{\zeta'_j; j=1,2,\dots\}$ is its dual basis. Then for any $n\ge 1$,

$$\begin{aligned}
|J_n|_{-q}^2 &= \sum_{j_1,\dots,j_n} |G_n(\zeta_{j_1},\dots,\zeta_{j_n})|^2 \\
&\le K^2\frac{1}{(n!)^2}(2aen)^n \sum_{j_1,\dots,j_n} |\zeta_{j_1}|_p^2\cdots|\zeta_{j_n}|_p^2 \\
&= K^2\frac{1}{(n!)^2}(2aen)^n\Big(\sum_{j=1}^{\infty}|\zeta_j|_p^2\Big)^n \\
&= K^2\frac{1}{(n!)^2}(2aen)^n\|i_{q,p}\|_{HS}^{2n}.
\end{aligned}$$

Now, we define a linear functional Φ on $(\mathcal{E})$ by

$$\varphi = \sum_{n=0}^{\infty} I_n(\theta_n) \longmapsto \langle\!\langle \Phi,\varphi\rangle\!\rangle = \sum_{n=0}^{\infty} n!\langle J_n,\theta_n\rangle, \tag{3}$$

where $J_0 = F(0)$. Note that $|J_0|\le K$. By the Schwarz inequality

$$\begin{aligned}
|\langle\!\langle \Phi,\varphi\rangle\!\rangle| &\le \sum_{n=0}^{\infty}\Big(\sqrt{n!}\,|J_n|_{-q}\Big)\Big(\sqrt{n!}\,|\theta_n|_q\Big) \\
&\le \Big(\sum_{n=0}^{\infty} n!\,|J_n|_{-q}^2\Big)^{1/2}\Big(\sum_{n=0}^{\infty} n!\,|\theta_n|_q^2\Big)^{1/2} \\
&\le K\|\varphi\|_q\Big(\sum_{n=0}^{\infty}\frac{1}{n!}(2aen)^n\|i_{q,p}\|_{HS}^{2n}\Big)^{1/2},
\end{aligned}$$

where we use the convention $0^0=1$. Since $n^n\le n!e^n$ for all $n\ge 0$, we have

$$|\langle\!\langle \Phi,\varphi\rangle\!\rangle| \le K\|\varphi\|_q\Big(\sum_{n=0}^{\infty}(2ae^2)^n\|i_{q,p}\|_{HS}^{2n}\Big)^{1/2}.$$

This shows that $\Phi\in(\mathcal{E}_q)^*$ for any q such that $2ae^2\|i_{q,p}\|_{HS}^2<1$ and we get the norm estimate for $\|\Phi\|_{-q}$ as given in the Theorem. To show that $S\Phi = F$, observe that from equation (2) we have

$$\langle J_n,\xi^{\otimes n}\rangle = G_n(\xi,\dots,\xi) = \frac{1}{n!}\frac{d^n}{dz^n}F(z\xi)\Big|_{z=0}.$$

Therefore, from equation (1) and the definition of Φ in (3),

$$\begin{aligned} S\Phi(\xi) &= \langle\langle \Phi, :e^{\langle\cdot,\xi\rangle}:\rangle\rangle \\ &= \sum_{n=0}^{\infty} \langle J_n, \xi^{\otimes n}\rangle \\ &= \sum_{n=0}^{\infty} \frac{1}{n!} \frac{d^n}{dz^n} F(z\xi)\Big|_{z=0} \\ &= F(\xi). \end{aligned}$$

Here in the last equality we have used the Taylor series of $F(z\xi)$ at $z = 1$.

3. Some remarks

Question 1. *Let F be a function on $\mathcal{E}_c$ satisfying conditions* (a) *and* (b) *in Theorem 1. Does it follow that the corresponding Hida distribution Φ is in $(\mathcal{E}_p)^*$ with the same p as given in condition* (b)*?*

The answer is negative in general. For example, let E be infinite dimensional and consider the function

$$F(\xi) = e^{-\frac{1}{2}\langle\xi,\xi\rangle}, \quad \xi \in \mathcal{E}_c. \tag{4}$$

This function F satisfies conditions (a) and (b) with $K = 1$, $a = \frac{1}{2}$, and $p = 0$. The corresponding Hida distribution Φ is a delta function [5]. We show that $\Phi \notin (L^2)$.

Suppose that $\Phi \in (L^2)$. Then $\Phi = \sum_{n=0}^{\infty} I_n(f_n), f_n \in E_c^{\widehat{\otimes} n}$, and

$$S\Phi(\xi) = \sum_{n-0}^{\infty} \langle f_n, \xi^{\otimes n}\rangle, \quad \xi \in \mathcal{E}_c. \tag{5}$$

It follows from equations (4) and (5) that for all $z \in \mathbb{C}$ and $\xi \in \mathcal{E}_c$,

$$\sum_{n=0}^{\infty} z^n \langle f_n, \xi^{\otimes n}\rangle = e^{-\frac{1}{2}z^2\langle\xi,\xi\rangle}.$$

This implies, in particular, that $f_2 \in E_c^{\widehat{\otimes} 2}$ should satisfy the following condition:

$$\langle f_2, \xi^{\otimes 2}\rangle = -\frac{1}{2}\langle\xi,\xi\rangle, \quad \forall \xi \in \mathcal{E}_c,$$

or equivalently,

$$\langle f_2, \xi \otimes \zeta\rangle = -\frac{1}{2}\langle\xi,\zeta\rangle, \quad \forall \xi, \zeta \in \mathcal{E}_c.$$

Let $\{\xi_j; j = 1, 2, \ldots\} \subset \mathcal{E}$ be an orthonormal basis for E. Then

$$|f_2|_0^2 = \sum_{j,k} |\langle f_2, \xi_j \otimes \xi_k\rangle|^2 = \sum_{j,k} \frac{1}{4}\delta_{j,k} = \infty.$$

Therefore, $f_2 \notin E^{\widehat{\otimes}2}$ and so $\Phi \notin (L^2)$.

Question 2. *Let F be a function on $\mathcal{E}_c$ satisfying conditions* (a) *and* (b) *in Theorem 1. Is the condition $2ae^2\|i_{q,p}\|_{HS}^2 < 1$ optimal for the corresponding Φ to belong to $(\mathcal{E}_q)^*$?*

The answer is negative in general. Consider the same function F in equation (4) for the special case $E = L^2(\mathbb{R})$ and $|h|_p = \left|(-D_u^2 + u^2 + 1)^p h\right|_0$. Note that the operator $-D_u^2+u^2+1$ has eigenvalues $2j$, $j = 1, 2, \ldots$. Hence $\|i_{q,0}\|_{HS}^2 = \sum_{j=1}^{\infty}(2j)^{-2q}$ and the condition $2ae^2\|i_{q,p}\|_{HS}^2 < 1$ with $a = \frac{1}{2}$ becomes

$$\frac{e^2}{2^{2q}} \sum_{j=1}^{\infty} \frac{1}{j^{2q}} < 1. \tag{6}$$

On the other hand, it follows from equation (4) that the corresponding Hida distribution Φ is given by

$$\varphi = \sum_{n=0}^{\infty} I_n(\theta_n) \longmapsto \langle\!\langle \Phi, \varphi \rangle\!\rangle = \sum_{n=0}^{\infty} (-1)^n \frac{(2n)!}{n! 2^n} \langle \tau^{\otimes n}, \theta_{2n} \rangle,$$

where $\tau \in \mathcal{S}'(\mathbb{R}^2)$ is given by $\langle \tau, \theta \rangle = \int \theta(u,u)\, du$. It has been shown in [5, 6] that

$$\|\Phi\|_{-q} = \left[\prod_{j=1}^{\infty} \left(1 - (2j)^{-4q}\right) \right]^{-1/4}.$$

Therefore, $\Phi \in (\mathcal{E}_q)^*$ if and only if $q > \frac{1}{4}$. Obviously, the number $q = 1/2$ does not satisfy inequality (6). Thus we can not conclude from Theorem 1 whether Φ belongs to $(\mathcal{S}_{1/2})^*$, while in fact it does since $1/2 > 1/4$.

Generalization. In [3] Kondratiev and Streit have introduced larger spaces of Hida distributions and generalized Theorem 1 to these spaces. Their construction is for the space $L^2(\mathbb{R})$ and the operator $-D_u^2 + u^2 + 1$, but it applies also to the case E and the family $\{|\cdot|_p; p = 1, 2, \ldots\}$ of norms in §1. More precisely, for $0 \leq \beta < 1, p = 1, 2, \ldots$, and $\varphi = \sum_{n=0}^{\infty} I_n(f_n) \in (L^2)$, define

$$\|\varphi\|_{p,\beta} = \left(\sum_{n=0}^{\infty} (n!)^{1+\beta} |f_n|_p^2 \right)^{1/2},$$

$$(\mathcal{E}_p)_\beta = \{\varphi \in (L^2);\ \|\varphi\|_{p,\beta} < \infty\}.$$

Let $(\mathcal{E})_\beta$ be the projective limit of $\{(\mathcal{E}_p)_\beta; p \geq 0\}$. Let $(\mathcal{E})_\beta^*$ and $(\mathcal{E}_p)_\beta^*$ denote the dual spaces of $(\mathcal{E})_\beta$ and $(\mathcal{E}_p)_\beta$, respectively. It can be shown [3, 6] that for any $\xi \in \mathcal{E}_{p,c}$,

$$\|:e^{\langle \cdot, \xi \rangle}:\|_{p,\beta} \leq 2^{\beta/2} \exp\left[(1-\beta) 2^{\frac{2\beta-1}{1-\beta}} |\xi|_p^{\frac{2}{1-\beta}}\right].$$

Thus $:e^{\langle\cdot,\xi\rangle}:\in(\mathcal{E})_\beta$ for any $\xi\in\mathcal{E}_c$. Hence the S-transform can be defined as in §1 for any $\Phi\in(\mathcal{E})_\beta^*$, i.e.,

$$S\Phi(\xi)=\langle\langle\Phi,:e^{\langle\cdot,\xi\rangle}:\rangle\rangle,\quad \xi\in\mathcal{E}_c,$$

where $\langle\langle\cdot,\cdot\rangle\rangle$ denotes the bilinear pairing of $(\mathcal{E})_\beta^*$ and $(\mathcal{E})_\beta$. The S-transform on $(\mathcal{E})_\beta^*$ is also injective. The proof of Theorem 1 can be modified to give a simple proof of the following Kondratiev–Streit characterization Theorem and the norm estimate. For details, see [6].

Theorem 2. *Let $\Phi\in(\mathcal{E})_\beta^*$. Then its S-transform $F=S\Phi$ satisfies the conditions:*

(a) *For any ξ and η in $\mathcal{E}_c$, the function $F(z\xi+\eta)$, $z\in\mathbb{C}$, is an entire function on the set $\mathbb{C}$.*

(b) *There exist nonnegative constants K, a, and p such that*

$$|F(\xi)|\leq K\exp\left[a|\xi|_p^{\frac{2}{1-\beta}}\right],\quad \forall\xi\in\mathcal{E}_c.$$

Conversely, suppose a function F defined on $\mathcal{E}_c$ satisfies the above two conditions. Then there exists a unique $\Phi\in(\mathcal{E})_\beta^*$ such that $F=S\Phi$. Moreover, for any number q such that $e^2\left(\frac{2a}{1-\beta}\right)^{1-\beta}\|i_{q,p}\|_{HS}^2<1$, the following inequality holds:

$$\|\Phi\|_{-q,-\beta}\leq K\left(1-e^2\left(\frac{2a}{1-\beta}\right)^{1-\beta}\|i_{q,p}\|_{HS}^2\right)^{-1/2},$$

where $\|\cdot\|_{-q,-\beta}$ is the norm on $(\mathcal{E}_q)_\beta^*$ and $i_{q,p}$ is the inclusion map from $\mathcal{E}_q$ into $\mathcal{E}_p$.

References

1. **Hida, T. et al.** (1993). *White Noise: An Infinite Dimensional Calculus.* Kluwer Academic Publishers.

2. **Kondratiev, Yu.G. and Streit, L.** (1992). A remark about a norm estimate for white noise distributions. *Ukrainian Math. J.* 44, 832–835.

3. **Kondratiev, Yu. G. and Streit, L.** (1993). Spaces of white noise distributions: Constructions, Descriptions, Applications. I. *Reports on Math. Phys.* 33, 341–366.

4. **Kubo, I. and Takenaka, S.** (1980). Calculus on Gaussian white noise I. *Proc. Japan Acad.* 56A, 376–380.

5. **Kubo, I. and Yokoi, Y.** (1989). A remark on the space of testing random variables in the white noise calculus. *Nagoya Math. J.* 115, 139–149.

6. **Kuo, H.-H.** (1995). *White Noise Distribution Theory.* Seoul National University. (In preparation).

7. **Obata, N.** (1994). *White Noise Calculus and Fock Space.* Lecture Notes in Math. **1577**, Springer-Verlag.

8. **Potthoff, J. and Streit, L.** (1991). A characterization of Hida distributions. *J. Functional Analysis* 101, 212–229.

9. **Treves, F.** (1967). *Topological Vector Spaces, Distributions and Kernels.* Academic Press.

Exploring Stochastic Laws. pp. 251–256.
Eds. A.V. Skorokhod and Yu.V. Borovskikh

Asymptotic Normality of an Estimator of an Infinite-dimensional Parameter in a Model with a C^1-smooth Regression Function

A.G. KUKUSH

Kiev State University, Vladimirskaya St. 64, 252001 Kiev, Ukraine

Received February 6, 1995

Abstract. A model with a C^1-smooth regression function in a Hilbert space is studied. A parametric set is infinite-dimensional and compact. An estimator analogous to the usual least-squares estimator is constructed. Under definite geometric restrictions on the compact set, asymptotic normality of the estimator is proved. These results generalize the finite-dimensional results of B. L. S. Prakasa Rao. In so doing to prove convergence of the finite-dimensional projections of the normalized estimator we use the weak convergence of the random field generated by a minimized functional.

Besides widely used parametric estimators when a parameter belongs to a finite-dimensional set, the problem of estimating infinite-dimensional parameters also has practical interest. Such are the problems of estimating transition functions, see Dorogovtsev (1993), of estimating a signal by observation of its perturbations with noise, see Nemirovsky et al. (1987), of estimating the intensity of a nonhomogeneous Poisson process.

The difficulties connected with the proof of asymptotic properties of similar estimators are discussed in Dorogovtsev (1993). A parametric set is a compact set in a Hilbert space H in the paper Dorogovtsev et al. (1987); the results of asymptotic normality of the estimators were carried over to the case of arbitrary compact sets in H in Kukush (1988). In these and other papers a regression function is assumed to be twice differentiable. Now consider only C^1-smooth regression functions. In so doing though we should put definite geometric restrictions on a parametric set. For example, an ellipsoid in H does not satisfy these restrictions. The obtained statements generalize the finite-dimensional results in Prakasa Rao (1986).

1. Notations and assumptions

Let H and B be real separable Hilbert spaces with norms $\|\cdot\|$, $|||\cdot|||$ respectively, the space B — infinite-dimensional, Θ — a compact set in B, $(\Omega, \mathcal{F}, P)$ — a fixed full probability space, and $\{\xi_n,\ n \geq 1\}$ — a sequence of i.i.d. H-valued random elements. For each $n \geq 1$ $f_n: \ \Theta \mapsto H$. In the sequel the spaces H and H^*, B and B^* will be identified.

Suppose that for a fixed but unknown $\vartheta_0 \in \Theta$ the first N terms of the sequence

$$y_n := f_n(\vartheta_0) + \xi_n, \qquad n \geq 1$$

are observed. From the observations it is required that we estimate the unknown value ϑ_0. Let

$$Q_N(\vartheta) := \sum_{n=1}^{N} \|y_n - f_n(\vartheta)\|^2, \qquad \vartheta \in \Theta.$$

As an estimator for ϑ_0 we take some random element $\hat{\vartheta}_N$ such that for every $\omega \in \Omega$ $Q_N(\hat{\vartheta}_N) = \min_{\vartheta\in\Theta} Q_N(\vartheta)$. If $f_n \in C(\Theta, H)$, $n \geq 1$, such a random element exists; see Dorogovtsev et al. (1987). Consider the following conditions.

(i) $E\|\xi_1\|^2 < \infty$, $E\xi_1 = \bar{0}$; the latter mean is the Bochner integral with respect to the measure P.

We denote by S the correlation operator of the element ξ_1;

$$\psi_N(\vartheta_1, \vartheta_2) := \frac{1}{N}\sum_{n=1}^{N} \|f_n(\vartheta_1) - f_n(\vartheta_2)\|^2, \qquad \vartheta_{1,2} \in \Theta.$$

(ii) $\forall\, r > 0: \ \underline{\lim}_{N\to\infty} \inf_{(|||u|||\geq r,\ u\in\Theta-\vartheta_0)} \psi_N(\vartheta_0 + u, \vartheta_0) > 0.$

(iii) $\exists\, \psi \in C(\Theta \times \Theta)$, $\psi(\vartheta, \vartheta) = 0$, $\vartheta \in \Theta$:

$$\overline{\lim_{N\to\infty}} \sup_{\vartheta_{1,2}\in\Theta} (\psi_N(\vartheta_1, \vartheta_2) - \psi(\vartheta_1, \vartheta_2)) \leq 0.$$

(iv) The vector ϑ_0 is the point of local convexity of Θ, i.e., for some neighbourhood $U(\vartheta_0)$ of this point the set $U(\vartheta_0) \cap \Theta$ is convex.

(v) There exists a linear and dense in B set L such that

$$\forall\, l \in L\ \exists\, \varepsilon > 0\ \exists\, r_l > 0\ \forall\, x \in B(\vartheta_0, r_l) \cap \Theta: \ x + \varepsilon l \in \Theta \cap U(\vartheta_0).$$

(vi) For some open in B set $U \supset \Theta$ the functions f_n, $n \geq 1$ can be continued on U so that the continuations f_n have Frechet derivatives on U; besides that in the sense of uniform operator convergence

$$\frac{1}{N}\sum_{n=1}^{N} f_n'^*(\vartheta_0) f_n'(\vartheta_0) \to V = V(\vartheta_0), \qquad N \to \infty,$$

and $N(V) = \{\bar{0}\}$, $R(V) = B$.

Let $\{e_i\}$ be a basis in B belonging to L from condition (v). Let

$$V_m x := \sum_{i=m+1}^{\infty} (x, e_i) e_i, \qquad m \geq 1,\ x \in B.$$

For a linear operator A denote its operator norm by $\|A\|$.

(vii) The next two relations are valid.

1) $\exists r_0 > 0\ \forall n \geq 1\ \forall \vartheta \in B(\vartheta_0, r_0)$:

$$\|f_n'(\vartheta) - f_n'(\vartheta_0)\| \leq \alpha_n(\vartheta_0) \cdot |||\vartheta - \vartheta_0|||,$$

and

$$\overline{\lim_{N\to\infty}} \frac{1}{N} \sum_{n=1}^{N} \alpha_n^2(\vartheta_0) < \frac{1}{\operatorname{tr}^2 S \cdot \|V^{-1}\|^2}.$$

2) $\forall m \geq m_0\ \exists r_m, N(m)\ \forall n \geq N(m)\ \forall v \in \Theta - \theta_0,\ v \in B(\bar{0}, r_m)$:

$$\|(f_n'(\vartheta_0 + v) - f_n'(\vartheta_0))\, V_m v\| \leq \alpha_{mn}(\vartheta_0) \cdot |||V_m v|||^2,$$

and

$$\underline{\lim}_{m\to\infty} \overline{\lim_{N\to\infty}} \frac{1}{N} \sum_{n=1}^{N} \alpha_{mn}^2(\vartheta_0) < \frac{1}{\operatorname{tr}^2 S \cdot \|V^{-1}\|^2}.$$

(viii) In the sense of convergence in nuclear operator topology

$$\frac{1}{N} \sum_{n=1}^{N} f_n'^*(\vartheta_0)\, S f_n'(\vartheta_0) \to T = T(\vartheta_0), \qquad N \to \infty.$$

(ix) $\|V \cdot V_m - V_m \cdot V\| \to 0,\ m \to \infty.$

(x) $\forall C > 0:\ \dfrac{1}{N} \sum_{n=1}^{N} \displaystyle\int_{\{|||f_n'^*(\vartheta_0)\,\xi_1||| > C\sqrt{N}\}} |||f_n'^*(\vartheta_0)\,\xi_1|||^2\, dP \to 0, \qquad N \to \infty.$

The last condition is analogous to the Lindeberg condition and together with conditions (i), and (viii) provide the convergence of the B-valued random elements

$$\gamma_N := \frac{1}{\sqrt{N}} \sum_{n=1}^{N} f_n'^*(\vartheta_0)\, \xi_n, \qquad N \geq 1$$

in distribution to a centered Gaussian element γ with correlation operator T.

2. Asymptotic normality

Theorem. *Assume the conditions (i)–(x). Then the normalized estimator*

$$\delta_N := \sqrt{N}\,(\hat{\vartheta}_N - \vartheta_0) \to V^{-1}\gamma, \qquad N \to \infty$$

in distribution in B.

Sketch of the proof. a) Conditions (i)–(iii) are analogous to the conditions of consistency of the estimator, see Ivanov (1991) and Shao (1992). By means of these conditions, so as it was literally done in the doctoral thesis of A. V. Ivanov for the one-dimensional case, we prove the strong consistency of the estimator, i. e. convergence of $\hat{\vartheta}_N$ to ϑ_0 a.s. by the norm in B.

b) Condition (iv) and the consistency imply that for large N with large probability, $\hat{\vartheta}_N$ belongs to the convex part of the parametric set $U(\vartheta_0) \cap \Theta$. The function

$$q(t) := Q_N(\hat{\vartheta}_N + t(\vartheta_0 - \hat{\vartheta}_N)), \qquad t \in [0,1],$$

is correctly defined with large probability and reaches its minimum at $t = 0$. Hence

$$\langle Q'_N(\hat{\vartheta}_N), \vartheta_0 - \hat{\vartheta}_N \rangle \geq 0,$$

and condition (vi) and the boundedness in probability of the sequence $\{\gamma_N\}$ imply the boundedness in probability of $\{\delta_N\}$. We follow here the arguments from Dorogovtsev et al. (1987).

c) Similarly to the paper Kukush (1988) we prove the equality

$$\lim_{m\to\infty} \overline{\lim_{N\to\infty}}\, \mathbf{P}\{|||V_m \delta_N||| > \varepsilon\} = 0, \tag{1}$$

that together with the boundedness in probability of $\{\delta_N\}$ imply relative compactness of the distributions of $\{\delta_N\}$. To prove (1) for a fixed $m \geq 1$ we consider the function

$$q_m(t) := Q_N(\hat{\vartheta}_N + tV_m(\vartheta_0 - \hat{\vartheta}_N)), \qquad t \in [0,1].$$

It attains its minimum at $t = 0$. Hence

$$q'_m(0) = \langle Q'_N(\hat{\vartheta}_N), V_m(\vartheta_0 - \hat{\vartheta}_N)\rangle \geq 0.$$

This inequality, the conditions (vi) and (ix), and relative compactness of the distributions of $\{\gamma_N\}$ imply (1).

d) The operator V is continuously invertible and by letting $\kappa = V^{1/2}\vartheta$ we reduce the proof to the case when V is the unit operator I. Assume that $B \times B$-valued vectors $(\gamma_N; \delta_N)$ converge in the distribution to some random vector $(\gamma_\infty; \delta_\infty)$ where

γ_∞ and γ have the same distribution. In the sequel we will prove that δ_∞ and $V^{-1}\gamma$ have identical distributions. For this it's sufficient to prove the convergence

$$\pi_m \delta_N \to \pi_m V^{-1}\gamma, \qquad N \to \infty$$

in distribution for each $m \geq 1$ where $\pi_m := I - V_m$.

To prove the convergence of the finite-dimensional projections we consider for each fixed $\tau > 0$ the random field

$$\Phi_N(\varphi) := Q_N\Big(\pi_m \vartheta_0 + \frac{\varphi}{\sqrt{N}} + V_m \hat{\vartheta}_N\Big) - Q_N(\vartheta_0),$$

$$\varphi \in \pi_m B, \ |||\varphi||| \leq \tau.$$

In consequence of the consistency and condition (v) it is correctly defined for large N with large probability. From condition (vii) and the convergence of $(\gamma_N; \delta_N)$ to $(\gamma_\infty; \delta_\infty)$ similarly to paper Prakasa Rao (1986) we deduce the convergence of Φ_N to the field Φ_∞ in distribution in the space of continuous functions of a finite number of variables. The field Φ_∞ equals

$$\Phi_\infty(\varphi) = \|\varphi\|^2 + 2\,(\varphi, \pi_m \gamma) + \Delta$$

where Δ does not depend upon φ. Since the limiting field has its only point of minimum $\varphi_\infty = -\pi_m \gamma$, the point of minimum of Φ_N that equals

$$\varphi_N = \sqrt{N}\,\pi_m(\hat{\vartheta}_N - \vartheta_0) = \pi_m \delta_N$$

converge in distribution to ϕ_∞. So for $V = I$ we have $\pi_m \delta_N \to \pi_m \gamma$, hence in the general case

$$\pi_m \delta_N \to \pi_m V^{-1}\gamma, \qquad N \to \infty.$$

e) Then for each subsequence $\{\delta_{N'}\}$ there exists an imbedded subsequence of numbers $\{N''\}$ such that the vectors $(\gamma_{N''}, \delta_{N''})$ converge in distribution to $B \times B$. From d) it follows that $\delta_{N''} \to V^{-1}\gamma$ in distribution. Then because of the relative compactness of $\{\delta_N\}$ we obtain $\delta_N \to V^{-1}\gamma$ in distribution.

References

Dorogovtsev, A.Ya. (1993). On asymptotic normality of MLS estimators of infinite-dimensional parameters. — *Ukr. Math. J.* **45**, 44–53. (Russian).

Dorogovtsev, A.Ya., Zerek, N. and Kukush, A.G. (1987). Weak convergence of an infinite-dimensional parameter to a normal distribution. *Teor. Veroyatnost. i Mat. Statist.* **37**, 45–51; English transl. in *Theory Probab. and Math. Statist.* (1988).

Kukush, A.G. (1988). Convergence in distribution of the normalized estimator of an infinite-dimensional parameter of regression. *Doklady Akademii Nauk Ukr. SSR,* ser. A, **5**, 11–14. (Russian).

Nemirovsky, A.S. and Khasminsky, R.Z. (1987). Nonparametric estimating of the functionals of the derivatives of the signal observed in white noise. *Problemy Peredachy Informatsii*, **23**, 27–38. (Russian).

Prakasa Rao, B.L.S. (1986). Weak convergence of the least squares random field in the smooth case. *Statistics and Decisions*, **4**, 363–377.

Ivanov, A.V. (1991). The theory of estimating parameters of nonlinear regression models. Author's summary of his Doctoral Dissertation. Kiev. (Russian).

Shao, J. (1992). Consistency of least-squares estimator and its jackknife variance estimator in nonlinear models. *Canad. J. of Statist.*, **20**, 415–428.

Exploring Stochastic Laws. pp. 257–262.
Eds. A.V. Skorokhod and Yu.V. Borovskikh

On the efficiency of the least squares estimator of regression coefficients of a random field observed on a sphere

N.N. LEONENKO, M.I. YADRENKO AND L.M. IL'ICHEVA

Kiev State University, Department of Mechanics and Mathematics, ul. Vladimirskaya 64, 252601 Kiev-17, Ukraine

Received January 25, 1995

Abstract. In this paper we study the efficiency of the LSE relative to BLUE of regression coefficients of homogeneous isotropic random fields, observed on a sphere.

The question of the asymptotic efficiency of the least square estimator (LSE) of regression coefficients relative to the best linear unbiased estimator (BLUE) of a stationary time series or stationary random processes with continuous and bounded spectral density was considered in Grenander (1954), Grenander and Rosenblatt (1957), Ibragimov and Rozanov (1970) and others.

If, however, the spectral densities have some singularities, then the loss of efficiency can be significant. LSE does not lose much efficiency compared to BLUE (see, for example, Adenstedt (1976), Samarov and Taqqu (1988), Yajima (1988), Yajima (1989), Künsch et al. (1993), Yadrenko (1980), Ivanov and Leonenko (1989), and others).

The problem of estimating regression coefficients of homogeneous and isotropic random fields was considered in Yadrenko (1980), Ivanov and Leonenko (1989). In this paper we use the results of books Yadrenko (1980), Ivanov and Leonenko (1989) to study the efficiency of the LSE relative to BLUE of regression coefficients of homogeneous isotropic random fields, observed on a sphere.

Denote by R^n a real Euclidean space of dimension $n \geq 2$. Let $s(r) = \{x \in R^n : |x| = r\}$ be a sphere of radius $r > 0$, (ρ, u), $\rho \geq 0$,

$$u = (\theta_1, \ldots, \theta_{n-2}, \varphi) \in s(1), \qquad 0 \leq \theta_j < \pi,\ j = 1, \ldots, n-2,\ 0 \leq \varphi < 2\pi$$

— spherical coordinates of point x, $S_m^l(u)$—real orthonormalized spherical harmonics of degree m, where

$$h(m,n) = (2m+n-2)\left[(m+n-3)!/(n-2)!\,m!\right]$$

is the number of such harmonics.

Suppose that the random field

$$\xi(\rho,u) = a\,g(\rho,u) + \eta(\rho,u) \tag{1}$$

is observed on the sphere $s(\rho)$ in R^n, where a is an unknown parameter, $g(\rho,u)$ is a known function, and $\eta(\rho,u)$ is a homogeneous isotropic in a wide sense measurable mean-square continuous random field with zero mean and correlation function

$$B(\rho) = E\,\eta(\rho_1,u_1)\,\eta(\rho_2,u_2), \qquad \rho = |x-y|,\ x = (\rho_1,u_1),$$

$$y = (\rho_2,u_2), \quad \rho_i \geq 0, \quad u_i \in s(1), \quad i = 1,2$$

where

$$\rho = \sqrt{\rho_1^2 + \rho_2^2 - 2\,\rho_1\,\rho_2\cos\gamma}, \quad cos\gamma = \frac{(x,y)}{|x|\cdot|y|}.$$

It is known that correlation function $B(\rho)$ allows the spectral representation:

$$B(\rho) = 2^{(n-2)/2}\,\Gamma\left(\frac{n}{2}\right)\int_0^\infty \frac{J_{(n-2)/2}(\lambda\rho)}{(\lambda\rho)^{(n-2)/2}}\,dG(\lambda),$$

where $J_\nu(z)$ is a Bessel function of the first kind of order $\nu > -\frac{1}{2}$ and $G(\lambda)$ is a bounded nondecreasing function.

Later the following functions will be important:

$$b_m(r) = 2^{n-1}\Gamma\left(\frac{n}{2}\right)\pi^{\frac{n}{2}}\int_0^\infty J^2_{m+\frac{n-2}{2}}(\lambda r)\,(\lambda r)^{2-n}\,dG(\lambda), \qquad m = 0,1,2,\ldots.$$

Remark 1. The asymptotic behaviour of $b_m(r)$ as $r \to \infty$ can be found from the Abelian and Tauberian theorems for integral transforms of Hankel type (see, for example, Theorem 5 Yadrenko (1980), Lemma 1.5.4 Leonenko and Ivanov (1986)).

For example, if for some $\gamma \in (-1, n-2)$, $G'(\lambda) = |s(1)|\,\lambda^\gamma h(\lambda)$ where $h(\lambda)$ is continuous in a neighborhood of zero, $h(0) \neq 0$, and $h(\lambda)$ is bounded on $[0,\infty)$, then from Lemma 1.5.4 Ivanov and Leonenko (1989), we obtain as $r \to \infty$:

$$b_m(r) = (2\pi)^n\,h(0)\,C_1(n,m,\gamma)\,r^{-1-\gamma}(1+o(1))$$

where $|s(1)| = 2\pi^{\frac{n}{2}}/\Gamma\left(\frac{n}{2}\right)$ is the area of the surface of the unit sphere $s(1)$ and

$$C_1(n,m,\gamma) = \frac{\Gamma(n-2-\gamma)\,\Gamma\big((2m+\gamma+1)/2\big)}{2^{n-2-\gamma}\,\Gamma^2\big((n-1-\gamma)/2\big)\,\Gamma\big((2m+2n-4-\gamma)/2\big)}.$$

Let $g_m^l(r) = \int_{s(1)} g(r,u)\, S_m^l(u)\, \sigma(du)$ where $\sigma(\cdot)$ is Lebesgue measure on the sphere. Suppose that

$$\sum_{m=0}^{\infty} \sum_{l=1}^{h(m,n)} \frac{[g_m^l(r)]^2}{b_m(r)} < \infty.$$

Theorem 1 (Yadrenko (1980)) shows that BLUE a_r^* of a parameter based on the observations $\xi(r,u) = \xi(x)$, $x \in s(r)$ is

$$a_r^* = \frac{\displaystyle\int_{s(1)} l(r,u)\, \xi(r,u)\, \sigma(du)}{\displaystyle\sum_{m=0}^{\infty} \sum_{l=1}^{h(m,n)} \frac{[g_m^l(r)]^2}{b_m(r)}}$$

where

$$l(r,u) = \sum_{m=0}^{\infty} \sum_{l=1}^{h(m,n)} \frac{g_m^l(u)\, S_m^l(u)}{b_m(r)}.$$

The variance of estimate a_r^* has the form

$$\mathrm{Var}\, a_r^* = \left[\sum_{m=0}^{\infty} \sum_{l=1}^{h(m,n)} \frac{[g_m^l(r)]^2}{b_m(r)} \right]^{-1}. \tag{2}$$

Since the BLUE often has a complicated form, statistics considers the LSE for a, i.e., the value a which minimizes the functional:

$$Q(a) = \int_{s(r)} [\xi(r,u) - a g(r,u)]^2\, \sigma(du)$$

where $\sigma(\cdot)$ is a Lebesgue measure on the sphere $s(r)$.

The LSE has the following form:

$$\hat{a}_r = \frac{\displaystyle\int_{s(r)} \xi(r,u)\, g(r,u)\, \sigma(du)}{\displaystyle\int_{s(r)} g^2(r,u)\, \sigma(du)}.$$

Using some results from books Yadrenko (1980), Ivanov and Leonenko (1989) we derive the expression for the variance of a LSE:

$$\mathrm{Var}\, \hat{a}_r = \frac{\displaystyle\sum_{m=0}^{\infty} \sum_{l=1}^{h(m,n)} A_{m,l}(r)\, b_m(r)}{\displaystyle\sum_{m=0}^{\infty} \sum_{l=1}^{h(m,n)} [g_m^l(r)]^2}, \tag{3}$$

where

$$A_{m,l}(r) = [g_m^l(r)]^2 \left[\sum_{m=0}^{\infty} \sum_{l=1}^{h(m,n)} [g_m^l(r)]^2 \right]^{-1}.$$

Clearly,

$$A_{m,l}(r) \geq 0, \quad \sum_{m=0}^{\infty} \sum_{l=1}^{h(m,n)} A_{m,l}(r) = 1,$$

thus for r, a discrete probability distribution $\{A_{m,l}(r)\}$, $(m,l) \in T$, on the set $T = \{0,1,\ldots\} \times \{1,2,\ldots,h(m,n)\}$ is defined. If random field (1) has a correlation function B and spectral function G, the efficiency of the LSE is defined as

$$\mathrm{eff}(\mathrm{r},\mathrm{B}) = \frac{\mathrm{Var}\,\mathrm{a}_\mathrm{r}^*}{\mathrm{Var}\,\hat{\mathrm{a}}_\mathrm{r}}.$$

The asymptotic efficiency is $\mathrm{eff}(\infty,\mathrm{B}) = \lim_{\mathrm{r}\to\infty} \mathrm{eff}(\mathrm{r},\mathrm{B})$. It is obvious that $0 \leq \mathrm{eff}(\mathrm{r},\mathrm{B}) \leq 1$, $0 \leq \mathrm{eff}(\infty,\mathrm{B}) \leq 1$.

Theorem 1. *The efficiency* $\mathrm{eff}(\mathrm{r},\mathrm{B})$ *of LSE* $\hat{a}_r$ *is*

$$\begin{aligned}\mathrm{eff}(\mathrm{r},\mathrm{B}) &= \frac{\left(\sum_{m=0}^{\infty} \sum_{l=1}^{h(m,n)} [g_m^l(r)]^2\right)^2}{\left(\sum_{m=0}^{\infty} \sum_{l=1}^{h(m,n)} \frac{[g_m^l(r)]^2}{b_m(r)}\right)\left(\sum_{m=0}^{\infty} \sum_{l=1}^{h(m,n)} [g_m^l(r)]^2\, b_m(r)\right)} \\ &= \left[\left\{\sum_{m=0}^{\infty} \sum_{l=1}^{h(m,n)} A_{m,l}(r)\, b_m(r)\right\}\left\{\sum_{m=0}^{\infty} \sum_{l=1}^{h(m,n)} \frac{A_{m,l}(r)}{b_m(r)}\right\}\right]^{-1} \end{aligned} \tag{4}$$

Proof. Result (4) is obtained from (2) and (3).

From (4) we receive the following statement.

Theorem 2. *Let random field* (1) *be observed on* $s(r)$. *Suppose that discrete distribution* $\{A_{m,l}(r)\}$, $(m,l) \in T$ *is concentrated only in one point* $(m_0,l_0) \in T$ *for every* $r > 0$ *(that is* $A_{m_0,l_0}(r) = 1$ *and remainder* $A_{m,l}(r) = 0$*). Then* $\mathrm{eff}(\mathrm{r},\mathrm{B}) = 1$ *for every correlation function* B *and every* $r > 0$.

In particular if

$$g(r,u) = \Psi_1(r)\,\Psi_2(u), \qquad r \geq 0, \quad u \in s(1) \tag{5}$$

then

$$g_m^l(r) = \Psi_1(r)\,\tilde{g}_m^l, \quad \tilde{g}_m^l = \int_{s(1)} \Psi_2(u)\, S_m^l(u)\, \sigma(du),$$

and

$$\mathrm{eff}(\mathrm{r},\mathrm{B}) = \frac{\left(\sum_{m=0}^{\infty} \sum_{l=1}^{h(m,n)} [\tilde{g}_m^l]^2\right)^2}{\left(\sum_{m=0}^{\infty} \sum_{l=1}^{h(m,n)} \frac{[\tilde{g}_m^l]^2}{b_m(r)}\right)\left(\sum_{m=0}^{\infty} \sum_{l=1}^{h(m,n)} [\tilde{g}_m^l]^2 b_m(r)\right)}$$

$$= \left[\left(\sum_{m=0}^{\infty} \sum_{l=1}^{h(m,n)} \tilde{A}_{m,l} b_m(r) \right) \left(\sum_{m=0}^{\infty} \sum_{l=1}^{h(m,n)} \frac{\tilde{A}_m^l}{b_m(r)} \right) \right]^{-1} \tag{6}$$

where

$$\tilde{A}_{m,l} = \frac{[\tilde{g}_m^l]^2}{\sum_{m=0}^{\infty} \sum_{l=1}^{h(m,n)} [\tilde{g}_m^l]^2} \geq 0, \quad \sum_{m=0}^{\infty} \sum_{l=1}^{h(m,n)} \tilde{A}_{m,l} = 1. \tag{7}$$

Thus a discrete probability distribution $\{\tilde{A}_{m,l}\}, (m,l) \in T$ does not depend on $r > 0$.

Theorem 3. *Let random field* (1) *be observed on* $s(r)$ *and let the regression function* $g(r,u)$ *satisfy condition* (5). *Suppose that discrete probability distribution* $\{\tilde{A}_{m,l}\}$, $(m,l) \in T$ *(see* (7)*) is concentrated only in one point* $(m_0, l_0) \in T$ *(that is* $\tilde{A}_{m_0,l_0} = 1$ *and remainder* $\tilde{A}_{m,l} = 0$*). Then* $\text{eff}(r, B) = 1$ *for every correlation function* B *and every* $r > 0$.

Theorem 4. *Let random field* (1) *be observed on a sphere* $s(r)$. *Suppose that for correlation function* B *and spectral function* G *there exists a function* $\mu(r) > 0$ *such that the limit:*

$$\lim_{r\to\infty} [b_m(r)\,\mu(r)] = b \in (0,\infty)$$

exists and b *does not depend on* m. *If there exists*

$$\lim_{r\to\infty} A_{m,l}(r) = \bar{A}_{m,l}, \qquad (m,l) \in T,$$

then asymptotic efficiency

$$\text{eff}(\infty, B) = 1.$$

Theorem 5. *Let random field* (1) *be observed on a sphere* $s(r)$ *and let the regression function* $g(r,u)$ *satisfy condition* (5). *Suppose, that for correlation function* B *and spectral function* G *there exists a function* $\mu(r) > 0$ *such that the limit*

$$\lim_{r\to\infty} [b_m(r)\,\mu(r)] = b \in (0,\infty)$$

exists and b *does not depend on* m. *Then asymptotic efficiency*

$$\text{eff}(\infty, B) = 1.$$

Remark 2. Let $g(r,u) = p(r)$ be a radial function. Then $g_0^1(r) = p(r)\sqrt{|s(1)|}$ and the remainder $g_m^l(r)$ vanishes. Hence, $A_{0,1}(r) = \tilde{A}_{0,1} = 1$, the remainder

$A_{m,l}(r) = A_{m,l} = 0$ and the discrete probability distribution is concentrated in a point $(m_0, l_0) = (0, 1)$. We obtain from Theorem 2 that $\text{eff}(\text{r}, \text{B}) = 1$. Let (5) hold and $\Psi_2(u) = S_{m_0}^{l_0}(u)$, then discrete probability distributions are concentrated in one point $(m_0, l_0) \in T$.

We obtain from Theorem 3 that $\text{eff}(\text{r}, \text{B}) = 1$. This is very much in contrast to the theory of efficiency for random processes.

Theorem 6. *Let random field* (1) *be observed on a sphere* $s(r)$ *and let there exist*

$$\lim_{r \to \infty} A_{m,l}(r) = \bar{A}_{m,l}, \qquad (m, l) \in T$$

such that $\bar{A}_{m_0,l_0} = 1$ *for some* $(m_0, l_0) \in T$ *and remainder* $\bar{A}_{m,l} = 0$. *Then asymptotic efficiency* $\text{eff}(\infty, \text{B}) = 1$ *for every correlation function* B *and every* $r > 0$.

References

Adenstedt, R.K. (1976). On large sample estimation for the mean of a stationary random sequences. *Annals of Statistics* **2**, 1095–1107.

Ibragimov, U.A. and Rozanov, Yu.A. (1970). *Gaussian random processes.* Nauka, Moscow, 384 p.

Grenander, U. (1954). On the estimation of regression coefficients in the case of autocorrelated disturbance. *Annals of Mathematical Statistics* **25**, 252–272.

Grenander, U. and Rosenblatt M. (1957). *Statistical Analysis of Stationary Time Series*, Wiley, New York.

Künsch, H, Beran, J. and Hampel, F. (1993). Contrast under long-range correlations. *Annals of statistics* **21**, 943–964.

Ivanov, A.V. and Leonenko, N.N. (1989). *Statistical Analysis of Random Fields.* Kluwer Academic Publishers, Dordreht, 224 p.

Samarov, A. and Taqqu, M.S. (1988). On the efficiency of the sample mean in long-memory noise. *Journal of Time Series Analysis* **9**, 191–200.

Yajima, Y. (1988). On the estimation of a regression model with long-memory stationary errors. *Annals of Statistics* **16**, 791–807.

Yajima, Y. (1989). Asymptotic properties of the LSE in a regression model with long-memory stationary errors. *Annals of Statistics* **19**, 158–177.

Yadrenko, M.I. (1983). *Spectral Theory of Random Fields.* Optimization Software Inc., New York.

Exploring Stochastic Laws. pp. 263–273.
Eds. A.V. Skorokhod and Yu.V. Borovskikh

Large deviation theorems in hypotheses testing problems

YU.N. LIN'KOV

Institute of Applied Mathematics and Mechanics, National Academy of Sciences of Ukraine, Roza Luxemburg St., 74, Donetsk, 340114, Ukraine

Received April 8, 1995

Abstract. Large deviation theorems for logarithm of the likelihood ratio in a scheme of general statistical experiments are considered. On the basis of these general theorems, the theorem about the rates of decrease of the error probabilities for Neyman–Pearson test is proved. Application of this theorems are considered.

1. Introduction

Large deviation theorems for the likelihood ratio play an important role in the asymptotical investigation of the rate of decrease of the 1st and 2nd type errors probabilities in the hypotheses testing problems (see, for example, Chernoff (1956), Bahadur (1971), Birgé (1981), Borovkov and Mogul'skij (1992), Lin'kov (1995)). The initial study of this subject was devoted to observations of i.i.d. random variables. In this scheme, the exponent of the 2nd type error probabilities for Neyman–Pearson test, when the exponent of the 1st type error probabilities is given, was been obtained by Birgé (1981).

In this paper we consider large deviation theorems for the likelihood ratio in a scheme of general statistical experiments. On the basis of these theorems we investigate the exponents of the 1st and 2nd type errors probabilities for Neyman–Pearson test. We explain the latest results in this area of mathematics for general statistical experiments. Then we apply these general results to particular statistical experiments generated by concrete observations. In Section 2 the Hellinger integral of order ε and Neyman–Pearson test are introduced. Large deviation theorems for logarithm of the likelihood ratio are considered in Section 3. In Section 4 the theorem about the rates of decrease of the error probabilities for Neyman–Pearson test are proved. In Section 5 two examples are considered where the Theorems from Sections 3 and 4 can be applied.

2. Preliminaries

Let $(X^t, \mathfrak{B}^t, P^t, \tilde{P}^t)_{t\in R_+}$ be a family of binary statistical experiments generated by the observations ξ^t of arbitrary character (see, for example, (Lin'kov, 1993)). Denote by H^t and $\tilde{H}^t$ two statistical hypotheses according to which a distribution of the observation ξ^t is defined by the measures P^t and $\tilde{P}^t$, respectively. Let $Q^t = 2^{-1}(P^t + \tilde{P}^t)$ be a probability measure on $\mathfrak{B}^t$ and let $\zeta_t = dP^t/dQ^t$ and $\tilde{\zeta}_t = d\tilde{P}^t/dQ^t$ be finite versions of Radon–Nikodym derivatives. We denote by $z_t = \tilde{\zeta}_t/\zeta_t$ a likelihood ratio setting $0/0 = 0$ for definiteness. We introduce the Hellinger integral $H_t(\varepsilon)$ of order ε for the measures $\tilde{P}^t$ and P^t, putting

$$H_t(\varepsilon) = H(\varepsilon; \tilde{P}^t, P^t) = \begin{cases} P^t(\tilde{\zeta}_t > 0) & \text{for } \varepsilon = 0, \\ \tilde{P}^t(\zeta_t > 0) & \text{for } \varepsilon = 1, \\ E_Q^t \tilde{\zeta}_t^{\varepsilon} \zeta_t^{1-\varepsilon} & \text{for } \varepsilon \neq 0 \text{ and } \varepsilon \neq 1, \end{cases}$$

where E_Q^t means an expectation with respect to the measure Q^t. We denote

$$\varepsilon_-^{(t)} = \inf\{\varepsilon\colon H_t(\varepsilon) < \infty\}, \quad \varepsilon_+^{(t)} = \sup\{\varepsilon\colon H_t(\varepsilon) < \infty\}.$$

Obviously, $H_t(\varepsilon) = E^t z_t^{\varepsilon}$ for all $\varepsilon \in (\varepsilon_-^{(t)}, \varepsilon_+^{(t)})$, $\varepsilon \neq 0$ and $\varepsilon \neq 1$, where E^t means an expectation with respect to P^t.

To test the hypotheses H^t and $\tilde{H}^t$ we use the Neyman–Pearson test δ_t with the level α_t:

$$\delta_t = I(\Lambda_t > d_t) + q_t I(\Lambda_t = d_t), \tag{1}$$

where $\Lambda_t = \ln z_t$, $I(A)$ is the indicator of the set A and $d_t \in [-\infty, \infty]$ and $q_t \in [0,1]$ are parameters of the test δ_t defined by the condition $E^t \delta_t = \alpha_t$. Note that $\alpha(\delta_t) = E^t \delta_t$ is the 1st type error probability for the test δ_t. Denote by $\beta_t = \beta(\delta_t) = \tilde{E}^t(1-\delta_t)$ the 2nd type error probability for the test δ_t, where $\tilde{E}^t$ means an expectation with respect to $\tilde{P}^t$.

Below we shall consider large deviation theorems for Λ_t as $t \to \infty$ and then we shall apply these theorems to investigate the rate of decrease of $\alpha(\delta_t)$ and $\beta(\delta_t)$ as $t \to \infty$.

3. Large deviation theorem

We introduce the following condition:

H. For all $\varepsilon \in (-\infty, \infty)$, there exists the following limit

$$\lim_{t\to\infty} \psi_t^{-1} \ln H_t(\varepsilon) = \kappa(\varepsilon),$$

where $\psi_t \to \infty$ as $t \to \infty$ and $\kappa(\varepsilon)$ is a proper convex function such that $\kappa(\varepsilon)$ is differentiable on $(\varepsilon_-, \varepsilon_+)$ and $\gamma_- < \gamma_+$ where

$$\varepsilon_- = \inf\{\varepsilon\colon \kappa(\varepsilon) < \infty\}, \quad \varepsilon_+ = \sup\{\varepsilon\colon \kappa(\varepsilon) < \infty\},$$

$$\gamma_- = \lim_{\varepsilon \downarrow \varepsilon_-} \kappa'(\varepsilon), \quad \gamma_+ = \lim_{\varepsilon \uparrow \varepsilon_+} \kappa'(\varepsilon).$$

It's easy to see that $\varepsilon_- \leq 0$ and $\varepsilon_+ \geq 1$. In addition, if $\varepsilon_- < 0$ then $\kappa(0) = 0$, and if $\varepsilon_+ > 1$ then $\kappa(1) = 0$. If $\varepsilon_- < 0$ then the derivative $\gamma_0 = \kappa'(0)$ is defined, and if $\varepsilon_+ > 1$ then the derivative $\gamma_1 = \kappa'(1)$ is defined too. We denote the Legendre–Fenchel transform of the function $\kappa(\varepsilon)$ by $I(\gamma) = \sup_\varepsilon(\gamma\varepsilon - \kappa(\varepsilon))$, Rockafellar (1970) and Ellis (1985). Under the condition H we have $I(\gamma) = \gamma\,\varepsilon(\gamma) - \kappa(\varepsilon(\gamma))$, where $\varepsilon(\gamma)$ is any solution of the equation $\kappa'(\varepsilon) = \gamma$, $\gamma \in (\gamma_-, \gamma_+)$.

Introduce the notation

$$\Gamma_0 = \gamma_0\, I(\varepsilon_- < 0) + \gamma_-\, I(\varepsilon_- = 0),$$

$$\Gamma_1 = \gamma_1\, I(\varepsilon_+ > 1) + \gamma_+\, I(\varepsilon_+ = 1).$$

The following theorem gives two large deviation theorems for Λ_t as $t \to \infty$, the first one is under the hypothesis H^t and the second one is under the hypothesis $\tilde{H}^t$.

Theorem 1. *Let the condition H be satisfied. Then in the case $\Gamma_0 < \gamma_+$ for all $\gamma \in (\Gamma_0, \gamma_+)$*

$$\begin{aligned} &\lim_{t\to\infty} \psi_t^{-1} \ln P^t(\psi_t^{-1}\Lambda_t > \gamma) \\ &= \lim_{t\to\infty} \psi_t^{-1} \ln P^t(\psi_t^{-1}\Lambda_t \geq \gamma) = -I(\gamma), \end{aligned} \tag{2}$$

and in the case $\gamma_- < \Gamma_1$ for all $\gamma \in (\gamma_-, \Gamma_1)$

$$\begin{aligned} &\lim_{t\to\infty} \psi_t^{-1} \ln \tilde{P}^t(\psi_t^{-1}\Lambda_t < \gamma) \\ &= \lim_{t\to\infty} \psi_t^{-1} \ln \tilde{P}^t(\psi_t^{-1}\Lambda_t \leq \gamma) = -I(\gamma) + \gamma, \end{aligned} \tag{3}$$

where $I(\gamma) = \gamma\,\varepsilon(\gamma) - \kappa(\varepsilon(\gamma))$ and $\varepsilon(\gamma)$ is any solution of the equation $\kappa'(\varepsilon) = \gamma$.

Proof. If $\varepsilon < 0$ and $\Gamma_0 < \gamma_+$ in the condition H then the proof of relation (2) comes from Theorem II.6.3, Ellis (1985). For a detailed proof of (2) in this case see Lin'kov and Medvedeva (1995b). If $\varepsilon_- = 0$ in condition H then the proof of (2) is analogous to the case $\varepsilon_- < 0$ with corresponding modifications (see Lin'kov and Medvedeva (1995b)).

If $\varepsilon_+ > 1$ and $\kappa(\varepsilon)$ is a strictly convex function then the relation (3) is proved by Lin'kov (1993). In the general case relation (3) is proved by Lin'kov and Medvedeva (1995b). The proof of Theorem 1 is complete.

Remark 1. If $\varepsilon_- < 0$ and $\gamma_- < \gamma_0$ in condition H then for all $\gamma \in (\gamma_-, \gamma_0)$

$$\begin{aligned} &\lim_{t\to\infty} \psi_t^{-1} \ln P^t(\psi_t^{-1}\Lambda_t < \gamma) \\ &= \lim_{t\to\infty} \psi_t^{-1} \ln P^t(\psi_t^{-1}\Lambda_t \leq \gamma) = -I(\gamma), \end{aligned} \tag{4}$$

(see Lin'kov and Medvedeva (1995a, b)). If $\varepsilon_+ > 1$ and $\gamma_1 < \gamma_+$ in condition H then for all $\gamma \in (\gamma_1, \gamma_+)$

$$\begin{aligned} &\lim_{t\to\infty} \psi_t^{-1} \ln \tilde{P}^t(\psi_t^{-1}\Lambda_t > \gamma) \\ &\quad = \lim_{t\to\infty} \psi_t^{-1} \ln \tilde{P}^t(\psi_t^{-1}\Lambda_t \geq \gamma) = -I(\gamma) + \gamma, \end{aligned} \tag{5}$$

(see Lin'kov and Medvedeva (1995a, b)).

Remark 2. The inequality $\Gamma_0 < \gamma_+$ is a restriction only in the case $\varepsilon_- < 0$. Also, the inequality $\gamma_- < \Gamma_1$ is a restriction only in the case $\varepsilon_+ > 1$. If $\varepsilon_- = 0$ then $\Gamma_0 = \gamma_-$ and we have the large deviation theorem (2) for all $\gamma \in (\gamma_-, \gamma_+)$, but we do not have the large deviation theorem of type (4). If $\varepsilon_+ = 1$, then $\Gamma_1 = \gamma_+$ and so we have the large deviation theorem (3) under the alternative $\tilde{H}^t$ for all $\gamma \in (\gamma_-, \gamma_+)$ but we do not have the large deviation theorem of type (5).

4. The rate of decrease of the error probabilities for Neyman–Pearson tests

Let δ_t be a Neyman–Pearson test with the level α_t for the testing of the hypotheses H^t and $\tilde{H}^t$ defined by equality (1). The following theorem describes the rates of decrease of the error probabilities for Neyman–Pearson test under condition H.

Theorem 2. *Let condition H be satisfied with the additional restriction $\Gamma_0 < \Gamma_1$. Then the following statements hold:*

1) *for all $a \in (I(\Gamma_0),\ I(\Gamma_1))$*

$$\lim_{t\to\infty} \psi_t^{-1} \ln \alpha_t = -a \quad \Longleftrightarrow \quad \lim_{t\to\infty} \psi_t^{-1} \ln \beta_t = -b(a), \tag{6}$$

where $b(a) = a - \gamma(a) \in (I(\Gamma_1) - \Gamma_1, I(\Gamma_0) - \Gamma_0)$ and $\gamma(a)$ is a unique solution of the equation $I(\gamma) = a$ with respect to $\gamma \in (\Gamma_0, \Gamma_1)$;

2) *for all $a \in [-\kappa(0), I(\Gamma_0)]$*

$$\lim_{t\to\infty} \psi_t^{-1} \ln \alpha_t = -a \quad \Longrightarrow \quad \limsup_{t\to\infty} \psi_t^{-1} \ln \beta_t \leq \Gamma_0 - I(\Gamma_0), \tag{7}$$

and for all $a \in [I(\Gamma_1), \infty]$

$$\lim_{t\to\infty} \psi_t^{-1} \ln \alpha_t = -a \quad \Longrightarrow \quad \liminf_{t\to\infty} \psi_t^{-1} \ln \beta_t \geq \Gamma_1 - I(\Gamma_1); \tag{8}$$

3) *for all $b \in [-\kappa(1), I(\Gamma_1) - \Gamma_1]$*

$$\lim_{t\to\infty} \psi_t^{-1} \ln \beta_t = -b \quad \Longrightarrow \quad \limsup_{t\to\infty} \psi_t^{-1} \ln \alpha_t \leq -I(\Gamma_1), \tag{9}$$

and for all $b \in [I(\Gamma_0) - \Gamma_0, \infty]$

$$\lim_{t\to\infty} \psi_t^{-1} \ln \beta_t = -b \quad \Longrightarrow \quad \liminf_{t\to\infty} \psi_t^{-1} \ln \alpha_t \geq -I(\Gamma_0). \tag{10}$$

Proof. The proof of the equivalence (6) can be found in Lin'kov and Medvedeva (1995b). Here we prove only Statements 2 and 3.

Let $\lim_{t\to\infty} \psi_t^{-1} \ln \alpha_t = -a$. If $a \in [-\kappa(0), I(\Gamma_0)]$ then for all $\lambda > \Gamma_0$ there exists $t_0 = t_0(\lambda)$ such that $\psi_t^{-1} d_t < \lambda$ for all $t > t_0$. Then for all $\varepsilon \in (0,1)$ and $t > t_0$ we have

$$\begin{aligned}\psi_t^{-1} \ln \beta_t &\leq \psi_t^{-1} \ln \tilde{P}^t(\Lambda_t \leq d_t) \leq \psi_t^{-1} \ln \tilde{P}^t(\psi_t^{-1} \Lambda_t < \lambda) \\ &\leq (1-\varepsilon)\,\lambda + \psi_t^{-1} \ln H_t(\varepsilon).\end{aligned}$$

From here, we obtain

$$\limsup_{t\to\infty} \psi_t^{-1} \ln \beta_t \leq (1-\varepsilon)\,\lambda + \kappa(\varepsilon). \tag{11}$$

Taking into account the equality $I(\Gamma_0) = -\kappa(0+)$ and tending to the limit in the inequality (11) as $\varepsilon \to 0$ and $\lambda \to \Gamma_0$, the implication (7) is validated.

Now, if $a \in [I(\Gamma_1), \infty]$ then for all $\lambda < \Gamma_1$ there exists $t_1 = t_1(\lambda)$ such that $\psi_t^{-1} d_t > \lambda$ for all $t > t_1$. Then we have $\beta_t \geq \tilde{P}^t(\psi_t^{-1} \Lambda_t < \lambda)$ for all $t > t_1$. From this inequality and relation (3) we obtain

$$\liminf_{t\to\infty} \psi_t^{-1} \ln \beta_t \geq \lambda - I(\lambda). \tag{12}$$

Tending to the limit in the inequality (12) as $\lambda \to \Gamma_1$, implication (8) is validated.

Relations (9) and (10) in Statement 3 can be proved by analogy with the proof of relations (7) and (8) in Statement 2. Therefore this proof is omited. Theorem 2 is proved.

Remark 3. In Lin'kov and Medvedeva (1995b) the following inequality is proved under the condition H

$$\limsup_{t\to\infty} \psi_t^{-1} \ln \alpha_t \leq \kappa(0). \tag{13}$$

Therefore if there exists the limit $\lim_{t\to\infty} \psi_t^{-1} \ln \alpha_t = -a$ then $a \geq -\kappa(0)$. By analogy with inequality (13) we can prove the following inequality

$$\limsup_{t\to\infty} \psi_t^{-1} \ln \beta_t \leq \kappa(1).$$

Therefore if there exists the limit $\lim_{t\to\infty} \psi_t^{-1} \ln \beta_t = -b$ then we have $b \geq -\kappa(1)$. Thus the implication (8) in Statement 2 for all $a \in [I(\Gamma_1), \infty]$ has the form

$$\begin{aligned}\lim_{t\to\infty} \psi_t^{-1} \ln \alpha_t = -a \quad \Longrightarrow \quad \kappa(1) &\geq \limsup_{t\to\infty} \psi_t^{-1} \ln \beta_t \\ &\geq \liminf_{t\to\infty} \psi_t^{-1} \ln \beta_t \geq \Gamma_1 - I(\Gamma_1).\end{aligned}$$

We note that $\Gamma_1 - I(\Gamma_1) = \kappa(1-)$. Therefore if $\kappa(1-) = \kappa(1)$ then from here we obtain the following known result

$$\lim_{t\to\infty} \psi_t^{-1} \ln \alpha_t = -a \quad \Longrightarrow \quad \lim_{t\to\infty} \psi_t^{-1} \ln \beta_t = \kappa(1)$$

for all $a \in [I(\Gamma_1), \infty]$ (see Lin'kov and Medvedeva (1995b)). By analogy with the above we obtain for all $b \in [I(\Gamma_0) - \Gamma_0, \infty]$ that implication (10) has the form

$$\begin{aligned} \lim_{t\to\infty} \psi_t^{-1} \ln \beta_t = -b \quad \Longrightarrow \quad -I(\Gamma_0) \; &\le \; \liminf_{t\to\infty} \psi_t^{-1} \ln \alpha_t \\ &\le \; \limsup_{t\to\infty} \psi_t^{-1} \ln \alpha_t \le \kappa(0). \end{aligned}$$

But $I(\Gamma_0) = -\kappa(0+)$. Therefore if $\kappa(0+) = \kappa(0)$ then from here we obtain the following known result

$$\lim_{t\to\infty} \psi_t^{-1} \ln \beta_t = -b \quad \Longrightarrow \quad \lim_{t\to\infty} \psi_t^{-1} \ln \alpha_t = \kappa(0)$$

for all $b \in [I(\Gamma_0) - \Gamma_0, \infty]$ (see Lin'kov and Medvedeva (1995b)).

5. Examples

Here we consider two examples where the conditions of Theorems 1 and 2 are satisfied. Other examples can be found in Birgé (1981), Lin'kov and Diallo (1993), Lin'kov (1993b, 1994, 1995), Lin'kov and Medvedeva (1995a, b, c, d).

Example 1. Let $\xi^t = (\xi_{t1}, \xi_{t2}, \ldots, \xi_{tt})$ where $\xi_{t1}, \xi_{t2}, \ldots, \xi_{tt}$ are independent random variables under the hypothesis H^t and under the hypothesis $\tilde{H}^t$ and let the distribution law of ξ_{ti} under the hypothesis H^t (respectively $\tilde{H}^t$) be exponential with the density $\lambda_{ti} \exp(-\lambda_{ti}(x - b_{ti}))\, I(x \ge b_{ti})$ (respectively $\tilde{\lambda}_{ti} \exp(-\tilde{\lambda}_{ti}(x - \tilde{b}_{ti}))\, I(x \ge \tilde{b}_{ti})$) where $\lambda_{ti} > 0$, $\tilde{\lambda}_{ti} > 0$, $b_{ti} \in (-\infty, \infty)$ and $\tilde{b}_{ti} \in (-\infty, \infty)$. It is easy to show that

$$\begin{aligned} \varepsilon_-^{(t)} &= -\min_{1\le i\le t} \left[\left(\left(\frac{\tilde{\lambda}_{ti}}{\lambda_{ti}} \vee 1 \right) - 1 \right)^{-1} I(\tilde{b}_{ti} \le b_{ti}) \right], \\ \varepsilon_+^{(t)} &= \min_{1\le i\le t} \left[1 - \left(\frac{\tilde{\lambda}_{ti}}{\lambda_{ti}} \wedge 1 \right) I(\tilde{b}_{ti} \ge b_{ti}) \right]^{-1} \end{aligned}$$

and for all $\varepsilon \in (\varepsilon_-^{(t)}, \varepsilon_+^{(t)})$

$$\begin{aligned} \ln H_t(\varepsilon) &= \sum_{i=1}^{t} \left\{ \varepsilon \ln \frac{\tilde{\lambda}_{ti}}{\lambda_{ti}} - \ln\left[\varepsilon \left(\frac{\tilde{\lambda}_{ti}}{\lambda_{ti}} - 1 \right) + 1 \right] \right\} + \\ &\quad + \varepsilon \ln H_t(1) + (1 - \varepsilon) \ln H_t(0), \end{aligned}$$

where

$$\begin{aligned} \ln H_t(1) &= -\sum_{i=1}^{t} \tilde{\lambda}_{ti} (b_{ti} - \tilde{b}_{ti})\, I(b_{ti} > \tilde{b}_{ti}), \\ \ln H_t(0) &= -\sum_{i=1}^{t} \lambda_{ti} (\tilde{b}_{ti} - b_{ti})\, I(\tilde{b}_{ti} > b_{ti}). \end{aligned}$$

Here $a \vee b = \max(a, b)$ and $a \wedge b = \min(a, b)$.

Let $\lambda_{ti} = \lambda$, $\tilde{\lambda}_{ti} = \tilde{\lambda}$, $b_{ti} = b$ and $\tilde{b}_{ti} = \tilde{b}$ for all $i = 1, 2, \ldots, t$ and $t = 1, 2, \ldots$. Then the condition H is satisfied with $\psi_t = t$,

$$\varepsilon_- = -\Big(\Big(\frac{\tilde{\lambda}}{\lambda} \vee 1\Big) - 1\Big)^{-1} I(\tilde{b} \le b), \quad \varepsilon_+ = \Big[1 - \Big(\frac{\tilde{\lambda}}{\lambda} \wedge 1\Big) I(\tilde{b} \ge b)\Big]^{-1}$$

and for all $\varepsilon \in (\varepsilon_-, \varepsilon_+)$

$$\kappa(\varepsilon) = \varepsilon \ln \frac{\tilde{\lambda}}{\lambda} - \ln\Big[\varepsilon\Big(\frac{\tilde{\lambda}}{\lambda} - 1\Big) + 1\Big] + \varepsilon\, \kappa(1) + (1 - \varepsilon)\, \kappa(0),$$

where

$$\kappa(1) = -\tilde{\lambda}(b - \tilde{b})\, I(b > \tilde{b}), \quad \kappa(0) = -\lambda(\tilde{b} - b)\, I(\tilde{b} > b).$$

If $\tilde{\lambda} = \lambda$ and $\tilde{b} \ne b$ then $\gamma_- = \gamma_+$ and hence the condition H is not satisfied.

Let $\tilde{\lambda} \ne \lambda$ and $\tilde{\lambda}/\lambda = \rho$. Suppose that $\tilde{b} > b$. Then $\varepsilon_- = 0$, $\varepsilon_+ = (1 - (\rho \wedge 1))^{-1}$ and

$$\kappa(\varepsilon) = \varepsilon \ln \rho - \ln[\varepsilon(\rho - 1) + 1] - (1 - \varepsilon)(\tilde{b} - b)\lambda.$$

In this case $\gamma_- = -(\rho - 1 - \ln \rho) + (\tilde{b} - b)\lambda$ for all $\rho \in (0, \infty)\backslash\{1\}$ and $\gamma_+ = \infty$ for all $\rho \in (0, 1)$ and $\gamma_+ = \ln \rho + (\tilde{b} - b)\lambda$ for all $\rho \in (1, \infty)$. In addition, $\Gamma_0 = \gamma_-$ and $\Gamma_1 = \gamma_1 = (\rho^{-1} - 1 - \ln \rho^{-1}) + (\tilde{b} - b)\lambda$. It is easy to show that

$$I(\gamma) = (z(\gamma) - 1 - \ln z(\gamma)) + (\tilde{b} - b)\lambda,$$

where

$$z(\gamma) = (\rho - 1)^{-1}\big(\ln \rho + (\tilde{b} - b)\lambda - \gamma\big).$$

If $0 < \rho < 1$ then $\gamma > \gamma_-$ only for $z(\gamma) > 1$. Let z_a be a solution of the equation $z - 1 - \ln z + (\tilde{b} - b)\lambda = a$ with respect to $z \in (1, \infty)$ for all $a \in (I(\Gamma_0), I(\Gamma_1)) = ((\tilde{b} - b)\lambda, \gamma_1)$. Then $\gamma(a) = \ln \rho + (1 - \rho)z_a + (\tilde{b} - b)\lambda$ and hence

$$b(a) = a - \ln \rho - (1 - \rho)\, z_a - (\tilde{b} - b)\, \lambda.$$

If $1 < \rho < \infty$ then $\gamma > \gamma_-$ only for $z(\gamma) \in (0, 1)$. Let $\tilde{z}_a$ be a solution of the equation $z - 1 - \ln z + (\tilde{b} - b)\lambda = a$ with respect to $z \in (0, 1)$ for all $a \in (I(\Gamma_0), I(\Gamma_1)) = ((\tilde{b} - b)\lambda, \gamma_1)$. Then $\gamma(a) = \ln \rho - (\rho - 1)\tilde{z}_a + (\tilde{b} - b)\lambda$ and hence

$$b(a) = a - \ln \rho + (\rho - 1)\, \tilde{z}_a - (\tilde{b} - b)\, \lambda.$$

If $\tilde{b} < b$ and $\rho \ne 1$ then $\varepsilon_- = -((\rho \vee 1) - 1)^{-1}$, $\varepsilon_+ = 1$ and

$$\kappa(\varepsilon) = \varepsilon \ln \rho - \ln[\varepsilon\, (\rho - 1) + 1] - \varepsilon\, (b - \tilde{b})\, \tilde{\lambda}.$$

This case can be considered by analogy with the case $\tilde{b} > b$.

If $\tilde{b}_{ti} = b_{ti}$ for all $i = 1, 2, \dots, t$ and $t = 1, 2, \dots$ then we obtain the case which is considered in Lin'kov and Medvedeva (1995b).

If $\tilde{\lambda}_{ti} = \lambda_{ti}$ for all $i = 1, 2, \dots, t$ and $t = 1, 2, \dots$ then the condition H is not satisfied.

Example 2. We consider the autoregressive process

$$\xi_i = \theta\, \xi_{i-1} + w_i, \quad i = 1, 2, \dots,$$

where $\xi_0 = 0$, $\theta \in (-\infty, \infty)$ is an unknown parameter and $w_1, w_2, \dots$ are independent standard normal variables. Then the distribution law of $\xi^t = (\xi_1, \xi_2, \dots, \xi_t)$ depends on the parameter θ and it is defined by the measure P^t_θ. Let $P^t = P^t_\theta$ and $\tilde{P}^t = \tilde{P}^t_{\tilde{\theta}}$ where $|\theta| < 1$ and $\tilde{\theta} \neq \theta$. It is easy to show that $H_t(\varepsilon) = (\det A_t(\varepsilon))^{-1/2}$ for all $\varepsilon \in (\varepsilon_-^{(t)}, \varepsilon_+^{(t)})$ where $A_t(\varepsilon) = (a_{ij}(\varepsilon))$ is a matrix of order t such that $a_{ij}(\varepsilon) = 0$ when $|i-j| > 1$, $a_{ij}(\varepsilon) = d(\varepsilon)$ when $|i-j| = 1$, $a_{ii}(\varepsilon) = c(\varepsilon)$ for all $i = 1, 2, \dots, t-1$ and $a_{tt}(\varepsilon) = 1$. Here $d(\varepsilon) = -\theta - \varepsilon(\tilde{\theta} - \theta)$ and $c(\varepsilon) = 1 + \theta^2 + \varepsilon\,(\tilde{\theta}^2 - \theta^2)$. It is easy to show that for all $t \geq 3$

$$\det A_t(\varepsilon) = \big(\alpha(\varepsilon) - \beta(\varepsilon)\big)^{-1} \big((1 - \beta(\varepsilon))\, \alpha^t(\varepsilon) - (1 - \alpha(\varepsilon))\, \beta^t(\varepsilon)\big), \tag{14}$$

where

$$\alpha(\varepsilon) = \frac{1}{2}\Big(c(\varepsilon) + \sqrt{c^2(\varepsilon) - 4\, d^2(\varepsilon)}\Big),$$
$$\beta(\varepsilon) = \frac{1}{2}\Big(c(\varepsilon) - \sqrt{c^2(\varepsilon) - 4\, d^2(\varepsilon)}\Big).$$

Let

$$\varepsilon_1 = -\frac{(1+\theta)^2}{(\tilde{\theta} - \theta)(\tilde{\theta} + \theta + 2)}, \quad \varepsilon_2 = \frac{(1-\theta)^2}{(\tilde{\theta} - \theta)(2 - \tilde{\theta} - \theta)},$$

where ε_1 is defined only when $\tilde{\theta} + \theta \neq -2$, and ε_2 is defined only when $\tilde{\theta} + \theta \neq 2$. If $\tilde{\theta}$ does not depend on t then for all $\varepsilon \in (-\infty, \infty)$ there exists the limit

$$\lim_{t \to \infty} \frac{1}{2} \ln H_t(\varepsilon) = \kappa(\varepsilon),$$

where $\kappa(\varepsilon) = -2^{-1} \ln \alpha(\varepsilon)$ for all $\varepsilon \in (\varepsilon_-, \varepsilon_+)$ when $1 - \theta\tilde{\theta} = 0$, for all $\varepsilon \in [\varepsilon_-, \varepsilon_+]$ when $1 - \theta\tilde{\theta} \neq 0$ and $|\tilde{\theta}| \leq 1$, for all $\varepsilon \in [\varepsilon_-, \varepsilon_+)$ when $1 - \theta\tilde{\theta} \neq 0$ and $|\tilde{\theta}| > 1$, $\kappa(\varepsilon) = 0$ for $\varepsilon = \varepsilon_+$ when either $1 - \theta\tilde{\theta} = 0$ or $1 - \theta\tilde{\theta} \neq 0$ and $|\tilde{\theta}| > 1$, and $\kappa(\varepsilon) = \infty$ for all another ε; here ε_- and ε_+ have the forms

$$\varepsilon_- = \begin{cases} \varepsilon_1\, I(\tilde{\theta} > \theta) + \varepsilon_2\, I(\tilde{\theta} < \theta) & \text{if } 1 - \theta\tilde{\theta} > 0, \\ \varepsilon_2\, I(\tilde{\theta} > \theta) + \varepsilon_1\, I(\tilde{\theta} < \theta) & \text{if } 1 - \theta\tilde{\theta} \leq 0, \end{cases}$$

$$\varepsilon_+ = \begin{cases} \varepsilon_2\, I(\tilde{\theta} > \theta) + \varepsilon_1\, I(\tilde{\theta} < \theta) & \text{if } |\tilde{\theta}| \leq 1, \\ 1 & \text{if } |\tilde{\theta}| > 1, \end{cases}$$

(the proof can be found in Lin'kov and Medvedeva (1995c)). We note that $\varepsilon_+ > 1$ if $|\tilde\theta| < 1$ and $\varepsilon_+ = 1$ if $|\tilde\theta| \geq 1$. In addition, $\kappa(1) = \kappa(1-) = 0$ if $|\tilde\theta| = 1$ and $\kappa(1-) < \kappa(1) = 0$ if $|\tilde\theta| > 1$. The calculation of the functions $I(\gamma)$ and $b(a)$ when $\tilde\theta$ does not depend on t, can be found in Lin'kov and Medvedeva (1995d).

Now let $\tilde\theta$ be dependent on t such that $\tilde\theta = \theta_t \to \theta$ as $t \to \infty$. Suppose that $t\,\Delta_t^2 \to \infty$ as $t \to \infty$ where $\Delta_t = \theta_t - \theta$. From the equality (14) we have

$$\ln H_t(\varepsilon) = -\frac{t}{2}\ln\alpha(\varepsilon) - \frac{1}{2}\ln\frac{1-\beta(\varepsilon)-(1-\alpha(\varepsilon))(\beta(\varepsilon)/\alpha(\varepsilon))^t}{\alpha(\varepsilon)-\beta(\varepsilon)} \tag{15}$$

for all $\varepsilon \in (\varepsilon_-^{(t)}, \varepsilon_+^{(t)})$. From Lin'kov and Medvedeva (1995c) we known that $\varepsilon_-^{(t)} \uparrow \varepsilon_-$ and $\varepsilon_+^{(t)} \downarrow \varepsilon_+$ as $t \to \infty$ when θ_t does not depend on t. But $\varepsilon_1 \to -\infty$ and $\varepsilon_2 \to \infty$ as $t \to \infty$ when $\Delta_t > 0$, and $\varepsilon_1 \to \infty$ and $\varepsilon_2 \to -\infty$ as $t \to \infty$ when $\Delta_t < 0$. Therefore $\varepsilon_-^{(t)} \to -\infty$ and $\varepsilon_+^{(t)} \to \infty$ as $t \to \infty$. It is easy to show that for all $\varepsilon \in (-\infty, \infty)$

$$\alpha(\varepsilon) = 1 + \frac{\varepsilon\,(1-\varepsilon)}{1-\theta^2}\,\Delta_t^2 + O(\Delta_t^3), \qquad t \to \infty, \tag{16}$$

and hence

$$\ln\alpha(\varepsilon) = \frac{\varepsilon\,(1-\varepsilon)}{1-\theta^2}\,\Delta_t^2\,(1 + O(\Delta_t)), \qquad t \to \infty. \tag{17}$$

By analogy, we obtain that for all $\varepsilon \in (-\infty, \infty)$

$$\beta(\varepsilon) = \theta^2 + 2\theta\varepsilon\Delta_t + O(\Delta_t^2), \qquad t \to \infty. \tag{18}$$

From (16) and (18), we have for all $\varepsilon \in (-\infty, \infty)$

$$\alpha(\varepsilon) - \beta(\varepsilon) = 1 - \theta^2 - 2\theta\varepsilon\Delta_t + O(\Delta_t^2), \qquad t \to \infty. \tag{19}$$

Now, from (16), (18) and (19), we obtain that for all $\varepsilon \in (-\infty, \infty)$

$$\ln\frac{1-\beta(\varepsilon)-(1-\alpha(\varepsilon))(\beta(\varepsilon)/\alpha(\varepsilon))^t}{\alpha(\varepsilon)-\beta(\varepsilon)} = O(\Delta_t^2), \qquad t \to \infty. \tag{20}$$

Finally, from (15), (17) and (20) we obtain that the condition H is satisfied with $\psi_t = t\,\Delta_t^2$, $\kappa(\varepsilon) = -2^{-1}\,(1-\theta^2)^{-1}\,\varepsilon\,(1-\varepsilon)$, $\varepsilon_- = -\infty$, $\varepsilon_+ = \infty$, $\gamma_- = -\infty$ and $\gamma_+ = \infty$. It's easy to show that $\gamma_0 = -(2(1-\theta^2))^{-1}$, $\gamma_1 = (2(1-\theta^2))^{-1}$ and

$$I(\gamma) = (8(1-\theta^2))^{-1}\,(1 + 2\gamma\,(1-\theta^2))^2.$$

Hence $\gamma(a) = (2a/(1-\theta^2))^{1/2} - (2(1-\theta^2))^{-1}$ for all $a \in (0, \infty)$. Thus, we have for all $a \in (0, \gamma_1)$

$$b(a) = \left((2a)^{1/2} - (2(1-\theta^2))^{-1/2}\right)^2.$$

References

Bahadur, R.R. (1971). *Some Limit Theorem in Statistics.* SIAM, Philadelphia.

Birgè, L. (1981). Vitesses maximales de dècroissance des erreurs et tests optimaux associes. *Z. Wahrscheinlichkeitstheorie und ver. Getiete* **55**(2), 261–273.

Borovkov, A.A. and Mogul'skij, A.A. (1992). *Large Deviations and Statistical Hypotheses Testing.* Nauka, Novosibirsk. (In Russian).

Chernoff, H. (1956). Large sample theory: parametric case. *Ann. Math. Statist.* **27**(1), 1-22.

Ellis, R.S. (1985). *Entropy, Large Deviations and Statistical Mechanics.* Springer-Verlag, Berlin—Heidelberg—New York.

Lin'kov, Yu.N. (1993a). *Asymptotical Methods of Statistics for Stochastic Processes.* Naukova Dumka, Kiev. (In Russian).

Lin'kov, Yu.N. (1993b). Large deviations in testing problems of counting processes. *Ukr. Math. J.* **45**(11), 1514–1521. (In Russian).

Lin'kov, Yu.N. (1994). Limit theorems for the local density of measures in the hypotheses testing problems of counting processes. *Probability Theory and Mathematical Statistics.* Proceedings of the Sixth Vilnius Conference (Vilnius, 28 June-3 July, 1993). B. Grigelionis et al. (eds). TEV, Vilnius; VSP, Utrecht, 497–515.

Lin'kov, Yu.N. (1995). Limit theorems for the likelihood ratio in the hypotheses testing problems. *Random Operators and Stochastic Equations.* **3**(1), 23–40.

Lin'kov, Yu. N. and Diallo, M. S. (1993). *Les propriètès asymptotiques de la densite locale des mesures pour les processes à accroissements indèpendants.* Prepublication 93. 06. Inst. Math. Appl. et Mecanique, Donetsk.

Lin'kov, Yu. N. and Medvedeva, M. I. (1995a). Large deviation theorems in testing problems of two simple hypotheses. *Ukr. Math. J.* **47**(2), 227–235. (In Russian).

Lin'kov, Yu.N. and Medvedeva, M.I. (1995b). Large deviation theorems for the logarithm of the likelihood ratio. *Teoria imovirnostej ta matematichna statistica.* **53**. (In Ukrainian, in press).

Lin'kov, Yu.N. and Medvedeva, M.I. (1995c). Large deviation theorems in the testing problems of the normal autoregressive processes. *Theory of Random Processes* **1**(17). (in Russian, in press).

Lin'kov, Yu.N. and Medvedeva, M.I. (1995d). On a method of calculation of the rate of decrease of the error probabilities for Neumann–Pearson test in testing autoregressive processes. *Theory of Random Processes.* **1**(17). (In Russian, in press).

Rockafellar, R.T. (1970). *Convex analysis.* Princeton University Press, New Jersey, Princeton.

Exploring Stochastic Laws. pp. 275–284.
Eds. A.V. Skorokhod and Yu.V. Borovskikh

Quasi-diffusion measure-valued processes and a limit theorem for jump measure-valued Markov processes

R.YA. MAIDANIUK[1] AND A.V. SKOROKHOD[2]

[1] *Kiev Transport Academy, Ukraine*

[2] *Institute of Mathematics of National Academy of Sciences of Ukraine, Michigan State University*

Received March 15, 1995

Abstract. The wide class of continuous measure-valued Markov processes which may be called quasi-diffusion is introduced. The limit theorem on the convergence of jump measure-valued Markov processes to the quasi-diffusion one is considered.

1. Introduction

Measure-valued processes were mainly considered within the theory of branching processes [4], but they may be used to describe randomly moving systems of particles, in particular, in the study of transport systems [5].

The goals of this article are to introduce the wide class of continuous measure-valued processes which can be called *quasi-diffusion measure-valued processes* [3], and to consider the limit theorem on the convergence of jump measure-valued Markov processes to the quasi-diffusion one.

In this paper we use the general theory of Markov processes [1] and martingale methods [2]. But the main tool which is used here are stochastic differential equations for complex systems [3].

2. Continuous measure-valued processes

Let X be a metric compact set, B its Borelian σ-algebra, C the space of continuous functions $f : X \to R$, and M the set of finite measures on B. We introduce in M a metric which generate the weak convergence of measures and denote by B_M the set of Borelian subsets of M. Consider Markov processes in the measurable space (M, B_M). We use the following notation: $\{\Omega, F\}$ is the probability space on which the process is determined, $\{F_t^s \subseteq F,\ \ 0 \le s \le t \le +\infty\}$ are σ-algebras connected with the Markov process. We denote its trajectories by $\mu(t) = \mu(t,\omega) : R_+ \times \Omega \to M$. $P_{s,m}$ is the probability on F corresponding to a process on the interval $[s,\infty]$ with initial value $\mu(s) = m$,

$$P(s,m,t,B) = P_{s,m}(\mu(t) \in B), \quad 0 \le s \le t, \ \ m \in M, \ \ B \in B_M$$

is the transition probability of the process. Denote

$$\langle f, m\rangle = \int f(x) m(dx),$$

where f is a measurable bounded function $X \to R$.

Preliminary assumptions

1. $\mu(t,\omega)$ is continuous in $t \in R_+$ for all $\omega \in \Omega$.

2. For any fixed $s \in R_+$ the filtration $\{F_t^s,\ \ t \ge s\}$ is right continuous and quasi left continuous (this is equivalent to the continuity of all F_t^s-right continuous martingales).

3. There exists a linear subset $D \subset C$ which is dense in C in the norm $\|f\| = \sup_{x \in X} |f(x)|$ and for $f \in D$ the function $\langle f, \mu_t\rangle$ is a semimartingale.

Assumption 3 implies the representation

$$\langle f, \mu_t\rangle = \mu(f,t) + a(f,t) + \langle f, \mu_0\rangle \tag{1}$$

where $\mu(f,t)$ is a continuous martingale, $a(f,t) : D \times R_+ \times \Omega \to R$ is an adapted continuous function of bounded variation, it is called *the compensator* of the semimartingale $\langle f, \mu_t\rangle$. $\mu(f,t)$ and $a(f,t)$ are linear in $f \in D$.

Introduce the function $b(f_1,f_2,t,\omega) = b(f_1,f_2,t) : D^2 \times R_+ \times \Omega \to R_+$ which is the mutual characteristic of the martingales $\mu(f_1,t)$ and $\mu(f_2,t)$. It is the bilinear form on D^2 adapted to filtration (F_t^0). It can be proved (see [3], Chapter 1, §1) that $a(f,t)$ and $b(f_1,f_2,t)$ are additive functionals of the Markov process μ. Functions $a(f,t)$ and $b(f_1,f_2,t)$ allow us to determine the transition probabilities of the process in the same way as the diffusion coefficients allow us to determine the distributions of the diffusion process.

Additional assumption

4. The functions $a(f,t)$ and $b(f_1,f_2,t)$ can be represented in the form:

$$a(f,t) \ = \ \int_0^t A(s,f,\mu(s))ds \tag{2}$$

$$b(f_1,f_2,t) \ = \ \int_0^t B(s,f_1,f_2,\mu(s))ds \tag{3}$$

where the function

$$A(s, f, m) : R_+ \times D \times M \to R$$

is continuous in $s \in R_1$ and $m \in M$, and linear in f; $B(s, f_1, f_2, m) : R_+ \times D^2 \times M \to R$ is a function continuous in $s \in R_+, m \in M$ and linear in f_1 and f_2. Denote by $D(R_+ \times M)$ the set of functions having the form

$$G(t, m) = g(t,\ \langle f_1, m\rangle,\ \ldots,\ \langle f_n, m\rangle) \tag{4}$$

where $f_1, \ldots, f_n \in D$ and $g \in C^{(2)}(R_+ \times R^n)$. Here $C^{(2)}(\ )$ is the space of continuous and bounded functions with continuous, bounded derivatives up to the second order. Obviously

$$D(R_+ \times M) \subset C(R_+ \times M),$$

where $C(R_+ \times M)$ is the space of continuous bounded functions $R_+ \times M \to R$.

It follows from Ito's formula that if $G \in D(R_+ \times M)$ then $G(t, \mu(t))$ is a semimartingale with a compensator of the form $\int_0^t LG(s, \mu(s))ds$, where

$$\begin{aligned} LG(s,\ m) \ &= \ \frac{\partial}{\partial s} g(s, \langle f_1,\ m\rangle,\ \ldots,\ \langle f_n, m\rangle) \\ &+ \sum_{i=1}^{n} \frac{\partial}{\partial z_i} g(s, \langle f_1,\ m\rangle,\ \ldots,\ \langle f_n,\ m\rangle) A(s,\ f_1,\ m) \\ &+ \frac{1}{2} \sum_{i,j=1}^{n} \frac{\partial^2}{\partial z_i \partial z_j} g(s, \langle f_1,\ m\rangle,\ \ldots,\ \langle f_n,\ m\rangle) B(s,\ f_i,\ f_j,\ m) \end{aligned} \tag{5}$$

and the function $g(s, z_1, \ldots, z_n)$ is connected with G by relation (4). It can be proved that $D(R_+ \times M)$ is dense in $C(R_+ \times M)$.

Theorem 1. *Let $G \in D(R_+ \times M)$ and let it satisfy the conditions*

a) $\sup_m |G(t, m)| \to 0, \quad t \to \infty;$ b) $\int_0^\infty \sup_m |LG(t, m)| dt < \infty.$

Then

$$G(t, m) = -\int_t^\infty T_s^t L(G(s, m)) ds \tag{6}$$

where T_s^t, $t \le s$ is a system of linear operators on the space $B(M)$ of B_M-measurable bounded functions $M \to R$,

$$T_s^t F(m) = \int F(m') P(t, m, s, dm'). \tag{7}$$

Proof. The process

$$G(t,\mu(t)) - \int_0^t LG(s,\mu(s))ds$$

is a martingale and therefore for $t < u$

$$\begin{aligned} E(G(u,\mu(u)) - G(t,\mu(t))/F_t^0) &= E\left(\int_t^u LG(s,\mu(s)ds/F_t^0)\right) \\ &= \int_t^u T_s^t LG(s,\mu(t))ds. \end{aligned}$$

Formula (6) is a consequence of this relation if $u \to \infty$.

Remark 1. Suppose that the set of functions of the form

$$\{F(s,m) = L(G(s,m)),\quad G \in D(R_+ \times M)\}$$

is a total subset of $C(R_+ \times M)$. Then if

$$\int_{R_+\times M} F(s,m)\nu_1(ds,dm) = \int_{R_+\times M} F(s,m)\nu_2(ds,dm)$$

for all such F where ν_1, ν_2 are measures on $R_+ \times M$, then $\nu_1 = \nu_2$. Now operator L which is determined by the functions A, B determines the transition probability of the process. This is a consequence of formula (6).

Remark 2. Denote by SM the set of semimartingale functions for the process $\mu(t)$: $F(t,m) \in SM$ if $F(t,m) \in C(R_+ \times M)$ and $F(t,\mu(t))$ is a semi-martingale, and denote by $\widetilde{SM} \subset SM$ the set of those $F(t,m)$ for which $F(t,\mu(t))$ has an absolutely continuous characteristic. It can be proved that for $F_k \in \widetilde{SM}$, $k = 1,\dots,n$ and $g(t,\ z_1,\ \dots,\ z_n) \in C^{(2)}(R_+ \times R^n)$ the function $g(t, F_1(t,m),\ \dots,\ F_n(t,m))$ belongs to $\widetilde{SM}$. So $\mu(t)$ is a quasi diffusion process (see[3], p.5). The second additional assumption concerns the factorization representation of the function $B(s,f_1,f_2,m)$.

5. There exists a measurable space (Y, B_Y) and a finite measure $\lambda(dy)$ on σ-algebra B_Y and a function $C(s,\ f,\ m,\ Y) : R_+ \times D \times M \times Y \to R$ which is continuous in $s \in R_+, m \in M$, linear in $f \in D$ and measurable in $y \in Y$

$$\int |C(s,\ f,\ m,\ y)|^2 \lambda(dy) < \infty$$

so that

$$B(s,\ f_1,\ f_2,\ m) = \int C(s,\ f_1,\ m,\ y)C(s,\ f_2,\ m,\ y)\lambda(dy). \tag{8}$$

Remark 3. Function C that satisfies relation (8) can be constructed (see [3], p.9, Theorem 1). The assumption concerns the properties of C.

3. Stochastic differential equations for measure-valued quasi diffusion process

We consider the Gaussian measure $W(dt,\ dy)$ on $R_+ \times Y$ with the properties:

I) $\quad EW(dt,\ dy) = 0, \quad EW^2(dt,\ dy) = dt\lambda(dy)$ (9)

II) if $S_1, \ldots, S_m \in B_{R_+} \otimes B_Y$ and $S_i \cap S_j = \emptyset$ for $i \neq j$ then $W(S_1), \ldots, W(S_m)$ are independent Gaussian random variables. Let the M-valued continuous process $\mu(t)$ be adapted to the filtration (W_t) generated by the measure W and let it satisfy the relation

$$d\langle f, \mu(t)\rangle = A(t, f, \mu(t))dt + \int_Y C(t, f, \mu(t), y)W(dt, dy) \tag{10}$$

where C is related to B by formula (8). Then $\mu(t)$ satisfies condition 4). Therefore relation (10) may be considered as the stochastic differential equation for the process $\mu(t)$.

Let $\varphi \in D(R_+ \times M)$. We now introduce the functions

$$\begin{aligned} A(t, \varphi, m) &= \frac{\partial}{\partial t}\varphi(t,\ \langle f_1, m\rangle,\ \ldots, \langle f_n, m\rangle) \\ &+ \sum_{k=1}^{n} \frac{\partial}{\partial z_k}\varphi_n(t, \langle f_1, m\rangle, \ldots, \langle f_n, m\rangle)A(t, f_k, m) \\ &+ \frac{1}{2}\sum_{k,j}^{n} \frac{\partial^2}{\partial z_k \partial z_j}\varphi_n(t, \langle f_1, m\rangle, \ldots, \langle f_n, m\rangle)B(t, f_k, f_j, m), \end{aligned} \tag{11}$$

$$C(t, \varphi, m, y) = \sum_{k=1}^{m} \frac{\partial}{\partial z_k}\varphi(t, \langle f_1, m\rangle, \ldots, \langle f_n, m\rangle)C(t, f_k, m, y) \tag{12}$$

where $\varphi(t, m) = \varphi(t, \langle f_1, m\rangle, \ldots, \langle f_n, m\rangle)$, $f_1, \ldots, f_n \in D$, $\varphi(t, z, \ldots, z_n) \in C^{(2)}(R_+ \times R_n)$. Equation (10) and Ito's formula imply the relation

$$d\varphi(t, \mu(t)) = A(t, \varphi, \mu(t))dt + \int_Y C(t, \varphi, \mu(t), y)W(dt, dy) \tag{13}$$

which is the general form of the stochastic differential equation for $\mu(t)$.

The assumptions concerning the coefficients of the stochastic differential equation. Introduce the set $D_1(M)$ as the subset of $D(R_+ \times M)$ of those functions $F(t, m) = F(m)$ which are constants in t and for which

$$\sup_{z_1, \ldots, z_n} \left(|\varphi(z_1, \ldots, z_n)| + \sum_{k=1}^{n} |\varphi'_{z_k}(z_1, \ldots, z_n)| + \sum_{k,j=1}^{n} |\varphi''_{z_k, z_j}(z_1, \ldots, z_n)| \right) \leq 1,$$

where $\varphi(t, \langle f_1, m\rangle, \ldots, \langle f_n, m\rangle) = F(m)$ and $\|f_i\| \leq 1,\ f_i \in D$.

6. There exists a constant $K > 0$ for which

$$\sup_{\varphi \in D_1(M)} E\left(|A(t, \varphi, \mu_1) - A(t, \varphi, \mu_2)|^2 + \int_Y |C(t, \varphi, \mu_1, y) - C(t, \varphi, \mu_2, y)|^2 \lambda(dy) \right) \leq$$

$$\leq K \sup_{\varphi \in D_1(M)} |\varphi(\mu_1) - \varphi(\mu_2)|^2$$

for any pair of M-valued random variables μ_1 and μ_2.

7. Denote by D^+ the set of nonnegative functions from D. Suppose that $D = D^+ - D^+$ and that for any $f \in D^+$ the functions

$$\frac{A(t,f,m)}{\langle f,m\rangle} \quad \text{and} \quad \frac{B(t,f,f,m)}{\langle f,m\rangle^2}$$

are bounded.

Remark 4. Let $\mu(t)$ satisfy equation (9). Then by using Ito's formula for $f \in D^+$ we have:

$$\begin{aligned} d\log\langle f,\mu(t)\rangle &= \left(A(t,f,\mu(t))\langle f,\mu(t)\rangle^{-1} - \frac{1}{2}B(t,f,f,\mu(t))\langle f,\mu(t)\rangle^{-2}\right) dt \\ &\quad + \int C(t,f,\mu(t),y)\langle f,\mu(t)\rangle^{-1} W(dt,dy) \\ &= R(t)dt + \int S(t,y)W(dt,dy). \end{aligned}$$

So,

$$\langle f,\mu(t)\rangle = \langle f,\mu(0)\rangle \exp\left\{\int_0^t R(u)du + \int_0^t S(u,y)W(du,dy)\right\}$$

and $\langle f,\mu(t)\rangle \geq 0$ if $\langle f,\mu(0)\rangle \geq 0$.

Theorem 2. *Let the coefficients of equation* (13) *satisfy conditions* 6 *and* 7. *Then for any initial value* $\mu(0) = m$ *there exists a* W_t*-measurable solution of equation* (13). *This solution is unique* (*i.e. any two solutions coincide with probability* 1), *and it is continuous with probability* 1. *Moreover,* $\mu(t)$ *is a Markov process with transition probability* $P(s,m,t,A) = P\{\mu_{s,m}(t) \in A\}$, *where* $\mu_{s,m}(t)$ *is the solution of equation* (13) *on the interval* $[s,\infty]$ *with the initial condition* $\mu_{s,m}(s) = m$.

Proof of uniqueness. Let $\mu(t)$ and $\tilde{\mu}(t)$ be two solutions with $\mu(0) = \tilde{\mu}(0)$. For $\varphi \in D_1(M)$ we can write

$$\varphi(\mu(t)) - \varphi(\tilde{\mu}(t)) = \int_0^t [A(s,\varphi,\mu(s)) - A(s,\varphi,\tilde{\mu}(s))]ds +$$

$$+ \int_0^t \int_Y [C(s,\varphi,\mu(s),y) - C(s,\varphi,\tilde{\mu}(s),y)]W(ds,dy),$$

therefore

$$E(\varphi(\mu(t)) - \varphi(\tilde{\mu}(t)))^2 \leq 2(1+t)\int_0^t E\Big(|A(s,\varphi,\mu(s)) - A(s,\varphi,\tilde{\mu}(s))|^2 +$$

$$+ \int_Y |C(s,\varphi,\mu(s),y) - C(s,\varphi,\tilde{\mu}(s),y)|^2\lambda(dy)\Big)\, ds \leq$$

$$\leq 2K(1+t)\int_0^t \sup_{\varphi\in D_1(M)} E|\varphi(\mu(s))-\varphi(\tilde{\mu}(s))|^2 ds. \tag{14}$$

Denote $\alpha(t)=\sup_{\varphi\in D_1(M)} E|\varphi(\mu(s))-\varphi(\tilde{\mu}(s))|^2$. Now (14) implies the inequality

$$\alpha(t)\leq 2K(1+t)\int_0^t \alpha(s)ds.$$

Since $\alpha(t)\geq 0$, so $\alpha(t)=0$.

Existence of the solution. Let $\{x_1, x_2, \ldots, x_2\}$ be the set of the different points of X. Let

$$m_{x_1,\ldots,x_n}(B)=n^{-1}\sum I\{x_k\in B\}.$$

We assume that $\mu(t)$ is a probability measure for any probability initial measure m. Note that $\mu^*(t,B)=\mu(t,B)/\mu(t,X)$ satisfies this property. If the function $f(x)=1$ belongs to D then $\mu^*(t)$ satisfies all the conditions which satisfy $\mu(t)$.

Set

$$\begin{aligned} A(t,f,m_{x_1,\ldots,x_n}) &= A^{(n)}(t,f,x_1,\ldots,x_n)\\ B(t,f_1,f_2,m_{x_1,\ldots,x_n}) &= B^{(n)}(t,f_1,f_2,x_1,\ldots,x_n)\\ C(t,f,m_{x_1,\ldots,x_n}) &= C^{(n)}(t,f,x_1,\ldots,x_n). \end{aligned}$$

Condition 6 implies the inequality

$$\sup_{f\in D,\|f\|\leq 1}\Big[\Big|A^{(n)}(t,f_k,x_1,\ldots,x_n)-A^{(n)}(t,f,x_1,\ldots,x_n)\Big|^2+$$

$$+\int_Y \Big|C^{(n)}(t,f_k,x_1,\ldots,x_n)-C^{(n)}(t,f,x_1,\ldots,x_n)\Big|^2\lambda(dy)\Big]\leq$$

$$\leq K\sup_{f\in D,\|f\|\leq 1}\left(\frac{1}{n}\sum_{k=1}^{n}(f(x_k)-f(\tilde{x}_k))\right)^2. \tag{15}$$

For any $\varepsilon>0$ we can find the functions

$$A^{(n,N)}(t,f,z_1,\ldots,z_N):\quad R_+\times D\times R^N\to R$$

$$C^{(n,N)}(t,f,z_1,\ldots,z_N,y):\quad R_+\times D\times R^N\times Y\to R$$

and $f_k\in D\cap\{\|f\|\leq 1\}$, $k-1,\ldots,N$ for which

$$\Big|A^{(n,N)}(t,f,\langle f_1,m_{x_1,\ldots,x_n}\rangle,\ldots,\langle f_N,m_{x_1,\ldots,x_n}\rangle)-A^{(n)}(t,f,x_1,\ldots,x_n)\Big|<\varepsilon,$$

$$\int\Big|C^{(n,N)}(t,f,\langle f_1,m_{x_1,\ldots,x_n}\rangle,\ldots,y)-C^{(n)}(t,f,x_1,\ldots,x_n,y)\Big|^2\lambda(dy)<\varepsilon^2$$

for $f\in D\cap\{\|f\|\leq 1\}$, and additionally

$$A^{(n,N)}(t,f,\langle f_1,m_{x_1,\ldots,x_n}\rangle,\ldots),\quad C^{(n,N)}(t,f,\langle f_1,m_{x_1,\ldots,x_n}\rangle,\ldots,y)$$

satisfy inequality (15).

We consider the system of stochastic differential equations

$$d\langle f_k, \mu^{(n,N)}(t)\rangle = A^{(n,N)}(t, f_k, \langle f_1, \mu^{(n,N)}\rangle, \dots, \langle f_N, \mu^{(n,N)}(t)\rangle)dt+$$

$$+\int_Y C^{(n,N)}\big(t, f_k, \langle f_1, \mu^{(n,N)}(t)\rangle, \dots, \langle f_N, \mu^{(n,N)}(t)\rangle, y\big)W(dt, dy), \tag{16}$$

$k = 1, \dots, N$. It follows from (15) that the coefficients of the system satisfy Lipschitz's condition. Therefore this system has a unique solution for any initial values. Using system $\langle f_k, \mu^{(n,N)}(t)\rangle$ for some sequence f_k which is dense in D, $n \to \infty$, $N \to \infty$ we can construct the system function $\langle f, \mu(t)\rangle f \in D$ which satisfies relation (10) (compare this with the results of [3], Chapter 1, §3). Considering remark 4 we can prove that $\mu(t)$ is a measure. Its uniqueness implies the Markov property of the solution and the continuity of $\langle f, \mu(t)\rangle$, $f \in D$ implies the continuity of $\mu(t)$.

4. The convergence of jump measure-valued processes to quasi diffusion processes

Consider the family of jump homogeneous Markov processes $\{\mu_\varepsilon(t), \varepsilon > 0\}$ in the space (M, B_M) with generators G_ε:

$$G_\varepsilon \varphi(m) = a_\varepsilon(m) \int (\varphi(m') - \varphi(m))\Pi_\varepsilon(m, dm') \tag{17}$$

$\varphi \in C(M)$, here $a_\varepsilon(m) > 0$, $a_\varepsilon(m) \in C(M)$, $\Pi_\varepsilon(m, dm')$ is a measure on B_M for fixed m and

$$\int \varphi(m')\Pi_\varepsilon(m, dm') \in C(M) \quad \text{if} \quad \varphi \in C(M).$$

Suppose that $a_\varepsilon(m) \to +\infty$ and $\int \varphi(m')\Pi_\varepsilon(m, dm') \to \varphi(m)$ if $\varepsilon \to 0$, $\varphi \in C(M)$. The problem is to investigate the asymptotic behavior of $\mu_\varepsilon(t)$ as $\varepsilon \to 0$.

Example. Let $\Lambda_\varepsilon \subset X$ be a finite set, $M(\delta, \Lambda_\varepsilon)$ — the set of measures m for which $m(X - \Lambda_\varepsilon) = 0$ and, $\frac{1}{\delta}m(\{x\})$ — a nonegative integer number. We consider the measures from $M(\delta, \Lambda_\varepsilon)$ as the distributions of the finite number of particles with mass δ. Suppose that $\mu_\varepsilon(t) \in M(\delta, \Lambda_\varepsilon)$ and $\mathrm{var}(\mu_\varepsilon(t) - \mu_\varepsilon(t-)) = 2\delta$, $\mu_\varepsilon(t, X)$ is a constant. This means that at the times of jumps only one of the particles changes its position. Let $\Delta_x(B) = I_{\{x \in B\}}$, which is a measure from M. Denote by $\Pi_\varepsilon(m, x, x')$ the probability of the transition of the particle from the point $x \in \Lambda_\varepsilon$ to the point $x' \in \Lambda_\varepsilon$ in the state $m \in M(\delta, \Lambda_\varepsilon)$. Then

$$G_\varepsilon \varphi(m) = a_\varepsilon(m) \sum_{x, x' \in \Lambda_\varepsilon} [\varphi(m - \delta\Delta_x + \delta\Delta_{x'}) - \varphi(m)]\Pi_\varepsilon(m, x, x') \tag{18}$$

where $a_\varepsilon^{-1}(m)$ is the average time of the system in state m. Note that such a process may be considered as a transport process (see for example [5]). Denote by N_ε the number of the points of Λ_ε and suppose that Λ_ε is an ε-net in X. We need to investigate the behavior of the system as $\varepsilon \to 0$, $N_\varepsilon \to \infty$, $\delta = \delta(\varepsilon) \to 0$.

Note that

$$\varphi(\mu_\varepsilon(t)) - \int_0^t G_\varepsilon \varphi(\mu_\varepsilon(s))ds \tag{19}$$

is a martingale with the square characteristic

$$\int_0^t [G_\varepsilon \varphi^2(\mu_\varepsilon(s)) - 2\varphi(\mu_\varepsilon(s))G_\varepsilon \varphi(\mu_\varepsilon(s))]ds \tag{20}$$

(see [3], Chapter 1, §1).

Introduce the functions

$$\begin{aligned} A_\varepsilon(f,m) &= G_\varepsilon \varphi_f(m), \\ B_\varepsilon(f_1,f_2,m) &= G_\varepsilon[\varphi_{f_1}(m)\varphi_{f_2}(m)] - \langle f_1,m\rangle G_\varepsilon \varphi_{f_1}(m) - \langle f_2,m\rangle G_\varepsilon \varphi_{f_2}(m). \end{aligned}$$

Here $\varphi_f(m) = \langle f,m\rangle$.

Theorem 3. *Let there exist a linear dense set $D \subset C(M)$ and functions $A(f,m)$, $B(f_1,f_2,m)$ which satisfy conditions 5, 6, 7 such that*

$$A_\varepsilon(f,m) \to A(f,m), \quad B_\varepsilon(f_1,f_2,m) \to B(f_1,f_2,m)$$

uniformly in f,m,f_1,f_2 if $\|f\| \le 1$, $\|f_1\| \le 1$, $\|f_2\| \le 1$, $f_1,f_2 \in D$ and $m(X) \le c$, where $c > 0$ is an arbitrary constant. Suppose that $\mu_\varepsilon(0)$ converges to a measure m_0. Then distribution of the process $\mu_\varepsilon(t)$ converges to the distribution of the solution of equation (10) with initial value $\mu(0) = m_0$.

The proof of the Theorem may be obtained in the same way as the proof of Theorem 2 in [3] on page 40.

References

1. **Dynkin, E.B.** (1965). *Markov processes.* Springer-Verlag, Berlin—Heidelberg—New York.

2. **Dellacherie, C. and Meyer, P.A.** (1975, 1980, 1983, 1987). *Probabilités et Potentiel.* Hermann, Paris.

3. **Skorokhod, A.V.** (1988). *Stochastic equations for complex systems.* Reidel Publishing Company.

4. Dynkin, E.B. (1994). *An introduction to branching measure-valued processes.* American Mathematical Society.

5. Maidaniuk, R.Y. and Skorokhod, A.V. (1995). On a class of Markov processes on graphs. *Theory of Probability and Mathematical Statistics.* (In press).

Exploring Stochastic Laws. pp. 285–291.
Eds. A.V. Skorokhod and Yu.V. Borovskikh

Asymptotic behavior of solutions of stochastic equations with periodic coefficients

S.YA. MAKHNO

Institute of Applied Mathematics and Mechanics, National Academy of Sciences of Ukraine, R. Luxemburg St 74, 340114 Donetsk, Ukraine.

Received April 15, 1995

Abstract. We consider the asymptotic behavior of solutions of diffusion stochastic equations with periodic coefficients, as time tends to infinity. Asymptotic normality and the logarithmic asymptotic of large deviations, and the law of iterated logarithms are studied.

1. Introduction

Let E_d be a Euclidean space with dimension d and $(\cdot,\cdot)$ the scalar product of its elements, ∇ denotes gradient, and "$'$" means conjugation. We denote by $\mathbb{P}_{l,k}(T,X)$ the space of periodic functions $f(t,x)$ with period T in $t \in E_1$ and with period X in all components x_i, $i=\overline{1,d}$, l times continuously differentiable on t and k times continuously differentiable on x, $x \in E_d$. The mean quantity of periodic function $f(t,x)$ is denoted by $\langle f\rangle$.

Let $(\Omega,\mathcal{F},\mathcal{F}_t,P)$, $t\geq 0$, be a probability space with filtering $\mathcal{F}_t$, and let E denote mathematical expectation.

Let $\xi(t)$ be a solution of the stochastic equation

$$\xi(t)=x+\int_0^t b(s,\xi(s))\,ds+\int_0^t \sigma(s,\xi(s))\,dw(s). \tag{1}$$

In equation (1), $(w(t),\mathcal{F}_t)$ is an m-dimensional standard Wiener process, $b(s,x)$ are d-vectors, and $\sigma(s,x)$ is a $d\times m$-matrix. Let $a(s,x)=\sigma(s,x)\sigma'(s,x)$. The following conditions are imposed:

1. The functions $b_i(s,x)\in\mathbb{P}_{1,1}(T,X)$, $a_{ij}(s,x)\in\mathbb{P}_{2,2}(T,X)$, $i,j=\overline{1,d}$.
2. The matrix $a(t,x)$ is uniformly non-degenerate.

Let

$$Lf(s,x)=\frac{\partial f(s,x)}{\partial s}+\frac{1}{2}\Big(a(s,x)\nabla,\nabla\Big)f(s,x)+\Big(b(s,x),\nabla f(s,x)\Big),$$

$$L^*f(s,x) = -\frac{\partial f(s,x)}{\partial s} + \frac{1}{2}\sum_{i,j=1}^{d} \frac{\partial^2}{\partial x_i \partial x_j}\Big(a_{ij}(s,x)\, f(s,x)\Big) - \\ -\sum_{i=1}^{d} \frac{\partial}{\partial x_i}\Big(b_i(s,x)\, f(s,x)\Big).$$

Lemma 1. (*Bensoussan et al.* (1978)). *Let conditions* 1 *and* 2 *be satisfied. The unique solution* $p(s,x) \in \mathbb{P}_{1,2}(T,X)$ *of the equation*

$$L^*p = 0, \qquad \langle p \rangle = 1,$$

exists. If $f(t,x) \in \mathbb{P}_{1.1}(T,X)$ *and* $\langle fp \rangle = 0$ *then the unique solution* $u(s,x) \in \mathbb{P}_{1,2}(T,X)$ *of the equation*

$$Lu = f, \qquad \langle u \rangle = 0, \tag{2}$$

exists.

Let $\lambda_i = \langle b_i p \rangle$, where λ is a vector with coordinates λ_i, $i = \overline{1,d}$. Denote by $v_k(s,x)$ the solution of equation (2) when

$$f(s,x) = \lambda_k - b_k(s,x), \qquad k = \overline{1,d},$$

where $v(s,x)$ is a vector-function with coordinates $v_k(s,x)$. Put

$$\hat{\sigma}_{ij}(s,x) = \sigma_{ij}(s,x) + \sum_{l=1}^{d} \sigma_{lj}(s,x)\, \frac{\partial v_i(s,x)}{\partial x_l}, \qquad i = \overline{1,d},\ j = \overline{1,m},$$

where $\hat{\sigma}(s,x)$ is a matrix with elements $\hat{\sigma}_{ij}(s,x)$,

$$h_{ij}(s,x) = \Big(\hat{\sigma}(s,x)\, \hat{\sigma}'(s,x)\Big)_{ij}, \quad H_{ij} = \langle h_{ij} p \rangle, \qquad i,j = \overline{1,d},$$

where H is a matrix with elements H_{ij}. Applying Ito's formula to the function $v_k(s,x)$ and process $\xi(t)$:

$$\begin{aligned} \int_0^t b_k(s,\xi(s))\, ds &= \lambda_k t + v_k(0,x) - v_k(t,\xi(t)) \\ &\quad + \int_0^t \Big(\nabla v_k(s,\xi(s)), \sigma(s,\xi(s))\, dw(s)\Big). \end{aligned}$$

From this and (1) we obtain

$$\xi(t) - \lambda t = x + v(0,x) - v(t,\xi(t)) + \int_0^t \hat{\sigma}(s,\xi(s))\, dw(s). \tag{3}$$

We shall introduce parameter ε, which has a small value. Let

$$\xi^\varepsilon(t) = \varepsilon\xi\left(\frac{t}{\varepsilon^2}\right), \quad \text{and} \quad w^\varepsilon(t) = \varepsilon w\left(\frac{t}{\varepsilon^2}\right).$$

From (1) and (3) we have

$$\xi^\varepsilon(t) = \varepsilon x + \frac{1}{\varepsilon}\int_0^t b\left(\frac{s}{\varepsilon^2}, \frac{\xi^\varepsilon(s)}{\varepsilon}\right) ds + \int_0^t \sigma\left(\frac{s}{\varepsilon^2}, \frac{\xi^\varepsilon(s)}{\varepsilon}\right) dw^\varepsilon(s), \tag{4}$$

$$\xi^\varepsilon(t) - \frac{\lambda t}{\varepsilon} = \varepsilon\left[x + v(0,x) - v\Big(\frac{t}{\varepsilon^2}, \frac{\xi^\varepsilon(t)}{\varepsilon}\Big)\right] + \int_0^t \hat{\sigma}\Big(\frac{s}{\varepsilon^2}, \frac{\xi^\varepsilon(s)}{\varepsilon}\Big)\, dw^\varepsilon(s). \tag{5}$$

2. Limit theorems

Theorem 1. *Let conditions* 1 *and* 2 *be satisfied. The random variable*

$$\frac{\xi(t)}{\sqrt{t}} - \lambda\sqrt{t}$$

converges in a distribution to the Gaussian random variable with parameter $(0, H)$ *as* $t \to \infty$.

Proof. Obviously, it is sufficient to investigate the random variable $\xi^{\varepsilon}(1) - \lambda/\varepsilon$ as $\varepsilon \to 0$. The random process $\xi^{\varepsilon}(t) - \frac{\lambda t}{\varepsilon}$ weakly converges to the random process $H^{1/2}w(t)$. It follows from the theorems of Makhno (1991).

For any positive function $\varphi(t) \to \infty$ as $t \to \infty$ the random process

$$\frac{1}{\varphi(t)}\left(\frac{\xi(t)}{\sqrt{t}} - \lambda\sqrt{t}\right) \to 0, \qquad t \to \infty$$

is in probability. This follows from Theorem 1 and so we can study the logarithmic asymptotic for the large deviation for this function.

We shall use the terminology of Freidlin and Wentzell (1984). Let $k(t)$ be a positive function which tends to infinity as $t \to \infty$, and $I(x)\colon E_d \to [0, \infty]$ is lower semi-continuous, and moreover, for any $a \geq 0$, the set

$$\{x \in E_d\colon I(x) \leq a\}$$

is compact.

A sequence of probability measures P^t on Borel σ-field $\mathcal{B}(E_d)$ is said to obey the large deviations principle if the following conditions hold:

a) For any open set $G \subset \mathcal{B}(E_d)$

$$\varliminf_{t\to\infty} k^{-1}(t) \ln P^t(G) \geq -\inf\{I(x), x \in G\};$$

b) For any closed set $F \subset \mathcal{B}(E_d)$

$$\varlimsup_{t\to\infty} k^{-1}(t) \ln P^t(F) \leq -\inf\{I(x), x \in F\}.$$

Function $k(t)\, I(x)$ is called a rate function.

Theorem 2. *Let conditions* 1 *and* 2 *be satisfied and non-negative function* $\psi(t)$ *be such that*

$$\lim_{t\to\infty} \frac{\psi(t)}{\sqrt{t}} = \infty, \quad \lim_{t\to\infty} \frac{\psi(t)}{t} = 0.$$

Then a family of probability measures

$$P^t(A) = P\left\{\frac{\xi(t) - \lambda t}{\psi(t)} \in A\right\}$$

obeys the large deviation principle and

$$k(t) = \frac{\psi^2 t}{t}, \quad I(x) = \frac{1}{2}(H^{-1}x, x).$$

Proof. We denote

$$M^t(\theta) = \ln E \exp\left\{\left(\theta, \frac{\xi(t) - \lambda t}{\psi(t)}\right)\right\}, \qquad \theta \in E_d.$$

According to the results of Freidlin and Wentzell (1984) it is sufficient to prove that,

$$\lim_{t\to\infty} \frac{t}{\psi^2(t)} M^t\left(\frac{\psi^2(t)}{t}\theta\right) = \frac{1}{2}(H\theta, \theta). \tag{6}$$

Let $\psi_\varepsilon = \psi(1/\varepsilon^2)$ and

$$N^\varepsilon(\theta) = \varepsilon^{-2}\psi_\varepsilon^{-2} \ln E \exp\left\{\varepsilon\psi_\varepsilon\left(\theta, \xi^\varepsilon(1) - \frac{\lambda}{\varepsilon}\right)\right\},$$

then (6) is equivalent to the following condition

$$\lim_{\varepsilon\to 0} N^\varepsilon(\theta) = \frac{1}{2}(H\theta, \theta). \tag{7}$$

Using (5) we have

$$\begin{aligned} N^\varepsilon(\theta) &= \varepsilon^{-2}\psi_\varepsilon^{-2} \ln E \exp\left\{\varepsilon^2\psi_\varepsilon\left(\theta, x + v(0,x) - v\left(\frac{1}{\varepsilon^2}, \frac{\xi^\varepsilon(1)}{\varepsilon}\right)\right)\right. \\ &\quad \left. + \varepsilon\psi_\varepsilon \int_0^1 \left(\theta, \hat\sigma\left(\frac{s}{\varepsilon^2}, \frac{\xi^\varepsilon(s)}{\varepsilon}\right) dw^\varepsilon(s)\right)\right\} \\ &= \varepsilon^{-2}\psi_\varepsilon^{-2} \ln E \exp\left\{\varepsilon^2\psi_\varepsilon\left(\theta, x + v(0,x) - v\left(\frac{1}{\varepsilon^2}, \frac{\xi^\varepsilon(1)}{\varepsilon}\right)\right)\right. \\ &\quad + \frac{\varepsilon^2\psi_\varepsilon^2}{2}\int_0^1 \left(h\left(\frac{s}{\varepsilon^2}, \frac{\xi^\varepsilon(s)}{\varepsilon}\right)\theta, \theta\right) ds + \varepsilon\psi_\varepsilon \int_0^1 \left(\theta, \hat\sigma\left(\frac{s}{\varepsilon^2}, \frac{\xi^\varepsilon(s)}{\varepsilon}\right) dw^\varepsilon(s)\right) \\ &\quad \left. - \frac{\varepsilon^2\psi_\varepsilon^2}{2}\int_0^1 \left(h\left(\frac{s}{\varepsilon^2}, \frac{\xi^\varepsilon(s)}{\varepsilon}\right)\theta, \theta\right) ds\right\}. \end{aligned} \tag{8}$$

Let $z(s,x)$ be the solution of (2) where

$$f(s,x) = \frac{1}{2}(H\theta,\theta) - \frac{1}{2}\left(h(s,x)\theta,\theta\right).$$

Applying Ito's formula to the function $\varepsilon^2 z\left(\frac{s}{\varepsilon^2}, \frac{x}{\varepsilon}\right)$ and the process $\xi^\varepsilon(t)$ from (4), we obtain

$$\begin{aligned} \frac{1}{2}\int_0^1 \left(h\left(\frac{s}{\varepsilon^2}, \frac{\xi^\varepsilon(s)}{\varepsilon}\right)\theta, \theta\right) ds &= \frac{1}{2}(H\theta,\theta) + \varepsilon^2\left[z(0,x) - z\left(\frac{1}{\varepsilon^2}, \frac{\xi^\varepsilon(1)}{\varepsilon}\right)\right] + \\ &\quad + \varepsilon\int_0^1 \left(\nabla Z\left(\frac{s}{\varepsilon^2}, \frac{\xi^\varepsilon(s)}{\varepsilon}\right), \sigma\left(\frac{s}{\varepsilon^2}, \frac{\xi^\varepsilon(s)}{\varepsilon}\right) dw^\varepsilon(s)\right). \end{aligned}$$

Then from (8) we have

$$\begin{aligned}
N^\varepsilon(\theta) &= \frac{1}{2}(H\theta,\theta) + \varepsilon^{-2}\psi_\varepsilon^{-2}\ln E\exp\Big\{\varepsilon^2\psi_\varepsilon\Big(\theta, x + v(0,x) - v\Big(\frac{1}{\varepsilon^2},\frac{\xi^\varepsilon(1)}{\varepsilon}\Big)\Big)\\
&\quad + \varepsilon^4\psi_\varepsilon^2\Big[z(0,x) - z\Big(\frac{1}{\varepsilon^2},\frac{\xi^\varepsilon(1)}{\varepsilon}\Big)\Big]\\
&\quad - \varepsilon^4\psi_\varepsilon^3\int_0^1\Big(\sigma\Big(\frac{s}{\varepsilon^2},\frac{\xi^\varepsilon(s)}{\varepsilon}\Big)\theta\hat{\sigma}'\Big(\frac{s}{\varepsilon^2},\frac{\xi^\varepsilon(s)}{\varepsilon}\Big),\nabla z\Big(\frac{s}{\varepsilon^2},\frac{\xi^\varepsilon(s)}{\varepsilon}\Big)\Big)\,ds\\
&\quad - \frac{\varepsilon^6\psi_\varepsilon^4}{2}\int_0^1\Big(a\Big(\frac{s}{\varepsilon^2},\frac{\xi^\varepsilon(s)}{\varepsilon}\Big)\nabla z\Big(\frac{s}{\varepsilon^2},\frac{\xi^\varepsilon(s)}{\varepsilon}\Big),\nabla Z\Big(\frac{s}{\varepsilon^2},\frac{\xi^\varepsilon(s)}{\varepsilon}\Big)\Big)\,ds\Big\}\\
&\quad \times\exp\Big\{\int_0^1\Big(\varepsilon\psi_\varepsilon\theta'\hat{\sigma}\Big(\frac{s}{\varepsilon^2},\frac{\xi^\varepsilon(s)}{\varepsilon}\Big)\\
&\qquad\qquad + \varepsilon^3\psi_\varepsilon^2\sigma'\Big(\frac{s}{\varepsilon^2},\frac{\xi^\varepsilon(s)}{\varepsilon}\Big)\nabla z\Big(\frac{s}{\varepsilon^2},\frac{\xi^\varepsilon(s)}{\varepsilon}\Big), dw^\varepsilon(s)\Big)\\
&\quad - \frac{1}{2}\int_0^1\Big|\varepsilon\psi_\varepsilon\theta'\hat{\sigma}\Big(\frac{s}{\varepsilon^2},\frac{\xi^\varepsilon(s)}{\varepsilon}\Big) + \varepsilon^3\psi_\varepsilon^2(\sigma'\nabla z)\Big(\frac{s}{\varepsilon^2},\frac{\xi^\varepsilon(s)}{\varepsilon}\Big)\Big|^2 ds\Big\}. \qquad (9)
\end{aligned}$$

Since the mathematical expectation of the second exponent equals one then from (9) we have

$$\Big|N^\varepsilon(\theta) - \frac{1}{2}(H\theta,\theta)\Big| \le \text{const}\,\Big|\frac{1}{\psi_\varepsilon} + \varepsilon + \varepsilon^2\psi_\varepsilon + \varepsilon^4\psi_\varepsilon^2\Big| \to 0, \qquad \varepsilon \to 0.$$

Therefore (7) and Theorem 1 are proved.

We now consider the law of iterated logarithms.

Lemma 2. *Let conditions* 1 *and* 2 *be valid and* $g(t,x) \in \mathbb{P}_{1,1}(T,X)$. *Then*

$$P\Big\{\lim_{t\to\infty}\frac{1}{t}\int_0^t g(s,\xi(s))\,ds = \langle gp\rangle\Big\} = 1.$$

Proof. Denote by $r(t,x)$ the solution of equation (2) with $f(t,x) = \langle gp\rangle - g(t,x)$. Applying Ito's formula to the function $r(t,x)$ and to the process $\xi(t)$, we have

$$\begin{aligned}
\frac{1}{t}\int_0^t g(s,\xi(s))\,ds &= \langle gp\rangle + \frac{1}{t}[r(0,x) - r(t,\xi(t))]\\
&\quad + \frac{1}{t}\int_0^t\Big(\nabla r(s,\xi(s)),\sigma(s,\xi(s))\,dw(s)\Big).
\end{aligned}$$

It is sufficient to prove that;

$$P\Big\{\lim_{t\to\infty}\frac{1}{t}\int_0^t\Big(\nabla r(s,\xi(s)),\sigma(s,\xi(s))\,dw(s) = 0\Big)\Big\} = 1. \qquad (10)$$

Put

$$M = \sup_{t,x}\Big(a(t,x)\nabla r(t,x), \nabla r(t,x)\Big),$$

$$\xi_k = \sup_{t\geq 2^k}\frac{1}{t}\left|\int_0^t\Big(\nabla r(s,\xi(s)), \sigma(s,\xi(s))\,dw(s)\Big)\right|.$$

According to Doob's estimate we have for any $\varepsilon > 0$

$$P\{\xi_k > \varepsilon\} \leq \sum_{m=k}^{\infty} P\left\{\sup_{2^m\leq t\leq 2^{m+1}}\left|\int_0^t\Big(\nabla r(s,\xi(s)), \sigma(s,\xi(s))\Big)\,dw(s)\right| > 2^k\varepsilon\right\}$$
$$\leq \frac{2M}{\varepsilon^2}\sum_{m=k}^{\infty}\frac{1}{2^m} \to 0, \qquad k\to\infty.$$

From this we obtain (10) when the random variable ξ_k increases. The lemma is proved.

Theorem 3. *If conditions* 1 *and* 2 *are valid then the following statements hold:*

1. $P\left\{\overline{\lim_{t\to\infty}}\frac{\xi_i(t)-\lambda_i t}{\sqrt{2H_{ii}t\ln\ln t}} = 1\right\} = 1, \qquad i=\overline{1,d};$
2. $P\left\{\underline{\lim}_{t\to\infty}\frac{\xi_i(t)-\lambda_i t}{\sqrt{2H_{ii}t\ln\ln t}} = -1\right\} = 1, \qquad i=\overline{1,d};$
3. $P\left\{\overline{\lim_{t\to\infty}}\frac{|\xi(t)-\lambda t|}{\sqrt{2\,\mathrm{Sp}\,H\,t\ln\ln t}} = 1\right\} = 1.$

Proof. Let

$$\mu_i(t) = \int_0^t\sum_{j=1}^{m}\hat\sigma_{ij}(s,\xi(s))\,dw_j(s), \qquad i=\overline{1,d};$$

$$\tau_i(t) = \int_0^t h_{ii}(s,\xi(s))\,ds, \qquad i=\overline{1,d}.$$

We have $\tau_i(t)\uparrow\infty$ as $t\to\infty$ with probability one. Let $\nu_i(t)$ be an inversion function to the function $\tau_i(t)$. By Theorem 1.3.4 of Gikhman and Skorokhod (1979) the process $\mu_i(\nu_i(t)) = \tilde w_i(t)$ is a Wiener process. According to Lemma 2 we have with probability one

$$\lim_{t\to\infty}\frac{\tau_i(t)\ln\ln\tau_i(t)}{t\ln\ln t} = H_{ii}. \tag{11}$$

Using the law of iterated logarithms for the Wiener process and (11) we obtain

$$\overline{\lim_{t\to\infty}}\frac{\mu_i(t)}{\sqrt{2H_{ii}\,t\ln\ln t}} = \overline{\lim_{t\to\infty}}\frac{\tilde w_i(\tau_i(t))}{\sqrt{2\tau_i(t)\ln\ln\tau_i(t)}}\sqrt{\frac{\tau_i(t)\ln\ln\tau_i(t)}{H_{ii}\,t\ln\ln t}} = 1. \tag{12}$$

By analogy,

$$\underline{\lim}_{t\to\infty}\frac{\mu_i(t)}{\sqrt{2H_{ii}\,t\ln\ln t}} = \underline{\lim}_{t\to\infty}\frac{\tilde w_i(\tau_i(t))}{\sqrt{2\tau_i(t)\ln\ln\tau_i(t)}}\sqrt{\frac{\tau_i(t)\ln\ln\tau_i(t)}{H_{ii}\,t\ln\ln t}} = -1. \tag{13}$$

Statements 1 and 2 follow from (3), (12) and (13). Statement 3 is a consequence of statements 1 and 2. The Theorem is now proved.

References

Bensoussan, A., Lions, J.-L. and Papanicolaou, G. (1978). *Asymptotic Analysis for Periodic Structures.* North-Holland, Amsterdam.

Freidlin, M.I. and Wentzell, A.D. (1984). *Random Perturbations of Dynamical Systems.* Springer, New York, etc.

Gikhman, I.I. and Skorokhod, A.V. (1979). *The Theory of Stochastic Processes III.* Springer–Verlag, New-York, Berlin, Heidelberg.

Makhno, S.Ya. (1991). On the behavior of the solution of a stochastic equation with unbounded drift. *J. Soviet Math. (JOSMAR)*, **53**(1), 62–65.

Exploring Stochastic Laws. pp. 293–305.
Eds. A.V. Skorokhod and Yu.V. Borovskikh

Trajectories of homogeneous and isotropic random flows have no fractal properties

A.A. MALYARENKO

National Academy of Sciences of Ukraine, Institute of Mathematics, Kiev, Ukraine

Received April 16, 1995

Abstract. Let $\xi(g)$ be a nondegenerate separable homogeneous and isotropic Gaussian random flow on the group SU(2). We prove that the Hausdorff dimension of its image may take the values 2 or 3, and the Hausdorff dimension of its graph is always equal to 3 almost surely.

1. Introduction

Let $(\Omega, \mathcal{F}, \mathbf{P})$ be a probability space, $\mathcal{H}$ a finite-dimensional complex Hilbert space, and $L^2(\Omega, \mathcal{F}, \mathbf{P}; \mathcal{H})$ a Hilbert space of $\mathcal{H}$-valued random vectors ξ with $\mathbf{E}\|\xi\|^2_{\mathcal{H}} < \infty$. Let $\mathcal{X} = \mathcal{G}/\mathcal{K}$ be a homogeneous space of a locally compact topological group $\mathcal{G}$ with respect to its compact subgroup $\mathcal{K}$. A continuous mapping $\xi(x) : \mathcal{X} \mapsto L^2(\Omega, \mathcal{F}, \mathbf{P}; \mathcal{H})$ will be called a *random flow*. We will suppose that group $\mathcal{G}$ has a countable base of neighborhoods of its unit, and $\mathbf{E}\xi(x) = 0$.

Definition 1. (Yaglom(1960)). A random flow $\xi(x)$ is called *left homogeneous* if

$$B(gx, gy) := \mathbf{E}\xi(gx)\xi^*(gy) = B(x, y), \ x, y \in \mathcal{X}, \ g \in \mathcal{G}.$$

Let $\mathcal{M}$ be a compact subgroup in the group $\mathrm{Aut}_{\mathcal{K}}\mathcal{G}$ of the automorphisms of group $\mathcal{G}$ preserving the subgroup $\mathcal{K}$. Let U be the irreducible unitary representation of group $\mathcal{M}$ in the Hilbert space $\mathcal{H}$.

Definition 2. (Malyarenko (1994)). The random flow $\xi(x)$ is called *isotropic* (with respect to the group $\mathcal{M}$ and the representation U) if

$$B(mx, my) = U(m)B(x, y)U^{-1}(m), \ m \in \mathcal{M}, \ x, y \in \mathcal{X}.$$

Let (X, ρ) be a metric space. Let $\varphi : (0, +\infty) \mapsto (0, +\infty)$ be a continuous strictly increasing function satisfying the condition

$$\lim_{t \downarrow 0} \varphi(t) = 0.$$

Then the *outer Hausdorff measure* $\varphi - \mu$ of a Borel set $A \subset X$ is defined as (Turbin and Pratsevytyi (1992)):

$$\varphi - \mu(A) = \lim_{\varepsilon \downarrow 0} \inf \sum_j \varphi(d(A_j^\varepsilon)),$$

where the infimum is taken over all countable collections of subsets $\{A_j^\varepsilon\}$ in X whose union covers A and for which the diameter of each set A_j^ε is not greater then ε.

Definition 3. (Turbin and Pratsevytyi (1992)). The *Hausdorff dimension* of a set $A \subset X$ is determined as

$$\dim_H A = \sup\{\alpha : \ t^\alpha - \mu(A) = \infty\}.$$

Let $\mathcal{G} = \mathrm{SU}(2)$ be the group of second order unimodular unitary matrices, let $\mathcal{K} = \{e\}$ and $\mathcal{M} = \mathrm{SU}(2)$ act on itself by inner automorphisms. According to (Vilenkin (1968)) any irreducible unitary representation U of the group $\mathcal{M}$ corresponds to the integer or half-integer number $u = 0, 1/2, 1, \ldots$. The complex dimension of the space $\mathcal{H}$ is equal to $2u + 1$, so its real dimension is equal to $4u + 2$.

Now we define *incremental variance functions:*

$$\sigma_j^2(g) := \mathbf{E}|\xi_j(g) - \xi_j(e)|^2, \ 1 \le j \le 4u + 2.$$

Theorem. *Let $\xi(g)$ be a separable Gaussian homogeneous and isotropic random flow on* SU(2). *Let the following condition be satisfied:*

$$\inf_{g \in \mathcal{G} \setminus \{e\}} \frac{\det \mathbf{E}(\xi(g) - \xi(e))(\xi(g) - \xi(e))^*}{\prod_{j=1}^{4u+2} \sigma_j^2(g)} > 0.$$

Then we have, with probability 1,

$$\begin{aligned} \dim_H \Im\xi &= \begin{cases} 2, & U \equiv 1 \\ 3, & U \not\equiv 1 \end{cases} \\ \dim_H \mathrm{Gr}\,\xi &\equiv 3. \end{aligned}$$

Here $\Im\xi$ is the image of the random flow $\xi(g)$, and $\mathrm{Gr}\,\xi$ is its graph:

$$\begin{aligned} \Im\xi &= \{\xi(g), \ g \in \mathrm{SU}(2)\} && \subset \mathcal{H} \\ \mathrm{Gr}\,\xi &= \{(g, \xi(g)), \ g \in \mathrm{SU}(2)\} && \subset \mathrm{SU}(2) \times \mathcal{H}. \end{aligned}$$

We exclude from consideration the degenerate case $\xi(g) \equiv 0$. So the Hausdorff dimension of either $\Im\xi$ and $\mathrm{Gr}\,\xi$ is an integer number and these sets are not *fractals* (Turbin and Pratsevytyi (1992)). The Theorem will be proved in the next section.

2. Proof of the Theorem

Since $\mathcal{K}$ is a compact subgroup, there exists a $\mathcal{G}$-invariant measure μ on $\mathcal{X}$ (Federer (1969), 2.7.12). Since $\mathcal{G}$ has a countable base of neighborhoods of its unit and $\mathcal{K}$ is a compact subgroup, there exists a $\mathcal{G}$-invariant metric ρ on $\mathcal{X}$ (Hewitt and Ross (1963), 8.14(α)).

Let $V(r)$ be a μ-measure of the closed ball of radius r (clearly it does not depend on the ball's location). We assume that there exist a real constant $\gamma > 1/2$ and a positive constant C such that

$$V(r) \geq Cr^{\gamma}, \; r > 0. \tag{1}$$

Let π be a canonical projection of $\mathcal{G}$ onto $\mathcal{X}$. Let $o = \pi(e) \in \mathcal{X}$ and

$$K_r = \{x \in \mathcal{X} : \; \rho(x, o) \leq r\}.$$

Lemma 1. *Let space $\mathcal{X}$ satisfy* (1) *and let functions $p(u)$ and $\psi(u)$, defined on $[-r, r]$ and $(-\infty, +\infty)$ respectively, satisfy*

$$\begin{aligned} p(r) &= p(-r) \downarrow 0 \quad , \; r \downarrow 0, \\ \psi(r) &= \psi(-r) \uparrow \infty \quad , \; r \uparrow \infty. \end{aligned}$$

Let ψ be a convex function, and $f(x)$ a continuous real-valued function on $\mathcal{X}$ for which

$$\int_K \int_K \psi\left(\frac{f(x) - f(y)}{p(r(K))}\right) d\mu(x)\, d\mu(y) \leq B, \tag{2}$$

for all closed balls K containing in K_r with radius $r(K)$. Then for all $x, y \in K_r$

$$|f(x) - f(y)| \leq 8 \int_0^{\rho(x,y)} \psi^{-1}(BC^{-2}u^{-2\gamma})\, dp(u).$$

Proof. Let x, y be two points in K_r. We construct a sequence of closed balls K_m having center in x as:

$$\begin{aligned} K_0 &= \{z \in \mathcal{X} : \; \rho(x, z) \leq \rho(x, y)\}; \\ p(r(K_m)) &= \frac{1}{2} p(r(K_{m-1})). \end{aligned} \tag{3}$$

Let K be any closed ball in $\mathcal{X}$. We denote:

$$f_K = (\mu(K))^{-1} \int_K f(x)\, d\mu(x), \; r_m = r(K_m).$$

It follows from (2), the convexity of ψ and Jencen's inequality that

$$\begin{aligned} \psi\left(\frac{f_{K_m} - f_{K_{m-1}}}{p(r_{m-1})}\right) &\leq [V(r_m)V(r_{m-1})]^{-1} \int_{K_m} \int_{K_m} \psi\left(\frac{f(x) - f(y)}{p(r_{m-1})}\right) d\mu(x)\, d\mu(y) \\ &\leq B[V(r_m)V(r_{m-1})]^{-1}. \end{aligned}$$

Inverting ψ in the above inequality gives

$$|f_{K_m} - f_{K_{m-1}}| \leq p(r_{m-1})\psi^{-1}(B[V(r_m)V(r_{m-1})]^{-1}). \tag{4}$$

It follows from (3) that

$$p(r_{m-1}) = 4(p(r_m) - p(r_{m+1})),$$

and it follows from (1) that $r \in [r_{m-1}, r_m]$ implies

$$C^{-2}r^{-2\gamma} \geq [V(r_m)V(r_{m-1})]^{-1}.$$

Now substituting into (4) gives to

$$|f_{K_m} - f_{K_{m-1}}| \leq 4\int_{r_{m+1}}^{r_m} \psi^{-1}(BC^{-2}r^{-2\gamma})\, dp(r).$$

Summing these inequalities for $m \geq 1$ yields

$$\limsup_{m\to\infty} |f_{K_m} - f_{K_0}| \leq 4\int_0^{r_0} \psi^{-1}(BC^{-2}r^{-2\gamma})\, dp(r).$$

Since f is continuous we have from the last inequality

$$|f(x) - f_{K_0}| \leq 4\int_0^{\rho(x,y)} \psi^{-1}(BC^{-2}r^{-2\gamma})\, dp(r).$$

The same inequality holds true with x replaced by y so the Lemma is proved.

Remark 1. This Lemma is a generalization of (Adler (1981), Lemma 3.3.3).

Now let $\xi(x)$ be a real-valued zero mean Gaussian random field on $\mathcal{X}$ with the continuous covariance function $B(x, y)$. Since the measure μ is positive on non-empty open subsets of K_r, we may apply *Merser's Theorem* (Berezansky et al. (1990)) and obtain:

$$B(x, y) = \sum_{j=1}^{\infty} \lambda_j \varphi_j(x)\varphi_j(y), \tag{5}$$

where φ_j are eigenfunctions of the integral operator

$$(A\varphi)(x) = \int_{K_r} B(x, y)\varphi(y)\, d\mu(y),$$

and λ_j are the corresponding eigenvalues. Series (5) converges absolutely and uniformly on $K_r \times K_r$.

Now let

$$\sigma^2(x, y) = \sum_{j=1}^{\infty} \lambda_j(\varphi_j(x) - \varphi_j(y))^2,$$

$$p(r) = \max\{\sigma(x, y) : \ \rho(x, y) \leq |r|\sqrt{\gamma}\}$$

and let ξ_j be a standard Gaussian orthonormal sequence. We set

$$\xi^{(m)}(x) = \sum_{j=1}^{m} \sqrt{\lambda_j}\varphi_j(t)\xi_j. \tag{6}$$

Lemma 2. *Suppose, that*

$$\int_0^1 (-\log r)^{1/2}\, dp(r) < \infty. \tag{7}$$

Then

$$|\xi^{(m)}(x) - \xi^{(m)}(y)| \leq 16\sqrt{\gamma}(\log C^{-2}B)^{1/2} p(\rho(x,y)) + $$

$$+\ 16\sqrt{2\gamma}\int_0^{\rho(x,y)} (-\log r)^{1/2}\, dp(r), \tag{8}$$

where

$$B(\omega) = \sup_m \int_{K_r}\int_{K_r} \exp\left\{\frac{\xi_m(x,\omega) - \xi_m(y,\omega)}{2\gamma^{1/2}p(\gamma^{-1/2}\rho(x,y))}\right\}^2 d\mu(x)\, d\mu(y)$$

and

$$\mathbf{E}|B(\omega)|^2 \leq 4\left(\frac{2\gamma}{2\gamma-1}\right)^{1/2}. \tag{9}$$

Moreover, $\xi^{(m)}(x)$ converges uniformly on K_r and are almost surely equicontinuous.

Proof. Let η_m be a sequence of random variables defined as

$$\eta_m = \exp\frac{1}{2}\left[\frac{\xi^{(m)}(x) - \xi^{(m)}(y)}{2\gamma^{1/2}p(\gamma^{-1/2}\rho(x,y))}\right]^2 - \exp\frac{1}{2}\left[\frac{\zeta_m}{2\sqrt{\gamma}}\right]^2,$$

where

$$\zeta_m = \frac{\xi^{(m)}(x) - \xi^{(m)}(y)}{p(\gamma^{-1/2}\rho(x,y))}.$$

It is easy to see that ζ_m have a zero mean Gaussian distribution with variance $\sigma_m^2 < 1$. Then we can show by direct computation that $\{\eta_m\}$ is a submartingale and

$$\mathbf{E}\eta_m^2 = \left(1 - \frac{\sigma_m^1}{2\gamma}\right)^{-1/2} \leq \left(1 - \frac{1}{2\gamma}\right)^{1/2} = \left(\frac{2\gamma}{2\gamma-1}\right)^{1/2}.$$

It immediately follows from (Gikhman and Skorokhod (1974), chapter 2, § 2, (17)), that

$$\mathbf{E}\{\max_{j\leq m}\eta_j^2\} \leq 4\mathbf{E}\eta_j^2 \leq 4\left(\frac{2\gamma}{2\gamma-1}\right)^{1/2}.$$

Integrating over $K_r \times K_r$ and using Fubini's Theorem gives us

$$\mathbf{E}\int_{K_r}\int_{K_r} \max_{j\leq m} \exp\left[\frac{\xi^{(j)}(x) - \xi^{(j)}(y)}{2\gamma^{1/2}p(\gamma^{-1/2}\rho(x,y))}\right]^2 d\mu(x)\, d\mu(y) \leq 4\left(\frac{2\gamma}{2\gamma-1}\right)^{1/2}.$$

According to (Gikhman and Skorokhod (1974), chapter 2, § 2, corollary to theorem 1) and the monotone convergence Theorem we obtain (9) by letting $m \uparrow \infty$. Now (9) implies almost sure finiteness of $B(\omega)$ so that we have, with probability one, for $m \geq 1$

$$\int_{K_r}\int_{K_r} \exp\left[\frac{\xi^{(m)}(x)-\xi^{(m)}(y)}{2\gamma^{1/2}p(\gamma^{-1/2}\rho(x,y))}\right]^2 d\mu(x)\,d\mu(y) \leq B(\omega) < \infty.$$

Now we apply Lemma 1 to obtain (8).

Finally, we can write

$$\int_{K_r}\left[B(x,x)-\sum_{k=1}^{m}\lambda_k\right] d\mu(x) = \int_{K_r}\left[B(x,x)-\sum_{k=1}^{m}\lambda_k\varphi_k^2(x)\right] d\mu(x)$$

which tends to 0 as $m \to \infty$ according to Mercer's Theorem. Hence

$$\mathbf{E}\sum_{k=1}^{\infty}\lambda_k\xi_k^2 = \sum_{k=1}^{\infty}\lambda_k = \int_{K_r} B(x,x)\,d\mu(x) < \infty$$

implies that

$$\sum_{k=1}^{\infty}\lambda_k\xi_k^2(\omega) < \infty$$

for almost all ω. So for $m > m'$

$$\int_{K_r} |\xi^{(m)}(x,\omega)-\xi^{(m')}(x,\omega)|^2\,d\mu(x) = \sum_{k=m'}^{m}\lambda_k\xi_k^2(\omega) \to 0 \text{ a. s.}$$

It means that $\xi^{(m)}(x,\omega)$ almost surely uniformly converge and from (8) we have that $\xi^{(m)}$ are almost surely equicontinuous.

Remark 2. This result is a generalization of Garsia's one (Garsia (1971)).

Lemma 3. *Let $\xi(x)$, $x \in K_r$ be a separable real valued, zero-mean Gaussian random field with continuous covariance function $B(x,y)$ which has Mercer's expansion* (5). *Let condition* (1) *hold true. Let*

$$\xi_j := \lambda_j^{-1/2}\int_{K_r}\xi(x)\varphi_j(x)\,d\mu(x). \tag{10}$$

Then sequence (6) *converges uniformly for $x \in K_r$ to $\xi(x)$ with probability 1, and the modulus of continuity $\omega_\xi(\delta)$ of the sample functions of ξ satisfies the inequality*

$$\omega_\xi(\delta) \leq 16\sqrt{\gamma}[\log C^{-2}B]^{1/2}p(\delta) + 16\sqrt{2}\gamma\int_0^\delta(-\log r)^{1/2}\,dp(r),$$

where B is a positive random variable satisfying (9).

To prove the Lemma we need to prove that (10) is an orthonormal Gaussian sequence. But it is a trivial computation.

Now we assume that the random field from Lemma 3 is left homogeneous and has an incremental variance function $\sigma^2(x)$. Let $0 \le \beta \le 1$.

Definition 4. (Adler (1981)). A random field $\xi(x)$ has *index* β if

$$\beta = \sup\{\alpha : \ \sigma(x) = o(\rho^\alpha(x, o))\} = \inf\{\alpha : \ \rho^\alpha(x, o) = o(\sigma(x))\}.$$

Lemma 4. *The trajectories of index-β Gaussian random field satisfy Hölder condition of order α for any $\alpha < \beta$.*

Proof. This is a generalization of (Adler (1981), Theorem 8.3.2). Their proofs are identical. According to the definition of $p(u)$ and the last definition one can easily show that there exists a finite constant $C_1 > 0$ such that for any $\beta > 0, \ \varepsilon > 0$

$$p(u)|u|^{-(\beta-\varepsilon)} < C_1, \ u \to 0.$$

Thus, as $\delta \to 0$,

$$(-\log u)^{1/2} p(u)\Big|_0^\delta \le C_1 \delta^{\beta-\varepsilon/2},$$

$$\int_0^\delta (-\log u)^{-1/2} u^{-1} p(u)\, du \le \int_0^\delta u^{-1+\beta-\varepsilon}\, du = (\beta-\varepsilon)^{-1}\delta^{\beta-\varepsilon}.$$

But according to Lemma 2 there exists a finite, positive number C_2 and an almost sure finite, positive random variable B such that, with probability 1,

$$\omega_\xi(\delta) \le Bp(\delta) + C_2 \int_0^\delta (-\log u)^{1/2}\, dp(u).$$

The last inequalities guarantee the possibility of integration by parts on the right hand side, and we obtain

$$\begin{aligned} \omega_\xi(\delta) &= Bp(\delta) + C_2 (-\log u)^{1/2} p(u)\Big|_0^\delta \\ &\quad + (1/2) C_2 \int_0^\delta (-\log u)^{-1/2} u^{-1} p(u)\, du \\ &\le A\delta^{\beta-\varepsilon/2}. \end{aligned}$$

This completes the proof.

Let (X, ρ) be a compact metric space. The ε*-covering* of X is a system of sets which covers X and the diameter of any set less or equal to 2ε. Let $N(\varepsilon)$ be the minimal quantity of sets in the ε-covering of X.

Definition 5. (Kolmogorov and Tihomirov (1959)). *The lower Kolmogorov's dimension of* X is defined as

$$\underline{\dim}_K X = \liminf_{\varepsilon \downarrow 0} \frac{\log_2 N(\varepsilon)}{\log_2 \varepsilon^{-1}}.$$

Lemma 5. *Let* (X, ρ) *be a compact metric space and* $N := \underline{\dim}_K X < \infty$. *Let* $\mathbf{F} : X \mapsto \mathbf{R}^d$ *and the* j*-th component of* $\mathbf{F}$ *satisfy a uniform Hölder condition of order* α_j, $0 < \alpha_j \leq 1$. *Let the coordinates be numbered such that* $0 < \alpha_1 \leq \ldots \leq \alpha_d \leq 1$.

Then we have

$$\dim_H \Im \mathbf{F} \leq \min \left\{ d, \frac{N + \sum_{j=1}^d (\alpha_d - \alpha_j)}{\alpha_d} \right\}, \tag{11}$$

$$\dim_H \operatorname{Gr} \mathbf{F} \leq \min \left\{ \frac{N + \sum_{j=1}^d (\alpha_d - \alpha_j)}{\alpha_d}, N + \sum_{j=1}^d (1 - \alpha_j) \right\}.$$

Proof. We will modify the proof of (Cuzick (1978)) of the case $X = [0,1]^N$. Since $\Im \mathbf{F} \subset \mathbf{R}^d$, then $\dim_H \Im \mathbf{F} \leq d$. According to definition 5, there exists a sequence $\varepsilon_n \downarrow 0$ and ε_n-coverings of X, each of which contains $\varepsilon_n^{-N}(1 + o(\varepsilon_n))$ elements. By Hölder's condition the image of each of these elements is contained in a d-dimensional interval, the length of whose edges is proportional to $\varepsilon_n^{\alpha_j}$, $1 \leq j \leq d$. Each of these intervals can be divided into $\varepsilon_n^{-\sum_{j=1}^d (\alpha_d - \alpha_j)}$ cubes with the length of edges proportional to $\varepsilon_n^{-\alpha_d}$. So the Hausdorff dimension of the image is no more then the infimum of all λ for which

$$\varepsilon_n^{-N}(1 + o(\varepsilon_n)) \varepsilon_n^{-\sum_{j=1}^d (\alpha_d - \alpha_j)} \varepsilon_n^{-\lambda \alpha_d} \to 0, \ n \to \infty.$$

But this is true for all

$$\lambda > \frac{N + \sum_{j=1}^d (\alpha_d - \alpha_j)}{\alpha_d}$$

and so (11) is proved.

A graph $\operatorname{Gr} \mathbf{F}$ can be covered by $\varepsilon_n^{-N}(1 + o(\varepsilon_n))$ sets, each of which is a Cartesian product of the subset of X having diameter $2\varepsilon_n$ and a d-dimensional interval with edges proportional to $\varepsilon_n^{-\alpha_j}$. From one side, we can cover each of these sets by similar sets, for which the first multiplier has diameter $\varepsilon_n^{-\alpha_d}$, and the second one is a d-dimensional cube with edge of the same length. The quantity of those sets is equal to $\varepsilon_n^{-\sum_{j=1}^d (\alpha_d - \alpha_j)}(1 + o(\varepsilon_n))$, and, proceeding as before, we find

$$\dim_H \operatorname{Gr} \mathbf{F} \leq \frac{N + \sum_{j=1}^d (\alpha_d - \alpha_j)}{\alpha_d}.$$

From the other side, we can cover each of these sets by similar sets, for which the diameter of the first multiplier and the length of the edge of the second one are equal to ε_n^{-1}, and the number of those sets is equal to $\varepsilon_n^{-\sum_{j=1}^{d}(1-\alpha_j)}(1+o(\varepsilon_n))$. Similar arguments prove the inequality

$$\dim_H \operatorname{Gr} \mathbf{F} \le N + \sum_{j=1}^{d}(1-\alpha_j),$$

so the proof is complete.

Let ν be a probability measure on (X, ρ). Similarly as in (Adler (1981)) we define the *energy integral*:

$$I_\alpha(\nu) = \int_X \int_X \rho^{-\alpha}(x,y)\, d\nu(x)\, d\nu(y), \ \alpha > 0.$$

Definition 6. (Adler (1981)). The *capacity dimension* of X is defined as follows:

$$\dim_C x = \sup\{\alpha > 0 : \ \exists\ \nu : \ I_\alpha(\nu) < \infty\} \ (\sup \emptyset = 0).$$

Lemma 6. *Let T be a compact subset in $\mathcal{X}$ satisfying the condition*

$$\underline{\dim}_K T = \dim_C T := N.$$

Let $\xi(x)$ be a d-dimensional Gaussian separable left homogeneous random flow on $\mathcal{X}$, each of its components $\xi_j(x)$ has an index β_j, satisfying

$$0 < \beta_1 \le \ldots \le \beta_d \le 1.$$

Let the following condition be satisfied

$$\inf_{x \ne o} \frac{\det \mathbf{E}(\xi(x) - \xi(o))(\xi(x) - \xi(o))^*}{\prod_{j=1}^{d} \sigma_j^2(x)} > 0. \tag{12}$$

Then we have, with probability 1

$$\dim_H \Im\xi = \min\left\{d, \frac{N + \sum_{j=1}^{d}(\beta_d - \beta_j)}{\beta_d}\right\}, \tag{13}$$

$$\dim_H \operatorname{Gr} \xi = \min\left\{\frac{N + \sum_{j=1}^{d}(\beta_d - \beta_j)}{\beta_d}, N + \sum_{j=1}^{d}(1-\beta_j)\right\}. \tag{14}$$

Proof. This Lemma was proven in (Cuzick (1978)) for the case $T = [0,1]^N$. We will modify Cuzick's proof for our case.

According to Lemmas 4 and 5 the right hand sides of (13) and (14) are almost surely upper bounds to their left hand sides so we must prove only the reverse inequalities. Suppose that the right hand side of (13) is equal to 1. Then we have $d = 1$ and almost sure trajectories continuity implies $\dim_H \Im\xi \geq 1$ with probability 1.

Now we assume that the right hand side of (13) is strictly greater then 1. At first we will show that the inequality

$$1 < \alpha < \min\left\{d, \frac{N + \sum_{j=1}^{d}(\beta_d - \beta_j)}{\beta_d}\right\}$$

implies that there exists a probability measure ν on T such that

$$\int_T \int_T \mathbf{E}\|\xi(x) - \xi(y)\|_{\mathbf{R}^d}^{-\alpha} \, d\nu(x)\, d\nu(y) < \infty. \tag{15}$$

Since $\xi(x)$ is left homogeneous we must only prove that

$$\int_{gT} \mathbf{E}\left(\sum_{j=1}^{d}(\xi_j(x) - \xi_j(o))^2\right)^{-\alpha/2} d\nu(gx) < \infty.$$

Here g is any element in $\mathcal{G}$, satisfying the condition $gy = o$. Now we will make a change of variables $\eta_j = \sigma_j^{-1}(x)(\xi_j(gx) - \xi_j(o))$. According to (12) the density of η is bounded and so we must prove

$$\int_T \int_{\mathbf{R}^d} \left\{\sum_{j=1}^{d}(t_j\sigma_j(x))^2\right\} dt\, d\nu(x) < \infty.$$

According to the definition of β, we have $\sigma_j(x) \geq K\rho^{\gamma_j}(x, o)$ for $\rho(x, o) < \delta$ for some $\delta > 0, K > 0$, where $\gamma_j > \beta_j$ for $\beta_j < 1$ and $\gamma_j = 1$ when $\beta_j = 1$ and $\gamma_1 \leq \gamma_2 \leq \ldots \leq \gamma_d \leq 1$. So we must prove

$$\int_T \rho^{-\alpha\gamma_1}(x, o) \int_{\mathbf{R}^d} \left[t_1^2 + \sum_{j=2}^{d}(t_j\rho^{\gamma_j - \gamma_1}(x, o))^2\right]^{-\alpha/2} dt\, d\nu(x) < \infty. \tag{16}$$

We can integrate over t_1 to obtain that the left hand side of (16) is less then a constant multiplied by

$$\int_T \rho^{-\alpha\gamma_1}(x, o) \int_{\mathbf{R}^{d-1}} \left[\sum_{j=2}^{d}(t_j\rho^{\gamma_j - \gamma_1}(x, o))^2\right]^{-\frac{\alpha+1}{2}} dt\, d\nu(x).$$

Now we can continue this process up to the last integral with respect to y_d. It will be finite if $\alpha < d$, so the left hand side of (16) will be less then constant multiplied by

$$\int_T \rho^{-\alpha\gamma_1 + (\gamma_2 - \gamma_1)(-\alpha+1) + \cdots + (\gamma_d - \gamma_{d-1})(-\alpha+d-1)}(x, o)\, d\nu(x)$$

and according to definition 6 this integral is finite for all

$$\alpha < \frac{N + \sum_{j=1}^{d}(\gamma_d - \gamma_j)}{\gamma_d}.$$

Letting $\gamma_j \downarrow \beta_j$ proves (15). Now let τ be the image of the measure ν under the transformation $x \mapsto \xi(x)$. We obtain

$$\int_{\Im\xi}\int_{\Im\xi} \|u-v\|^{-\alpha}\, d\tau(u)\, d\tau(v) < \infty,$$

i. e. $\dim_C \Im\xi \geq \alpha$. But according to (Adler (1981), Lemma 8.8.3) we have $\dim_H A \geq \dim_C A$ for any metric compact A (Adler formulates this Lemma for $A \subset \mathbf{R}^N$, but his proof does not depend on this fact). So formula (13) is proved.

If $\dim_H \Im\xi < d$ then we have

$$\dim_H \Im\xi = \frac{N + \sum_{j=1}^{d}(\beta_d - \beta_j)}{\beta_d} \leq \dim_H \mathrm{Gr}\,\xi \leq \frac{N + \sum_{j=1}^{d}(\beta_d - \beta_j)}{\beta_d}.$$

The first equality is a consequence of (13), the second one is trivial and the third one is a consequence of Lemmas 4 and 5.

When $\dim_H \Im\xi = d$ we finish the proof as above using the integral

$$\int_T\int_T \mathbf{E}\{[\|\xi(x)-\xi(y)\|^2 + \rho^2(x,y)]^{-\alpha/2}\}\, d\nu(x)\, d\nu(y).$$

Lemma 7. *All the components of the homogeneous and isotropic random flow on the group* SU(2) *have index* 1.

Proof. Using the main Theorem of (Malyarenko (1994)) we can show that the random flow $\xi(g)$ has the form

$$\xi_\zeta(g) = \sum_{H=|\zeta|}^{u}\sum_{\lambda=H}^{\infty}\sum_{\varepsilon=-H}^{H}\sum_{\gamma=-\lambda}^{\lambda} T_\gamma^\varepsilon(\lambda; g) Z_{\zeta\varepsilon\gamma}^{H}(\lambda).$$

Here $T_\gamma^\varepsilon(\lambda; g)$ are the matrix elements of the irreducible unitary representation $T_\lambda(g)$, $\lambda = H, H+1, \ldots$, and $Z_{\zeta\varepsilon\gamma}^{H}(\lambda)$ are zero mean Gaussian random variables satisfying the condition

$$\mathbf{E}Z_{\zeta\varepsilon\gamma}^{H}(\lambda)Z_{\beta\zeta\gamma'}^{H'}(\lambda) = \delta_\lambda^{\lambda'}\delta_\gamma^{\gamma'}\delta_H^{H'}\langle U,\zeta;U,\beta|2H,\zeta+\beta\rangle\times$$

$$\times\langle\lambda,\zeta;\lambda,\varepsilon|2H,\zeta+\varepsilon\rangle\nu_H(\lambda),$$

where the numbers in angular brackets are *Klebsch–Gordan coefficients* of the group SU(2) (Vilenkin (1968)) and $\nu_H(\lambda)$ are arbitrary non-negative numbers satisfying the condition

$$\sum_{\lambda=H}^{\infty} \nu_H(\lambda) < \infty, \quad H \leq u. \tag{17}$$

Let $(\varphi, \psi, \vartheta)$ be *Euler's angles* on the group SU(2) (Vilenkin (1968)). Then the invariant metric ρ will have the form

$$\rho(\varphi,\psi,\vartheta;0,0,0) = \arccos\left(\cos\frac{\vartheta}{2}\cos\frac{\varphi+\psi}{2}\right).$$

According to (Gikhman and Skorokhod (1974)) the covariance of both the real and imaginary part of the complex Gaussian vector $\xi_\zeta(g)$, $-u \le \zeta \le u$ has the form $(1/2)\mathbf{E}|\xi_\zeta(g)|^2$. So we must prove the equality

$$\begin{aligned}\lim_{\varphi,\psi,\vartheta\to 0}[\rho(\varphi,\psi,\vartheta;0,0,0)]^{-2\alpha}\textstyle\sum_{H=|\zeta|}^{u}\langle U,\zeta;U,\zeta|2H,2\zeta\rangle\sum_{\lambda=H}^{\infty}\times\\ \times\textstyle\sum_{\varepsilon=-H}^{H}\langle\lambda,\varepsilon;\lambda,\varepsilon|2H,2\varepsilon\rangle(1-\Re T_\varepsilon^\varepsilon(\lambda;\varphi,\psi,\vartheta)\nu_H(\lambda))=0,\ \alpha<1.\end{aligned} \tag{18}$$

The Clebsch–Gordan coefficient and matrix elements are elements of unitary matrices so their modulus are less or equal to 1. Using condition (17) and Weierstrass criterion we conclude that the series on the right hand side of (18) converges uniformly. So we can calculate the last limit term by term. To prove the Lemma we must show that

$$A(\alpha) := \lim_{\varphi,\psi,\vartheta\to 0}\vartheta^{-2\alpha}(1-\Re T_\varepsilon^\varepsilon(\lambda;\varphi,\psi,\vartheta)) = 0,\ \alpha<1.$$

But according to (Vilenkin (1968)) we have

$$1-\Re T_\varepsilon^\varepsilon(\lambda;\varphi,\psi,\vartheta) = 1-\cos(\varepsilon(\varphi+\psi))\cot^{2\varepsilon}\frac{\vartheta}{2}\sum_{k=\varepsilon}^{\lambda}\frac{(\lambda+k)!}{(\lambda-k)!(k+\varepsilon)!}\sin^{2k}\frac{\vartheta}{2},$$

and so we obtain

$$1-\Re T_\varepsilon^\varepsilon(\lambda;\varphi,\psi,\vartheta) \sim \frac{\varepsilon(\lambda+\varepsilon)!}{2(\lambda-\varepsilon)!}\vartheta^2$$

and

$$A(\alpha) = \lim_{\vartheta\to 0}\vartheta^{-2\alpha}\frac{\varepsilon(\lambda+\varepsilon)!}{2(\lambda-\varepsilon)!}\vartheta^2 = 0.$$

To prove the Theorem we must show that $\underline{\dim}_K \mathrm{SU}(2) = \dim_C \mathrm{SU}(2)$ and the group SU(2) satisfies condition (1). The first statement is obvious and the second one is proved by a direct computation using the fact that the invariant measure on SU(2) has the form (Vilenkin (1968))

$$d\mu = \frac{1}{16\pi^2}\sin\vartheta\, d\vartheta\, d\varphi\, d\psi.$$

References

Adler, R.J. (1981). *The Geometry of Random Fields.* Wiley, New York.

Berezansky, Yu.M., Us, G.F. and Sheftel', Z.G. (1990). *Functional Analysis.* Vyshcha Shkola Publishing, Kiev. (In Russian, to appear in English in Birkhäuzer, Zürich).

Cuzick, J. (1978). Some local properties of Gaussian vector fields. *Ann. Prob.* **6**(6), 984–994.

Federer, G. (1969). *Geometric Measure Theory.* Springer, Berlin.

Garsia, A. (1971). Continuity properties of Gaussian processes with multidimensional time parameter. *Proc. 6th Berkeley Symp. Math. Stat. Prob., Univ. of California Press, Berkeley* **2**, 369–374.

Gikhman, I.I. and Skorokhod, A.V. (1974). *The Theory of Stochastic Processes, Vol. 1* Springer, Berlin.

Hewitt, E. and Ross, K.A. (1963). *Abstract Harmonic Analysis, Vol. 1* Springer, Berlin.

Kolmogorov, A.N. and Tihomirov, V.M. (1959). ε-entropy and ε-capacity of sets in function spaces. *Uspehi Mat. Nauk,* **14**, No. 2(86), 3–86. (In Russian).

Malyarenko, A.A. (1994). Spectral decomposition of homogeneous and isotropic random flows on homogeneous space. *Asymptotical Analysis of Random Evolutions,*Institute of Mathematics Ukr. Ac. Sci., Kiev, 190–209.

Turbin, A.F. and Pratsevytyi, N.V. (1992). *Fractal Sets, Functions, Distributions.* Naukova Dumka Publishing, Kiev. (In Russian).

Vilenkin, N.Ya. (1968). *Special functions and the theory of group representations.* AMS, Providence, R.I.

Yaglom, A.M. (1960). Second-order homogeneous random fields. *Proc. 4th Berkeley Sympos. Math. Stat. and Prob. Univ. of California Press, Berkeley.* **2**, 593-622.

Exploring Stochastic Laws. pp. 307–328.
Eds. A.V. Skorokhod and Yu.V. Borovskikh

The weak convergence of point fields on a plane to a point field with conditionally independent increments

JU.S. MISHURA

Kiev University, Department of Mechanics and Mathematics, ul. Vladimirskaya 64, 252601 Kiev-17, Ukraine

Received February 1, 1995

Abstract. The functional limit theorem for point fields in a plane is established, where the conditions for convergence are formulated in terms of weakly predictable compensators and the limit field is a point field with conditionally independent increments. In the first part, the method of stochastic exponents in the plane is developed and used for the demonstration of the weak convergence of finite-dimensional distributions. In the second part the conditions of weak compactness and weak convergence are established.

We use the notions introduced in Mishura (1987, 1979, 1990).

1. The convergence of finite-dimensional distributions of point fields

Let $(\Omega, \mathfrak{F}, P)$ be a probability space, and $\{\mathfrak{F}_t, t \in R^2_+\}$ be σ-fields satisfying conditions (F1)–(F3) Cairoli and Walsh (1975). Denote by X the space of point fields

$$x = \{x_t(\omega), t \in R^2_+, \omega \in \Omega\},$$

i.e. such step functions that $x_t(\omega) \in N \cup \{0\}$ and $\Box x_t \in \{0, 1\}$. We consider such point fields that satisfy the condition

$$\text{(A)} \qquad x_t = M_t + A_t \qquad (1)$$

where M_t is a local strong martingale (Gushchin and Mishura (1992)), and A_t is a weakly predictable field (Gushchin (1982)), the compensator of x. It is well known that under condition (F4) Cairoli and Walsh (1975) any point field satisfies condition (A). If in addition $Ex_t < \infty$ then M_t is a strong martingale.

Definition 1. *Point field x is called a field with conditionally independent increments with respect to σ-field G if for any $s \le t$*

$$E(\exp\{i\lambda x]s,t]\}/\mathfrak{F}_s^*) = E(\exp\{i\lambda x]s,t]\}/G), \quad \mathfrak{F}_t^* = \bigvee_{s \not> t} \mathfrak{F}_s.$$

Theorem 1. *Let the point field x satisfy condition* (A), *A is a G-measurable function, $G \subset \mathfrak{F}_0$. Then x is a field with G-conditionally independent increments and*

$$\begin{aligned} G_s^t(x,\lambda) :&= E(\exp\{i\lambda x]s,t]\}/\mathfrak{F}_s^*) = \\ &= \exp\{aA]s,t]\} \cdot \prod_{s<z\le t} (1 + a\,\Box A_z)\exp\{-a\,\Box A_z\} \end{aligned} \tag{2}$$

where $a = e^{i\lambda} - 1$.

Proof. Let $F\colon R \to C$ be a continuous function, x be a point field, and $s \in R_+^2$ be fixed. Let

$$\bar{x}_t = x]s,t] = x_t - x_{t_1 s_2} - x_{s_1 t_2} + x_s.$$

Then $\bar{x}_s = \bar{x}_{t_1 s_2} = \bar{x}_{s_1 t_2} = 0$ for $t_i \ge s_i$, $i = 1,2$, and

$$F(\bar{x}_t) = F(0) + \sum_{s<u\le t} \Box F(\bar{x}_u), \tag{3}$$

where

$$\Box F(\bar{x}_u) = F(\bar{x}_u) - F(\bar{x}_u^{+-}) - F(\bar{x}_u^{-+}) + F(\bar{x}_u^{--}).$$

In particular, if $F(x) = \exp\{i\lambda x\}$, then

$$\begin{aligned} \Box\, F(\bar{x}_u) = \exp\{i\lambda \bar{x}_u^{--}\}\Big[&\exp\{i\lambda\Delta^1\bar{x}_{u-}\}\exp\{i\lambda\Delta^2\bar{x}_{u-}\}(\exp\{i\lambda\,\Box\bar{x}_u\} - 1) + \\ &+ (\exp\{i\lambda\Delta^1\bar{x}_{u-}\} - 1)(\exp\{i\lambda\Delta^2\bar{x}_{u-}\} - 1)\Big], \end{aligned} \tag{4}$$

where $\Delta^1\bar{x}_{u-} = x_u^{+-} - x_u^{--}$, $\Delta^2 x_{u-} = x_u^{-+} - x_u^{--}$. Denote $\delta_u^j = \exp\{i\lambda\Delta^j\bar{x}_u\}$, $\delta_{u-}^j = \exp\{i\lambda\Delta^j\bar{x}_{u-}\}$, $j = 1,2$.

We transform δ_u^j using the generalized one-parameter Ito formula

$$\begin{aligned} \delta_u^i &= 1 + \sum_{s_k<z_k\le t_k} \Big[\exp\{i\lambda\Delta^j\bar{x}_{u_j z_k}\} - \exp\{i\lambda\Delta^j\bar{x}_{(u_j,z_k)-}\}\Big] = \\ &= 1 + \sum_{s_k<z_k\le t_k} a\exp\{i\lambda\Delta^j\bar{x}_{(u_j,z_k)-}\}\,\Box x_{u_j z_k} = \\ &= 1 + a\int_{s_k}^{t_k} \delta_{u-}^j\, d_k(\Delta^j x_u), \qquad j = 1,2,\ k = 3 - j \end{aligned} \tag{5}$$

(if $j = 2$, $k = 1$ then $(u_j, t_k) = (t_1, u_2)$).

Substitute (4) and (5) into (3):

$$\exp\{i\lambda\bar{x}_t\} = 1 + \int_{]s,t]} \exp\{i\lambda\bar{x}_{u-}\}\, d\bar{x}_u^\lambda + \int_{]s,t]} \exp\{i\lambda\bar{x}_{u-}\}\, d_1 x_u^{\lambda,1,s}\, d_2 x_u^{\lambda,2,s}, \tag{6}$$

where

$$\bar{x}_{u-} = \bar{x}_u^{--}, \quad \bar{x}_u^\lambda = a\int_{]s,u]} \delta_{z-}^1 \delta_{z-}^2 \, dx_z, \quad x_u^{\lambda,j,s} = a\int_{]s,u]} \delta_{z-}^j \, dx_z. \tag{7}$$

Consider now formula (1). All the fields in it are locally integrable. Let $\{M_t^n, D_n\}$, $\{A_t^n, D_n\}$ and $\{x_t^n, D_n\}$, $n \geq 1$ be the localizers of M, a and x, respectively; then $x_t^n = M_t^n + A_t^n$ and all the fields in this representation are integrable, $(y_t - y_t^n)I_{D_n}(t) = 0$, $y = M, A, x$.

Put

$$v_t^n = \int_{[0,t]} I_{D_n}(z)\, dx_z = \int_{[0,t]} I_{D_n}(z)\, dx_z^n.$$

In addition field v_t^n is point and integrable. Evidently, $v_t^n = \tilde{M}_t^n + \tilde{A}_t^n$, where

$$\tilde{M}_t^n = \int_{[0,t]} I_{D_n}(z)\, dM_z^n = \int_{[0,t]} I_{D_n}(z)\, dM_z,$$

$$\tilde{A}_t^n = \int_{[0,t]} I_{D_n}(z)\, dA_z^n = \int_{[0,t]} I_{D_n}(z)\, dA_z,$$

the field $\tilde{M}^n$ is a strong martingale.

We use the Ito formula similar to (6) for v^n, which is true because v^n is a point field:

$$\exp\{i\lambda \bar{v}_t^n\} = 1 + \int_{]s,t]} \exp\{i\lambda \bar{v}_{u-}^n\}\, dv_u^{n,\lambda} + \int_{]s,t]} \exp\{i\lambda \bar{v}_{u-}^n\}\, d_1 v_u^{n,\lambda,1,s} d_2 v_u^{n,\lambda,2,s}, \tag{8}$$

where $v_u^{n,\lambda}$ and $v_u^{n,\lambda,j,s}$ can be obtained from (7) when we substitute v^n instead of x into the right-hand side. Now let now $\delta_z^{n,j} = \exp\{i\lambda\Delta^j v_z^n\}$,

$$A_u^{n,\lambda} = a\int_{]s,u]} \delta_{z-}^{n,1}\delta_{z-}^{n,2}\, d\tilde{A}_z^n = \int_{]s,u]} \delta_{z-}^{n,1}\delta_{z-}^{n,2} I_{D_n}(z)\, dA_z,$$

$$A_u^{n,\lambda,j,s} = a\int_{]s,u]} \delta_{z-}^{n,j} I_{D_n}(z)\, dA_z.$$

Then

$$E(\exp\{i\lambda \bar{v}_t^n\}/\mathfrak{F}_s^*) = 1 + E\Big(\int_{]s,t]} \exp\{i\lambda \bar{v}_{u-}^n\}\, dA_u^{n,\lambda}/\mathfrak{F}_s^*\Big) +$$

$$+ E\Big(\int_{]s,t]} \exp\{i\lambda \bar{v}_{u-}^n\}\, d_1 A_u^{n,\lambda,1,s} d_2 A_u^{n,\lambda,2,s}/\mathfrak{F}_s^*\Big).$$

From the G-measurability of A

$$E\Big(\int_{]s,t]} \exp\{i\lambda \bar{v}_{u-}^n\}\, dA_u^{n,\lambda}/\mathfrak{F}_s^*\Big) = a\int_{]s,t]} E(\exp\{i\lambda \bar{v}_{u-}^n\}\, \delta_{u-}^{n,1}\delta_{u-}^{n,2} I_{D_n}(u)/\mathfrak{F}_s^*)\, dA_u.$$

All the functions under the sign of the integral are bounded by 1, so therefore

$$\lim_{n\to\infty} \int_{]s,t]} E(\exp\{i\lambda \bar{v}_{u-}^n\}\, \delta_{u-}^{n,1}\delta_{u-}^{n,2} I_{D_n}(u)/\mathfrak{F}_s^*)\, dA_u =$$

$$= \int_{]s,t]} E(\exp\{i\lambda \bar{x}_{u-}\}\, \delta_{u-}^1\delta_{u-}^2/\mathfrak{F}_s^*)\, dA_u =$$

$$= \lim_{n\to\infty} \int_{]s,t]} E(\exp\{i\lambda \bar{v}_{u-}^n\}\, \delta_{u-}^{n,1}\delta_{u-}^{n,2}/\mathfrak{F}_s^*)\, dA_u. \tag{9}$$

Further,

$$E(\exp\{i\lambda\bar{v}^n_{u-}\}\,\delta^{n,1}_{u-}\delta^{n,2}_{u-}/\mathfrak{F}^*_s) =$$
$$= \lim_{\substack{z\uparrow u, z<u\\ p_j\downarrow u_j}} E\Big(\exp\{i\lambda\bar{v}^n_z\}E((\exp\{i\lambda(\bar{v}^n_{p_1 z_2}-\bar{v}^n_z)/\mathfrak{F}^*_{z_1 s_2})\times$$
$$\times E(\exp\{i\lambda(\bar{v}^n_{z_1 p_2}-\bar{v}^n_z)\}/\mathfrak{F}^*_{s_1 z_2})/\mathfrak{F}^*_s\Big). \tag{10}$$

Denote

$$\mathfrak{F}^*_{u_1-,s_2} = \bigvee_{z_1<u_1}\mathfrak{F}^*_{z_1,s_2}, \qquad \mathfrak{F}^*_{s_1,u_2-} = \bigvee_{z_2<u_2}\mathfrak{F}^*_{s_1 z_2}.$$

Then

$$\lim_{z_1\uparrow u_1, p_1\downarrow u_1} E(\exp\{i\lambda(\bar{v}^n_{p_1 z_2}-\bar{v}^n_z)\}/\mathfrak{F}^*_{z_1 s_2}) = E\{\delta^{n,1}_u/\mathfrak{F}^*_{u_1-,s_2}\}. \tag{11}$$

Present $\delta^{n,1}_u$ similarly to (5):

$$\delta^{n,1}_u = 1 + a\int_{s_2}^{u_2}\delta^{n,1}_{(u_1,z_2)-}d_2(\Delta^1\bar{v}^n_{u_1 z_2}) =$$
$$= 1 + a\int_{s_2}^{u_2}\delta^{n,1}_{(u_1,z_2)-}d_2(\Delta^1\tilde{M}^n_{u_1 z_2}) + a\int_{s_2}^{u_2}\delta^{n,1}_{(u_1,z_2)-}d_2(\Delta^1\tilde{A}^n_{u_1 z_2}).$$

Note that $|M^n_t| \le x^n_T + A^n_T$ if $t \le T$, and x^n_T, A^n_T are integrable variables. Therefore for any points (z_1,p_2), (z_1,t_2) such that $u_1 < z_1 \le T_1$, $s_2 \le p_2 \le t_2 \le T_2$

$$E(\Delta^1 M^n_{z_1 t_2} - \Delta^1 M^n_{z_1 p_2}/\mathfrak{F}^*_{u_1-,p_2}) = \lim_{z_1'\downarrow z_1, p_1\uparrow u_1} E(M^n](z_1,p_2),(z_1',t_2)]/\mathfrak{F}^*_p) = 0.$$

Thus

$$E(\delta^{n,1}_u/\mathfrak{F}^*_{u_1-,s_2}) = 1 + a\int_{s_2}^{u_2}E(\delta^{n,1}_{(u_1,z_2)-}I_{D_n}(u_1,z_2)/\mathfrak{F}^*_{u_1-,s_2})\,d_2(\Delta^1 A_{u_1 z_2}).$$

Taking the limit we obtain

$$E(\delta^1_u/\mathfrak{F}^*_{u_1-,s_2}) = 1 + a\int_{s_2}^{u_2}E(\delta^1_{(u_1,z_2)-}/\mathfrak{F}^*_{u_1-,s_2})\,d_2(\delta^1 A_{u_1 z_2}). \tag{12}$$

Equality (12) means that the stochastic process $\beta_{u_2} = E(\delta^1_u/\mathfrak{F}^*_{u_1-,s_2})$ satisfies the usual one-parameter Dolean's equation. This equation has a unique solution that can be presented as

$$\sigma^1_u = E(\delta^1_u/\mathfrak{F}^*_{u_1-,s_2}) =$$
$$= \exp\{-a(\Delta^1\bar{A}_u - \Delta^1\bar{A}_{u_1 s_2})\}\prod_{s_2<z_2\le u_2}(1+a\,\Box A_{u_1 z_2})\exp\{-a\,\Box A_{u_1 z_2}\}. \tag{13}$$

Similarly,

$$\sigma^2_u = E(\delta^2_u/\mathfrak{F}^*_{s_1,u_2-}) =$$
$$= \exp\{-a(\Delta^2\bar{A}_u - \Delta^2\bar{A}_{s_1 u_2})\}\prod_{s_1<z_1\le u_1}(1+a\,\Box A_{z_1 u_2})\exp\{-a\,\Box A_{z_1 u_2}\}. \tag{14}$$

Finally, by considering (10) and (11) we obtain

$$E(\exp\{i\lambda\bar{x}_{u-}\}\,\delta^1_{u-}\delta^2_{u-}/\mathfrak{F}^*_s) = E(\exp\{i\lambda\bar{x}_{u-}\}/\mathfrak{F}^*_s)\sigma^1_{u-}\sigma^2_{u-},$$

where σ^i_u, $i = 1, 2$ are equal to the right-hand sides of (13) and (14), respectively.

Having transformed similarly the second integral on the right-hand side of (8) and taking into account (9) we obtain

$$E(\exp\{i\lambda\bar{x}_t\}/\mathfrak{F}^*_s) = \lim_{n\to\infty} E(\exp\{i\lambda\bar{v}^n_t\}/\mathfrak{F}^*_s) =$$
$$= 1 + \int_{]s,t]} E(\exp\{i\lambda\bar{x}_{u-}\}/\mathfrak{F}^*_s)\,dB^\lambda_u + \int_{]s,t]} E(\exp\{i\lambda\bar{x}_{u-}\}/\mathfrak{F}^*_s)\,d_1B^{1,\lambda}_u\,d_2B^{2,\lambda}_u,$$

where

$$B^\lambda_u = a\int_{]s,t]} \sigma^1_{u-}\sigma^2_{u-}\,dA_u, \qquad B^{j,\lambda}_u = a\int_{]s,t]} \sigma^j_{u-}\,dA_u.$$

Fields B^λ and $B^{j,\lambda}$ satisfy the relations

$$B^{j,\lambda}_u = B^\lambda_u - a\int_{]s,t]} \sigma^j_{u-}(\sigma^{3-j}_{u-} - 1)\,dA_u,$$
$$\Delta^j B^{3-j,\lambda}_u = \sigma^{3-j}_u - 1 = a\int_{s_{3-j}}^{u_{3-j}} \sigma^{3-j}_{(u_j,z_{3-j})-}\,d_{3-j}(\Delta^j A)_{u_j z_{3-j}}.$$

Therefore

$$B^{j,\lambda}_u = B^\lambda_u - ((3-j)\text{-}]B^{j,\lambda}]\circ B^{3-j,\lambda}) \tag{16}$$

(The notion of the integrals $(i\text{-}]B]\circ C)$ was introduced in Mishura (1990)). Relations (15) and (16) mean that the random function $\phi_t := E(\exp\{i\lambda\bar{x}_t\}/\mathfrak{F}^*_s)$ satisfies the two-parameter Dolean's equation with respect to the fields of bounded variation (Mishura (1988)). This equation has a unique solution that is assigned by any of the following equalities

$$\phi_t = \exp\{B^{1,\lambda}_t\}\prod_{0<s_1\le t_1}(1+\Delta^1 B^{1,\lambda}_{s_1t_2})\exp\{-\Delta^1 B^{1,\lambda}_{s_1t_2}\} =$$
$$= \exp\{B^{2,\lambda}_t\}\prod_{0<s_2\le t_2}(1+\Delta^2 B^{2,\lambda}_{t_1s_2})\exp\{-\Delta^2 B^{2,\lambda}_{t_1,s_2}\}.$$

We use the first from these equalities.

$$\prod_{0<s_1\le t_1}(1+\Delta^1 B^{1,\lambda}_{s_1t_2}) = \prod_{0<s_1\le t_1}(1+\Delta^1 B^{1,\lambda}_{(s_1,t_2)-})(1+\Box B^{1,\lambda}_{s_1,t_2}(1+\Delta^1 B^{1,\lambda}_{(s_1,t_2)-})^{-1}).$$

Further, from Mishura (1988)

$$\Box B^{1,\lambda}_{s_1t_2} = \Box B^\lambda_{s_1t_2}(1+\Delta^2 B^{2,\lambda}_{(s_1,t_2)-})^{-1}.$$

In addition,

$$\Box B^\lambda_{s_1t_2} = a\sigma^1_{(s_1,t_2)-}\sigma^2_{(s_1,t_2)-}\Box A_{s_1t_2}, \qquad 1+\Delta^i B^{i,\lambda}_{(s_1,t_2)-} = \sigma^i_{(s_1,t_2)-}.$$

So,

$$\square B^{1,\lambda}_{s_1,t_2}(1+\Delta^1 B^{1,\lambda}_{(s_1,t_2)-})^{-1} = a\,\square A_{s_1 t_2}.$$

Now, taking into account the presentation (13) for σ^1_u, we have

$$\prod_{s_1<z_1\le t_1}(1+\Delta^1 B^{1,\lambda}_{(z_1,t_2)-}) =$$

$$= \prod_{s_1<z_1\le t_1}\exp\{a\Delta^1\bar{A}_{z_1 t_2}-\Delta^1\bar{A}_{z_1 s_2})\}\prod_{s<z\le t}(1+a\,\square A_z)\exp\{-a\,\square A_z\},$$

finally,

$$\prod_{s_1<z_1\le t_1}(1+\Delta^1 B^{1,\lambda}_{z_1,t_2}) =$$

$$= \prod_{s_1<z_1\le t_1}\exp\{a(\Delta^1\bar{A}_{z_1 t_2}-\Delta^1\bar{A}_{z_1 s_2})\}\prod_{s<z\le t}(1+a\,\square A_z)\exp\{-a\,\square A_z\}.$$

Further, in accordance with Mazziotto et al (1981) any field B of bounded variation has a decomposition

$$B_t = B^c_t + B^d_t + B^{cd}_t + B^{dc}_t, \tag{17}$$

where B^c is continuous, B^d is a purely discontinuous field, and B^{cd} and B^{dc} are continuous in one coordinate and purely discontinuous in another.

Now,

$$\bar{B}^{1,\lambda}_t = \sum_{s_1<z_1\le t_1}\Delta^1\bar{B}^{1,\lambda}_{z_1 t_2} = \bar{B}^{1,\lambda,cd}_t + \bar{B}^{1,\lambda,c}_t.$$

But

$$\bar{B}^{1,\lambda,cd}_t + \bar{B}^{1,\lambda,c}_t = a\int_{]s,t]}\sigma^1_{u-}\,d(A^c+A^{cd})_u = a(A^c+A^{cd})]s,t] = a\bar{A}^c_t + a\bar{A}^{cd}_t,$$

because

$$\int_{]s,t]}(\sigma^1_{u-}-1)\,d(A^c+A^{cd})_u = \int_{]s,t]}\left[a\int_{s_2}^{u_2}\sigma^1_{(u_1,z_2)-}\,d_2(\Delta^1 A)_{u_1 z_2}\right]d(A^c+A^{cd})_u = 0.$$

The last equality is true because the integrand differs from zero in no more than a countable number of vertical lines.

Therefore

$$\phi_t = \exp\{a\bar{A}_t\}\prod_{s<z\le t}(1+a\,\square A_z)\exp\{-a\,\square A_z\}. \tag{18}$$

The right-hand side of (18) is G-measurable, thus $\phi_t = E(\exp\{i\lambda\bar{x}_t\}/G)$, i.e., the field x has G-conditionally independent increments.

Now $\lambda = i\mu$, $\mu \in R$, $a = e^{-\mu} - 1$. Note that the compensator has such a modification for which $\square A_t \leq 1$. Then $1 + a\square A_z = 1 - (1 - e^{-\mu})\square A_z > 0$. Further we let $1 - e^{-\mu} = b$. Transform the discontinuous part of ϕ_t:

$$\prod_{0<z\leq t} (1 - b\square A_z)\exp\{b\square A_z\} =$$

$$= \exp\Big\{\sum_{0<z\leq t} (\ln(1 - b\square A_z) + b\square A_z)\Big\} = \exp\Big\{\sum_{k=2}^{\infty}(-b)^k/k \sum_{0<z\leq t} (\square A_z)^k\Big\} \quad (19)$$

Further we denote S as the set $S_1 \times S_2$, where S_i, $i = 1,2$ are dense in R_+. We formulate the following result for non-random functions without including a proof.

Lemma 1. *Let $\{A_t^n, A_t,\ n \geq 1,\ t \in R_+^2\}$ be non-random, non-decreasing real functions. Under the following conditions*

1) $A_t^n \to A_t, \quad n \to \infty,\ t \in S;$

2) $\sum_{0<z\leq t} (\square A_z^n)^2 \to \sum_{0<z\leq t} (\square A_z)^2, \quad n \to \infty,\ t \in S,$

for any $k \geq 3$

$$\sum_{0<z\leq t} (\square A_z^n)^k \to \sum_{0<z\leq t} (\square A_z)^k, \qquad t \in S.$$

Lemma 2. *Let $\{A_t^n, A_t,\ n \geq 1,\ t \in R_+^2\}$ be the sequence of non-decreasing random fields on the same probability space. If for any $t \in S$ $A_t^n \xrightarrow{P} A_t$,*

$$\sum_{0<s\leq t} (\square A_s^n)^2 \xrightarrow{P} \sum_{0<s\leq t} (\square A_s)^2, \qquad n \to \infty,$$

then for any $k \geq 3$, $t \in S$

$$\sum_{0<s\leq t} (\square A_s^n)^k \xrightarrow{P} \sum_{0<s\leq t} (\square A_s)^k, \qquad n \to \infty.$$

Theorem 2. *Let $(x_t^n, \mathfrak{F}_t^n,\ n \geq 1,\ t \in R_+^2)$ be the sequence of point fields with compensators $(A_t^n, \mathfrak{F}_t^n,\ n \geq 1)$, and $x \in X$ be a point field with G-measurable compensator A. Under the following conditions*

3) $G \subset \bigcap_{n\geq 1} \mathfrak{F}_0^n;$

4) $A_t^n \xrightarrow{P} A_t, \quad \sum_{0<s\leq t} (\square A_s^n)^2 \xrightarrow{P} \sum_{0<s\leq t} (\square A_s)^2, \qquad n \to \infty,\ t \in S,$

we have

$$G_0^t(x^n, \lambda) \xrightarrow{P} G_0^t(x, \lambda), \qquad n \to \infty,\ t \in S,\ \lambda \in R.$$

Proof of lemma 2 and theorem 2 is similar to the one-parameter case (Liptser, Shiryayev (1986)) if we take into account formulas (2) and (19) and lemma 1.

Further we denote $\mathfrak{M}_{\text{loc}}$ ($s\mathfrak{M}_{\text{loc}}$) as the class of local (strong) martingales. Let the point $T \in S$ be fixed.

Consider such stopping times $\sigma \geq 0$ and $\tau \geq 0$ that for any $t_1 \in [0, T_1]$ the set $\{\sigma \leq t_1\} \in \mathfrak{F}_{t_1 T_2}$ and for any $t_2 \in [0, T_2]$ the set $\{\tau \leq t_2\} \in \mathfrak{F}_{T_1 t_2}$. We call such stopping times σ and τ as $\mathfrak{F}_{t_1 T_2}$- ($\mathfrak{F}_{T_1 t_2}$-) stopping times, respectively.

We now introduce the notation

$$\zeta_t^n = \exp\{i\lambda x_t^n\}(G_0^t(x^n, \lambda))^{-1}.$$

Lemma 3. *Let $\lambda \in R_0 = R \setminus \{\pi + 2k\pi,\ k \in Z\}$. Then there exist the unions $\{\sigma_{nk},\ n \geq 1, k \geq 1\}$ and $\{\tau_{nk},\ n \geq 1, k \geq 1\}$ of $\mathfrak{F}_{t_1 T_2}$- and $\mathfrak{F}_{T_1 t_2}$-stopping times that satisfy the conditions:*

5) *for any $n \geq 1$, $k \geq 1$ $\tau_{nk+1} \geq \tau_{nk}$, $\sigma_{nk+1} \geq \sigma_{nk}$;*

6) *there exist constants c_k, $k \geq 1$ such that $|\zeta_t^{nk}| \leq c_k$ a.s.;*

7) $E(\zeta_t^{nk}/\mathfrak{F}_0^n) = 1$, $\zeta_t^{nk} = \zeta^n_{t_1 \wedge \sigma_{nk}, t_2 \wedge \tau_{nk}}$.

If, in addition, the following condition holds

(B) $$G_0^t(x^n, \lambda) \xrightarrow{P} G_0^t(x, \lambda), \qquad n \to \infty,\ t \in S,\ \lambda \in R_0$$

and $T \in S$, then

8) $$\overline{\lim_{n\to\infty}} P\{\sigma_{nk} < T_1,\ |G_0^t(x,\lambda)| > 1/k\} = \overline{\lim_{n\to\infty}} P\{\tau_{nk} < T_2,\ |G_0^t(x,\lambda)| > 1/k\} = 0, \quad k \geq 0.$$

Proof. Index n at the beginning will be omitted.

1) We know that

$$\square A_t = E\{\square x_t / \hat{\mathfrak{F}}_t\},$$

where $\hat{\mathfrak{F}}_t = (\bigvee_{s<t} \mathfrak{F}_s^*) \cap \mathfrak{F}_t$. So

$$E\{\exp\{i\lambda \square x_t\} / \hat{\mathfrak{F}}_t\} = 1 + a\, E\{\square x_t / \hat{\mathfrak{F}}_t\} = 1 + a \square A_t,$$

therefore $|1 + a \square A_t| \leq 1$ (a.s.). Now,

$$\begin{aligned} |G_0^t(x,\lambda)(G_0^{(s_1,t_2)}(x,\lambda))^{-1}| &= |\exp\{a\alpha\}| \prod_{(s_1,0)<z\leq t} |1 + a \square A_z| \\ &\leq \left|\exp\left\{-\sin^2 \frac{\alpha\lambda}{2}\right\}\right| \\ &\leq 1, \end{aligned}$$

$$\alpha = (A^c + A^{cd} + A^{dc})_{t_1 T_2} - (A^c + A^{cd} + A^{dc})_{s_1 T_2} \geq 0.$$

Therefore

$$|G_0^t(x,\lambda)| \le |G_0^{(s_1,t_2)}(x,\lambda)| \le 1, \tag{20}$$

if $s_1 \le t_1$, $\lambda \in R$.

Besides it,

$$|1 + a\,\Box A_t|^2 = 4\sin^2\frac{\lambda}{2}\cdot p_1^2 - 4\sin^2\frac{\lambda}{2}\cdot p_1 + 1,$$

where $p_1 = P(\Box x_t = 1/\hat{\mathfrak{F}}_t)$.

It means that $|1 + a\,\Box A_t| \ge |\cos\frac{\lambda}{2}| > 0$, $\lambda \in R_0$. Therefore the field ζ_t is defined for any $t \in R_+^2$.

2) Equalities (5) and (6) mean that $\exp\{i\lambda x_t\}$ satisfies two-parameter Dolean's equation with respect to the fields of bounded variation (Mishura (1988)). This equation has a unique solution that can be presented as

$$\exp\{i\lambda x_t\} = \exp\{x_t^{1,\lambda,0}\} \prod_{0<s_1\le t_1} (1 + \Delta^1 x_{s_1 t_2}^{1,\lambda,0}) \exp\{-\Delta^1 x_{s_1 t_2}^{1,\lambda,0}\} =$$
$$= \prod_{0<s_1\le t_1} (1 + \Delta^1 x_{s_1 t_2}^{1,\lambda,0})$$

(the field $x_t^{1,\lambda,0}$ was introduced in Mishura (1988)). Transforming the last expression similarly to (18) we obtain

$$\exp\{i\lambda x_t\} = \prod_{0<s\le t} (1 + a\,\Box x_s).$$

Taking into account the equality

$$A_t - \sum_{0<s\le t} \Box A_s = (A^c + A^{cd} + A^{dc})_t,$$

we obtain

$$\zeta_t = \prod_{0<s\le t} (1 + a\,\Box M_s(1 + a\,\Box A_s)^{-1}) \exp\{-a(A^c + A^{cd} + A^{dc})_t\},$$

where field $M = x - A \in s\mathfrak{M}_{\mathrm{loc}}$.

3) Let us now transform the field ζ_t with the help of the generalized one-parameter Ito formula. For this purpose denote ζ_t as a product $\zeta_t = \zeta_t^1\zeta_t^2$, where

$$\zeta_t^1 = \prod_{0<s\le t} (1 + a\,\Box M_s(1 + a\,\Box A_s)^{-1}),$$
$$\zeta_t^2 = \exp\{-a(A^c + A^{cd} + A^{dc})_t\},$$

and note that for fixed t_2 the product of functions mentioned as the processes in t_1 can be written in the form

$$\zeta_t = \zeta_t^1\zeta_t^2 = \zeta^1_{(0,t_2)}\zeta^2_{(0,t_2)} + \int_0^{t_1} (\zeta^1_{s_1 t_2})^{-+}\, d_1\zeta^2_{s_1 t_2} +$$
$$+ \int_0^{t_1} (\zeta^2_{s_1 t_2})^{-+}\, d_1\zeta^1_{s_1 t_2} + \sum_{0<s_1\le t_1} \Delta^1\zeta^1_{s_1 t_2}\Delta^1\zeta^2_{s_1 t_2} =$$
$$= 1 - a\int_0^{t_1} \zeta^{-+}_{s_1 t_2}\, d^1(A^c + A^{cd})_{s_1 t_2} +$$
$$+ \sum_{0<s_1\le t_1} \zeta^{-+}_{s_1 t_2}(\exp\{-a\Delta^1 A^{dc}_{s_1 t_2}\} - 1) \prod_{0<s_2\le t_2} (1 + a\,\square M_s(1 + a\,\square A_s)^{-1}) +$$
$$+ \sum_{0<s_1\le t_1} \zeta^{-+}_{s_1 t_2}\Big(\prod_{0<s_2\le t_2} (1 + a\,\square M_s(1 + a\,\square A_s)^{-1}) - 1\Big). \qquad (21)$$

Consider now the following processes in t_2 under fixed s_1

$$\xi_{s_1 t_2} = \exp\{-a\Delta^1 A^{dc}_{s_1 t_2}\} - 1, \quad \eta_{s_1 t_2} = \prod_{0<s_2\le t_2} (1 + a\,\square M_s(1 + a\,\square A_s)^{-1})$$

and transform their product using again the one-parameter Ito formula

$$(\xi\eta)_{s_1 t_2} = -a\int_0^{t_2} \eta_s^{+-} \exp\{-a\Delta^1 A_s^{dc}\}\, d^2(\Delta^1 A^{dc})_s +$$
$$+ \sum_{0<s_2\le t_2} \xi_s^{+-}\, a\,\square M_s(1 + a\,\square A_s)^{-1}. \qquad (22)$$

In addition,

$$\eta_{s_1 t_2} - 1 = \sum_{0<s_2\le t_2} \eta_s^{+-}\, a\,\square M_s(1 + a\,\square A_s)^{-1}. \qquad (23)$$

Substituting (22) and (23) into (21) we obtain

$$\zeta_t = 1 - a\int_0^{t_1} \zeta^{-+}_{s_1 t_2}\, d^1(A^c + A^{cd})_{s_1 t_2} - a\int_{[0,t]} \zeta^{-+}_{s_1 t_2}\eta_s^{+-} \exp\{-a\Delta^1 A_s^{dc}\}\, dA_s^{dc} +$$
$$+ a\int_{[0,t]} \zeta^{-+}_{s_1 t_2}\eta_s^{+-}(\xi_s + 1)(1 + a\,\square A_s)^{-1}\, d(x_s - A_s^d) =$$
$$= 1 + a\int_{[0,t]} \zeta^{-+}_{s_1 t_2}\eta_s^{+-}(\xi_s + 1)(1 + a\,\square A_s)^{-1}\, d(x_s - A_s^d) -$$
$$- a\int_{[0,t]} \zeta^{-+}_{s_1 t_2}\eta_s^{+-} \exp\{-a\Delta^1 A_s^{dc}\}\, A_s^{dc} - a\int_{[0,t]} \zeta^{-+}_{s_1 t_2}\, d(A^c + A^{cd})_s.$$

Note that for any finite function ϕ_s

$$\int_{[0,t]} \phi_s\, dA_s^{dc} - \int_{[0,t]} \phi_s(1 + a\,\square A_s)^{-1}\, dA_s^{dc} = \sum_{s:\,\square A_s\ne 0} \phi_s(1 + a\,\square A_s)^{-1}\square A_s^{dc} = 0,$$

because $\square A_s^{dc} = 0$ and the sum have no more than a countable number of items. Therefore

$$\int_{[0,t]} \zeta^{-+}_{s_1 t_2}\eta_s^{+-} \exp\{-a\Delta^1 A_s^{dc}\}\, dA_s^{dc} =$$
$$= \int_{[0,t]} \zeta^{-+}_{s_1 t_2}\eta_s^{+-} \exp\{-a\Delta^1 A_s^{dc}\}(1 + a\,\square A_s)^{-1}\, dA_s^{dc}.$$

We prove similarly that

$$\int_{[0,t]} \zeta_{s_1 t_2}^{-+} d(A^c + A^{cd})_s = \int_{[0,t]} \zeta_{s_1 t_2}^{-+} \eta_s^{+-} \exp\{-a\Delta^1 A_s^{dc}\}(1 + a\,\Box A_s)^{-1} d(A^c + A^{cd})_s.$$

Finally,

$$\zeta_t = 1 + a \int_{[0,t]} \zeta_{s_1 t_2}^{-+} \eta_s^{+-} \exp\{-a\Delta^1 A_s^{dc}\}(1 + a\,\Box A_s)^{-1} dM_s =$$

$$= 1 + a \int_{[0,t]} \zeta_{s_1 t_2}^{-+} \eta_s^{+-} \exp\{-a\Delta^1 A_s^{dc}\} dM_s.$$

4) Let $\{D_r^n,\ r \geq 1\}$ be a sequence of weakly predictable stopping sets that localize M^n, $M^{n,r} \in s\mathfrak{M}$ are the localized fields. Consider the sequence of the following stopping times:

$$\sigma_{nk} = \inf\{t_1 > 0 \colon \sup_{t_2 \leq T_2} |\zeta_t^n| \geq k\} \wedge T_1,$$

$$\tau_{nk} = \inf\{t_2 > 0 \colon \sup_{t_1 \leq T_1} |\zeta_t^n| \geq k\} \wedge$$

$$\wedge \inf\{t_2 > 0 \colon \sup_{t_1 \leq T_1} |\eta_t^n \exp\{-a\Delta^1 A_t^{n,dc}\}| \geq 2k + 1\} \wedge T_2.$$

It is evident that σ_{nk} are $\mathfrak{F}_{t_1 T_2}^n$-stopping times, τ_{nk} are $\mathfrak{F}_{T_1 t_2}^n$-stopping times and statement (5) holds. Let

$$\beta_t^{nk} = \beta_{t_1 \wedge \sigma_{nk}, t_2 \wedge \tau_{nk}}^n,$$

so now we have

$$\begin{aligned} |\zeta_t^{nk}| &\leq 3k + |\Box \zeta_t^{nk}| \\ &= 3k + |a|\,|(\zeta_t^{nk})^{+-}|\,|\eta_t^{nk}|\,|\exp\{-a\Delta^1 (A^{dc})_t^{nk}\}|\,|\Box M_t^{nk}| \\ &\leq 3k + |a|(2k+1)2k\,|(1 + a\,\Box A_t^{nk})^{-1}| \\ &\leq 3k + 4|a|k(2k+1)\,|\cos \lambda/2|^{-1} \\ &= c_k, \end{aligned} \tag{25}$$

i.e. statement 6) holds.

5) Now we form the fields

$$\zeta_t^{n,r,k} = 1 + a \int_{[0,(t_1 \wedge \sigma_{nk}, t_2 \wedge \tau_{nk})]} (\zeta_{s_1 t_2}^n)^{-+} \eta_s^n \exp\{-a\Delta^1 A_s^{n,dc}\} I_{D_r^n}(s)\, dM_s^n =$$

$$= 1 + a \int_{[0,t]} I_{[0,\sigma_{nk}]}(s_1) I_{[0,\tau_{nk}]}(s_2) (\zeta_{s_1 t_2}^n)^{-+} \eta_s^n \exp\{-a\Delta^1 A_s^{n,dc}\} I_{D_r^n}(s)\, dM_s^{n,r}.$$

We shall establish that $E(\zeta_t^{n,r,k}/\mathfrak{F}_0^n) = 1$. Really, the fields $I_{[0,\sigma_{nk}]}(t_1)$, $t_1 \leq T_1$ and $(\zeta_t^n)^{-+}$, $t_1 \leq T_1$ are continuous from the left consequently 1-predictable (Gushchin (1982)), the field

$$I_{[0,\tau_{nk}]}(t_2), \quad t_2 \leq T_2$$

is continuous from the right consequently 2-predictable, the fields $\eta_s^n \exp\{-a\Delta^1 A_s^{n,dc}\}$ and $I_{D_r^n}(s)$ are weakly predictable. Therefore it is sufficient to establish that for any bounded and 1- (2-) predictable fields ϕ_t^1, $\phi_{t_2}^2$ and any bounded weakly predictable field χ_t , $t \in R_+^2$

$$E\Big\{\int_{[0,t]} \phi_{s_1 t_2}^1 \phi_{s_2}^2 \chi_s \, dM_s^{n,r} / \mathfrak{F}_0^n\Big\} = 0.$$

For this purpose it is sufficient to consider simple bounded functions ϕ^i. For simple functions

$$\phi_{s_1 t_2}^1 = \sum_l \phi_l^1 I\{s_1 \in (s_1^l, s_1^{l+1}]\}$$

and

$$\phi_{s_2}^2 = \sum_m \phi_m^2 I\{s_2 \in (s_2^m, s_2^{m+1}]\}$$

we have

$$E\Big\{\int_{[0,t]} \phi_{s_1 t_2}^1 \phi_{s_2}^2 \chi_s \, dM_s^{n,r} / \mathfrak{F}_0^n\Big\} = E\Big(\sum_{l,m} \phi_l^1 \phi_m^2 E\Big(\int_{\square_{lm}} \chi_s \, dM_s^{n,r} / \mathfrak{F}_{lm}^{*,n}\Big) / \mathfrak{F}_0^n\Big) = 0$$

($\square_{lm} = (s_1^l, s_1^{l+1}] \times (s_2^m, s_2^{m+1}]$, $\mathfrak{F}_{lm}^{*,n} = \mathfrak{F}_{(s_1^l, s_2^m)}^{*,n}$). Now, similarly to (25) we obtain that $|\zeta_t^{n,k,r}| \le c_k$. Therefore, taking into account (24), we obtain for any $n \ge 1$, $k \ge 1$

$$E(\zeta_t^{nk} / \mathfrak{F}_0^n) = \lim_{r \to \infty} E(\zeta_t^{n,k,r} / \mathfrak{F}_0^n) = 1,$$

i.e., statement 7) holds.

6) If condition (B) is satisfied then from (20) we obtain

$$\begin{aligned}
&\overline{\lim_{n\to\infty}} P\Big\{\sigma_{nk} < T_1, |G_0^T(x,\lambda)| > \frac{1}{k}\Big\} = \\
&= \overline{\lim_{n\to\infty}} P\Big\{\sup_{t<T} |\zeta_t^n| \ge k,\ |G_0^T(x,\lambda)| > \frac{1}{k}\Big\} = \\
&= \overline{\lim_{n\to\infty}} P\Big\{|G_0^T(x^n,\lambda)| \le \frac{1}{k},\ |G_0^T(x,\lambda)| > \frac{1}{k}\Big\} = 0.
\end{aligned}$$

Now note that

$$|\eta_{s_1,t_2}^n \exp\{-a\Delta^1 A_{s_1 t_2}^{dc,n}\}| \le 1 + |\Delta^1 \zeta_{s_1 t_2}^n| \, |(\zeta_{s_1 t_2}^n)^{-+}|^{-1} \le 1 + |\Delta^1 \zeta_{s_1 t_2}^n|.$$

So

$$\begin{aligned}
P\Big\{\tau_{nk} < T_2,\ |G_0^t(x,\lambda)| > \frac{1}{k}\Big\} &\le P\Big\{|G_0^T(x^n,\lambda)| \le \frac{1}{k},\ |G_0^T(x,\lambda)| > \frac{1}{k}\Big\} + \\
&+ P\Big\{\sup_{s\le T} |\zeta_s^n| \ge k,\ |G_0^T(x,\lambda)| > \frac{1}{k}\Big\} \to 0, \quad n \to \infty,
\end{aligned}$$

and therefore 8) holds.

We now introduce the following notations: $x^n \xrightarrow{d} x$ (G-stable) means that for any finite subset $\{t^1, \ldots, t^m\} \subset S$, any bounded and continuous on R^m function $h\colon R^m \to C$ and any bounded G-measurable random variable ξ

$$E\xi h(x^n_{t^1}, \ldots, x^n_{t^m}) \to E\xi h(x_{t^1}, \ldots, x_{t^m}), \qquad n \to \infty.$$

Theorem 3. *Under the following conditions*

(C) 1) $G \subset \bigcap_{n \geq 1} \mathfrak{F}^n_0$;

2) $G^t_0(x^n, \lambda) \xrightarrow{P} G^t_0(x, \lambda), \qquad n \to \infty,\ t \in S,\ \lambda \in R$;

3) *the compensator A of x is a G-measurable function,*

$$x^n \xrightarrow{d} x, \qquad n \to \infty \quad (G\text{-stable}).$$

Proof. 1) To start with we shall prove that under condition (C)

$$E(\exp\{i\lambda x^n_t\}/\mathfrak{F}^n_0) \xrightarrow{P} G^t_0(x, \lambda), \qquad t \in S, \tag{26}$$

for $\lambda \in R_0$. Using lemma 3 we obtain (let $t \leq T$, $t \in S$, $T \in S$, $G^{nk}_0 = G^{(t_1 \wedge \sigma_{nk}, t_2 \wedge \tau_{nk})}_0$). Then for any $k \geq 1$

$$\begin{aligned}
&\overline{\lim_{n\to\infty}}\, E\,|E(\exp\{i\lambda x^n_t\}/\mathfrak{F}^n_0) - G^t_0(x,\lambda)|\, I\Big\{|G^T_0(x,\lambda)| > \frac{1}{k}\Big\} \leq \\
&\quad \leq \overline{\lim_{n\to\infty}}\, E\,|E(\exp\{i\lambda x^n_t\} - G^t_0(x,\lambda)\zeta^{nk}_t/\mathfrak{F}^n_0)|\, I\Big\{|G^T_0(x,\lambda)| > \frac{1}{k}\Big\} \leq \\
&\quad \leq \overline{\lim_{n\to\infty}}\, E\,|\exp\{i\lambda x^n_t\} - G^t_0(x,\lambda)\zeta^{nk}_t|\, I\Big\{|G^T_0(x,\lambda)| > \frac{1}{k}\Big\} \leq \\
&\quad \leq \overline{\lim_{n\to\infty}}\, E\,|1 - G^{nk}_0(x,\lambda)(G^{nk}_0(x^n,\lambda))^{-1}|\, I\Big\{|G^T_0(x,\lambda)| > \frac{1}{k}\Big\} \leq \\
&\quad \leq c_k \overline{\lim_{n\to\infty}}\, E\,|G^t_0(x^n,\lambda) - G^t_0(x,\lambda)|\, I\Big\{\sigma_{nk} \geq T_1,\ \tau_{nk} \geq T_2,\ |G^T_0(x,\lambda)| > \frac{1}{k}\Big\} + \\
&\qquad + 2c_k \overline{\lim_{n\to\infty}} \Big(P\Big\{\sigma_{nk} < T_1,\ |G^T_0(x,\lambda)| > \frac{1}{k}\Big\} + \\
&\qquad\qquad + P\Big\{\tau_{nk} < T_2,\ |G^T_0(x,\lambda)| > \frac{1}{k}\Big\}\Big) = 0.
\end{aligned} \tag{27}$$

Now,

$$\begin{aligned}
&\lim_{k\to\infty} \overline{\lim_{n\to\infty}}\, E\,|E(\exp\{i\lambda x^n_t\}/\mathfrak{F}^n_0) - G^t_0(x,\lambda)|\, I\Big\{|G^T_0(x,\lambda)| \leq \frac{1}{k}\Big\} \leq \\
&\qquad \leq 2 \lim_{k\to\infty} P\Big\{|G^T_0(x,\lambda)| \leq \frac{1}{k}\Big\} = 0.
\end{aligned} \tag{28}$$

Statement (26) follows from (27) and (28).

2) It follows from (26) that for $\lambda = \pi/2$

$$E(\exp\{i\lambda x^n_t\}/\mathfrak{F}^n_0) \to E(\exp\{i\lambda x_t\}/G),$$

i.e.,

$$P\{x_t^n = 2k+1,\ k \in Z/\mathfrak{F}_0^n\} \to P\{x_t = 2k+1,\ k \in Z/G\}.$$

Therefore for $\lambda = \pi + 2k\pi$, $k \in Z$

$$E(\exp\{i\lambda x_t^n\}/\mathfrak{F}_0^n) = P\{x_t^n = 2k,\ k \in Z/\mathfrak{F}_0^n\} - P\{x_t^n = 2k+1,\ k \in Z/\mathfrak{F}_0^n\} \to$$
$$\to P\{x_t = 2k,\ k \in Z/G\} - P\{x_t = 2k+1,\ k \in Z/G\} = G_0^t(x,\lambda).$$

It means that (26) holds for any $\lambda \in R$.

3) Now let $t_{jl} = (t_1^j, t_2^l)$, $1 \le j \le J$, $1 \le l \le L\}$ be a collection of the points from S, and $\{\lambda_{jl}, 1 \le j \le J, 1 \le l \le L\}$ be a collection of real numbers. We shall now prove the relation

$$E\xi \exp\Big\{i\sum_{j=1}^{J}\sum_{l=1}^{L}\lambda_{jl}x_{t_{jl}}^n\Big\} \to E\xi \exp\Big\{i\sum_{j=1}^{J}\sum_{l=1}^{L}\lambda_{jl}x_{t_{jl}}\Big\}, \tag{29}$$

ξ is G-measurable , $|\xi| \le C$.

We introduce for this purpose the following notations

$$x_{jl} = x_{t_{jl}}, \quad \Box_{jl} = \Box_{t_{jl}}x_{t_{j+1l+1}} = x]t_{jl}, t_{j+1l+1}],$$
$$G_{jl}(x,\lambda) = G_{t_{jl}}^{t_{j+1l+1}}(x,\lambda), \quad \mathfrak{F}_{jl}^{*,n} = \mathfrak{F}_{t_{jl}}^{*,n},$$
$$\Box_{j0}x = x_{j+1,1} - x_{j1}, \quad \Box_{0l}x = x_{1l+1} - x_{1l},$$
$$\Box_{00}x = x_{11}, \quad \Lambda_{jl} = \sum_{j' \le j, l' \le l}\lambda_{j'l'}$$

and note that the relation

$$E(\exp\{i\lambda\,\Box_{jl}x^n\}/\mathfrak{F}_{jl}^{*,n}) \xrightarrow{P} G_{jl}(x,\lambda), \qquad n \to \infty \tag{30}$$

can be proved similarly to (26).

Now it is evident that

$$\sum_{j=1}^{J}\sum_{l=1}^{L}x_{jl}\lambda_{jl} = \sum_{j=0}^{J-1}\sum_{l=0}^{L-1}\Box x_{jl}\Lambda_{j+1l+1},$$

whence

$$E\xi \exp\Big\{i\sum_{j=1}^{J}\sum_{l=1}^{L}x_{jl}\lambda_{jl}\Big\} = E\xi\prod_{j=0}^{J-1}\prod_{l=0}^{L-1}G_{jl}(x,\Lambda_{j+1l+1}). \tag{31}$$

We renumber the points t_{jl} in such a way that they became linearly ordered, namely

$$\tau_1 = t_{11}, \tau_2 = t_{12}, \ldots, \tau_L = t_{1L},$$
$$\tau_{L+1} = t_{21}, \tau_{L+2} = t_{22}, \ldots, \tau_{2L} = t_{2L},$$
$$\tau_{2L+1} = t_{31}, \ldots, \tau_{JL} = t_{JL}.$$

Renumber in the same way Λ_{jl}, $G_{jl}(x,\lambda)$, $\mathfrak{F}^*_{jl}$, $\square_{jl}x$. Then using (30) and (31), we obtain

$$\lim_{n\to\infty}\left|E\xi\left(\exp\left\{i\sum_{j=1}^{J}\sum_{l=1}^{L}x^n_{jl}\lambda_{jl}\right\}-\exp\left\{i\sum_{j=1}^{J}\sum_{l=1}^{L}x_{jl}\lambda_{jl}\right\}\right)\right|$$
$$=\lim_{n\to\infty}\left|\sum_{m=0}^{JL}E\xi\exp\left\{i\sum_{m'=0}^{m-1}\square_{m'}x^n\Lambda_{m'}\right\}\times\right.$$
$$\left.\times\prod_{m'=m+1}^{JL}G_{m'}(x,\Lambda_{m'})\Big(E(\exp\{i\Lambda_m\square_m x^n\}/\mathfrak{F}^*_m)-G_m(x,\Lambda_m)\Big)\right|$$
$$\leq(JL+1)\lim_{n\to\infty}\sum_{m=0}^{JL}E\left|E(\exp\{i\Lambda_m\square_m x^n\}/\mathfrak{F}^*_m)-G_m(x,\Lambda_m)\right|=0,$$

i.e., equality (29) holds and the proof therefore follows.

The next result follows from Theorems 2 and 3.

Theorem 4. *Let the conditions of Theorem 2 be satisfied. Then* $x^n\xrightarrow{d}x$, $n\to\infty$ *(G-stable).*

2. Weak compactness and weak convergence of point fields

Consider the field $M=x-A\in s\mathfrak{M}_{\text{loc}}$ from formula (1). As far as the field M has bounded jumps, it follows from Theorem 3 (Gushchin, Mishura (1992)) that $M\in s\mathfrak{M}^2_{\text{loc}}$. Therefore it has characteristics, i.e., such a unique non-decreasing weakly predictable field $\langle M\rangle$ that $M^2-\langle M\rangle\in\mathfrak{M}_{\text{loc}}$.

Theorem 5. *The characteristics* $\langle M\rangle$ *equals*

$$\langle M\rangle_t=A_t-\sum_{0<s\leq t}(\square A_s)^2.$$

Proof. We apply the generalized Ito formula from Mishura (1990) to the field M_t^2:

$$M_t^2-2\int_{[0,t]}M_s^{--}\,dM_s+2\int_{[0,t]}d_1M_s\,d_2M_s+$$
$$+\sum_{0<s_1\leq t_1}(\Delta^1M_{s_1t_2})^2+\sum_{0<s_2\leq t_2}(\Delta^2M_{t_1s_2})^2-\sum_{0<s\leq t}(\square M_s)^2.$$

Now

$$\sum_{0<s_1\le t_1}(\Delta^1 M_{s_1t_2})^2=\sum_{0<s_1\le t_1}(\Delta^1 x_{s_1t_2})^2+\sum_{0<s_1\le t_1}(\Delta^1 A_{s_1t_2})^2-2\sum_{0<s_1\le t_1}\Delta^1 x_{s_1t_2}\Delta^1 A_{s_1t_2},$$

$$\sum_{0<s_1\le t_1}(\Delta^1 x_{s_1t_2})^2=\sum_{0<s_1\le t_1}\Big(\sum_{0<s_2\le t_2}\square x_s\Big)^2=\sum_{0<s\le t}\square x_s+\sum_{0<s\le t}\square x_s\Delta^1 x_{s-}=$$
$$=x_t+2\int_{[0,t]}\Delta^1 x_{s-}\,dx_s=M_t+A_t+2\int_{[0,t]}\Delta^1 A_{s-}\,dA_s+$$
$$+2\int_{[0,t]}\Delta^1 M_{s-}\,dA_s+2\int_{[0,t]}\Delta^1 A_{s-}\,dM_s+2\int_{[0,t]}\Delta^1 M_{s-}\,dM_s.$$

We use the generalized one-parameter Ito formula to calculate the sum

$$\sum_{0<s_1\le t_1}(\Delta^1 A_{s_1t_2})^2=2\sum_{0<s_1\le t_1}\int_0^{t_2}\Delta^1 A_{s-}\,d_2(\Delta^1 A)_s+$$
$$+\sum_{0<s\le t}((\Delta^1 A_s)^2-(\Delta^1 A_{s-})^2-2\Delta^1 A_{s-}\square A_s)=$$
$$=2\int_{[0,t]}\Delta^1 A_{s-}\,dA_s+\sum_{0<s\le t}(\square A_s)^2,$$

similarly,

$$\sum_{0<s_1\le t_1}\Delta^1 x_{s_1t_2}\Delta^1 A_{s_1t_2}=$$
$$=\sum_{0<s_1\le t_1}\int_0^{t_2}\Delta^1 x_{s-}\,d_2(\Delta^1 A)_s+\sum_{0<s_1\le t_1}\int_0^{t_2}\Delta^1 A_{s-}\,d_2(\Delta^1 x)_s+\sum_{0<s\le t}\square A_s\square x_s=$$
$$=\int_{[0,t]}\Delta^1 A_{s-}\,dA_s+\int_{[0,t]}\Delta^1 M_{s-}\,dA_s+\int_{[0,t]}\Delta^1 A_{s-}\,dA_s+\int_{[0,t]}\Delta^1 A_{s-}\,dM_s+$$
$$+\sum_{0<s\le t}(\square A_s)^2+\sum_{0<s\le t}\square A_s\square M_s.$$

It follows from these equalities that

$$\sum_{0<s\le t_1}(\Delta^1 M_{s_1t_2})^2=A_t-\sum_{0<s\le t}(\square A_s)^2+M_t+2\int_{[0,t]}(\Delta^1 M_{s-}-\square A_s)\,dM_s.$$

Similarly,

$$\sum_{0<s_2\le t_2}(\Delta^2 M_{t_1s_2})^2=A_t-\sum_{0<s\le t}(\square A_s)^2+M_t+\int_{[0,t]}(\Delta^2 M_{s-}-\square A_s)\,dM_s.$$

Further,

$$\sum_{0<s\le t}(\square M_s)^2=\sum_{0<s\le t}(\square x_s)^2+\sum_{0<s\le t}(\square A_s)^2-2\sum_{0<s\le t}\square x_s\square A_s=$$
$$=A_t+M_t-\sum_{0<s\le t}(\square A_s)^2-2\int_{[0,t]}\square A_s\,dM_s.$$

It means that

$$M_t^2 = A_t - \sum_{0<s\leq t} (\square A_s)^2 + M_t + 2\int_{[0,t]} (\Delta^1 M_{s-} + \Delta^2 M_{s-} - \square A_s)\, dM_s + \\ + 2\int_{[0,t]} M_s^{--}\, dM_s + \int_{[0,t]} d_1 M_s\, d_2 M_s.$$

It is evident that the fields

$$\Big\{M_t,\quad \int_{[0,t]} (\Delta^1 M_{s-} + \Delta^2 M_{s-} - \square A_s)\, dM_s,\quad \int_{[0,t]} M_s^{--}\, dM_s,\quad \text{and}\quad \int_{[0,t]} d_1 M_s\, d_2 M_s\Big\}$$

belong to $\mathfrak{M}_{\text{loc}}$. It means that the difference

$$M_t^2 - \Big(A_t - \sum_{0<s\leq t} (\square A_s)^2\Big)$$

belongs to $\mathfrak{M}_{\text{loc}}$, and therefore the proof follows.

The following two lemmas are proved similarly to lemma 3 from Mishura (1987).

Lemma 4. *Let the field $M \in s\mathfrak{M}^2_{loc}$, $|\square M| \leq C$. Then for any $\alpha > 0$ and $\beta > 0$, $t \in R^2_+$*

$$P\{\sup_{0<s\leq t} |M_s| > \alpha\} \leq P\{\langle M\rangle_t > \beta\} + 16\alpha^{-2} E(\langle M\rangle_t \wedge (2\beta + C^2)).$$

Lemma 5. *Let τ be a $\mathfrak{F}_{t_1 T_2}$-stopping time, $M \in s\mathfrak{M}^2_{loc}$ and $|\square M| \leq C$. Then for any $\alpha, \beta, \delta > 0$*

$$P\Big\{\sup_{\substack{\tau_1 \leq t_1 \leq \tau_1+\delta \\ 0<t_2\leq T_2}} |M_{(\tau_1+t_1,t_2)} - M_{(\tau_1,t_2)}| > \alpha\Big\} \leq \\ \leq 2^{10} P\Big\{\langle M\rangle_{(\tau_1+\delta,T_2)} - \langle M\rangle_{(\tau_1,t_2)} > \beta\Big\} + \\ + 16\alpha^{-2} E\Big((\langle M\rangle_{(\tau_1+\delta,T_2)} - \langle M\rangle_{(\tau_1,T_2)}) \wedge (2\beta + C^2)\Big).$$

Lemma 5 can be formulated "in a symmetric way" for $\mathfrak{F}_{T_1 t_2}$-stopping times.

Now we shall formulate the result that concerns the weak compactness of local strong martingales.

Theorem 6. *Let $\{M^n, n \geq 1\} \subset s\mathfrak{M}^2_{loc}$, $|\square M^n| \leq C$ and the following conditions be fulfilled:*

9) $\langle M^n\rangle_t \xrightarrow{P} B_t$, $t \in S_B$, $n \to \infty$, *B_t is non-decreasing field,*

$$S_B = \Big\{t \in R^2_+ : P\{B \text{ is continuous in the point } t\} = 1\Big\}.$$

10) *the sequence of measures corresponding to the fields* $\langle M^n\rangle$, $n \geq 1$ *is weakly compact in the space* (D_T, J_T) *(Mishura (1979)) for any* $T \in S_B$.

Then the sequence of measures corresponding to the fields M^n, $n \geq 1$ *is weakly compact in* (D_T, J_T) *for any* $T \in S_B$.

Proof. Let $T \in S_B$. Then it follows from condition 9) that

$$\lim_{C\to\infty} \overline{\lim_{n\to\infty}} P\{\langle M^n\rangle_T > C\} = 0$$

and it follows from lemma 4 that

$$\lim_{C\to\infty} \overline{\lim_{n\to\infty}} P\{\sup_{s\leq T} |M_s^n| > C\} = 0. \tag{32}$$

Further, it follows from 10) that the measures corresponding to the process $\{\langle M^n\rangle_{t_1T_2}$, $t_1 \leq T_1$, $n \geq 1\}$ are weakly compact. Then, from the criterion of weak compactness (Rebolledo (1979)) we obtain for any $\varepsilon > 0$ and any $\mathfrak{F}_{t_1T_2}$-stopping times τ_1

$$\lim_{\delta\downarrow 0} \overline{\lim_{n\to\infty}} P\Big\{\sup_{\tau_1\leq t_1\leq\tau_1+\delta} (\langle M^n\rangle_{(\tau_1+t_1,T_2)} - \langle M^n\rangle_{(\tau_1,T_2)}) > \varepsilon\Big\} = 0. \tag{33}$$

In turn, it follows from (33) and lemma 5 that

$$\lim_{\delta\downarrow 0} \overline{\lim_{n\to\infty}} P\Big\{\sup_{\substack{\tau_1\leq t_1\leq\tau_1+\delta\\ 0\leq t_2\leq T_2}} |M^n_{(t_1+\tau_1,t_2)} - M^n_{(\tau_1,t_2)}| > \varepsilon\Big\} = 0. \tag{34}$$

and weak compactness of measures corresponding to $\{M^n,\ n \geq 1\}$ in (D_T, J_T) follows from (32), (34) and lemma 1 Mishura (1987).

We now formulate sufficient conditions for the convergence of non-random, non-decreasing functions in (D_T, J_T) (we denote this convergence as $\xrightarrow{D_T,J_T}$).

Lemma 6. *Under the following conditions*

11) $A_t^n \to A_t$, $n \to \infty$, $t \in S_A \cap [0,T]$, $T \in S_A$;

12) *a)* $\sum_{0<s_1\leq t_1} (\Delta^1 A^n_{s_1T_2})^2 \to \sum_{0<s_1\leq t_1} (\Delta^1 A_{s_1T_2})^2$, $n \to \infty$, $(t_1, T_2) \in S_A \cap [0,T]$;

b) $\sum_{0<s_2\leq t_2} (\Delta^2 A^n_{(T_1t_2)})^2 \to \sum_{0<s_2\leq t_2} (\Delta^2 A_{(T_1t_2)})^2$, $n \to \infty$, $(T_1, t_2) \in S_A \cap [0,T]$

where S_A *is the set of continuity of* A,

$$A^n \xrightarrow{D_T,J_T} A, \qquad n \to \infty.$$

Proof. If conditions 11) and 12)a are fulfilled then it follows from Jacod, Memin (1980) that the sequence of one parameter functions

$$\{A^n_{t_1T_2},\ t_1 \leq T_1,\ n \geq 1\}$$

converges in J-topology on the segment $[0, T_1]$. It means that

$$\lim_{\delta\downarrow 0}\overline{\lim_{n\to\infty}}\,\Delta^1(A^n_{\cdot T_2},\delta)=0$$

where

$$\Delta^1(A^n_{\cdot T_2},\delta)=\sup_{\delta\le u_1\le T_1-\delta}\min(A^n_{(u_1+\delta,T_2)}-A^n_{(u_1,T_2)},A^n_{(u_1,T_2)}-A^n_{(u_1-\delta,T_2)})+ \\ +(A^n_T-A^n_{(T_1-\delta,T_2)}+A^n_{(\delta,T_2)}).$$

Note that the functions A^n are non-decreasing, so

$$\Delta^1(A^n_{\cdot T_2},\delta)=\gamma^1(A^n,\delta):=\sup_{\substack{\delta\le u_1\le T_1-\delta\\ 0<t_2\le T_2}}\min(A^n_{(u_1+\delta,t_2)}-A^n_{(u_1,t_2)},\ A^n_{(u_1,t_2)}-A^n_{(u_1-\delta,t_2)})+ \\ +\sup_{0\le t_2\le T_2}(A^n_{(T_1,t_2)}-A^n_{(T_1-\delta,t_2)}+A^n_{(\delta,t_2)}).$$

Therefore

$$\lim_{\delta\downarrow 0}\overline{\lim_{n\to\infty}}\,\gamma^1(A^n,\delta)=0. \tag{35}$$

Similarly, it follows from conditions 11) and 12)b that

$$\lim_{\delta\downarrow 0}\overline{\lim_{n\to\infty}}\,\gamma^2(A^n,\delta)=0 \tag{36}$$

where $\gamma^2(A^n,\delta)$ can be obtained from $\gamma^1(A^n,\delta)$ by a symmetric change of coordinates. According to Theorem 2 Mishura (1987), equalities (35) and (36) satisfy the convergence $A^n \xrightarrow{D_T,J_T} A$, $n\to\infty$.

Let us now formulate a stochastic analogy of lemma 6.

Lemma 7. *Let $\{A^n_t, A_t,\ t\in R^2_+, n\ge 1\}$ be non-decreasing fields, and let the following conditions be satisfied:*

13) $A^n_t \xrightarrow{P} A_t$, $n\to\infty$, $t\in S_A\cap[0,T]$, $t\in S_A$;

14) $\sum\limits_{0<s_1\le t_1}(\Delta^1 A^n_{s_1t_2})^2 \xrightarrow{P} \sum\limits_{0<s_1\le t_1}(\Delta^1 A_{s_1t_2})^2$, $\sum\limits_{0<s_2\le t_2}(\Delta^2 A^n_{t_1s_2})^2 \xrightarrow{P} \sum\limits_{0<s_2\le t_2}(\Delta^2 A_{t_1s_2})^2$, $n\to\infty$, $t\in S_A\cap[0,T]$, $T\in S_A$.

Then

$$A_n \xrightarrow{D_T,J_T} A, \qquad n\to\infty.$$

Lemma 8. *Let $\{A^n_t,\ n\ge 1,\ A_t,\ t\in R^2_+\}$ be non-decreasing fields, and let the following conditions be satisfied:*

15) $A^n_t \xrightarrow{P} A_t$, $t\in S_A\cap[0,T]$, $t\in S_A$, $n\to\infty$;

16) $\sum\limits_{0<s\le t}(\square A^n_s)^2 \xrightarrow{P} \sum\limits_{0<s\le t}(\square A_s)^2$, $\int_{[0,T]}\Delta^i A^n_{s-}\,dA^n_s \xrightarrow{P} \int_{[0,T]}\Delta^i A_{s-}\,dA_s$, $n\to\infty$, $t\in S_A\cap[0,T]$, $i=1,2$, $T\in S_A$.

Then the sequence of measures that correspond to the fields

$$\left\{A_t^n, \quad \sum_{0<s\le t} (\Box A_s^n)^2, \quad A_t^n - \sum_{0<s\le t} (\Box A_s^n)^2, \quad n \ge 1\right\}$$

are weakly compact in (D_T, J_T).

Proof. Note that

$$\sum_{0<s_1\le t_1} (\Delta^1 A_{s_1 t_2}^n)^2 = \sum_{0<s\le t} (\Box A_s^n)^2 + \int_{[0,t]} \Delta^1 A_{s-}^n \, dA_s^n.$$

Therefore the weak convergence of measures corresponding to the fields $\{A_t^n, n \ge 1\}$ follows from lemma 7. Denote $B_t^n = \sum_{0<s\le t}(\Box A_s^n)^2$, $C_t^n = A_t^n - B_t^n$. Then it is easy to verify that

$$\gamma^i(B^n, \delta) \le A_T^n \cdot \gamma^i(A^n, \delta),$$
$$\gamma^i(C^n, \delta) \le (1 + A_T^n)\, \gamma^i(A^n, \delta),$$

and under conditions 15) and 16)

$$\lim_{\delta\downarrow 0} \overline{\lim_{n\to\infty}} P\{\gamma^i(E^n, \delta) > \varepsilon\} = 0, \qquad \varepsilon > 0,\ E = B, C,\ i = 1, 2,$$

and then taking into account Theorem 2 Mishura (1979) we obtain $E^n \xrightarrow{D_T, J_T} E$, $n \to \infty$, $E = B, C$, so consequently the measures that correspond to these fields are weakly compact.

The next statement follows from Theorems 3, 5, 6 and lemma 8.

Theorem 7. *Let* $\{x_t^n,\ \mathfrak{F}_t^n,\ n \ge 1,\ t \in R_+^2\}$ *be the sequence of point fields with compensators* $\{A_t^n, \mathfrak{F}_t^n, n \ge 1, t \in R_+^2\}$, *and* x *be the point field with compensator* A, *that is a* G*-measurable function. Under condition 1 of Theorem 3 and conditions 15)–16) of lemma 8 for any* $T \in S_A$, *the sequence of measures corresponding to fields* x^n *weakly converges to the measure that corresponds to field* x, *and in addition*

$$x^n \xrightarrow{d} x, \qquad n \to \infty \quad (G\text{-stable.})$$

Proof. Under conditions 15)–16) of lemma 8, and by virtue of Theorems 5 and 6, the sequence of measures corresponding to the fields $\{x^n, n \ge 1\}$ is weakly compact. Now the convergence

$$x^n \xrightarrow{d} x, \qquad n \to \infty \quad (G\text{-stable.})$$

is guaranteed by Theorem 3 to satisfy the weak convergence of finite-dimensional distributions of the fields $\{x^n, n \ge 1\}$ whence we obtain weak convergence of measures.

Remark 1. The results obtained generalize the corresponding functional limit theorem for point processes (Kabanov et al (1980), Liptser, Shiryayev (1986)) and also partial cases of the convergence of point fields that were considered in Ivanoff (1985), Brown et al (1986), Banys (1985) and Roitgartz (1991).

References

Banys, R. (1985). Convergence of random measures and point processes on the plane. *Prob. and Math. Stat.* **5**(2), 211–219.

Brown, T.C., Ivanoff, B.G. and Weber, N.C. (1986). Poisson convergence in two dimensions with application to row and column exchangeable arrays. *Stoch. Proc. and Appl.* **23**(2), 307–318.

Cairoli, R. and Walsh, J.B. (1975). Stochastic integrals in a plane. *Acta Mathem.* **134**, 1–2, 111-183.

Gushchin, A.A. (1982). On the general theory of random fields on a plane. *Russian Math. Surveys.* **37**, 53–74.

Gushchin, A.A. and Mishura, Yu.S. (1992). The Davis inequalities and the Gundy decomposition for two-parameter strong martingales. III. *Theor. Probability and Math. Stat.* **40**, 75–85. (In Russian).

Ivanoff, B.G. (1985). Poisson convergence for point processes on a plane. *J. Austral. Math. Soc.* **139**, 253–269.

Jacod, J. and Memin, J. (1980). *Un nouveau critere de compacite relative pour une suite de processus.* Preprint.

Kabanov, Yu.M., Liptser, R.Sh. and Shiryayev, A.N. (1980). Some limit theorems for simple point processes (a martingale approach). *Stochastics.* **3**(3), 203–216.

Liptser, R.Sh. and Shiryayev, A.N. (1986). *Theory of martingales.* Moscow, Nauka. (In Russian).

Mazziotto, G., Merzbach, E. and Szpirglas, F. (1981). Discontinuites des processus croissants et martingales a variation integrale. *Lect. Notes Math.* **863**, 59–83.

Mishura, Yu.S. (1979). On the convergence of random fields in the J-topology. *Theor. Probability and Math. Stat.* **17**, 111–119.

Mishura, Yu.S. (1987). Sufficient conditions for relative compactness of measures corresponding to two-parameter strong martingales. *Theor. Probability and Math. Stat.* **34**, 117–125.

Mishura, Yu.S. (1988). Exponential formulas and Dolean's equation for discontinuous two-parameter processes. *Theor. Probab. and Appl.* **33**(2), 412–417. (In Russian).

Mishura, Yu.S. (1990). Conditions for the existence of stochastic integrals with respect to weak semimartingales, and the generalized Ito formula for semimartingales on a plane. *Theor. Probability and Math. Stat.* **40**, 75–85.

Roitgartz, A. D. (1991). Limit theorems for point random fields defined on a plane. *Ukr. Math. Journ.* **43**(1), 118–123. (In Russian).

Exploring Stochastic Laws. pp. 329–341.
Eds. A.V. Skorokhod and Yu.V. Borovskikh

Interpolation of vector-valued stochastic processes

M.P. MOKLYACHUK

Department of Mechanics and Mathematics, Kiev University

Received January 10, 1995

Abstract. We consider the problem of the linear mean square optimal estimation for the functional

$$A_T\xi = \int_0^T \langle a(t), \xi(t) \rangle \, dt$$

of a Hilbert space valued stationary stochastic process $\xi(t)$, $t \in \mathbf{R}^1$ from observations of the process $\xi(t)+\eta(t)$ for $t \in \mathbf{R}^1 \setminus [0,T]$, where $\eta(t)$ is an uncorrelated with $\xi(t)$ Hilbert space valued stationary stochastic process. Formulas are derived for computing the mean square error and the spectral characteristic of the optimal estimate of the functional $A_T\xi$. The minimax spectral characteristic and the least favorable spectral densities are found for various classes of spectral densities.

1. Introduction

The classical Kolmogorov method of linear interpolation, extrapolation and filtering of weak stationary stochastic processes (see, for example, Kolmogorov (1986), and Rosanov (1967)) may be employed under the condition that spectral densities of stochastic processes are known. In practice, however, the problem of the estimation of unknown values of stochastic process arises when the spectral density is not known exactly. To solve the problem, the parametric or non-parametric estimate of the unknown spectral density is found. Then the classical method is applied provided that the estimate of the density is correct. This procedure can result in a significant increase in the value of the error as Vastola and Poor have demonstrated with the help of some examples, Vastola and Poor (1983). For this reason it is necessary to search for the estimate of the unknown value of the stochastic process that has the least value for the error for all densities from a certain class of spectral densities. Such an approach to the problem of interpolation, extrapolation and filtering of stationary stochastic processes has been taken into consideration by many investigators, such as Vastola and Poor (1983) – Moklyachuk (1993e). A survey of results in minimax (robust) methods of data processing can be found in the paper Kassam and Poor (1985). The paper Taniguchi (1981) was the first one where the minimax interpolation problem for the ε-pollution model was investigated. The relation of the minimax interpolation problem with the problem of robust hypothesis testing is shown in Kassam (1982). In papers Franke (1984) – Kassam (1982) the minimax interpolation problem is investigated with the help of the convex optimization methods. In papers Moklyachuk (1981) – Moklyachuk (1993e) the minimax interpolation, extrapolation and filtering problems are investigated for linear functionals of the unknown values of stationary stochastic processes. In this article it is shown that convex optimization methods can be applied to determine the least favorable spectral densities and minimax (robust) spectral characteristic of the optimal estimate of linear functional of a Hilbert space valued stationary stochastic process.

2. The classical method of linear interpolation

Denote by X a separable Hilbert space with an orthonormal basis $\{e_k\colon k = 1, 2, \ldots\}$. Stationary stochastic processes $\xi(t)$ and $\eta(t)$ with values in X have spectral densities $f(\lambda)$ and $g(\lambda)$ if the correlation functions $B_\xi(s)$ and $B_\eta(s)$ can be represented in the form

$$\langle B_\xi(s)\, e_k, e_j\rangle = \frac{1}{2\pi}\int_{-\infty}^{\infty} e^{i\lambda s}\langle f(\lambda)\, e_k, e_j\rangle\, d\lambda, \qquad k, j = 1, 2, \ldots,$$
$$\langle B_\eta(s)\, e_k, e_j\rangle = \frac{1}{2\pi}\int_{-\infty}^{\infty} e^{i\lambda s}\langle g(\lambda)\, e_k, e_j\rangle\, d\lambda, \qquad k, j = 1, 2, \ldots.$$

For almost all $\lambda \in \mathbf{R}^1$ the spectral densities $f(\lambda)$ and $g(\lambda)$ are nuclear operators with integrable nuclear norms (see Kallianpur and Mandrekar (1965)).

Denote by $\mathbf{K}(f+g)$ a set of all $k \in \mathbf{N}$ such that $f_k(\lambda)+g_k(\lambda)$, $f_k(\lambda) = \langle f(\lambda)e_k, e_k\rangle$, $g_k(\lambda) = \langle f(\lambda)e_k, e_k\rangle$ satisfy the minimality condition:

$$\int_{-\infty}^{\infty} \frac{|\gamma(\lambda)|}{f_k(\lambda) + g_k(\lambda)}\, d\lambda < \infty$$

for some nontrivial function of the exponential type $\gamma(\lambda) = \int_0^T \alpha(t)\, e^{it\lambda}\, dt$. We will consider the estimation problem in the case when $\mathbf{K}(f + g) \neq \emptyset$. This condition is necessary and sufficient in order that the mean square error of the optimal linear estimate of the unknown value of the process $\xi(t)$ is not equal to zero (Kolmogorov (1986), Rosanov (1967)).

Let the sequence $a(t)$ that determines the functional

$$A_T\xi = \int_0^T a(t)\, \xi(t)\, dt = \sum_{k=1}^{\infty} \int_0^T a_k(t)\, \xi_k(t)\, dt,$$

satisfy the condition

$$\sum_{k=1}^{\infty} \int_0^T |a_k(t)|\, dt < \infty, \qquad \sum_{k=1}^{\infty} \int_0^T t\, |a_k(t)|^2\, dt < \infty. \tag{1}$$

Under this condition $\mathbf{E}\,|A_T\xi|^2 < \infty$.

The mean square error of a linear estimate $\hat{A}_T\xi$ of the functional $A_T\xi$ is determined by the spectral characteristic $h(\lambda)$ of the estimate and spectral densities $f(\lambda)$ and $g(\lambda)$ of processes $\xi(t)$ and $\eta(t)$. The value of the mean square error is

$$\begin{aligned}\Delta(h;f,g) &= \mathbf{E}\left|A_T\xi - \hat{A}_T\xi\right|^2 \\ &= \sum_{k=1}^{\infty}\frac{1}{2\pi}\int_{-\infty}^{\infty}\left[\left|A_k(\lambda)-\hat{h}_k(\lambda)\right|^2 f_k(\lambda) + \left|\hat{h}_k(\lambda)\right|^2 g_k(\lambda)\right]d\lambda,\end{aligned}$$

where

$$A_k(\lambda) = \int_0^T a_k(t)\,e^{it\lambda}\,dt.$$

The spectral characteristic $h(\lambda) = \left\{\hat{h}_k(\lambda)\colon k = 1,2,\ldots\right\}$ of the estimate $\hat{A}_T\xi$ belongs to the subspace $\mathbf{L}_2^{-T}(f+g)$ of the space $\mathbf{L}_2(f+g)$ generated by the functions

$$\hat{h}_k(\lambda) = \int_{\mathbf{R}\setminus[0,T]} h_k(t)\,e^{it\lambda}\,dt,$$

$$\sum_{k=1}^{\infty}\int_{-\infty}^{\infty}\left|\hat{h}_k(\lambda)\right|^2 (f_k(\lambda)+g_k(\lambda))\,d\lambda < \infty.$$

The spectral characteristic $h(f,g)$ of the optimal linear estimate of the functional $A_T\xi$ minimizes the mean square error. With the help of the classical Kolmogorov method (Kolmogorov (1986), Rosanov (1967)) we can derive the following formulas for computing the value $\Delta(f,g)$ of the error and the spectral characteristic $h(f,g)$ of the optimal linear estimate of the functional $A_T\xi$ under the condition that spectral densities $f(\lambda)$, and $g(\lambda)$ of processes $\eta(t)$ and $\xi(t)$ are known. In this case

$$\begin{aligned}\Delta(f,g) &= \sum_{k\in\mathbf{K}}\left\{\frac{1}{2\pi}\int_{-\infty}^{\infty}\frac{|A_k(\lambda)g_k(\lambda)+C_k(\lambda)|^2}{(f_k(\lambda)+g_k(\lambda))^2}f_k(\lambda)\,d\lambda\right. \\ &\quad\left.+\frac{1}{2\pi}\int_{-\infty}^{\infty}\frac{|A_k(\lambda)f_k(\lambda)-C_k(\lambda)|^2}{(f_k(\lambda)+g_k(\lambda))^2}g_k(\lambda)\,d\lambda\right\} \\ &= \sum_{k\in\mathbf{K}}\left[\int_0^T(\mathbf{B}_{kT}\,\mathbf{c}_k)(t)\,\bar{c}_k(t)\,dt + \int_0^T(\mathbf{R}_{kT}\,\mathbf{a}_k)(t)\,\bar{a}_k(t)\,dt\right] \\ &= \sum_{k\in\mathbf{K}}\left[\langle\mathbf{B}_{kT}\,\mathbf{c}_k,\mathbf{c}_k\rangle + \langle\mathbf{R}_{kT}\,\mathbf{a}_k,\mathbf{a}_k\rangle\right], \qquad (2)\end{aligned}$$

$$\begin{aligned}h_k(f,g) &= \frac{A_k(\lambda)\,f_k(\lambda)-C_k(\lambda)}{f_k(\lambda)+g_k(\lambda)} \\ &= A_k(\lambda) - \frac{A_k(\lambda)\,g_k(\lambda)+C_k(\lambda)}{f_k(\lambda)+g_k(\lambda)}, \qquad (3)\end{aligned}$$

where

$$C_k(\lambda) = \int_0^T c_k(t)\,e^{it\lambda}\,dt, \qquad c_k(t) = \left(\mathbf{B}_{kT}^{-1}\,\mathbf{D}_{kT}\,\mathbf{a}_k\right)(t),\ k\in\mathbf{K}(f+g),$$

$\langle \mathbf{a}, \mathbf{c} \rangle$ is the inner product and $\mathbf{B}_{kT}$, $\mathbf{D}_{kT}$, $\mathbf{R}_{kT}$ are operators in the space $L_2([0,T])$ that are determined by the relations

$$\left(\mathbf{B}_{kT}\,\mathbf{c}\right)(t) = \frac{1}{2\pi}\int_0^T c(u)\int_{-\infty}^{\infty} e^{i(u-t)\lambda}(f_k(\lambda)+g_k(\lambda))^{-1}\,d\lambda\,du,$$
$$\left(\mathbf{D}_{kT}\,\mathbf{c}\right)(t) = \frac{1}{2\pi}\int_0^T c(u)\int_{-\infty}^{\infty} e^{i(u-t)\lambda}f_k(\lambda)\,(f_k(\lambda)+g_k(\lambda))^{-1}\,d\lambda\,du,$$
$$\left(\mathbf{R}_{kT}\,\mathbf{c}\right)(t) = \frac{1}{2\pi}\int_0^T c(u)\int_{-\infty}^{\infty} e^{i(u-t)\lambda}f_k(\lambda)\,g_k(\lambda)\,(f_k(\lambda)+g_k(\lambda))^{-1}\,d\lambda\,du,$$

where $0 \leq t \leq T$.

Lemma 1. *Let $\xi(t)$ and $\eta(t)$ be uncorrelated stationary stochastic processes with spectral densities $f(\lambda)$ and $g(\lambda)$ such that $\mathbf{K}(f+g)\neq\emptyset$ and condition* (1) *be satisfied. The mean square error $\Delta(f,g)$ and the spectral characteristic $h(f,g)$ of the optimal linear estimate of the functional $A_T\xi$ from observations of the process $\xi(t)+\eta(t)$ for $t\in\mathbf{R}\setminus[0,T]$ can be computed by* (2) *and* (3).

Lemma 2. *Let $\xi(t)$ be a stationary stochastic process with spectral density $f(\lambda)$ such that $\mathbf{K}(f)\neq\emptyset$ and condition* (1) *be satisfied. The mean square error $\Delta(f)$ and the spectral characteristic $h(f)$ of the optimal linear estimate of the functional $A_T\xi$ from observations of the process $\xi(t)$ for $t\in\mathbf{R}\setminus[0,T]$ can be computed by the formulas*

$$\Delta(f) = \sum_{k\in\mathbf{K}}\frac{1}{2\pi}\int_{-\infty}^{\infty}|C_k(\lambda)|^2 f_k^{-1}(\lambda)\,d\lambda = \sum_{k\in\mathbf{K}}\left\langle \mathbf{B}_{kT}^{-1}\,\mathbf{a}_k,\,\mathbf{a}_k\right\rangle, \tag{4}$$
$$h_k(f) = A_k(\lambda) - C_k(\lambda)\,f_k^{-1}(\lambda), \tag{5}$$

where

$$C_k(\lambda) = \int_0^T \left(\mathbf{B}_{kT}^{-1}\,\mathbf{a}_k\right)(t)\,e^{it\lambda}\,dt,$$
$$\left(\mathbf{B}_{kT}\,\mathbf{c}\right)(t) = \frac{1}{2\pi}\int_0^T c(u)\int_{-\infty}^{\infty} e^{i(u-t)\lambda}f_k(\lambda)^{-1}\,d\lambda\,du.$$

3. Minimax (robust) approach to interpolation problem

Formulas (1)–(5) can be applied to compute the mean square error and the spectral characteristic of the optimal linear estimate of the functional $A_T\xi$ if the spectral densities $f(\lambda)$ and $g(\lambda)$ of processes $\eta(t)$ and $\xi(t)$ are known. In the case when the spectral densities are not known exactly, but sets $\mathcal{D}_f$ and $\mathcal{D}_g$ of possible spectral densities are given, we can apply the minimax (robust) approach to the problem of the estimation of the unknown values of stochastic processes and their transformations Vastola and Poor (1983) – Moklyachuk (1993e). With the help of this approach we can find an estimate that minimizes the mean square error for all spectral densities $f(\lambda)$, $g(\lambda)$ from a given class $\mathcal{D} = \mathcal{D}_f \times \mathcal{D}_g$ simultaneously.

Definition 1. *For a given class of spectral densities $\mathcal{D} = \mathcal{D}_f \times \mathcal{D}_g$ the spectral densities $f^0(\lambda) \in \mathcal{D}_f$ and $g^0(\lambda) \in \mathcal{D}_g$ are called the least favorable in $\mathcal{D} = \mathcal{D}_f \times \mathcal{D}_g$ for the optimal linear estimation of the functional $A_T\xi$ if*

$$\Delta(f^0, g^0) = \Delta\Big(h(f^0, g^0); f^0, g^0\Big) = \max_{(f,g)\in\mathcal{D}_f\times\mathcal{D}_g} \Delta\Big(h(f,g); f, g\Big).$$

Definition 2. *For a given class of spectral densities $\mathcal{D} = \mathcal{D}_f \times \mathcal{D}_g$ the spectral characteristic $h^0\lambda$ is called minimax (robust) if*

$$h^0\lambda \in H_{\mathcal{D}} = \bigcap_{(f,g)\in\mathcal{D}_f\times\mathcal{D}_g} L_2^{-T}(f+g),$$

$$\min_{h\in H_{\mathcal{D}}} \max_{(f,g)\in\mathcal{D}_f\times\mathcal{D}_g} \Delta\Big(h; f, g\Big) = \max_{(f,g)\in\mathcal{D}_f\times\mathcal{D}_g} \Delta\Big(h^0; f, g\Big).$$

Detailed analysis of relations (2)–(5) make it possible to conclude that the following statements are true.

Lemma 3. *Spectral densities $f^0(\lambda)$ and $g^0(\lambda)$ are the least favorable in the class $\mathcal{D} = \mathcal{D}_f \times \mathcal{D}_g$ for the optimal linear estimation of the functional $A_T\xi$ if $\mathbf{K}(f^0, g^0) \neq \emptyset$ and the Fourier transformation of the functions*

$$\Big(f_k^0(\lambda) + g_k^0(\lambda)\Big)^{-1}, \quad f_k^0(\lambda)\Big(f_k^0(\lambda) + g_k^0(\lambda)\Big)^{-1}, \quad f_k^0(\lambda)\, g_k^0(\lambda)\Big(f_k^0(\lambda) + g_k^0(\lambda)\Big)^{-1}$$

form the operators $\mathbf{B}_{kT}^0$, $\mathbf{D}_{kT}^0$, $\mathbf{R}_{kT}^0$ that determine a solution of the following conditional extremum problem

$$\max_{(f,g)\in\mathcal{D}_f\times\mathcal{D}_g} \sum_{k\in\mathbf{K}} \Big[\Big\langle \mathbf{D}_{kT}\, \mathbf{a}_k, \mathbf{B}_{kT}^{-1}\mathbf{D}_{kT}\, \mathbf{a}_k\Big\rangle + \Big\langle \mathbf{R}_{kT}\, \mathbf{a}_k, \mathbf{a}_k\Big\rangle\Big] =$$
$$= \sum_{k\in\mathbf{K}} \Big[\Big\langle \mathbf{D}_{kT}^0\, \mathbf{a}_k, (\mathbf{B}_{kT}^0)^{-1}\, \mathbf{D}_{kT}^0\, \mathbf{a}_k\Big\rangle + \Big\langle \mathbf{R}_{kT}^0\, \mathbf{a}_k, \mathbf{a}_k\Big\rangle\Big]. \qquad (6)$$

The minimax (robust) spectral characteristic $h^0 = h(f^0, g^0)$ can be computed by formula (3) if $h(f^0, g^0) \in H_{\mathcal{D}}$.

Lemma 4. *A spectral density $f^0(\lambda) \in \mathcal{D}_f$ is the least favorable in the class $\mathcal{D}_f$ for the optimal linear estimation of the functional $A_T\xi$ from observations of the process $\xi(t)$ for $t \in \mathbf{R} \setminus [0, T]$ if $\mathbf{K}(f^0) \neq \emptyset$ and the Fourier transformation of the functions $(f_k^0(\lambda))^{-1}$, $k = 1, 2, \ldots$ form the operators $\mathbf{B}_{kT}^0$, $k = 1, 2, \ldots$ that determine a solution of the conditional extremum problem*

$$\max_{f\in\mathcal{D}_f} \sum_{k\in\mathbf{K}} \Big\langle \mathbf{B}_{kT}^{-1}\, \mathbf{a}_k, \mathbf{a}_k\Big\rangle = \sum_{k\in\mathbf{K}} \Big\langle (\mathbf{B}_{kT}^0)^{-1}\, \mathbf{a}_k, \mathbf{a}_k\Big\rangle. \qquad (7)$$

The minimax (robust) spectral characteristic $h^0 = h(f^0)$ can be computed by formula (5) *if $h(f^0) \in H_{\mathcal{D}_f}$.*

The least favorable spectral densities $f^0(\lambda)$ and $g^0(\lambda)$ and the minimax (robust) spectral characteristic $h^0 = h(f^0, g^0)$ form a saddle point of the function $\Delta(h; f, g)$ on the set $H_{\mathcal{D}} \times \mathcal{D}$. The saddle point inequalities

$$\Delta\left(h^0; f, g\right) \leq \Delta\left(h^0; f^0, g^0\right) \leq \Delta\left(h; f^0, g^0\right), \qquad \forall h \in H_{\mathcal{D}},\ \forall f \in \mathcal{D}_f,\ \forall g \in \mathcal{D}_g$$

hold when $h^0 = h(f^0, g^0)$ and $h(f^0, g^0) \in H_{\mathcal{D}}$, and (f^0, g^0) is a solution of the conditional extremum problem

$$\sup_{(f,g)\in\mathcal{D}_f\times\mathcal{D}_g} \Delta\left(h(f^0, g^0); f, g\right) = \Delta\left(h(f^0, g^0); f^0, g^0\right), \tag{8}$$

where

$$\begin{aligned} \Delta\left(h(f^0, g^0); f, g\right) &= \sum_{k\in\mathbf{K}} \left\{ \frac{1}{2\pi} \int_{-\infty}^{\infty} \frac{|A_k(\lambda) g_k^0(\lambda) + C_k^0(\lambda)|^2}{(f_k^0(\lambda) + g_k^0(\lambda))^2} f_k(\lambda)\, d\lambda + \right. \\ &\left. + \frac{1}{2\pi} \int_{-\infty}^{\infty} \frac{|A_k(\lambda) f_k^0(\lambda) - C_k^0(\lambda)|^2}{(f_k^0(\lambda) + g_k^0(\lambda))^2} g_k(\lambda)\, d\lambda \right\}. \end{aligned}$$

The conditional extremum problem (8) is equivalent to the unconditional extremum problem:

$$\Delta_{\mathcal{D}}(f, g) = -\Delta\left(h(f^0, g^0); f, g\right) + \delta\left((f, g) \mid \mathcal{D}_f \times \mathcal{D}_g\right) \to \inf, \tag{9}$$

where $\delta\left((f, g) \mid \mathcal{D}_f \times \mathcal{D}_g\right)$ is the indicator function of the set $\mathcal{D} = \mathcal{D}_f \times \mathcal{D}_g$. The solution (f^0, g^0) of problem (9) is characterized by the condition $0 \in \partial\Delta_{\mathcal{D}}(f^0, g^0)$ Pshenichnyi (1971), where $\partial\Delta_{\mathcal{D}}(f^0, g^0)$ is the sub-differential of the convex functional $\Delta_{\mathcal{D}}(f, g)$ at the point (f^0, g^0).

This condition gives us a possibility to determine the least favorable spectral densities for concrete classes of spectral densities.

Lemma 5. *Let (f^0, g^0) be a solution of extremum problem* (9). *The spectral densities $f^0(\lambda)$ and $g^0(\lambda)$ are the least favorable in the class $\mathcal{D} = \mathcal{D}_f \times \mathcal{D}_g$ and the spectral characteristic $h^0 = h(f^0, g^0)$ is the minimax (robust) for the optimal linear estimate of the functional $A_T\xi$ if $h(f^0, g^0) \in H_{\mathcal{D}}$.*

4. Least favorable spectral densities in the class $\mathcal{D}_f^0 \times \mathcal{D}_g^0$

Let us consider the problem for the set of spectral densities $\mathcal{D} = \mathcal{D}_f^0 \times \mathcal{D}_g^0$, where

$$\begin{aligned} \mathcal{D}_f^0 &= \left\{ f \,\middle|\, \frac{1}{2\pi} \sum_{k=1}^{\infty} \int_{-\infty}^{\infty} f_k(\lambda)\, d\lambda \leq P_1 \right\}, \\ \mathcal{D}_g^0 &= \left\{ g \,\middle|\, \frac{1}{2\pi} \sum_{k=1}^{\infty} \int_{-\infty}^{\infty} g_k(\lambda)\, d\lambda \leq P_2 \right\}. \end{aligned}$$

Stochastic processes which have spectral densities from the class $\mathcal{D}_f^0$ satisfy the power constraint: $\mathbf{E}|\xi(t)|^2 \leq P_1$. Let the densities $f^0(\lambda) \in \mathcal{D}_f^0$ and $g^0(\lambda) \in \mathcal{D}_g^0$, $\mathbf{K}(f^0 + g^0) \neq \emptyset$ and the functions

$$h_{kf}(f^0, g^0) = \frac{\left|A_k(\lambda)\, g_k^0(\lambda) + C_k^0(\lambda)\right|}{f_k^0(\lambda) + g_k^0(\lambda)}, \tag{10}$$

$$h_{kg}(f^0, g^0) = \frac{\left|A_k(\lambda)\, f_k^0(\lambda) - C_k^0(\lambda)\right|}{f_k^0(\lambda) + g_k^0(\lambda)}, \tag{11}$$

$$k \in \mathbf{K}(f^0 + g^0),$$

be bounded. Under these conditions $\Delta\big(h(f^0, g^0); f, g\big)$ is a continuous linear functional in the space $L_1 \times L_1$ and

$$\partial\Delta_{\mathcal{D}_f^0 \times \mathcal{D}_g^0}(f^0, g^0) = -\partial\Delta\big(h(f^0, g^0); f^0, g^0\big) + \partial\delta\big((f^0, g^0) \mid \mathcal{D}_f^0 \times \mathcal{D}_g^0\big).$$

From the condition $0 \in \partial\Delta_{\mathcal{D}}(f^0, g^0)$ for $\mathcal{D} = \mathcal{D}_f^0 \times \mathcal{D}_g^0$ we find that the components of the least favorable densities $f^0 \in \mathcal{D}_f^0$, $g^0 \in \mathcal{D}_g^0$ satisfy the equations

$$\alpha_{k1}\left|A_k(\lambda)\, g_k^0(\lambda) + C_k^0(\lambda)\right| = \big(f_k^0(\lambda) + g_k^0(\lambda)\big), \tag{12}$$

$$\alpha_{k2}\left|A_k(\lambda)\, f_k^0(\lambda) - C_k^0(\lambda)\right| = \big(f_k^0(\lambda) + g_k^0(\lambda)\big), \tag{13}$$

where $\alpha_{k1} \geq 0$, $\alpha_{k2} \geq 0$. Note, that $\alpha_{k1} \neq 0$ if

$$\sum_{k \in \mathbf{K}} \frac{1}{2\pi} \int_{-\infty}^{\infty} f_k^0(\lambda)\, d\lambda = P_1, \tag{14}$$

and $\alpha_{k2} \neq 0$ if

$$\sum_{k \in \mathbf{K}} \frac{1}{2\pi} \int_{-\infty}^{\infty} g_k^0(\lambda)\, d\lambda = P_2. \tag{15}$$

Theorem 1. *Let spectral densities $f^0(\lambda)$ and $g^0(\lambda)$ belong to the set $\mathcal{D} = \mathcal{D}_f^0 \times \mathcal{D}_g^0$, $\mathbf{K}(f^0 + g^0) \neq \emptyset$ and assume the functions $h_{kf}(f^0, g^0)$ and $h_{kg}(f^0, g^0)$, $k \in \mathbf{K}(f^0 + g^0)$ computed by formulas* (10) *and* (11) *are bounded. The spectral densities $f^0(\lambda)$ and $g^0(\lambda)$ are the least favorable in the class $\mathcal{D} = \mathcal{D}_f^0 \times \mathcal{D}_g^0$ for the optimal linear estimation of the functional $A_T\xi$ if they satisfy relations* (12) *and* (13) *and determine a oolution of cxtrcmum problem* (6). *The minimax (robust) spectral characteristic of the optimal linear estimate of the functional is computed by formula* (3).

Theorem 2. *Let spectral density $f(\lambda)$ be known, assume density $g^0(\lambda)$ belongs to the set $\mathcal{D}_g^0$, $\mathbf{K}(f + g^0) \neq \emptyset$ and assume functions $h_{kg}(f, g^0)$, $k \in \mathbf{K}(f + g^0)$ computed by formula* (11) *are bounded. The spectral density $g^0(\lambda)$ is the least favorable in the class $\mathcal{D}_g^0$ for the optimal estimation of functional $A_T\xi$ if the components of the density satisfy the relations*

$$g_k^0(\lambda) = \max\left\{0, \alpha_{k2}\left|A_k(\lambda) f_k(\lambda) - C_{kT}^0(\lambda)\right| - f_k(\lambda)\right\}$$

and $(f(\lambda), g^0(\lambda))$ *determines a solution of extremum problem* (6). *The function* $h(f, g^0)$ *computed by formula* (3) *is the minimax (robust) spectral characteristic of the optimal linear estimate of the functional* $A_T\xi$.

Theorem 3. *Let spectral density* $f^0(\lambda)$ *belong to the set* $\mathcal{D}_f^0$, $\mathbf{K}(f^0) \neq \emptyset$ *and assume the functions* $h_{kf}(f^0)$, $k \in \mathbf{K}(f^0)$ *computed by formula* (5) *are bounded. The spectral density* $f^0(\lambda)$ *is the least favorable in the class* $\mathcal{D}_f^0$ *for the optimal estimation of the functional* $A_T\xi$ *from observations of the process* $\xi(t)$ *for* $t \in \mathbf{R} \setminus [0, T]$ *if it satisfies the equation*

$$f_k^0(\lambda) = \alpha_{k1} \left| C_k^0(\lambda) \right|$$

and determines a solution of extremum problem (7). *The minimax (robust) spectral characteristic of the optimal linear estimate of the functional is computed by formula* (5).

5. Least favorable spectral densities in the class $\mathcal{D} = \mathcal{D}_\varepsilon \times \mathcal{D}_v^u$

Let us consider the problem for the set of spectral densities $\mathcal{D} = \mathcal{D}_\varepsilon \times \mathcal{D}_v^u$,

$$\begin{aligned} D_\varepsilon &= \left\{ f \,\middle|\, f_k(\lambda) = (1-\varepsilon) f_k^1(\lambda) + \varepsilon w_k(\lambda), \ \frac{1}{2\pi} \sum_{k=1}^{\infty} \int_{-\infty}^{\infty} f_k(\lambda)\, d\lambda = P_1 \right\}, \\ \mathcal{D}_v^u &= \left\{ g \,\middle|\, v_k(\lambda) \leq g_k(\lambda) \leq u_k(\lambda); \ \frac{1}{2\pi} \sum_{k=1}^{\infty} \int_{-\infty}^{\infty} g_k(\lambda)\, d\lambda \leq P_2 \right\}, \end{aligned}$$

where the spectral densities $v(\lambda)$, $u(\lambda)$ and $f^1(\lambda)$ are known and the densities $u_k(\lambda)$, $k = 1, 2, \ldots$ are bounded. Class $\mathcal{D}_\varepsilon$ describes the "ε-pollution" model of stochastic processes. Class $\mathcal{D}_v^u$ describes the "band" model of stochastic processes.

If $f^0(\lambda) \in \mathcal{D}_\varepsilon$, $g^0(\lambda) \in \mathcal{D}_v^u$, $\mathbf{K}(f^0 + g^0) \neq \emptyset$ and functions $h_{kf}(f^0, g^0)$ and $h_{kg}(f^0, g^0)$, $k \in \mathbf{K}(f^0 + g^0)$ computed by formulas (10) and (11) are bounded, the condition $0 \in \partial\Delta_{\mathcal{D}}(f^0, g^0)$ for $\mathcal{D} = \mathcal{D}_\varepsilon \times \mathcal{D}_v^u$ is satisfied if components of the densities $f^0(\lambda)$ and $g^0(\lambda)$ satisfy the equations

$$\left| A_k(\lambda)\, g_k^0(\lambda) + C_k^0(\lambda) \right| = \left(f_k^0(\lambda) + g_k^0(\lambda) \right) \left(\varphi_k(\lambda) + \alpha_{k1}^{-1} \right), \tag{16}$$

$$\left| A_k(\lambda)\, f_k^0(\lambda) - C_k^0(\lambda) \right| = \left(f_k^0(\lambda) + g_k^0(\lambda) \right) \left(\gamma_{k1}(\lambda) + \gamma_{k2}(\lambda) + \alpha_{k2}^{-1} \right), \tag{17}$$

where $\gamma_{k1}(\lambda) \leq 0$ a.e. and $\gamma_{k1}(\lambda) = 0$ if $g_k^0(\lambda) \geq v_k(\lambda)$; $\gamma_{k2}(\lambda) \geq 0$ a.e. and $\gamma_{k2}(\lambda) = 0$ if $g_k^0(\lambda) \leq u_k(\lambda)$; $\varphi_k(\lambda) \leq 0$ a. e. and $\varphi_k(\lambda) = 0$ if $f_k^0(\lambda) \geq (1-\varepsilon) f_k^1(\lambda)$.

Theorem 4. *Let spectral densities* $f^0(\lambda)$ *and* $g^0(\lambda)$ *belong to the set* $\mathcal{D} = \mathcal{D}_\varepsilon \times \mathcal{D}_v^u$, $\mathbf{K}(f^0 + g^0) \neq \emptyset$ *and assume the functions* $h_{kf}(f^0, g^0)$ *and* $h_{kg}(f^0, g^0)$, $k \in \mathbf{K}(f^0 + g^0)$ *computed by formulas* (10) *and* (11) *are bounded. The spectral densities* $f^0(\lambda)$ *and* $g^0(\lambda)$ *are the least favorable in the class* $\mathcal{D} = \mathcal{D}_\varepsilon \times \mathcal{D}_v^u$ *for the optimal linear* *estimation of the functional $A_T\xi$ if they satisfy relations* (14)–(17) *and determine a solution of extremum problem* (6). *The minimax (robust) spectral characteristic of the optimal linear estimate of the functional is computed by formula* (3).

Theorem 5. *Let spectral density $f(\lambda)$ be known, assume density $g^0(\lambda)$ belongs to the set $\mathcal{D}_v^u$, $\mathbf{K}(f+g^0)\neq\emptyset$ and assume the functions $h_{kg}(f,g^0)$ and $k\in\mathbf{K}(f+g^0)$ computed by formula* (11) *are bounded. The spectral density $g^0(\lambda)$ is the least favorable in the class $\mathcal{D}_v^u$ for the optimal estimation of the functional $A_T\xi$ if components of the density satisfy the relations*

$$g_k^0(\lambda)=\max\left\{v_k(\lambda),\min\left\{u_k(\lambda),\ \alpha_{k2}\left|A_k(\lambda)f_k(\lambda)-C_k^0(\lambda)\right|-f_k(\lambda)\right\}\right\}$$

and $(f(\lambda),g^0(\lambda))$ determines a solution of extremum problem (6). *The function $h(f,g^0)$ computed by formula* (3) *is the minimax (robust) spectral characteristic of the optimal linear estimate of the functional $A_T\xi$.*

Theorem 6. *Let a spectral density $f^0(\lambda)$ belong to the set $\mathcal{D}_\varepsilon$, $\mathbf{K}(f^0)\neq\emptyset$ and assume the functions $h_{kf}(f^0)$, $k\in\mathbf{K}(f^0)$ computed by formula* (5) *are bounded. The spectral density $f^0(\lambda)$ is the least favorable in the class $\mathcal{D}_\varepsilon$ for the optimal estimation of the functional $A_T\xi$ from observations of the sequence $\xi(t)$ for $t\in\mathbf{R}\setminus[0,T]$ if the components $f_k^0(\lambda)$, $k\in\mathbf{K}(f^0)$ of the density satisfy the equation*

$$f_k^0(\lambda)=\max\left\{(1-\varepsilon)\,f_k^1(\lambda),\alpha_{k1}\left|C_{kT}^0(\lambda)\right|\right\}$$

and determine a solution of extremum problem (7). *The minimax (robust) spectral characteristic of the optimal linear estimate of the functional is computed by formula* (5).

6. Least favorable spectral densities in the class $\mathcal{D}_{2\delta_1}\times\mathcal{D}_{1\delta_2}$

Let the set of densities be of the form $\mathcal{D}=\mathcal{D}_{2\delta_1}\times\mathcal{D}_{1\delta_2}$,

$$\begin{aligned}\mathcal{D}_{2\delta_1}&=\left\{f\;\Big|\;\frac{1}{2\pi}\sum_{k=1}^{\infty}\int_{-\infty}^{\infty}\left|f_k(\lambda)-f_k^1(\lambda)\right|^2d\lambda\le\delta_1\right\},\\ \mathcal{D}_{1\delta_2}&=\left\{g\;\Big|\;\frac{1}{2\pi}\sum_{k=1}^{\infty}\int_{-\infty}^{\infty}\left|g_k(\lambda)-g_k^1(\lambda)\right|d\lambda\le\delta_2\right\},\end{aligned}$$

where $f_k^1(\lambda)$ and $g_k^1(\lambda)$ are known bounded spectral densities. The sets $\mathcal{D}_{2\delta_1}$ and $\mathcal{D}_{1\delta_2}$ describe the "δ-neighbourhood" models of stochastic processes in the spaces $\mathbf{L}_1$ and $\mathbf{L}_2$. If $f^0(\lambda)\in\mathcal{D}_{2\delta_1}$ and $g^0(\lambda)\in\mathcal{D}_{1\delta_2}$, $\mathbf{K}(f^0+g^0)\neq\emptyset$ and functions $h_{kf}(f^0,g^0)$ and $h_{kg}(f^0,g^0)$, $k\in\mathbf{K}(f^0+g^0)$ computed by formulas (10) and (11) are bounded, the condition $0 \in \partial\Delta_{\mathcal{D}}(f^0, g^0)$ for $\mathcal{D} = \mathcal{D}_{2\delta_1} \times \mathcal{D}_{1\delta_2}$ is satisfied if the components of the densities $f^0(\lambda)$ and $g^0(\lambda)$ satisfy the equations

$$\begin{aligned} \left|A_k(\lambda)g_k^0(\lambda) + C_k^0(\lambda)\right|^2 &= \left(f_k^0(\lambda) + g_k^0(\lambda)\right)^2 \left(f_k^0(\lambda) - f_k^1(\lambda)\right)\alpha_{k1}, \quad (18) \\ \left|A_k(\lambda)f_k^0(\lambda) - C_k^0(\lambda)\right| &= \left(f_k^0(\lambda) + g_k^0(\lambda)\right)\psi_k(\lambda)\alpha_{k2}, \quad (19) \\ & k \in \mathbf{K}(f^0 + g^0), \end{aligned}$$

where $|\psi_k(\lambda)| \leq 1$ a.e. and $\psi_k(\lambda) = \operatorname{sign}(g_k^0(\lambda) - g_k^1(\lambda))$ if $g_k^0(\lambda) \neq g_k^1(\lambda)$. Equations (18) and (19) with extremum condition (6) and conditions

$$\sum_{k=1}^{\infty} \frac{1}{2\pi} \int_{-\infty}^{\infty} \left|f_k(\lambda) - f_k^1(\lambda)\right|^2 d\lambda = \delta_1, \quad (20)$$

$$\sum_{k=1}^{\infty} \frac{1}{2\pi} \int_{-\infty}^{\infty} \left|g_k(\lambda) - g_k^1(\lambda)\right| d\lambda = \delta_2 \quad (21)$$

determine the least favorable spectral densities.

Theorem 7. *Let spectral densities $f^0(\lambda)$ and $g^0(\lambda)$ belong to the set $\mathcal{D}_{2\delta_1} \times \mathcal{D}_{1\delta_2}$, $\mathbf{K}(f^0+g^0) \neq \emptyset$ and assume the functions $h_{kf}(f^0, g^0)$ and $h_{kg}(f^0, g^0)$, $k \in \mathbf{K}(f^0+g^0)$ computed by formulas (10) and (11) are bounded. The spectral densities $f^0(\lambda)$ and $g^0(\lambda)$ are the least favorable in the class $\mathcal{D}_{2\delta_1} \times \mathcal{D}_{1\delta_2}$ for the optimal linear estimation of the functional $A_T\xi$ if they satisfy relations (18)–(21) and determine a solution of extremum problem (6). The minimax (robust) spectral characteristic of the optimal linear estimate of the functional is computed by formula (3).*

Theorem 8. *Let spectral density $f(\lambda)$ be known, assume density $g^0(\lambda)$ belongs to the set $\mathcal{D}_{1\delta_2}$, $\mathbf{K}(f+g^0) \neq \emptyset$ and assume the functions $h_{kg}(f, g^0)$, $k \in \mathbf{K}(f+g^0)$ computed by formula (11) are bounded. The spectral density $g^0(\lambda)$ is the least favorable in the class $\mathcal{D}_{1\delta_2}$ for the optimal estimation of the functional $A_T\xi$ if components of the density satisfy the relations*

$$g_k^0(\lambda) = \max\left\{g_k^1(\lambda), \alpha_{k2}\left|A_k(\lambda)f_k(\lambda) - C_k^0(\lambda)\right| - f_k(\lambda)\right\}$$

and $(f(\lambda), g^0(\lambda))$ determines a solution of extremum problem (6). The function $h(f, g^0)$ computed by formula (3) is the minimax (robust) spectral characteristic of the optimal linear estimate of the functional $A_T\xi$.

Theorem 9. *Let a spectral density $f^0(\lambda)$ belong to the set $\mathcal{D}_{2\delta_1}$, $\mathbf{K}(f^0) \neq \emptyset$ and assume the functions $h_{kf}(f^0)$, $k \in \mathbf{K}(f^0)$ computed by formula (5) are bounded. The spectral density $f^0(\lambda)$ is the least favorable in the class $\mathcal{D}_{2\delta_1}$ for the optimal estimation of the functional $A_T\xi$ from observations of the sequence $\xi(t)$ for $t \in \mathbf{R} \setminus [0, T]$ if the components $f_k^0(\lambda)$, $k \in \mathbf{K}(f^0)$ of the density satisfy the equation*

$$\left|C_k^0(\lambda)\right|^2 = \left(f_k^0(\lambda)\right)^2 \left(f_k^0(\lambda) - f_k^1(\lambda)\right)\alpha_{k1},$$

and determine a solution of extremum problem (7). *The minimax (robust) spectral characteristic of the optimal linear estimate of the functional is computed by formula* (5).

References

Kolmogorov, A. N. (1986). *Theory of Probability and Mathematical Statistics. Selected Papers.* Nauka, Moscow. (in Russian)

Rosanov, Yu. A. (1967). *Stationary Random Processes.* Holden-Day, San Francisco.

Vastola, K. S. and Poor, H.V. (1983). An analysis of the effects of spectral uncertainty on Wiener filtering, *Automatica*, **19**, 289–293.

Franke, J. (1984). On the robust prediction and interpolation of time series in the presence of correlated noise, *J. Time Series Analysis*, **5**, No 4, 227–244.

Franke, J. (1985). Minimax-robust prediction of discrete time series. *Z. Wahrsch. verw. Gebiete.* **68**, No 2, 337–364.

Franke, J. and Poor, H. V. (1984). Minimax-robust filtering and finite length robust predictors. *Robust and Nonlinear Time Series Analysis. Lecture Notes in Statist.* **26**, 87–126.

Hosoya, Y. (1978). Robust linear extrapolation of second order stationary processes. *Ann. Probab.* **6**, 574–584.

Kassam, S. A. (1982). Robust hypothesis testing and robust time series interpolation and regression. *J. Time Series Analysis.* **3**, 185–194.

Kassam, S. A. and Poor, H. V. (1985). Robust techniques for signal processing. A Survey. *Proc. IEEE.* **73**, 433–481.

Taniguchi, M. (1981). Robust regression and interpolation for time series. *J. Time Series Analysis.* **2**, 53–62.

Moklyachuk, M. P. (1981). Estimation of linear functionals of a stationary stochastic processes and a two-person zero-sum game, *Stanford Univ., Technical Report. No 169.*

Moklyachuk, M. P. (1982). On an antagonistic game and prediction of stationary random processes in a Hilbert space, *Theory Probab. Math. Stat.* **24**, 121–127.

Moklyachuk, M. P. (1990). Minimax extrapolation and autoregression-moving average processes. *Theory Probab. and Math. Stat.* **41** 77–84.

Moklyachuk, M. P. (1991a). Minimax filtering of linear transformations of stationary sequences. *Ukrainian Math. Journal.* **43**, No 1. 75–81.

Moklyachuk, M. P. (1991b). Extrapolation of transformations of random processes, perturbed by white noise. *Ukrainian Math. Journal* **43**, No 2, 188–194.

Moklyachuk, M. P. (1991c). Minimax extrapolation of random processes for models of ε-pollution. *Theory Probab. and Math. Stat* **42**, 113–121.

Moklyachuk, M. P. (1991d). Minimax filtering of stationary sequences with white noise. *Theory Probab. and Math. Statistics* **43**, 109–122.

Moklyachuk, M. P. (1992a). Minimax filtering of linear transformations of stationary processes. *Theory Probab. and Math. Statistics*, **44**, 93–101.

Moklyachuk, M. P. (1992b). On minimax extrapolation problem for vector-valued sequences perturbed by white noise. *Theory Probab. and Math. Stat.* **46**, 88–104.

Moklyachuk, M. P. (1992c). On stochastic equations which describe one-sided moving average processes and minimax filtering. *Random Operators and Stochastic Equations.* **1**, No 4, 329–343.

Moklyachuk, M. P. (1993a). On filtering problem for vector valued processes. *Theory Probab. and Math. Stat.* **47**, 104–118.

Moklyachuk, M. P. (1993b). Autoregressive stochastic sequences and minimax interpolation. *Theory Probab. and Math. Stat.* **48**, 135–146.

Moklyachuk, M. P. (1993c). Minimax filtering of vector-valued processes. *Ukrainian Math. Journal*, **45**, No 3, 389–397.

Moklyachuk, M. P. (1993d). Minimax-robust interpolation problem for discrete time series, In: *Evolutionary Stochastic Systems in Physics and Biology.* V.S. Korolyuk et al. (Eds). TVP/VSP, Moscow/Utrecht, 336–347.

Moklyachuk, M. P. (1993e). Minimax-robust interpolation of stationary stochastic processes, In: *New Trends in Probability and Statistics. Proceedings of the second Ukrainian–Hungarian conference.* M. Arato, M.I. Yadrenko (Eds). Moscow/Utrecht/Kiev, TVP/VSP/TBiMC, 183–193.

Kallianpur, G. and Mandrekar, V. (1965). Multiplicity and representation theory of purely non-deterministic stochastic processes, *Theor. Probab. Appl.* **10**, 614–644.

Pshenichnyi, B.N. (1971). *Necessary Conditions for an Extremum.* Marcel Dekker, New York, 1971.

Exploring Stochastic Laws. pp. 343–349.
Eds. A.V. Skorokhod and Yu.V. Borovskikh

Commutation Relations for Mixtures of Gaussian Measures

N.V. NORIN

Moscow Institute of Radio Engineering, Electronics, and Automation, pr. Vernadskogo 78, 117454, Moscow, Russia

Received February 15, 1995

Abstract. The aim of this paper is to establish commutation relations for a class of differentiable measures. This class M consists of mixtures of Gaussian measures of a special type. Note that infinite-dimensional Cauchy measures, Bessel and Riesz potentials, and spherically symmetric measures belong to M. Previously obtained by (Norin (1995a)) Itô-Wick, decomposition for such measures allows us to use explicit calculations as in the case of Gaussian measures.

1. Basic definitions

Let H and E be separable Hilbert spaces, H^m — an m-folded Hilbert tensor product of H, $V_m, G_m \in E \otimes H^m$, $|V_m|_{m,E}$ and $\langle V_m, G_m\rangle_{m,E}$ — the norm and inner product in $E \otimes H^m$ ($|V_m|_m \equiv |V_m|_{m,\mathbb{R}}$ and $\langle V_m, G_m\rangle_m \equiv \langle V_m, G_m\rangle_{m,\mathbb{R}}$), and S^m — the symmetrization operator in $E \otimes H^m$:

$$S^m V_m = \frac{1}{m!} \sum_{\pi \in S_m} e \otimes h_{\pi(1)} \otimes \cdots \otimes h_{\pi(m)}$$

where $V_m = e \otimes h_1 \otimes \cdots \otimes h_m$, $e \in E$, $h_i \in H$ (S_m is the permutation group of order m), and $E \otimes H^{m_s} = S^m(E \otimes H^m)$. Let ν be a σ-finite Borelian measure on $\mathbb{R}_+$ without an atom in zero and let $V_m: \mathbb{R}_+ \to E \otimes H^m: t \mapsto V_m(t)$ be a Borelian multivector field such that

$$\int_0^\infty |S^m V_m(t)|^2_{m,E}\, \nu(dt) < \infty.$$

The set of such fields is a direct integral of Hilbert space $E \otimes H^m$ over measurable space $(\mathbb{R}_+, \nu)$ (cf. Gelfand and Vilenkin (1964), ch. 1, add. to §4, or Dixmier (1969), p. 145); the direct integral is denoted by $\int_0^\infty \oplus (E \otimes H^m)\, d\nu \equiv H^m_{\nu,E}$ ($H^m_{\nu,\mathbb{R}} \equiv H^m_\nu$). The space $H^m_{\nu,E}$ has the structure of a Hilbert space with an inner product

$$(V_m, G_m)_{H^m_{\nu,E}} = m! \int_0^\infty \langle V_m(t), G_m(t)\rangle_{m,E}\, \nu(dt).$$

In a similar manner one can define the direct integral $H^{m_s}_{\nu,E}$ of Hilbert space $E\otimes H^{m_s}$.

Lemma 1. *i) The direct integral $H^m_{\nu,E}$ is canonically isomorphic to the tensor product $L^2(\mathbb{R}_+,\nu)\otimes E\otimes H^m$. Under this isometry the multivector field $f\cdot V_m \in H^m_{\nu,E}$ where $f\in L^2(\mathbb{R}_+,\nu)$, $V_m\in E\otimes H^m$, corresponds to element $f\otimes V_m\in L^2(\mathbb{R}_+,\nu)\otimes E\otimes H^m$;*

ii) $H^m_{\nu,E}$ is a separable Hilbert space;

iii) items i) and ii) will be true if we replace $H^m_{\nu,E}$ by $H^{m_s}_{\nu,E}$ and H^m by H^{m_s}.

Proof follows from (Dixmier (1969), p. 153). Note only that we can take the set of multivector fields $V_m(t)=\varphi(t)\,e_i\otimes h_{i_1}\otimes\cdots\otimes h_{i_m}$ as the countable basis in $H^m_{\nu,E}$. Here $\{\varphi_i\}$ is a basis in $L^2(\mathbb{R}_+,\nu)$, $\{e_i\}$ is a basis in E and $\{h_i\}$ is a basis in H.

The orthogonal direct sum $\mathcal{F}_{\nu,E}=\oplus_{m=0}^\infty H^{m_s}_{\nu,E}$, $H^{0_s}_{\nu,E}\equiv E\otimes L^2(\mathbb{R}_+,\nu)$ with inner product

$$(V,G)_{\mathcal{F}_{\nu,E}}=\sum_{m=0}^\infty m!\int_0^\infty \langle V_m(t),G_m(t)\rangle_{m,E}\,\nu(dt)$$

is called a (vector) *Fock space*. It is easy to see that

$$\mathcal{F}_{\nu,E}=E\otimes(\bigoplus_{m=0}^\infty H^{m_s}_\nu)\cong\bigoplus_{m=0}^\infty E\otimes H^{m_s}_\nu .$$

Let γ_t be a Gaussian Radon measure on a Borelian σ-algebra $\mathcal{B}(X)$ of locally convex space X, $H\subset X$ — the subspace of differentiability of this measure, and $J_{E,m}(t,V_{m,E}(t))$ — an E-valued logarithmic derivative of Gaussian measure γ_t along the direction of multivector field $V_{m,E}(t)$ (see Norin (1990, 1991) for details),

$$\eta(x)=\begin{cases}\lim\limits_{n\to\infty}\dfrac{1}{n}\sum\limits_{i=1}^n\langle h_i,x\rangle^2 & \text{if the limit is finite,}\\ 0 & \text{if otherwise,}\end{cases}$$

$B_t=\{x\in X\mid \eta(x)=t\}$.

Definition 1. *The mixture of Gaussian measures is a measure μ_ν on X, defined by the equality*

$$\mu_\nu(B)=\int_0^\infty\gamma_t(B)\,\nu(dt),\qquad \forall B\in\mathcal{B}(X).$$

Definition 2. *Borelian function*

$$H_{E,m}(x,V_m)=\begin{cases}(-1)^m t^{m/2}J_{E,m}(t,V_{m,E}(t))(x) & \text{if } x\in B_t,\\ 0 & \text{if } x\notin B_t \text{ for all } t\in\mathbb{R}_+,\end{cases}$$

is called an E-valued Hermite pseudopolynomial of order m (corresponding to the multivector field $V_{m,E}(t)$).

The following Theorem summarizes the properties of Hermite pseudopolynomials. Note that detailed investigations of these properties are contained in Norin (1995a, b).

Theorem 1. 1. *Let δ_{mn} be the Kronecker symbol. Then*

$$\int_X H_m(x, V_m)\, H_n(x, G_n)\, \mu_\nu(dx) = \delta_{nm}\, m! \int_0^\infty \langle S^m V_m(t), S^m G_m(t)\rangle_{m,E}\, \nu(dt);$$

2. $H_m(x, V_m) \in L^2(X, E, \mu_\nu) \Longleftrightarrow V_m \in H^m_{\nu,E}$;

3. *Mapping* $H_m \colon H^{m_s}_{\nu,E} \to L^2(X, E, \mu_\nu) \colon V_m \mapsto H_m(x, V_m)$ *is an isometry onto its image*;

4. *Let $\{e_j\}$ be an orthonormal basis in E and let $\{h_i\}$ be an orthonormal basis in H,*

$$V_m(t) = \sum_{i,j,i_1,\cdots,i_m=1}^{\infty} a_{i,j,i_1,\cdots,i_m}\, \varphi_i(t)\, e_j \otimes h_{i_1} \otimes \cdots \otimes h_{i_m} \in H^m_{\nu,E}.$$

Define the inner product $V_m \lrcorner\, h$, $h \in H$, with the help of equality

$$V_m(t) \lrcorner\, h = \sum_{i,j,i_1,\cdots,i_m=1}^{\infty} a_{i,j,i_1,\cdots,i_m}\, \varphi_i(t)\, \langle h, h_{i_1}\rangle\, e_j \otimes h_{i_2} \otimes \cdots \otimes h_{i_m} \in H^{m-1}_{\nu,E}.$$

Now if $V_m(t) \in E \bar{\otimes} (\bar{X}^*)^m$ for all t, then there exists a Frechét derivative of Hermite pseudopolynomial $H_m(x, V_m)$ along H, and

$$\langle h, D_H H_m(x, V_m)\rangle = (\eta(x))^{-1/2}\, m\, H_{m-1}(x, (S^m V_m) \lrcorner\, h); \tag{1}$$

5. *If a function $\varphi \colon X \to \mathbb{R}$ is measurable with respect to the minimal σ-algebra generated by sets B_t, $t \geq 0$, then*

$$H_m(x, \varphi \cdot V_m) = \varphi\, H_m(x, V_m). \tag{2}$$

Theorem 2. (Itô-Wick decomposition for mixtures of Gaussian measures). *The space $L^2(X, E, \mu_\nu)$ of E-valued square summable functionals on X is isometric to $\mathcal{F}_{\nu,E}$. This isometry is given by the mapping*

$$\sum_{m=0}^{\infty} V_m(\cdot) \longmapsto \sum_{m=0}^{\infty} H_{E,m}(x, V_m) : \mathcal{F}_{\nu,E} \to L^2(X, E, \mu_\nu).$$

The *stochastic derivative* of the functional $\varphi \in L^2(X, E, \mu_\nu)$ is defined by the equality (cf. Eq. (1))

$$D\varphi = (\eta(x))^{-1/2} \sum_{m=1}^{\infty} m\, H_{E,m-1}(x, S^m V_m) \tag{3}$$

where $\varphi = \sum_{m=0}^{\infty} H_{E,m}(x, S^m V_m)$ is the expansion of φ into a series of Hermite pseudopolynomials, and V_m is an element of $E \otimes H_\nu^{m,s}$.

Definition 3. *The extended stochastic integral of $E \otimes H$-valued functional $\varphi \in L^2(X, E \otimes H, \mu_\nu)$, $\varphi = \sum_{m=0}^{\infty} H_{E,m}(x, S^m V_m)$, is*

$$\mathcal{I}\,\varphi(x) \stackrel{\text{def}}{=} (\eta(x))^{-1/2} \sum_{m=0}^{\infty} H_{E,m+1}(x, S^m V_m).$$

It follows from Eq. (2) that $D(\eta^\alpha \varphi) = \eta^\alpha\, D(\varphi)$ and $\mathcal{I}(\eta^\alpha \varphi) = \eta^\alpha\, \mathcal{I}(\varphi)$.

2. Commutation relations

Lemma 2. *Let $\varphi \in L^2(X, E \otimes H, \mu_\nu)$, $\varphi = \sum_{m=0}^{\infty} H_{E\otimes H,m}(S^m V_{m+1})$, be such that*

$$\sum_{m=0}^{\infty} (m+1)^2\, m! \int_0^\infty t^{-2} \left|S^{m+1} V_{m+1}(t)\right|^2_{E\otimes H^{m+1}} \nu(dt) < \infty. \tag{4}$$

Then

$$[D, \mathcal{I}](\varphi) \stackrel{\text{def}}{=} D(\mathcal{I}\varphi) - \mathcal{I}(D\varphi) = \eta^{-1}\varphi.$$

Proof. It is easy to see that

$$\begin{aligned}
\mathcal{I}(D\varphi) &= \mathcal{I}\Big(\sum_{m=1}^{\infty} m\,\eta^{-1/2}\, H_{E\otimes H^2,m-1}(S^m V_{m+1})\Big) \\
&= \sum_{m=1}^{\infty} \eta^{-1} \left[\sum_{k=1}^{m} H_{E\otimes H,m}(S^{0,k}(S^m V_{m+1}))\right],
\end{aligned}$$

where operators $S^{0,k}$ are defined on multivectors $h_0 \otimes h_1 \otimes \cdots \otimes h_k \otimes \cdots \otimes h_m$ by the equality

$$S^{0,k}(h_0 \otimes h_1 \otimes \cdots \otimes h_k \otimes \cdots \otimes h_m) = h_k \otimes h_1 \otimes \cdots \otimes h_0 \otimes \cdots \otimes h_m$$

and are extended onto H^{m+1} by linearity and continuity. On the other hand,

$$\begin{aligned}
\mathcal{I}\varphi &= \sum_{m=0}^{\infty} \eta^{-1/2} H_{E\otimes H,m+1}(S^{m+1} V_{m+1}) \\
&= \sum_{m=0}^{\infty} (m+1)^{-1}\, \eta^{-1/2} \left[\sum_{k=1}^{m} H_{E,m+1}\big(S^{0,k}(S^m V_{m+1})\big) + H_{E,m+1}(S^m V_{m+1})\right].
\end{aligned}$$

Therefore

$$D(\mathcal{I}\varphi) = \sum_{m=0}^{\infty} \eta^{-1} \left[\sum_{k=1}^{m} H_{E,m}\big(S^{0,k}(S^m V_{m+1})\big) + H_{E,m}(S^m V_{m+1})\right],$$

and

$$D(\mathcal{I}\varphi) - \mathcal{I}(D\varphi) = \sum_{m=0}^{\infty} \eta^{-1} H_{E,m}(S^m V_{m+1}) = \eta^{-1}\varphi.$$

Condition (4) guarantees the validity of the above calculations.

Definition 4. *An (unbounded) operator*

$$\mathcal{L}: L^2(X, E, \mu_\nu) \to L^2(X, E, \mu_\nu): \varphi \longmapsto \mathcal{I}(D\varphi)$$

is called an Ornstein–Uhlenbeck operator.

Lemma 3. *Let $f \in L^2(X, E, \mu_\nu)$, $f = \sum_{m=0}^{\infty} H_{E,m}(S^m V_m^f)$, be such that*

$$\sum_{m=0}^{\infty} (m+1)^4 m! \int_0^{\infty} t^{-3} \left|S^{m+1} V_{m+1}^f(t)\right|^2_{E\otimes H^{m+1}} \nu(dt) < \infty. \tag{5}$$

Then

$$[D, \mathcal{L}](f) \stackrel{\text{def}}{=} D(\mathcal{L}f) - \mathcal{L}(Df) = \eta^{-1} Df.$$

Proof. First,

$$\begin{aligned} D(\mathcal{L}f) &= D\mathcal{I}\Big(\sum_{m=1}^{\infty} m\,\eta^{-1/2} H_{E\otimes H,m-1}(S^m V_m^f)\Big) \\ &= D\Big(\sum_{m=1}^{\infty} m\,\eta^{-1} H_{E,m}(S^m V_m^f)\Big) = \sum_{m=1}^{\infty} m^2 \eta^{-3/2} H_{E\otimes H,m-1}(S^m V_m^f), \end{aligned}$$

and by (5)

$$\begin{aligned} \|D(\mathcal{L}f)\|^2_{2,E\otimes H} &= \sum_{m=1}^{\infty} m^4 (m-1)! \int_0^{\infty} t^{-3} \left|S^m V_m^f\right|^2_{E\otimes H^m} \nu(dt) \\ &= \sum_{m=0}^{\infty} (m+1)^4 m! \int_0^{\infty} t^{-3} \left|S^{m+1} V_{m+1}^f\right|^2_{E\otimes H^{m+1}} \nu(dt) < \infty. \end{aligned}$$

On the other hand,

$$\begin{aligned} \mathcal{L}(Df) &= \mathcal{L}\Big(\sum_{m=1}^{\infty} m\,\eta^{-1/2} H_{E\otimes H,m-1}(S^m V_m^f)\Big) \\ &= \mathcal{I}\Big(\sum_{m=2}^{\infty} m(m-1)\,\eta^{-1} H_{E\otimes H^2,m-2}(S^m V_m^f)\Big) \\ &= \sum_{m=2}^{\infty} m(m-1)\,\eta^{-3/2} H_{E\otimes H,m-1}(S^m V_m^f), \end{aligned}$$

and

$$\begin{aligned} \|\mathcal{L}(Df)\|^2_{2,E\otimes H} &= \sum_{m=2}^{\infty} m^2(m-1)^2 (m-1)! \int_0^{\infty} t^{-3} \left|S^m V_m^f\right|^2_{E\otimes H^m} \nu(dt) \\ &= \sum_{m=1}^{\infty} (m+1)^2 m^2 m! \int_0^{\infty} t^{-3} \left|S^{m+1} V_{m+1}^f\right|^2_{E\otimes H^{m+1}} \nu(dt) \le \|D(\mathcal{L}f)\|^2_{2,E\otimes H}. \end{aligned}$$

Assume $\varphi = Df$, then $\varphi = \sum_{m=0}^{\infty} H_{H,m}(V_{m+1}^{\varphi})$, where $V_{m+1}^{\varphi} = (m+1)\eta^{-1/2}V_{m+1}$. It follows from (5) that condition (4) is fulfilled for φ, and, therefore, we can apply Lemma 2 to φ. Then

$$D(\mathcal{L}f) - \mathcal{L}(Df) = D\mathcal{I}D\,f - \mathcal{I}DD\,f = D\mathcal{I}\,\varphi - \mathcal{I}D\,\varphi = \eta^{-1}\varphi = \eta^{-1}Df.$$

Theorem 3. *Let $\varphi = \sum_{m=0}^{\infty} H_{E\otimes H,m}(S^m V_{m+1}) \in L^2(X, E\otimes H, \mu_\nu)$ be such that*

$$\sum_{m=0}^{\infty} (m+1)^2\,(m+1)! \int_0^{\infty} t^{-3} \left|S^{m+1}V_{m+1}(t)\right|^2_{E\otimes H^{m+1}} \nu(dt) < \infty. \tag{6}$$

Then

$$[\mathcal{L},\mathcal{I}]\,(\varphi) \stackrel{\text{def}}{=} \mathcal{L}\,(\mathcal{I}\varphi) - \mathcal{I}\,(\mathcal{L}\varphi) = \eta^{-1}\mathcal{I}\varphi. \tag{7}$$

Proof. It follows from Eq. (6) and Lemma 2 that the following calculations are correct

$$\begin{aligned}[\mathcal{L},\mathcal{I}]\,(\varphi) &= \mathcal{L}\,(\mathcal{I}\varphi) - \mathcal{I}\,(\mathcal{L}\varphi) = \mathcal{I}D\mathcal{I}\,\varphi - \mathcal{I}\mathcal{I}D\,\varphi \\ &= \mathcal{I}\,(D\mathcal{I}\,\varphi - \mathcal{I}D\,\varphi) = \eta^{-1}\mathcal{I}\varphi.\end{aligned}$$

Note that in the case of Gaussian measures, Eq. (7) is called the Strook's commutation formula (see Malliavin (1985)). On the other hand, this formula is a special case of commutation formulae for differentiable measures (Norin (1987)).

References

Dixmier, J. (1969). *Les algébres d'opérateurs dans l'espace hilbertien.* Gautier-Villars, Paris.

Gelfand, I.M. and Vilenkin, N. Ya. (1964). *Generalized Functions. Vol. 4.* Academic Press, New York.

Malliavin, P. (1984). Calcul des variations, intégrales stochastiques et complexe de Rham sur l'espace de Wiener. *C. R. Acad. Sc. Paris,* **299**(1), 347–350.

Norin, N.V. (1987). Two commutation formulas for extended stochastic integrals with respect to differentiable measures. In: *Ergodic Theory of Markov processes.* Abstracts of All-Union School-Seminar. Kyzyl. September, 1987. Kyzyl. p. 43.

Norin, N.V. (1990). Vector extended stochastic integrals in Sobolev spaces of Wiener functionals. In: *Proc. of Fifth Int. Vilnius Conf. on Probab. Theory and Math. Statist.* June, 1989. B. Grigelionis et al. (Eds) VSP/Mokslas, **2,** 244–256.

Norin, N.V. (1991). Abstract stochastic integral equations with vector extended stochastic integrals. *Matem. zametki,* **49**, 153–155. (In Russian).

Norin, N.V. (1995a). Itô-Wick decomposition for mixtures of Gaussian measures. In: *Frontiers in Pure and Applied Probability II: Proc. of the Fourth Russian-Finnish Symp. on Probab. Theory and Math. Statist.* (Moscow, October 3–8, 1993). A. N. Shiryaev et al. (Eds). TVP, Moscow. (In press).

Norin, N.V. (1995b). Differentiability properties of mixtures of Gaussian measures. In: *Statistics and Control for Stochastic processes V: Festschrift in Honour of Professor A. N. Shiryaev.* A. A. Novikov (Ed). TVP, Moscow. (In press).

Exploring Stochastic Laws. pp. 351–355.
Eds. A.V. Skorokhod and Yu.V. Borovskikh

A characterization of an exponential distribution based on renewals

A. OBRETENOV*

Bulgarian Academy of Sciences, Institute of Mathematics, P.O. Box 373, 1090 Sofia, Bulgaria

Received February 1, 1995

Abstract. It is shown that if the probability of renewals on the interval $[0,t)$ is random and depends on t, and if these probabilities are geometrically distributed then the randomizing function is exponentially distributed. Several statements are given about this property under certain assumptions concerning the randomizing function.

Let $X_1, X_2, \dots, X_n, \dots$ be non-negative, independent and identically distributed random variables with common distribution function $F(x)$. $G(t)$ denotes a distribution function o $[0,\infty)$. Assume that $F(0+) = G(0+) = 0$.

Let

$$N(t) = \max\{n: X_1 + X_2 + \cdots + X_n < t\}$$

and

$$p_k(t) = P(N(t) = k), \qquad k = 0, 1, 2, \dots .$$

Consider the randomized distribution

$$\int_0^\infty p_k(t)\, dG(t) = g_k, \qquad k = 0, 1, 2, \dots . \tag{1}$$

If $F(x)$ and $G(x)$ are exponential distributions it is easy to see that the distribution g_k from (1) is geometric. Moreover, if $F(x)$ is arbitrary, but $G(t) = 1 - \exp(-at)$, $0 < a < \infty$, then

$$\int_0^\infty p_k(t)\, dG(t) = h^k(a)\,[1 - h(a)], \qquad k = 0, 1, 2, \dots$$

where $h(s) = \int_0^\infty e^{-sx}\, dF(x)$, i.e., g_k is a geometric distribution.

We ask if the converse conclusion is true. In other words, assuming that for given $F(x)$, the probabilities g_k from (1) are $g_k = p^k q$, $k = 0, 1, 2, \dots, 0 < p < 1$,

* Work is supported by The Bulgarian Ministry of Sciences and Education, Grant No. MM17/91

$q = 1-p$, we wish to determine the distribution function $G(t)$. We shall be interested in the case when the distribution $G(t)$ is exponential. Our aim is to characterize the distribution $G(t)$ when $F(x)$ is given and $\{g_k\}$ is the geometric distribution. Galambos (1978), p. 143 has shown that if $F(x) = 1 - \exp(-\lambda x)$, $\lambda > 0$, and g_k are geometric, then $G(x)$ possesses an exponential distribution.

In the sequel we shall use the following notation: $L_2[0,\infty)$ will be a set of real functions, twice integrable on $[0,\infty)$; $\delta(x)$ will be the Dirac function: $\delta(x) = 0$, $x < 0$, $\delta(x) = 1$, $x \geq 0$. *NBU* (or *NWU*)—the class of distribution functions G which satisfy

$$\Delta(x_1, x_2) = \bar{G}(x_1 + x_2) - \bar{G}(x_1)\bar{G}(x_2) \geq 0 \quad (\text{or } \leq 0).$$

A straightforward result in this connection is the following one.

Proposition 1. *Let*

$$\int_0^\infty p_k(t)\, dG(t) = p^k q, \qquad k = 0, 1, 2, \ldots, \quad 0 < p < 1 \tag{2}$$

hold for all degenerate distributions $F(x) = \delta(x - x_0)$, $x_0 \in [0,\infty)$ *and for all* $k = 0$, 1, 2, ... *Then* $G(x)$ *is an exponential distribution.*

Actually, in this case equation (2) is $\bar{G}(kx_0) = \bar{G}^k(x_0)$, for every $x_0 \geq 0$. Setting $f(x) = \log \bar{G}(x)$, we get $f(kx_0) = k\, f(x_0)$ from which it follows that

$$f\left(\frac{k}{m}\, x_0\right) = \frac{k}{m}\, f(x_0)$$

for all integers k and m. Therefore $f(x)$ is a linear function and $\bar{G}$ is exponential.

If we are interested in solution G of (1) in a narrower class we may assume that (1) holds only for all $k = 0, 1, 2, \ldots$.

In this case we have the following proposition.

Proposition 2. *Let* (2) *be fulfilled for* $k = 0, 1, 2, \ldots$ *and for fixed* $F(x)$. *If the distribution function* $G(t)$ *belongs to the class NBU (or NWU), then* $G(t)$ *is an exponential distribution whenever* $\operatorname{supp} F(dx) = [0,\infty)$.

Proof. Define the measure $\mu(x_1, x_2)$ on $R_2 = [0,\infty) \times [0,\infty)$ as the product

$$\mu(dx_1, dx_2) = F(dx_1)\, F(dx_2).$$

Then equation (2) for $k = 2$ can be written as follow:

$$\int_{R_2} \bar{G}(x_1, x_2)\, \mu(dx_1, dx_2) = \int_{R_2} \bar{G}(x_1)\, \bar{G}(x_2)\, \mu(dx_1, dx_2). \tag{3}$$

This implies that

$$\int_{R_2} \Delta(x_1, x_2)\, \mu(dx_1, dx_2) = 0. \tag{4}$$

Since $G \in NBU$ (or NWU), we have $\Delta(x_1, x_2) \geq 0$ (≤ 0) and (4) yields $\Delta(x_1, x_2) = 0$. The last equality leads to the well-known Cauchy equation which gives the required statement.

Another method to obtain the result we are interested in is to assume that the unknown unction $\bar{G}(x)$ is an absolutely monotone unction. Such a unction by Bernstein's representation can be written in the ollowing integral orm

$$\bar{G} = \int_0^\infty e^{-yx}\, dU(y), \tag{5}$$

where $U(y)$ is a distribution unction.

Proposition 3. *If $\bar{G}(x)$ is an absolutely monotone function, and equation* (2) *holds for all integers k and for some $F(x)$, then $\bar{G}(x)$ is exponential.*

Proof. From (2), using (5), we get

$$\int_0^\infty h^k(y)\, dU(y) = \left[\int_0^\infty h(y)\, dU(y)\right]^k. \tag{6}$$

Since the Laplace-Stieltjes trans orm $h(y)$ is non-increasing there exists the inverse unction $h^{-1}(u) = y$ o $h(y)$. Putting $h(y) = u$ and changing the variable in (6), we obtain

$$\int_0^1 u^k\, dB(u) = a^k, \quad a = \int_0^\infty f(y)\, dU(y), \quad B(u) = 1 - U(h^{-1}(u)). \tag{7}$$

Since (7) is ulfilled orall $k = 0, 1, 2, \ldots$ the unction $B(u)$ is $B(u) = \delta(u - a)$ and the unction $U(y)$ will also be a Dirac unction at a point, say $y = y_0$. There ore $\bar{G}(x)$, defined by (5), will be an exponential unction.

The general result which we shall prove, is the ollowing one.

Theorem 1. *Suppose that* (2) *holds for an arbitrary, but fixed distribution $F(x)$ and for all $k = 0, 1, 2, \ldots$. Let the unknown function $G(x)$ be from the class of functions having a derivative and such that $x\, G'(x) \in L_2[0, \infty)$. Then $G(x) = 1 - \exp(-ax)$, $a > 0$.*

Proof. Denote

$$K(x, y) = \sum_0^\infty p_k\left(\frac{y}{1-y}\right) x^k, \quad \varphi(x) = G'\left(\frac{x}{1-x}\right)(1-x)^{-2},$$

$$f(x) = q(1 - px)^{-1}, \quad 0 \leq x \leq 1, \quad 0 \leq y \leq 1.$$

Multiplying equation (2) by x^k and summing with respect to k, we get

$$f(x) = \int_0^1 K(x,y)\,\varphi(y)\,dy. \tag{8}$$

Equality (8) is a Fredholm equation with given kernel K, given $f(x)$ and unknown function φ. Obviously

$$0 \le K(x,y) \le \sum_k p_k\left(\frac{y}{1-y}\right) = 1$$

and

$$\int_0^1 K^2(x,y)\,dy = A^2(x) \le 1,$$

therefore $K \in L_2(0,1)$. Let

$$K_R(x,y) = \int_0^1 K(x,z)\,K(y,z)\,dz.$$

Following the theory of integral equations (see Tricomi (1957), Chap. 3.16) we denote by λ_1^2, λ_2^2, ... the eigenvalues of the kernel K_R and by $\mu_1(x)$, $\mu_2(x)$, ... the corresponding eigenfunctions functions.

Let

$$f_k = \int_0^1 f(x)\,\mu_k(x)\,dx.$$

It is known that

$$f_k = \frac{1}{\lambda_k}\int_0^1 \varphi(x)\,\mu_k(x)\,dx, \tag{9}$$

and by Hilbert–Schmidt Theorem we have

$$f(x) = \sum_1^\infty p_k\,\mu_k(x), \tag{10}$$

where the series is uniformly convergent on (0, 1).

Since $\{\mu_k\}$ is an orthonormal system, it is closed. By Picard's Theorem about Fredholm integral equations of the first kind, equation (8) will have a unique solution if the series

$$\sum_1^\infty \lambda_k^2 f_k^2$$

converges.

With the help of (9), using Bessel's inequality, we get

$$\sum_1^\infty \lambda_k^2 f_k^2 \le \int_0^1 \varphi^2(x)\,dx. \tag{11}$$

Let $G_0(x)$ be a solution of (2) for which the conditions of our Theorem are fulfilled. Then

$$\varphi_0(x) = G_0'\left(\frac{x}{1-x}\right)(1-x)^{-2}$$

will be a solution of (8). At least one solution of this kind exists, and the exponential distribution is such a one.

Moreover,

$$\int_0^1 \varphi_0^2(x)\,dx = \int_0^1 G_0'^2\left(\frac{x}{1-x}\right)(1-x)^{-4}\,dx = \int_0^\infty G_0'^2(1+t)^2 < \infty$$

and comparing with (11) we conclude that the series $\sum_1^\infty \lambda_k^2 f_k^2$ converges. Therefore the unique solution of (2) is $G_0(x) = 1 - \exp\{-ax\}$, where constant a is determined by the equation $h(a) = p$.

All results can be reformulated as answers to the following problem:

Let $X_1, X_2, \ldots$ be non-negative i.i.d. random variables and equality

$$E\,\varphi(X_1 + X_2 + \cdots + X_n) = [E\,\varphi(X_1)]^n \tag{12}$$

holds for $n = 0, 1, 2, \ldots$. Determine the form of the function φ.

Note that (12) is a well-known property of the Laplace-Stieltjes transform in the case of identically distributed and independent random variables. The results given here show that under certain assumptions, the Laplace-Stieltjes transform only obeys property (12).

References

Galambos, J. (1978). *The Asymptotic Theory of Extreme Order Statistics.* Wiley, New York.

Tricomi, F.G. (1957). *Integral Equations.* Interscience Publishers, Inc., New York.

Exploring Stochastic Laws. pp. 357–366.
Eds. A.V. Skorokhod and Yu.V. Borovskikh

Planar random evolution with three directions

ENZO ORSINGHER AND ARISTIDE SAN MARTINI

Dipartimento di Statistica, Probabilita e Statistiche Applicate p. le A. Moro 5, 00185 Rome, Italy

Received January 30, 1995

Abstract. A planar random motion with three possible directions changing cyclically at Poisson-paced times is analysed and the equation governing the distribution of its current position is derived. When the angular amplitude between the three possible velocity directions is equal to $2\pi/3$, recurrence relationships for the position distributions (when the number of velocity switches is fixed) are worked out by solving suitable initial-value problems. An integral relationship for the Fourier transform of the distribution of the position at time t is presented.

1. Introduction

Planar random evolutions governed by hyperbolic equations have been considered and analysed in some recent papers (Kolesnik and Turbin (1990a, b), Orsingher and Kolesnik (1994)).

Planar random motions depend in a crucial way on the number of possible directions. This is because the order of the partial differential equations governing the transition law of the randomly moving particle is equal to the number of directions.

We observe that the minimal number of possible directions for a non - trivial planar motion is three (a two-direction motion yields a process where the set of possible positions at an arbitrary time t is just a segment). The equation governing the distribution can have a cumbersome form, even if the number of directions is as small as three, unless these are suitably chosen.

The changes of direction are governed by a homogeneous Poisson process with rate λ. In the papers by Kolesnik (1989) and by Kolesnik and Turbin (1990b) it is assumed that from each direction, after a Poisson event, the motion can switch to any of the other possible directions. Instead, we assume in this paper that the particle takes up all possible directions cyclically.

The paper by Kolesnik and Turbin (1990b) presents the nth order hyperbolic equation governing the transition function when the angles of the directions are $\theta = k\frac{2\pi}{n}$, $k = 0, 1, \dots, n-1$, and when, at Poisson events, switching from any direction to any other is possible.

So far, only in a special case with four orthogonal directions, it has been possible to solve the equations obtaining the explicit form of the transition function at time t (see Orsingher and Kolesnik (1994)).

We start now describing the three-direction random evolution dealt with in this paper.

The possible velocities are

$$\vec{v}_n = c(\cos\theta_n \vec{i} + \sin\theta_n \vec{j}), \qquad n = 1, 2, 3,$$

where

$$0 \leq \theta_1 \leq \frac{2}{3}\pi \leq \theta_2 \leq \frac{4}{3}\pi \leq \theta_3 \leq 2\pi. \tag{1.1}$$

At time $t = 0$ the particle starts moving from the origin with velocity $\vec{v}_n$ (chosen with probability 1/3) until the first Poisson event, when the particle's velocity switches to $\vec{v}_n + 1$ (of course $\vec{v}_{n+3i} = \vec{v}_n$, $i = 1, 2, \ldots$). At the second Poisson event the velocity changes again to $\vec{v}_n + 2$, and so on, repeating cyclically the velocities in this way.

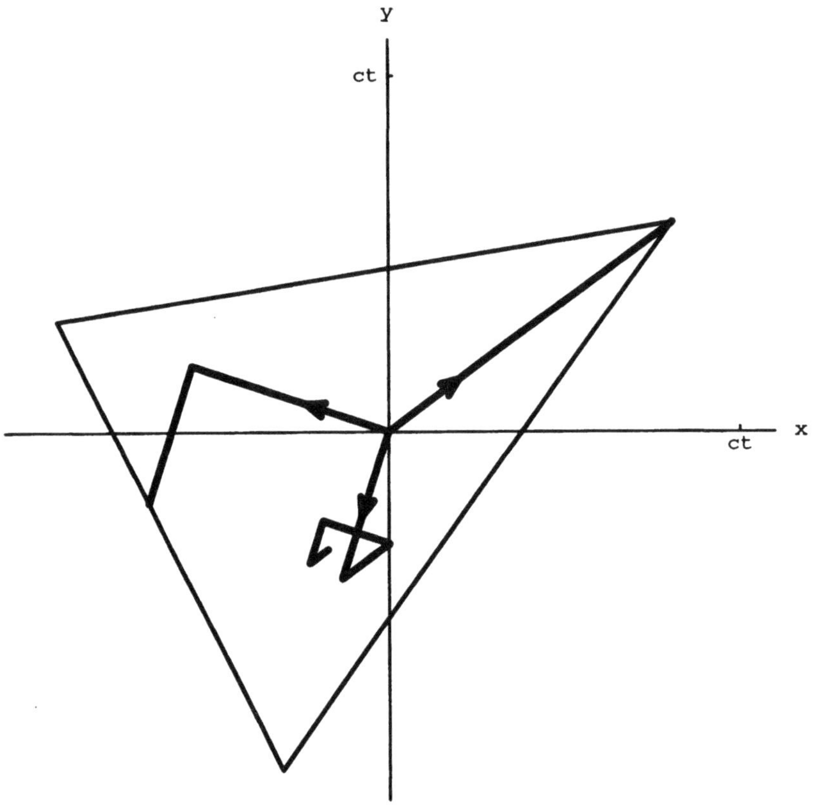

Figure 1. Three examples of paths with initial velocities $\vec{v}_1, \vec{v}_2, \vec{v}_3$, and with, respectively, $0, 1, 4$ Poisson events occurring before time t.
The arrows represent the possible velocities.

The constraint (1.1) ensures that both components of the velocity change sign at least once during each cycle.

The set of possible position of the particle at time t is the triangle T with vertices $(ct\cos\theta_n, ct\sin\theta_n)$, $n = 1,2,3$ (see Fig.1). It is clear that at time t, if no Poisson event has occurred, the particle is located on one of the vertices of T, depending on the initial direction. If exactly one change of direction happens before t, the particle reaches the boundary of T, vertices excluded. If more than one Poisson event occurs before t, the particle is located in the interior of T. Three examples of such paths are depicted in Fig.1. If $(X(t),\ Y(t))$, $t \geq 0$, denotes the position at time t of the moving particle, our interest is to study the distribution

$$P(x,y,t) = \Pr\{X(t) \leq x,\ Y(t) \leq y\}. \tag{1.2}$$

We present the hyperbolic third-order partial differential equation governing the distribution (1.2) and discuss its form when some particular assumptions on the angles θ_n, $n = 1,2,3$, are made.

In the special case where $\theta_1 = \pi/6$, $\theta_2 = 5\pi/6$, $\theta_3 = 3\pi/2$, we also study the following distributions

$$P_n^k(x,y,t) = \Pr\{X(t) \leq x,\ Y(t) \leq y,\ N(t) = k,\ D(t) = n\}, \quad n = 1,2,3 \tag{1.3}$$

where $k \geq 0$ and $N(t)$ denotes the number of speed changes until time t and $D(t)$ denotes the current velocity.

By solving suitable initial-value problems for the third - order equations governing (1.3), we are able to obtain recurrence relationships for the distributions P_n^k.

We would like to remark that the special case referred to above, possesses a fundamental advantage in that the triangular circuits, with sides parallel to the boundary, form equiprobability curves. This permits us to deal with partial differential equations depending on time t and only one space parameter v.

The random motions discussed here seem to us simplified but realistic models of physical phenomena like motions of particles in a turbulent medium.

2. Derivation of the general equation

We start considering the distributions of the triple $\big(X(t),\ Y(t),\ D(t);\quad t \geq 0\big)$, namely

$$P_n(x,y,t) = \Pr\{X(t) \leq x, Y(t) \leq y, D(t) = n\} = \sum_{k=0}^{+\infty} P_n^k(x,y,t), n = 1,2,3, \tag{2.1}$$

and the corresponding densities p_n.

Theorem 2.1. *The densities p_n of the distributions* (2.1) *are solutions of the following differential system*

$$\begin{aligned}\frac{\partial p_1}{\partial t} &= -c\cos\theta_1\frac{\partial p_1}{\partial x} - c\sin\theta_1\frac{\partial p_1}{\partial y} + \lambda(p_3 - p_1)\\ \frac{\partial p_2}{\partial t} &= -c\cos\theta_2\frac{\partial p_2}{\partial x} - c\sin\theta_2\frac{\partial p_2}{\partial y} + \lambda(p_1 - p_2)\\ \frac{\partial p_3}{\partial t} &= -c\cos\theta_3\frac{\partial p_3}{\partial x} - c\sin\theta_3\frac{\partial p_3}{\partial y} + \lambda(p_2 - p_3).\end{aligned} \tag{2.2}$$

Hint of proof. The particle is near the point (x, y) with velocity $\vec{v}_n$ at time $t + \Delta t$ provided that one of the following events occurs:

i) at time t the particle is near the point $(x - c\Delta t\cos\theta_n, y - c\Delta t\sin\theta_n)$ and no change of direction occurs in $(t, t + \Delta t]$;

ii) at time t the particle is near the point $(x - c\Delta t\cos\theta_{n-1}, y - c\Delta t\sin\theta_{n-1})$ and one change of direction occurs.

Remark 2.1. The case where the possible directions are $\theta_n = n\frac{2\pi}{3}$, $n = 1, 2, 3$, and from each direction the motion can switch to any of the other two, is studied in Kolesnik (1989).

Theorem 2.2. *The densities p_1, p_2, p_3 and their sum $p = p_1 + p_2 + p_3$ are solutions of the third-order hyperbolic equation*

$$\begin{aligned}&\left(\frac{\partial}{\partial t}+\lambda\right)^3 u - \lambda^3 u = -c\{\cos\theta_1+\cos\theta_2+\cos\theta_3\}\frac{\partial}{\partial x}\left(\frac{\partial}{\partial t}+\lambda\right)^2 u\\ &-c\{\sin\theta_1+\sin\theta_2+\sin\theta_3\}\frac{\partial}{\partial y}\left(\frac{\partial}{\partial t}+\lambda\right)^2 u - c^2\Big\{\sin\theta_1(\cos\theta_2+\cos\theta_3)+\\ &\quad+\sin\theta_2(\cos\theta_1+\cos\theta_3)+\sin\theta_3(\cos\theta_1+\cos\theta_2)\Big\}\frac{\partial^2}{\partial x\partial y}\left(\frac{\partial}{\partial t}+\lambda\right)u\\ &-c^2\{\cos\theta_1\cos\theta_2+\cos\theta_1\cos\theta_3+\cos\theta_2\cos\theta_3\}\frac{\partial^2}{\partial x^2}\left(\frac{\partial}{\partial t}+\lambda\right)u\\ &-c^2\{\sin\theta_1\sin\theta_2+\sin\theta_1\sin\theta_3+\sin\theta_2\sin\theta_3\}\frac{\partial^2}{\partial y^2}\left(\frac{\partial}{\partial t}+\lambda\right)u\\ &-c^3\cos\theta_1\cos\theta_2\cos\theta_3\frac{\partial^3 u}{\partial x^3} - c^3\sin\theta_1\sin\theta_2\sin\theta_3\frac{\partial^3 u}{\partial y^3}\\ &-c^3\{\sin\theta_1\cos\theta_2 cos\theta_3+\sin\theta_2\cos\theta_1\cos\theta_3+\sin\theta_3\cos\theta_1\cos\theta_2\}\frac{\partial^3 u}{\partial x^2\partial y}\\ &-c^3\{\cos\theta_1 sin\theta_2\sin\theta_3+\cos\theta_2\sin\theta_1\sin\theta_3+\cos\theta_3\sin\theta_1\sin\theta_2\}\frac{\partial^3 u}{\partial x\partial y^2}.\end{aligned} \tag{2.3}$$

Proof. Equation (2.3) is obtained by repeated derivations of equations (2.2) and suitable substitutions.

Equation (2.3) has a very complicated structure and we must restrict ourselves to particular cases in order to develop the investigation further.

Remark 2.2. In the special case where $\theta_1 = \theta$, $\theta_2 = \pi - \theta$, $\theta_3 = \frac{3\pi}{2}$, equation (2.3) reduces to

$$\begin{aligned}\left(\frac{\partial}{\partial t}+\lambda\right)^3 u - \lambda^3 u &= -c(2\sin\theta - 1)\frac{\partial}{\partial y}\left(\frac{\partial}{\partial t}+\lambda\right)^2 u + c^2\cos^2\theta\frac{\partial^2}{\partial x^2}\left(\frac{\partial}{\partial t}+\lambda\right)u \\ &\quad -c^2(\sin^2\theta - 2\sin\theta)\frac{\partial^2}{\partial y^2}\left(\frac{\partial}{\partial t}+\lambda\right)u + c^3\sin^2\theta\frac{\partial^3 u}{\partial y^3} \\ &\quad -c^3\cos^2\theta\frac{\partial^3 u}{\partial x^2\partial y}. \end{aligned} \tag{2.4}$$

We observe that the horizontal component of motion has a law satisfying the equation governing a one-dimensional motion with velocities $c\cos\theta, -c\cos\theta, 0$, taken up cyclically. This equation reads as following

$$\left(\frac{\partial}{\partial t}+\lambda\right)^3 u - \lambda^3 u = c^2\cos^2\theta\frac{\partial^2}{\partial x^2}\left(\frac{\partial}{\partial t}+\lambda\right)u$$

The Galilean transformation $x' = x, t' = t, y' = y + ct$ reduces equation (2.4) to the following one

$$\begin{aligned}\left(\frac{\partial}{\partial t'}+\lambda\right)\Big\{\left(\frac{\partial}{\partial t'}+\lambda\right)^2 &+ c^2(1+\sin\theta)^2\frac{\partial^2}{\partial y'^2} - \\ &- c^2\cos^2\theta\frac{\partial^2}{\partial x'^2} + \\ &+ 2c(1+\sin\theta)\frac{\partial}{\partial y'}\left(\frac{\partial}{\partial t}+\lambda\right)\Big\}u = \lambda^3 u. \end{aligned} \tag{2.5}$$

Equation (2.5) refers to the motion of the particle as observed in the frame of reference (x', y', t') in which the possible velocities are

$$\Big(c\cos\theta, c(1+\sin\theta)\Big),\ \Big(-c\cos\theta, c(1+\sin\theta)\Big),\ (0,0).$$

3. Special case

In this Section, we restrict ourselves to the case where $\theta_1 = \frac{\pi}{6}, \theta_2 = \frac{5\pi}{6}, \theta_3 = \frac{3\pi}{2}$, so that the set of possible positions at time t forms an equilateral with sides $\sqrt{3}ct$ long.

Equation (2. 3), and therefore equation (2. 4) also, reduce to

$$\left(\frac{\partial}{\partial t}+\lambda\right)^3 u - \lambda^3 u = \frac{3}{4}c^2\left(\frac{\partial}{\partial t}+\lambda\right)\left(\frac{\partial^2}{\partial x^2}+\frac{\partial^2}{\partial y^2}\right)u - \frac{3}{4}c^3\frac{\partial^3 u}{\partial x^2\partial y} + \frac{c^3\partial^3 u}{4\partial y^3}. \tag{3.1}$$

The transformation $u = e^{-\lambda t}\omega$ permits us to obtain a simpler equation which reads

$$\left(\frac{\partial^3}{\partial t^3} - \lambda^3\right)\omega = \frac{3}{4}c^2\frac{\partial}{\partial t}\left(\frac{\partial^2}{\partial x^2} + \frac{\partial^2}{\partial y^2}\right)\omega - \frac{3}{4}c^3\frac{\partial^3\omega}{\partial x^2\partial y} + \frac{c^3\partial^3\omega}{4\partial y^3}. \tag{3.2}$$

Our main result in this Section is stated in the following theorem.

Theorem 3.1. *On the triangular circuit made up of the lines*

$$\begin{aligned} \sqrt{3}x - y &= v, \\ \sqrt{3}x + y &= -v, \qquad 0 \le v \le ct, \\ 2y &= v \end{aligned} \tag{3.3}$$

equation (3.1) *reduces to the following one*

$$\left(\frac{\partial}{\partial t} + \lambda\right)^3 u - \lambda^3 u = 3c^2\left(\frac{\partial}{\partial t} + \lambda\right)\frac{\partial^2 u}{\partial v^2} + 2c^3\frac{\partial^3 u}{\partial v^3}, \tag{3.4}$$

and equation (3.2) *reduces to*

$$\frac{\partial^3\omega}{\partial t^3} - \lambda^3\omega = 3c^2\frac{\partial^3\omega}{\partial t\partial v^2} + 2c^3\frac{\partial^3\omega}{\partial y^3}. \tag{3.5}$$

Proof. It is sufficient to evaluate $\frac{\partial}{\partial x}$ and $\frac{\partial}{\partial y}$ in terms of $\frac{\partial}{\partial v}$ along each of the three segments in (3.3).

Remark 3.1. Theorem 3.1 states that the distribution p does not change along each triangular circuit (3.3). Thus

$$p(v,t)dv = \Pr\left\{(X(t), Y(t)) \in T_{v+dv} - T_v\right\},$$

where T_v is the triangle with vertices

$$\left(\frac{\sqrt{3}}{2}v, \frac{v}{2}\right), \quad \left(-\frac{\sqrt{3}}{2}v, \frac{v}{2}\right), \quad (0, -v)$$

and is drawn in Fig.2. An analogous result permitting us to pass from an equation in (x, y, t) to another one in (v, t) is not possible even for the case

$$\begin{aligned} \theta_1 &= \theta, \\ \theta_2 &= \pi - \theta, \\ \theta_3 &= \frac{3\pi}{2}. \end{aligned}$$

Remark 3.2. In the limit $\lambda \to +\infty$, $c \to +\infty$ and $\frac{c^2}{\lambda} \to 1$, equation (3.1) becomes the two-dimensional heat equation

$$\frac{\partial u}{\partial t} = \frac{1}{4}\left(\frac{\partial^2}{\partial x^2} + \frac{\partial^2}{\partial y^2}\right)u.$$

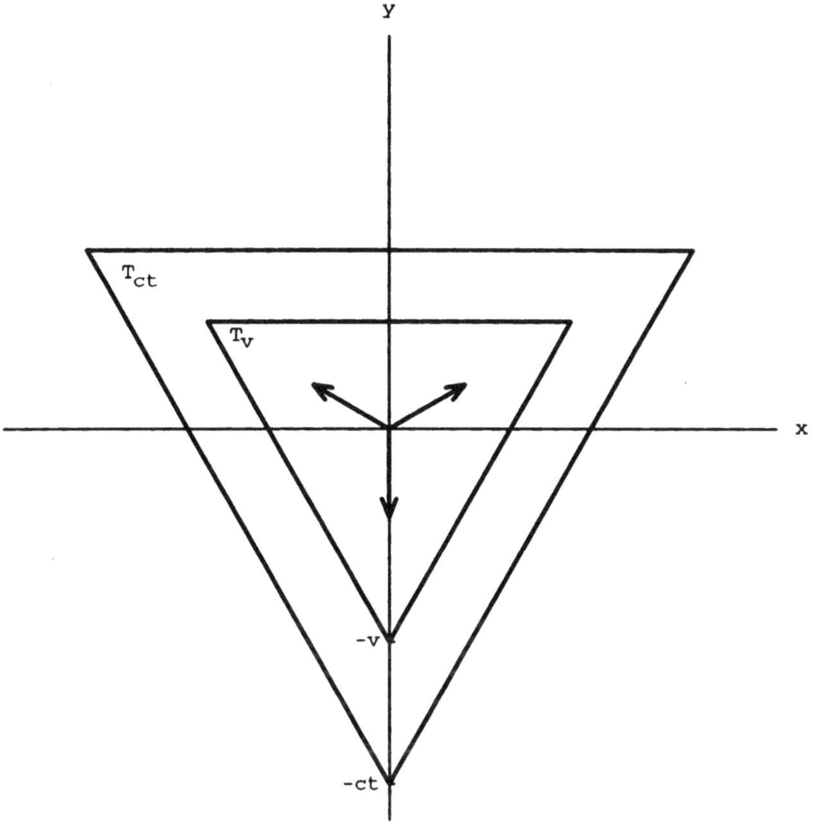

Figure 2. Along the boundary of the triangle T_v, the probability distribution of the particle's position at time t is constant. The three arrows are the possible velocities.

4. Recurrence relationships

The derivation of the distribution $p = p(v,t)$, $0 \leq v \leq ct$, by solving an initial-value problem for equation (3.4), poses daunting difficulties. This is because the general solution of (3.4) cannot be written down in a simple, closed form.

We therefore prefer another approach based on the analysis of the distributions $P_n^k(x,y,t)$, $n = 1,2,3$ and $k \geq 0$, defined in (1.3), for which we write down the governing differential system by suitable adapting the reasoning leading to (2.2). We are still assuming, as in Section 3, that $\theta_1 = \pi/6$, $\theta_2 = 5\pi/6$, $\theta_3 = 3\pi/2$.

Theorem 4.1. *The densities of distributions* (1.3), p_n^k, $n = 1,2,3$ *and* $k \geq 0$ *are solutions of the following differential-difference system.*

$$\begin{aligned}\frac{\partial p_1^k}{\partial t} &= -\frac{\sqrt{3}}{2}c\frac{\partial p_1^k}{\partial x} - \frac{c}{2}\frac{\partial p_1^k}{\partial y} + \lambda(p_3^{k-1} - p_1^k),\\ \frac{\partial p_2^k}{\partial t} &= \frac{\sqrt{3}}{2}c\frac{\partial p_2^k}{\partial x} - \frac{c}{2}\frac{\partial p_2^k}{\partial y} + \lambda(p_1^{k-1} - p_2^k),\\ \frac{\partial p_3^3}{\partial t} &= c\frac{\partial p_3^k}{\partial y} + \lambda(p_2^{k-1} - p_3^k).\end{aligned} \tag{4.1}$$

Proof. The same as in theorem 2.1, with the necessary restrictions on θ_n, $n = 1, 2, 3$.

Corollary 4.1. *By the same procedure as in Theorem 2.1, it can be proved that the densities p_n^k and $p^k = \sum_{i=1}^3 p_n^k$ are solutions of the following third-order differential-difference equation*

$$\begin{aligned}\left(\frac{\partial}{\partial t} + \lambda\right)^3 u^k &= \frac{3}{4}c^2\left(\frac{\partial}{\partial t} + \lambda\right)\left(\frac{\partial^2}{\partial x^2} + \frac{\partial^2}{\partial y^2}\right)u^k\\ &\quad -\frac{3}{4}c^3\frac{\partial^3 u^k}{\partial x^2 \partial y}\\ &\quad +\frac{c^3}{4}\frac{\partial^3 u^k}{\partial y^3} + \lambda^3 u^{k-3}, \quad k \geq 3.\end{aligned} \tag{4.2}$$

The reader should remark that summing up (4.2) with respect to $k \geq 0$, equation (3.4) emerges.

Theorem 4.2. *On the triangular circuit* (3.3), *equation* (4.2) *reduces to*

$$\left(\frac{\partial}{\partial t} + \lambda\right)^3 u^k = 3c^2\left(\frac{\partial}{\partial t} + \lambda\right)\frac{\partial^2 u^k}{\partial v^2} + 2c^3\frac{\partial^3 u^k}{\partial v^3} + \lambda^3 u^{k-3}, \quad k \geq 3. \tag{4.3}$$

Proof. The same as in theorem 3.1.

For us it is important to work with equation

$$\frac{\partial^3 \omega^k}{\partial t^3} = 3c^2\frac{\partial^3 \omega^k}{\partial t \partial v^2} + 2c^3\frac{\partial^3 \omega^k}{\partial v^3} + \lambda^3 \omega^{k-3}, \quad k \geq 3. \tag{4.4}$$

extracted from (4. 3) by means of the usual position $u^k = e^{-\lambda t}\omega^k$.

We now give a recurrence formula for the Fourier transforms

$$W^k = \int_{-\infty}^{+\infty} e^{i\beta v}\omega^k(v,t)dt, \quad k \geq 0. \tag{4.5}$$

Theorem 4.3. *For the Fourier transforms* (4.5) *the following recursive relationships hold*

$$W^k(\beta,t) = -\frac{\lambda^3}{9c^2\beta^2}\int_0^t W^{k-3}(\beta,s)\left[e^{-2ic\beta(t-s)} + e^{ic\beta(t-s)}\left\{3ic\beta(t-s)-1\right\}\right]ds, \quad (4.6)$$

$k \geq 3$.

Proof. The Fourier transforms (4.5) are solutions of the ordinary linear non-homogeneous differential equations

$$\frac{d^3W^k}{dt^3} = -3c^2\beta^2\frac{dW^k}{dt} + 2ic^3\beta^3W^k + \lambda^3W^{k-3}, \quad k \geq 3 \quad (4.7)$$

whose general solution reads

$$W^k(\beta,t) = Ae^{-2ic\beta t} + Be^{ic\beta t} + Cte^{ic\beta t} + \Gamma(\beta,t) \quad (4.8)$$

where A, B, C are arbitrary constants and Γ is a solution of the non-homogeneous equation (4.7). The function Γ can be obtained by means of the usual Lagrange technique after some calculations. Since functions $\omega^k = e^{\lambda t}p^k(v,t)$ satisfy the initial conditions

$$\omega^k(v,0) = \frac{\partial\omega^k}{\partial t}(v,0) = \frac{\partial^2\omega^k}{\partial t^2}(v,0) = 0,$$

we can conclude that $A = B = C = 0$, hence relationship (4.6) follows.

Theorem 4.4. *For the functions* $\omega^k(v,t) = e^{\lambda t}p^k(v,t), k \geq 3$, *the following recursive relationships are true*

$$\omega^k(v,t) = \frac{\lambda^3}{9c^2}\int_0^t ds\int_{v-c(t-s)}^{v+2c(t-s)}\{2c(t-s)+v-z\}\omega^{k-3}(z,s)dz. \quad (4.9)$$

Proof. We start taking the inverse Fourier transforms of (4.6), that is,

$$\begin{aligned}
\omega^k(v,t) &= \frac{1}{2\pi}\int_{-\infty}^{+\infty} e^{-i\beta v}W^k(\beta,t)d\beta \\
&= \frac{\lambda^3}{9c^2}\int_0^t ds\int_{-\infty}^{+\infty}\omega^{k-3}(z,s)dz\left\{\frac{1}{2\pi}\int_{-\infty}^{+\infty}\frac{e^{i\beta v+i\beta z}}{(-i\beta)^2}\left[e^{-2ic\beta(t-s)}+\right.\right. \\
&\quad \left.\left. + e^{ic\beta(t-s)}\{3ic\beta(t-s)-1\}\right]d\beta\right\} \\
&= \frac{\lambda^3}{9c^2}\int_0^t ds\int_{-\infty}^{+\infty}\omega^{k-3}(z,s)\left\{[z-\{v+2c(t-s)\}]I_{(v+2c(t-s),+\infty)}(z)+\right. \\
&\quad \left. -[z-\{v-c(t-s)\}]I_{(v-c(t-s),+\infty)}(z)+3c(t-s)I_{(v-c(t-s),+\infty)}(z)\right\}dz.
\end{aligned}$$

Elementary calculations straightforwardly lead to (4.9). In the last equality above, the following formula was used

$$\frac{1}{2\pi}\int_{-\infty}^{+\infty}\frac{e^{-i\beta z+i\beta a}}{(-i\beta)^k}d\beta = \frac{(z-a)^{k-1}}{(k-1)!}I_{a,+\infty}(z), k \geq 1.$$

The scrupulous reader can check with some care that (4.9) is indeed a solution of (4.4).

It is interesting to remark that the recursive formula (4.9) is similar to the well known D'Alembert solution of the forced wave equation.

By summing up relationships (4.9) with respect to k, $k \geq 3$, one obtains an integral equation for the component $\omega(v,t)$ of the distribution of the position at time t when the number of velocity changes is random. Similarly, by summing up relationships (4.6), one obtains an analogous integral equation for the Fourier transform $W(\beta, t)$.

References

Kolesnik, A.D. (1989). On a model of Markovian random evolution on a plane. In: *Analytical methods of investigation of the evolution of stochastic systems.* Inst. Math. Ukr. Acad. Sci., Kiev, pp. 55–61 (in Russian).

Kolesnik, A.D. and Turbin, A.F. (1990a). *Symmetric Markovian random evolutions on a plane.* Preprint 90.12, Inst. Math. Ukr. Acad. Sci. Kiev, pp. 39. (In Russian).

Kolesnik, A.D. and Turbin, A.F. (1990b). Symmetric random evolutions in R^2. *Dokl. Acad. Nauk Ukrain.* pp. 10–11. (In Russian).

Orsingher, E. and Kolesnik, A.D. (1994). The explicit probability law of a planar random motion governed by a fourth-order hyperbolic equation. (Submitted for publication).

Exploring Stochastic Laws. pp. 367–382
Eds. A.V. Skorokhod and Yu.V. Borovskikh

Applications of null-additive set functions in mathematics

ENDRE PAP

Institute of Mathematics, University of Novi Sad, Trg D. Obradovića 4, 21000 Novi Sad, Yugoslavia

Received January 18, 1995

Abstract. In this paper a wide class of non-additive set functions is presented. Null-additive set functions covers decomposable measures, submeasures, k-triangular set functions, maxitive measures, etc. Two groups of applications in mathematics are presented. First on the class of Diagonal Theorems (which are very powerful methods in Measure theory and Functional Analysis) and the second – applications in solving nonlinear differential and difference equations.

Introduction

Non-additive set functions occur also in classical measure theory, as for example sub-measures, but interest in them starts to grow rapidly with many different applications, as for example in decision theory, artificial intelligence, game theory, etc. (see the references at the end). Early traces of non-additive set functions goes back to Choquet (1953–1954), Aleksjuk (1968), Dobrakov (1974), Sugeno (1974), Drewnowski (1972). There are books specially devoted to non-additive set functions such as in Aumann and Shapley (1974), Wang and Klir (1992), Denneberg (1994), Pap (1995).

1. Null-additive set functions

Let R be a ring of subsets of the given set X.

Definition 1. We say that

a) m is *monotone* if $A, B \in R$ and $A \subset B$ imply $m(A) \leq m(B)$;

b) m is *continuous* from below if for every increasing sequence (A_i) of elements of R such that $\cup_{i \in N} A_i \in R$, we have $m(\cup_{i \in N} A_i) = \lim_{i \to \infty} m(A_i)$;

c) m is *exhaustive* if for every disjoint sequence (A_i) of elements from R we have $\lim_{i\to\infty} m(A_i) = 0$;

d) m is *order continuous* if for each sequence (E_n) such that $E_n \searrow \emptyset$ we have

$$\lim_{n\to\infty} m(E_n \bigcap Y) = 0$$

uniformly with respect to $Y \in R$;

e) let $k \in [1, +\infty), m$ is *σ-k-subadditive* if for every disjoint sequence (A_i) of elements of R such that $\cup_{i\in N} A_i \in R$, we have

$$m(\bigcup_{i\in N} A_i) \leq m(A_1) + k \sum_{i=2}^{+\infty} m(A_i).$$

Definition 2. A set function $m : R \to [0, \infty]$ with $m(\emptyset) = 0$ is called *null-additive*, if

$$m(A \cup B) = m(A)$$

whenever $A, B \in R$, $A \cap B = \emptyset$ and $m(B) = 0$.

For properties of null-additive set functions see Pap (1994a), Pap (1995), Wang (1984), Wang and Klir (1992).

Example 1. Let S be a triangular conorm (t-conorm), i.e. a binary operation on $[0, 1]$ such that it is associative, commutative and monotone with a neutral element 0. A set function $m : R \to [0, 1]$ is called an S-decomposable measure if $m(\emptyset) = 0$ and

$$m(A \cup B) = m(A)Sm(B)$$

whenever $A, B \in R$ and $A \cap B = \emptyset$ (Fine (1973), Sugeno (1974), Dubois and Prade (1982), Pap (1990a, b, 1991a, b, 1995)). m is a monotone null-additive set function.

Example 2. ($\oplus$-decomposable measure — Sugeno and Murofushi (1987), Maslov (1987), Pap (1990b, 1992, 1993, 1994b, 1995), Wang and Klir (1992), Klement and Weber (1991), Baccelli et al. (1992)). The operation $\oplus$ (pseudo-addition) is a function $\oplus : [0, \infty] \times [0, \infty] \to [0, \infty]$ which is commutative, nondecreasing, associative, continuous and has a zero element 0. For example $x \oplus y = (x^p + y^p)^{1/p}$ for a fixed $p > 0$.

A set function $m : R \to [0, \infty]$ is a $\oplus$-decomposable measure if $m(\emptyset) = 0$ holds, and

$$m(A \cup B) = m(A) \oplus m(B),$$

whenever $A, B \in R$ and $A \cap B = \emptyset$. It is obvious that m is a null-additive set function. There are further generalizations for operations, defined on $[a, b] \subset [-\infty, +\infty]$.

Example 3. (k-triangular set functions, Agafonova and Klimkin (1974), Pap (1976, 1982, 1986, 1987, 1995), Guariglia (1990)). A set function $m : R \to [0, \infty)$ is said to be *k-triangular* for $k \geq 1$ if $m(\emptyset) = 0$ and

$$m(A) - km(B) \leq m(A \cup B) \leq m(A) + km(B),$$

whenever $A, B \in R$, $A \cap B = \emptyset$. It is obvious that k-triangular set function is always null-additive, although it may not be monotone. We shall also call 1-triangular set functions *only triangular* set functions. Special triangular set functions are submeasures.

Definition 3. Let R be a σ-ring, a set function $m : R \to [0, \infty]$ with $m(\emptyset) = 0$ is *σ-null-additive* if for any sequence (B_i) of pairwise disjoint sets from R such that $A \cap B_i = \emptyset$ and $m(B_i) = 0$ we have

$$m(A \cup \bigcup_{i=1}^{\infty} B_i) = m(A).$$

Monotone set functions which vanish at the empty set were investigated by G. Choquet (1953–1954) under the name *capacities.* He introduced an integral, which generalize the Lebesgue integral. Sugeno (1974) introduced an integral on set $A \subset X$

$$(S) \int_A h(x) dm := \sup_{\alpha \in [0, +\infty]} \min[\alpha, m(A \cap \{x : h(x) \geq \alpha\})].$$

The importance of null-additive fuzzy measures are stressed by the following results of Wang (1984).

Theorem 1. *Whenever, $f_1 = f_2$ a.e.,*

$$(S) \int f_1 dm = (S) \int f_2 dm$$

holds, if and only if m is null-additive.

Theorem 2. *Whenever $f_n \to f$ a.e., and there exist n_0 and a constant $c' \leq (S)\int f dm$ such that $m(\{\sup_{n \geq n_0} f_n > c'\}) < \infty$, then $(S)\int f_n dm \to (S)\int f dm$, if and only if m is null-additive.*

Sugeno and Murofushi (1991a) (see also Pap (1995)) stressed the importance of null-additivity for Choquet integral.

2. Diagonal theorem for triangular set function

We shall present a version of a Diagonal Theorem for triangular set functions. Diagonal Theorems as extracts and generalizations of the classical "sliding hump" method are very useful in the proofs of the generalized fundamental theorems of Functional Analysis and Measure Theory (Antosik (1978), Antosik and Saeki (1992), Antosik and Shwartz (1995), Mikusiński (1970), Pap (1974, 1982, 1995), Shwartz (1992)).

2.1. Diagonal theorems

Let R be a ring subset of a set. Let $m : R \to [0, \infty]$ be a triangular set function, i.e. k-triangular set function for $k = 1$. For each set $A \subset X$, let

$$\hat{m}(A) = \sup\{m(B) : \ B \in R, \ B \subset A\}.$$

Definition 4. We say that m is *α-bounded* ($0 \leq \alpha < \infty$) if $\inf_n m(B_n) \leq \alpha$ for each disjoint sequence $\{B_n\}$ in R.

If $\{A_n\}$, is a sequence of sets and $I \subset N$, we write A_I for $\cup_{n \in I} A_n$.

We have by Antosik and Saeki (1992) the following theorem.

Theorem 3. (*Diagonal Theorem for triangular set functions*). *Let $\{m_n\}$ be a sequence of α-bounded triangular set functions on the ring $R (0 \leq \alpha < \infty)$, let $\{B_n\}$ be a disjoint sequence in R, and let $\{\delta_n\}$ be a sequence in $[0, \infty]$. Then either there exists an infinite set $I \subset N$ such that*

(i) $\hat{m}_n(B_I \backslash B_n) < \delta_n \quad (n \in I)$,

or there exists a finite set $J \subset N$ which satisfies (i) with $I = J$ and

(ii) $\hat{m}_n(B_j) \geq \delta_n - \alpha$ *for infinitely many $n \in N$.*

Corollary 1. (Webers Version of the Nikodym Boundedness Principle (1986)). *Let R be a σ-algebra, and let $\mathcal{T}$ be a set of α-bounded triangular set functions on R for $\alpha < \infty$, such that*

$$\sup\{m(B) : m \in \mathcal{T}\} < \infty$$

for each $B \in R$. Then $\mathcal{T}$ is uniformly bounded.

Corollary 2. *Let $\alpha, \beta, \gamma \in [0, \infty]$, and let $\{m_n\}$ be a sequence of α-bounded triangular set functions on the ring R.*

(i) *Suppose that for each disjoint sequence $\{B_n\}$ in R and each subsequence $\{m_{n_k}\}$ of $\{m_n\}$, we have*

$$\inf\{m_{n_k}(B_E) : k \in E \subset N, \ B_E \in R\} \leq \beta,$$

and

$$\underline{\lim}_{j\to\infty}\{\underline{\lim}_{k\to\infty} m_{n_k}(B_j)\} \leq \gamma \overline{\lim}_{n\to\infty} m_n(B_j).$$

Then m_n are uniformly $(\alpha+\beta+\gamma)$-bounded.

(ii) *If R is a σ-algebra and if*

$$\psi(A) = \overline{\lim}_{n\to\infty} m_n(A)$$

is β-bounded, then the sequence $\{m_n\}$ is uniformly $(\alpha+2\beta)$-bounded.

Let S be a semigroup with triangular functional f, i.e. $f(x)-f(y) \leq f(x+y) \leq f(x)+f(y)$ for $x, y \in S$. Let x_{ij} be an infinite matrix (indexed by $N \times N$) with entries in S. For each sequence $\{x_j\}$ in S and each $I \subset N$, write $f(\sum_{j\in I} x_j)$ for $\lim_{n\to\infty} \sup f(\sum_{j\in I\cap[1,n]} x_j)$.

We have by Mikusiński (1970) and Pap (1974)

Corollary 3. (Mikusiński, Antosik, Pap Diagonal Theorem). *Suppose*

(i) $\lim_{n\to\infty} f(x_{ij}) = 0 \quad (i \in N)$.

Then there exists an infinite set I and a set $J \subset I$ such that

(a) $\sum_{j\in I} f(x_{ij}) < \infty \quad (i \in N)$, *and*

(b) $f(\sum_{j\in J} x_{ij}) \geq f(x_{ij})/2 \quad (i \in I)$.

Proof. By (i), we may suppose that (x_{ij}), satisfy $\sum_{j\in N} f(x_{ij}) < \infty$ for each $i \in N$. Define

$$m_i(E) = f(\sum_{j\in E} x_{ij}) \quad (i \in N, E \subset N).$$

Then each m_i is continuous (hence exhaustive) triangular set function on $\mathcal{P}(N)$. If $I \subset N$ is any set such that $\hat{m}_i(I\backslash\{i\}) < m_i(\{i\})/2$ whenever $i \subset I$, then $i \in I$ implies

$$m_i(I) \geq m_i(\{i\}) - m_i(I\backslash\{i\}) \geq m_i(\{i\})/2.$$

Therefore the desired result follows by Diagonal Theorem 3 for triangular set functions with $\delta_i = m_i(\{i\})/2$ and $\alpha = 0$.

Let G be a *quasi-normed group,* i.e. a group endowed with a functional $|.|$ such that it satisfies the conditions: (i) $|0| = 0$; (ii) $|x| = |-x|$; (iii) $|x+y| \leq |x|+|y|$.

Corollary 4. (Antosik Diagonal Theorem). *Let $\beta, \gamma \in [0,\infty)$, and suppose that*

(i) $\lim_{j\to\infty} |x_{ij}| = 0 \quad (i \in N)$,

(ii) *for each finite set $F \subset N$, $\overline{\lim}_{j\to\infty} |\sum_{j\in F} x_{ij}| \leq \beta$, and that each principle infinite submatrix of (x_{ij}) has a further principle submatrix y_{ij} such that*

(iii) $\underline{\lim}_{i\to\infty} |\sum_{j\in N} y_{ij}| \leq \gamma$.

Then $\overline{\lim}_{i\to\infty} |x_{ij}| \leq \beta + \gamma$.

Proof. Taking an infinite submatrix of (x_{ij}), we may suppose that $\sum_{j\in N}|x_{ij}| < \infty$ for each $i \in N$ and $\lim_{i\to\infty}|x_{ii}|$ exists in $[0,\infty]$. Let m_i be the continuous triangular set function on $\mathcal{P}(N)$ defined as in the proof of the last theorem.

Fix any $\beta' > \beta$. Then (ii) implies that there exists no finite set $J \subset N$ such that $\hat{m}_i(J) \geq \beta'$ for infinitely many $i \in N$. So, by Diagonal Theorem 3 with $\alpha = 0$ and $\delta_i = \beta'$, there exists an infinite set $I \subset N$ such that $\hat{m}_i(I\backslash\{i\}) < \beta'$ for each $i \in I$. Then, $i \in I' \subset I$ implies

$$|x_{ii}| = m_i(\{i\}) \leq m_i(I') + m_i(I'\backslash\{i\}) < m_i(I') + \beta'.$$

Then by (iii) we obtain that $\lim|x_{ii}| \leq \gamma + \beta'$.

As an immediate consequence we obtain a Diagonal Theorem (which we will denote by DT) given in the following statement.

Theorem 4. *Let $(G, |\cdot|)$ be a quasi-normed group, and (x_{ij}) an infinite matrix in G such that for every increasing sequence $\{m_i\}$ in N there exists a subsequence $\{n_i\}$ of $\{m_i\}$ such that*

$$\lim_{i\to\infty} x_{n_i n_j} = 0 \quad \forall\, j \in N, \quad \text{and} \quad \lim_{i\to\infty}\sum_{j=1}^{\infty} x_{n_i n_j} = 0.$$

Then $\lim_{i\to\infty} x_{ii} = 0$.

2.2. Applications

We present some applications of Diagonal Theorems. We do not intend to formulate the theorems in full, but to present the method of the proofs.

The first two theorems are from Measure Theory.

Nikodym convergence. *Let $\{\mu_n\}_{n\in N}$ be a sequence of countable additive functions from a σ-algebra $\sum$ to a topological group G and $\{E_n\}$ — a sequence of pairwise disjoint sequence of sets E_n from $\sum$.*

If $\mu_n(E) \to 0$ for $E \in R$, then $\mu_n(E_n) \to 0$. Applying the DT to the matrix $\{\mu_i(E_j)\}$, we get the assertion.

Nikodym boundedness. *If G is a topological vector space and $\{\mu_n(E)\}$ is a bounded sequence for each $E \in \sum$, then $\{\mu_n(E_n)\}$ is a bounded sequence, for each pairwise disjoint sequence $\{E_n\}$ from $\sum$.*

Applying the DT to the matrix $\{\alpha_i\mu_i(E_j)\}$ where $\{\alpha_j\}$ is a scalar sequence converging to zero, we get the assertion.

Continuous Convergence. *Assume that f_n for $n \in N$ are additive and continuous mappings from a complete metric group G to a topological group $Y, x_n \in G$ for $n \in N$ and $x_n \to 0$. If $f_n(x) \to 0$ for each $x \in G$, then $f_n(x_n) \to 0$.*

We remark that due to the completeness of G, for each subsequence $\{v_n\}$ of $\{x_n\}$ there are a subsequence $\{u_n\}$ of $\{u_n\}$ and $v \in G$ such that $v_1 + v_2 + \ldots = v$. Applying the DT to the matrix $\{f_i(x_j)\}$, we get $f_i(x_i) \to 0$.

Equicontinuity. *If Y is a topological vector space and $\{f_n(x)\}$ is a bounded sequence for each $x \in G$, then $f_n(x_n) \to 0$, $\{x_n\}$ is the same as in the previous case.*

Let $\{k_n\}$ be a sequence of positive integers such that $k_n \to \infty$ and $k_n x_n \to 0$. Applying the DT to the matrix $\{k_i^{-1} f_i(k_j x_j)\}$ we get $f_n(x_n) \to 0$.

Joint continuity. *If G is a metric group and at the same time G is a linear space such that the multiplication ax is separately continuous, then the multiplication is jointly continuous.*

Assume that $\alpha_n \to 0$ and $x_n \to 0$. Applying the DT to the matrix $\{\alpha_j x_j\}$ we get $\alpha_i x_i \to 0$.

Weak boundedness. *Let G be a normed space, and let G' be the dual of $G, x_n \in G$ and $x_n^{'} \in G'$ for $n \in N$. We note that G' is a Banach space.*

If $\{\langle x_n, x' >\}$ is a bounded sequence for each $x' \in G$ and $\|x_n^{'}\| \leq 1$ for $n \in N$, then $\{\langle x_n, x_n^{'} >\}$ is a bounded sequence.

Applying the DT to the matrix $\{\langle \sqrt{\alpha_i} x_i, \sqrt{\alpha_i} x_j^{'} \rangle\}$, where $\{\alpha_i\}$ is a numerical sequence converging to zero, we get $\alpha_i \langle x_i, x_i^{'} \rangle \to 0$.

Weak $*$-boundedness. *If G is a Banach space, $\{\langle x, x_n^{'} \rangle\}$ is a bounded sequence for each $x \in X$ and $\|x_n\| \leq 1$ for $n \in N$, then $\{\langle x_n, x_n^{'} \rangle\}$ is a bounded sequence.*

Applying the DT to the matrix $\{\langle \sqrt{\alpha_j} x_j, \sqrt{\alpha_i} x_i \rangle\}$ where $\{\alpha_i\}$ is a numerical sequence converging to zero, we get $\alpha_i \langle x_i, x_i^{'} \rangle \to 0$.

Orlicz–Pettis. *If $\|x_n^{'}\| \leq 1$ for $n \in N$ and for each increasing sequence of positive integers k_j there exists an $x \in G$ such that*

$$\sum_{j=1}^{\infty} \langle x_{k_j}, x' \rangle = \langle x, x' \rangle$$

for each $x' \in X$, then $\langle x_n, x_n^{'} \rangle \to 0$.

We may assume that G is separable and $\langle x, x_n' \rangle \to \langle x, x' \rangle$ for each $x \in G$ and an x' in G'. Otherwise we would take a suitable subsequence of $\{x_n'\}$. Applying the DT to the matrix $\{\langle x_j, x_i' - x' \rangle\}$ we get $\langle x_i, x_i' - x' \rangle \to 0$, or equivalently, $\langle x_i, x_i' \rangle \to 0$ for $\langle x_i, x' \rangle \to 0$.

2.3. Adjoint Theorem

If (G, τ) is a topological vector space, a sequence $\{x_k\}$ in G is said to be a $\tau - K$ sequence if every subsequence of $\{x_k\}$ has a further subsequence $\{x_{n_k}\}$ such that the subseries $\sum_k x_{n_k}$ is τ-convergent to an element of G. A topological vector space (G, τ) is said to be K- space if every sequence which converges to 0 is a $\tau - K$ sequence. A subset A of G is said to be $\tau - K$ bounded if for every sequence $\{x_n\} \subseteq A$ and every scalar sequence $\{t_n\}$ which converges to 0, the sequence $\{t_n x_n\}$ is a $\tau - K$ sequence.

Throughout this subsection let E and F be Hausdorff locally convex topological vector spaces and $T : E \to F$ a linear mapping. The domain of the adjoint operator, T', is defined to be $D(T') = \{y' \in F' : y'T \in E'\}$ and $T' : D(T') \to E'$ is defined by $T'y' = y'T$.

Let $\sigma(E, E')$ and $\beta(E, E')$ denote weak and strong topology on E, respectively. Now we have the Adjoint Theorem (Pap (1985, 1995)).

Theorem 5. *T' is sequentially continuous with respect to the relative $\sigma(F', F)$ (weak) topology on $D(T')$ and the topology on E' of uniform convergence on $\sigma(E, E')$ — K convergent sequences. In particular, T' is bounded with respect to these topologies.*

3. Nonlinear differential and difference equations

3.1. g-derivative

Let g be the additive generator of the strict pseudo-addition $\oplus$ on $[a, b]$ such that g is continuously differentiable on (a, b). The corresponding pseudo-multiplication $\otimes$ will always be defined as

$$u \otimes v = g^{-1}(g(u)g(v)).$$

Let f and h be two functions defined on the interval $[c, b]$ and with values in $[a, b]$. Then, we define for any $x \in [c, d]$

$$(f \oplus h)(x) = f(x) \oplus h(x),$$

$$(f \otimes h)(x) = f(x) \otimes h(x)$$

and for any $\lambda \in [a, b]$

$$(\lambda \otimes f)(x) = \lambda \otimes f(x).$$

We have by Pap (1993)

Definition 5. If the function f is differentiable on (c, d) and has the same monotonicity as the function g, then the g-derivative of f at the point $x \in (c, d)$ is

$$\frac{d^{\oplus} f(x)}{dx} := g^{-1}\left(\frac{d}{dx} g(f(x))\right),$$

when the right part is meaningful.

Definition 6. Let $f : [c, d] \to [a, b]$ be a measurable function. Then

$$\int_{[c,d]}^{\oplus} f dx := g^{-1}\left(\int_c^d g(f) dx\right)$$

is the g-integral of f on the interval $[c, d]$.

We have introduced in Pap (1990) an integral with respect to a $\oplus$-decomposable measure. In particular, if $\oplus$ is a strict pseudo-addition with a monotone generator g we have for measurable function f and σ-$\oplus$-decomposable measure m

$$\int^{\oplus} f \oplus dm = g^{-1}\left(\int (g \circ f) dx\right).$$

3.2. Differential equations

Now we shall consider as an example the following ordinary differential equation of the second order for $p > 0$ and $n \in R^+$

$$y'' + (p-1)y^{-1}(y')^2 + (2npx^{-1} \quad x^{-np}y^p)y' + n(np-1)x^{-2}y = 0.$$

Introducing a special generator $g, g(x) = x^p$ for $p > 0$, and the corresponding pseudo-addition $\oplus$ and pseudo-multiplication $\otimes$

$$u \oplus v = (u^p + v^p)^{1/p}$$

and

$$u \otimes v = u \cdot v,$$

we can write the preceding equation in the following form

$$y \otimes \frac{d^{\oplus}}{dx}(y) = \frac{d^{(2)\oplus}}{dx^2}(x^n \otimes y).$$

We can now apply the g-calculus on this equation (see Pap (1995)).

We shall consider as an important nonlinear partial differential equation the Burgers equation (see Maslov (1987), Pap (1992, 1993, 1995))

$$\frac{\partial u}{\partial t} + 1/2(\frac{\partial u}{\partial x})^2 - c/2\frac{\partial^2 u}{\partial x^2} = 0$$

for $x \in R$ and $t > 0$ with the initial condition $u(x,0) = u_0(x)$, where c is a given positive constant.

Now consider the corresponding "pseudo-operations". We shall consider this equation by g-calculus with the generator $g(u) = e^{-u/c}$. Then the corresponding pseudo-addition is

$$u \oplus v := -c\ln(e^{-u/c} + e^{-v/c})$$

and the distributive pseudo-multiplication

$$u \otimes v := u + v.$$

Then for solutions u_1 and u_2 the function $(\lambda_1 \oplus u_1) \otimes (\lambda_2 \otimes u_2)$ is also a solution of Burger's equation.

Further details can be found in Pap (1995).

3.3. Difference equation

We shall consider for given $\alpha, \beta \in [a, b]$ the following difference equation (see Pap (1995))

$$c_{m,n}^{k+1} = \alpha \otimes c_{m-1,n}^{k} \oplus \beta \otimes c_{m,n-1}^{k} \qquad (*)$$

where $k = 0, 1, 2, \ldots; \quad m, n = 0, \pm 1, \pm 2, \ldots$ with the initial conditions

$$c_{m,n}^{0} = \begin{cases} 1, & n = 0,\ m \geq 0 \\ 1, & m = 0,\ n \geq 0 \\ 0, & \text{otherwise.} \end{cases}$$

Example 4. Let $\oplus = \min$ and $\otimes = +$ on $[-\infty, +\infty]$. Then we have $0 = +\infty$ and $1 = 0$. Taking $\alpha = \beta = 1 = 0$ equation $(*)$ reduces to the Bellman equation (see Maslov (1987))

$$c_{m,n}^{k+1} = \min\{c_{m-1,n}^{k}, c_{m,n-1}^{k}\}$$

where $k = 0, 1, 2, \ldots; \quad m, n = 0, \pm 1, \pm 2, \ldots$
with the initial conditions

$$c_{m,n}^{0} = \begin{cases} 0, & n = 0,\ m \geq 0 \\ 0, & m = 0,\ n \geq 0 \\ +\infty, & \text{otherwise.} \end{cases}$$

Example 5. Let $\oplus = \max$ and $\otimes = +$ on $[-\infty, +\infty)$. Then we have $\mathbf{0} = -\infty$ and $\mathbf{1} = 0$. Equation (*) reduces to

$$c_{m,n}^{k+1} = \max\{\alpha + c_{m-1,n}^k, \beta + c_{m,n-1}^k\}$$

where $k = 0, 1, 2, \ldots;\quad m, n = 0, \pm 1, \pm 2, \ldots,$
with the initial conditions

$$c_{m,n}^0 = \begin{cases} 0, & n = 0,\ m \geq 0 \\ 0, & m = 0,\ n \geq 0 \\ -\infty, & \text{otherwise.} \end{cases}$$

The limit behavior of the solution of (*) was examined in Maslov (1987) and Pap (1995).

References

Abraham, P. (1992). Saeki's improvement of the Vitali–Hahn–Saks–Nikodym theorem holds precisely for Banach spaces having cotype, Proc. Amer. Math. Soc. **116**, 171–179.

Agafonova, L.V. and Klimkin, V.M. (1974). Nikodym theorem for triangular set functions. *Siberian Math. J.* **15**, 669–674. (In Russian).

Aleksjuk, V.N. (1968). Two theorems on existence of quasibase for a family of quasimeasures. *Izv. vissh. utcheb. zaved.* **6**(73), 11–18. (In Russian).

Antosik, P. (1978). Sur les suites d'applications. *C. R. Acad. Sci., Paris.* **287 A**, 75–77.

Antosik, P. and Saeki, S. (1992). *Diagonal Theorem for measuroids and its Applications.* Preprint.

Antosik, P. and Shwartz, C. (1985). *Matrix Methods in Analysis.* Lecture Notes in Mathematics **1113**, Springer-Verlag, Heidelberg.

Aumann, R.J. and Shapley, L.S. (1974). *Values of Non-Atomic Games.* Princeton Univ. Press.

Baccelli, F., Cohen, G., Olsder, G.J. and Quadrat, J.P. (1992). *Synchronization and Linearity: an Algebra for Discrete Event Systems.* John Wiley and Sons, New York.

Berres, M. (1988). λ-additive measures on measure spaces. *Fuzzy Sets and Systems.* **27**, 159–169.

Billot, A. (1992). From fuzzy set theory of non-additive probabilities: How have economists reacted? *Fuzzy Sets and Systems.* **49**, 75–89.

Chateauneuf, A. (1987). Uncertainty aversion and risk aversion in models with nonadditive probabilities. In: *Risk, Decision and Rationality,* B.R. Munier (Ed.), D. Reidel Publ. Comp., pp.615–627.

Chateauneuf, A. (1991). On the use of capacities in modeling uncertainty aversion and risk aversion. *Journal of Math. Economics* **20**, 343–369.

Choquet, G. (1953–1954). Theory of capacities. *Ann. Inst. Fourier (Grenoble).* **5**, 131–292.

de Campos, L.M., Lamata, M.T. and Moral, S. (1990b). The concept of conditional fuzzy measure. *International Journal of Intelligent Systems.* **5**, 237–246.

Delgado, M. and Moral, S. (1989). Upper and lower fuzzy measures. *Fuzzy Sets and Systems.* **33**, 191–200.

Dellacherie, C. (1972). *Capacite et processus stochastiques.* (Ergeb. Math. Grenzgeb. 67). Springer-Verlag. Berlin–Heidelberg.

Denneberg, D. (1994). *Non-Additive Measure and Integral.* Kluwer Academic Publisher.

Dobrakov, I. (1974). *On submeasures - I.* Dissertationes Math. **112**. Warszawa.

Dobrakov, I. and Farkova, J. (1980). *On submeasures - II.* Math. Slovaca. **30**, 65–81.

Drewnowski, L. (1972). Topological rings of sets, continuous set functions, integration, I, II, III. *Bull. Acad. Polon. Sci. Ser. Math. Astronom. Phys.* **20**, 269–276, 277–286, 439–445.

Dubois, D. and Prade, H. (1980). *Fuzzy Sets and Systems: Theory and applications.* Academic Press, New York.

Dubois, D. and Prade, H. (1982). A class of fuzzy measures based on triangular norms. *Internat. J. Gen. System.* **8**, 43–61.

Dubois, D. and Prade, H. (1982). Evidence Measures Based on Fuzzy Information. *Automatica.* **21**, 547–562.

Falconer, K. (1990). *Fractal Geometry.* John Wiley and Sons, New York.

Fine, T.L. (1973). *Theories of Probability: An Examination of Foundations.* Academic Press, New York.

Fishburn, P.C. (1970). *Utility Theory for Decision Making.* John Wiley and Sons, New York.

Golan, J.S. (1992). The Theory of Semirings with Applications in Mathematics and Theoretical Computer Sciences. *Pitman Monog. and Surveys in Pure and Appl. Maths.* **54**, Longman, New York.

Grabish, M., Murofushi, T. and Sugeno, M. (1992). Fuzzy measure of fuzzy events defined by fuzzy integrals. *Fuzzy Sets and Systems* **50**, 293–313.

Guan, J.W. and Bell, D.A. (1991). *Evidence and its Applications, Vol. 1.* North Holland, New York.

Guariglia, E. (1990). On Dieudonne's Boundedness Theorem. *J. Math. Anal. Appl.* **145**, 447–454.

Huber, P.J. (1973). The use of Choquet Capacities in Statistics. *Bull. of Int. Institute of Statistics.* **45**, 181–191.

Huber, P.J. and Strassen, V. (1973). Minimax Tests and the Neyman–Pearson Lemma for Capacities. *The Annals of Statistics.* **1**, 251–263.

Klement, E.P. and Weber, S. (1991). Generalized measures. *Fuzzy Sets and Systems.* **40**, 375–394.

Klement, E.P. and Weber, S. (1994). An integral representation for decomposable measures of measurable functions. *Aequat. Math.* **47**, 255–262.

Klimkin, V.M. (1989). Uniform boundedness of a family of nonadditive set functions. *Mat. Sb.* **180**(3), 385–396. (In Russian).

Maslov, V.P. (1987). *Asymptotical methods for solving pseudo-differential equations.* Moscow. (In Russian).

Mikusiński, J. (1970). A theorem on vector matrices and its applications in measure theory and functional analysis. *Bull. Acad Polon. Sci. Ser. Math.* **18**, 193–196.

Murofushi, T. and Sugeno, M. (1989). An interpretation of fuzzy measures and the Choquet integral as an integral with respect to a fuzzy measure. *Fuzzy Sets and Systems.* **29**, 201–227.

Murofushi, T. and Sugeno, M. (1991a). A theory of fuzzy measures. Representation, the Choquet integral and null sets. *J. Math. Anal. Appl.* **159**, 532–549.

Murofushi, T. and Sugeno, M. (1991b). Fuzzy t- conorm integral with respect to fuzzy measures: Generalization of Sugeno integral and Choquet integral. *Fuzzy Sets and Systems.* **42**, 57–71.

Pap, E. (1974). A generalization of the Diagonal theorem on a block-matrix. *Mat. ves.* **11**(26), 66–71.

Pap, E. (1976). Uniform boundedness of a family of exhaustive set functions. *Mat. ves.* **13**(28), 319–326.

Pap, E. (1982). *Functional Analysis (Sequential convergences. Some principle of functional analysis) (with English Summary).* Novi Sad.

Pap, E. (1985). Functional analysis with K-convergence. *Proceedings of the Conference on Convergence, Bechyne, Czech. 1984*, Akademie-Verlag, Berlin, 245–250.

Pap, E. (1986). A generalization of a theorem Dieudonne for k-triangular set functions. *Acta Sci. Math.* **50**, 159–167.

Pap, E. (1987). The Vitali–Hahn–Saks theorems for k-triangular set functions. *Atti Sem. Mat. Fis. Univ. Modena.* **26**, 21–32.

Pap, E. (1990a). Lebesgue and Saks decomposition of $\perp$-decomposable measures. *Fuzzy Sets and Systems.* **38**, 345–353.

Pap, E. (1990b). Integral generated by decomposable measures. *Univ. u Novom Sadu Zd. Rad. Prirod.- Mat. Fak. Ser. Mat.* **20**(1), 135–144.

Pap, E. (1991a). On non-additive set functions *Atti. Sem. Mat. Fis. Univ. Modena* **39**, 345–260.

Pap, E. (1991b). Many dimensional Lyapunov's convexity theorem for E_λ-decomposable measure. *Univ. u Novom Sadu Zd. Rad. Prirod.- Mat. Fak. Ser. Mat.* **21**(1), 65–74.

Pap, E. (1992). Decomposable measures and applications on nonlinear partial differential equations. *Rend. del Circolo Mat. di Palermo.* **ser II-num 28**, 387–403.

Pap, E. (1993). g-calculus. *Univ. u N. Sadu, Zd. Rad. Prirod.- Mat.* **23**(1), 145–156.

Pap, E. (1994a). The range of null-additive fuzzy and non-fuzzy measures. *Fuzzy Sets and Systems.* **65**(1), 105–115.

Pap, E. (1994b). Decomposition of supermodular functions and □-decomposable measures. *Fuzzy Sets and Systems.* **65**(1), 71–83.

Pap, E. (1995). *Null-Additive Set Functions.* Kluwer, Dordrecht; Ister Science, Bratislava.

Schilkret, N. (1971). Maxitive measure and integration *Indag. Math.* **33**, 109–116.

Schmeidler, D. (1986). Integral Representation without Additivity. *Proc. Amer. Math. Soc.* **97**, 253-261.

Schmeidler, D. (1989). Subjective Probability and Expected Utility without Additivity. *Econometrica* **57**, 571–587.

Shafer, G. (1976). A Mathematical *Theory of Evidence.* Princeton University Press. Princeton. New Jersey.

Sugeno, M. (1974). *Theory of fuzzy integral and their applications.* Ph. D. Thesis, Tokyo Institute of Technology.

Sugeno, M. (1977). Fuzzy measures and fuzzy integrals: A survey, in Ed M.M. Gupta, G.N. Saridis and B.R. Gaines. *Fuzzy Automata and Decision Processes.* North-Holland, New York, 89–102.

Sugeno, M. and Murofushi, T. (1987). Pseudo-additive measures and integrals. *J. Math. Anal. Appl.* **122**, 197–222.

Swartz, C. (1992). *Introduction to Functional Analysis.* Marcel Dekker, Inc., New York.

Šipoš, J. (1979). Integral with respect to a pre-measure. *Math. Slovaca.* **29**, 141–155.

Wang, Z. (1984). The Autocontinuity of Set Function and the Fuzzy Integral. *J. Math. Anal. Appl.* **99**, 195–218.

Wang, Z. (1985). Asymptotical structural characteristics of fuzzy measure and their applications. *Fuzzy Sets and Systems.* **16**, 277–290.

Wang, Z. and Klir, G. (1992). *Fuzzy Measure Theory.* Plenum Press, New York and London.

Weber, H. (1986). Compactness in spaces of group valued contents, the Vitali–Hahn–Saks theorem and Nikodym's boundedness theorem. *Rocky Mountain J. Math.* **16**, 253–275.

Weber, S. (1984). $\perp$-decomposable measures and integrals for Archimedean t-conorm. *J. Math. Anal. Appl.* **101**, 114–138.

Exploring Stochastic Laws. pp. 383–393.
Eds. A.V. Skorokhod and Yu.V. Borovskikh

Gaussian processes via independence of linear forms

AGNIESZKA PLUCIŃSKA AND JACEK WESOŁOWSKI

Mathematical Institute, Warsaw University of Technology, Plac Politechniki 1, 00-661 Warsaw, Poland

Received January 23, 1995

Abstract. A simple property of the covariance matrix, which ensures that infinite random sequences with independent linear forms are Gaussian, is discovered. For continuous time stochastic processes it is simplified further by assuming some consistency conditions. The results are parallel to the known characterizations of Gaussian processes by Gaussian conditional structures of the second order.

1. Introduction

Let $X = (X_t)_{t\in T}$, $T = \{0, 1, \dots, N\}$, where N is finite (or infinite, or $T = [0, \infty)$), be a square integrable zero mean stochastic process defined on a probability space $(\Omega, \mathcal{F}, P)$ with a covariance function K. Additionally assume that for any $0 \leq t_1 < t_2 < \cdots < t_n$, $n = 1, 2, \dots$

$$K^{(n)}(t_1, \dots, t_n) = \det[K(t_i, t_j)]_{i,j=1,\dots,n} > 0, \tag{1}$$

i.e. $X_{t_1}, \dots, X_{t_n}$ are linearly independent functions on Ω.

We say that X is an ILF process (process with Independent Linear Forms) or it fulfils an ILF condition if for any $0 < t_1 < \cdots < t_n$, $n = 1, 2, \dots$, there exist real functions $c_{ij} = c_{ij}(t_1, \dots, t_n)$, $i = 1, \dots, j-1$, $j = 2, 3, \dots, n$, such that

$$X_{t_1}, \quad X_{t_2} - c_{1,2}X_{t_1}, \quad \dots, \quad X_{t_n} - c_{n-1,n}X_{t_{n-1}} - \cdots - c_{1,n}X_{t_1} \tag{2}$$

are independent. Our aim in this paper is to find a property of the covariance function K which, together with the ILF condition, sufficiently characterizes the Gaussian process. A suggestion that such assumptions may determine finite dimensional distributions of the process gives the Darmois–Skitovitch Theorem on characterizing the normal law by independence of two linear forms in independent r.v's, but only for those r.v's which are present in both linear forms — see for instance Feller (1971), Ch. III.

Observe that the ILF condition implies linearity of regression, i.e.

$$E(X_{t_n} \mid X_{t_1}, \dots, X_{t_{n-1}}) = c_{1,n} X_{t_1} + \dots + c_{n-1,n} X_{t_{n-1}}$$

and homoscedasticity, i.e.

$$\mathrm{Var}(X_{t_n} \mid X_{t_1}, \dots, X_{t_{n-1}}) = \mathrm{const}(t_1, \dots, t_n).$$

These two relations strengthened by adding in the conditions also future states are the core of the definition of Gaussian conditional structures of the second order (GCS_2) introduced in Wesołowski (1991). GCS_2 continuous time stochastic processes were first considered by Plucińska (1983), where it was proved that they are Gaussian if and only if the covariance function fulfills some conditions of the Lipschitz type. In other papers: Wesołowski (1984), Bryc (1985), Bryc and Szabłowski (1990), it was shown that essentially the continuity of the covariance function is sufficient to characterize the Gaussian process by the GCS_2 condition. However in the case of GCS_2 infinite sequences, Bryc and Plucińska (1985) needed an additional assumption on the form of the covariance function for characterizing the Gaussian measure. Namely, they assumed additionally that

$$c_{n-1,n} \neq 0 \tag{3}$$

for all $0 \leq t_1 < \dots < t_n$, $n = 2, 3, \dots$. Observe that this is essentially a condition imposed on the covariance matrix since

$$c_{i,n} = -\frac{K_{i,n}^{(n)}}{K^{(n-1)}}, \tag{4}$$

where $K_{i,n}^{(n)}$ is the cofactor of the element $K(t_i, t_n)$ in the matrix $[K(t_i, t_j)]_{i,j=1,\dots,n}$, see Plucińska (1984) for the proof of this formula.

In Section 2 we will first consider ILF discrete time processes and give an analogue of condition (3) for the covariance matrix which guarantees that the sequence is Gaussian. As an auxiliary result a characterization of the finite dimensional Gaussian measure is used. Then continuous time processes are investigated. It is proved that some consistency assumptions allows us to weaken the conditions imposed on the covariance matrix in the discrete time case. The proof of the finite dimensional version of the characterization involving laborious computations in terms of cofactors of the covariance matrix is given in Section 3. It is the technical core of the paper.

2. ILF processes with discrete and continuous time

The main result of this section is an analogue of the characterization obtained for GCS_2 random sequences with an additional assumption (3) in Bryc and Plucińska (1985). Instead of linearity of regression and homoscedasticity involving conditioning with respect to past and future states we assume the independence condition leading only to Gaussian second order conditional structures for conditioning with respect to the past.

Theorem 1. *Let $X = (X_n)_{n=1,2...}$ be a square integrable zero mean random sequence fulfilling* (1). *Assume that it has the ILF property. If for any $1 \leq t_1 < \cdots < t_n$, $t_i \in \{1, 2, \ldots\}$, $i = 1, 2, \ldots, n$, $n = 2, 3, \ldots$,*

$$K_{1,n}^{(n)} \neq 0 \tag{5}$$

then X is a Gaussian sequence.

The proof of this result is based on the following characterization of the finite dimensional Gaussian measure making use of a weaker version of the ILF condition. It is similar to the Darmois–Skitovitch Theorem, but without the independence of the components of the involved random vector. While a conjecture that some result of such a type holds true is quite natural since we increase the number of independent linear forms, instead of assuming independence among components, the main difficulty was to discover some convenient properties of the covariance matrix sufficient to obtain the characterization. Another version of the Darmois–Skitovitch Theorem, where weakening of the independence assumption was compensated by considering more than two linear forms, was given in Kagan (1988).

Proposition 1. *Let $(Y_1, \ldots, Y_m)$, $m > 2$, be a square integrable zero mean random vector with a covariance matrix K fulfilling* (1) *for any $n = 1, \ldots, m$. Assume that r.v's*

$$Y_1, \quad Y_2 - a_{1,2}Y_1, \quad \ldots, \quad Y_m - a_{m-1,m}Y_{m-1} - \cdots - a_{1,m}Y_1 \tag{6}$$

are independent for some real numbers $a_{i,j}$, $i = 1, \ldots, j-1$, $j = 1, \ldots, m$ and additionally the linear forms

$$Y_{m-1} - b_{m-2,m-1}Y_{m-2} - \cdots - b_{2,m-1}Y_2, \quad Y_m - b_{m-1,m}Y_{m-1} - \cdots - b_{2,m}Y_2 \tag{7}$$

are independent for some real numbers b_{ij}, $i = 1, \ldots, j-1$, $j = m-1, m$. If condition (5) *holds for all $n = 2, \ldots, m$ then $(Y_1, \ldots, Y_{m-1})$ is a Gaussian random vector.*

The proof of Proposition 1, which essentially is a quite laborious computation involving cofactors of the covariance matrix, is put off until Section 3. Now we will quickly show how Theorem 1 follows from the above result.

Proof of Theorem 1. Fix any integer $m > 2$ and any $0 \le t_1 < \cdots < t_m$, and let $Y_i = X_{t_i}$, $i = 1, \ldots, m$. Then by the ILF condition we have (6) with $a_{ij} = c_{ij}(t_1, \ldots, t_j)$, $i = 1, \ldots, j-1$, $j = 1, \ldots, m$ and (7) with $b_{ij} = c_{ij}(t_2, \ldots, t_j)$, $i = 2, \ldots, j-1$, $j = m-1, m$. Consequently by Proposition 1 the random vector $(X_{t_1}, \ldots, X_{t_{m-1}})$ is Gaussian for any $0 \le t_1 < \cdots < t_{m-1}$ and any integer $m > 2$. Hence all finite dimensional distributions of the random sequence X are Gaussian.

Now consider the case when $T = [0, \infty)$. Let us introduce a natural consistency condition of the following form: for any $0 \le t_1 < \cdots < t_n$, $n \ge 4$

$$\lim_{t_{n-1} \to t_{n-2}} E(X_{t_n} \mid X_{t_1}, \ldots, X_{t_{n-1}}) = E(X_{t_n} \mid X_{t_1}, \ldots, X_{t_{n-2}}), \tag{8}$$

where the limit is understood to be in a quadratic mean or in a probability one. A similar condition proved to be useful in the investigations of the memory of stochastic processes with linearity of regression lead in Plucińska (1988).

Additionally our considerations will be limited to ILF processes with the property: for any $n > 3$ and any $0 \le t_1 < \cdots < t_{n-1} < t_n$ the following implication holds true

$$c_{1,n}(t_1, \ldots, t_n) = 0 \quad \Rightarrow \quad c_{1,n}(t_1, \ldots, t_{n-2}, s, t_n) = 0, \qquad \forall s \in [t_{n-2}, t_{n-1}], \tag{9}$$

where c's are coefficients of linear forms and, at the same time, regression coefficients (see the formulas in Section 1). Consequently the above implication may be rewritten as

$$E(X_{t_n} \mid X_{t_1}, \ldots, X_{t_{n-1}}) = E(X_{t_n} \mid X_{t_2}, \ldots, X_{t_{n-1}})$$

which implies

$$E(X_{t_n} \mid X_{t_1}, \ldots, X_{t_{n-2}}, X_s) = E(X_{t_n} \mid X_{t_2}, \ldots, X_{t_{n-2}}, X_s), \qquad \forall s \in [t_{n-2}, t_{n-1}].$$

Let us point out that some interesting consequences of the vanishing of some regression coefficients in models with linear regression were discovered in Plucińska (1988).

For continuous time we have the following characterization

Theorem 2. *Let $X = (X_t)_{t \in [0,\infty)}$ be a square integrable zero mean stochastic process fulfilling* (1). *Assume that it has the ILF property and that consistency conditions* (8) *and* (9) *are satisfied. If for any $0 \le t_1 < t_2 < t_3$*

$$K_{1,n}^{(n)} \neq 0 \tag{10}$$

for $n = 2, 3$ then X is a Gaussian process.

If X is a Markov process then for any $0 \le r < s < t$

$$K(s,s)K(r,t) = K(r,s)K(s,t)$$

and consequently $K_{1,3}^{(3)} = 0$, hence the method we develop here is not applicable in that case. Observe that for the ILF Markov process fulfilling (10) for $n = 2$ there is a real function $g(t) = K(0,t)K^{-1}(t,t)$ such that $(X_t g^{-1}(t))_{t\in[0,\infty)}$ is a process with independent increments. In a recent paper by Plucińska (1994) it was shown that if in addition linearity of regression with respect to the σ-field generated not only by the past but also by one future state of the process holds, then the process is Gaussian or $(X_t g^{-1}(t))_{t\in[0,\infty)}$, where g is a real positive function with independent increments. Additionally it was assumed that $E(X_t^4) < \infty$, for all $t \in [0,\infty)$. It remains an open question if only these two types of stochastic processes have the ILF property in the case when assumption (10) or the linearity of regression with respect also to future states are replaced by, for instance, continuity of the covariance function. This question seems to be quite natural for continuous time processes since the continuity of the covariance function ensures a characterization of the Gaussian process by the GCS_2 condition — see Section 1.

From the proof given below it follows easily that in the case of continuous covariance function it is sufficient to impose condition (10) only for (t_1, t_2, t_3) from a dense set in $[0,\infty)^3$.

Proof of Theorem 2. By Theorem 1 it is sufficient to prove that assumption (5) from this Theorem holds. We apply induction with respect to n. For $n = 2, 3$, it is obvious. Now assume that (5) holds for some $m \geq 3$, i.e. for any $0 \leq t_1 < \cdots < t_m$ and let $K_{1,m+1}^{(m+1)} = 0$ for some $0 \leq t_1 < \cdots < t_{m+1}$. We will prove that it is impossible. The ILF assumption implies

$$\begin{aligned} E(X_{t_{m+1}} \mid X_{t_1}, \ldots, X_{t_m}) &= c_{1,m+1}(t_1, \ldots, t_{m+1})X_1 + \cdots \\ &\quad + c_{m,m+1}(t_1, \cdots, t_{m+1})X_{t_m}, \end{aligned} \tag{11}$$

and by (8)

$$\begin{aligned} \lim_{t_m \to t_{m-1}} E(X_{t_{m+1}} \mid X_{t_1}, \ldots, X_{t_m}) &= E(X_{t_{m+1}} \mid X_{t_1}, \ldots, X_{t_{m-1}}) \\ &= c_{1,m+1}(t_1, \ldots, t_{m-1}, t_{m+1})X_{t_1} + \cdots \\ &\quad + c_{m-1,m+1}(t_1, \ldots, t_{m-1}, t_{m+1})X_{t_{m-1}}. \end{aligned} \tag{12}$$

On the other hand by (4) the coefficient $c_{1,m+1}(t_1, \ldots, t_{m+1})$ in (11) equals zero. Consequently

$$E(X_{t_{m+1}} \mid X_{t_1}, \ldots, X_{t_m}) = E(X_{t_{m+1}} \mid X_{t_2}, \ldots, X_{t_m}).$$

Hence by (8) and (9)

$$\begin{aligned} &\lim_{t_m \to t_{m-1}} E(X_{t_{m+1}} \mid X_{t_1}, \ldots, X_{t_m}) = \\ &= c_{2,m+1}(t_2, \ldots, t_{m-1}, t_{m+1})X_{t_2} + \cdots + c_{m-1,m+1}(t_2, \ldots, t_{m-1}, t_{m+1})X_{t_{m-1}} \end{aligned}$$

which contradicts (12) since by the induction assumption

$$c_{1,m+1}(t_1,\ldots,t_{m-1},t_{m+1}) \neq 0.$$

3. Finite dimensional case

This section contains a proof of the main technical result, i.e. Proposition 1. It involves laborious calculations in terms of cofactors of the covariance matrix. One of the tricks we will often use is the following formula for cofactors of a symmetric matrix $K = [K(i,j)]_{i,j=1,\ldots,m}$

$$K^{(m)}K^{(m)}_{i,j,n,n} + K^{(m)}_{i,n}K^{(m)}_{j,n} = K^{(m)}_{i,j}K^{(m)}_{n,n} \tag{13}$$

for any $i \neq n$, $j \neq n$, $i,j,n \in \{1,\ldots,m\}$, $m = 2,3,\ldots$, where $K^{(m)} = \det(K)$, $K^{(m)}_{i,j}$ is a cofactor of the element $K(i,j)$ in the matrix K and $K^{(m)}_{i,j,n,n}$ is a cofactor of the element $K(i,j)$ in a matrix created from K by deleting the n-th row and n-th column. This formula is taken from Plucińska (1984).

Fix an arbitrary integer m and introduce for a covariance $m \times m$ matrix K quantities similar to those from (4)

$$c_{j,k}(1,\ldots,j,k) = -\frac{K^{(j,k)}_{j,k}}{K^{(j,k)}_{k,k}},$$

where $K^{(j,k)}_{j,k}$ is the cofactor of the element $K(j,k)$ in the matrix created from the matrix K by deleting all the rows and columns of numbers $j+1,\ldots,k-1,k+1,\ldots,m$, $j<k$, $K^{(k,k)}_{k,k} = 1$, $k = 1,\ldots,m$. The following identity will be used in the course of the proof later in this section.

Lemma 1. *For any* $j = 1,\ldots,m-1$

$$\sum_{k=j}^{m-1} c_{k,m}(1,\ldots,m)c_{j,k}(1,\ldots,j,k) = c_{j,m}(1,\ldots,j,m). \tag{14}$$

Proof. Consider the Gaussian random vector $(V_1,\ldots,V_m)$ with the covariance matrix K. Then

$$E(V_m \mid V_1,\ldots,V_j) = \sum_{i=1}^{j} c_{i,m}(1,\ldots,j,m)V_i.$$

On the other hand

$$\begin{aligned} E(V_m \mid V_1,\ldots,V_j) &= E(E(V_m \mid V_1,\ldots,V_{m-1}) \mid V_1,\ldots,V_j) \\ &= E\Big(\sum_{i=1}^{m-1} c_{i,m}(1,\ldots,m)V_i \mid V_1,\ldots,V_j\Big) \end{aligned}$$

$$= \sum_{i=1}^{j} c_{i,m}(1,\ldots,m)V_i + \sum_{i=j+1}^{m-1} c_{i,m}(1,\ldots,m)\, E(V_i \mid V_1,\ldots,V_j)$$

$$= \sum_{i=1}^{j} c_{i,m}(1,\ldots,m)V_i + \sum_{i=j+1}^{m-1} c_{i,m}(1,\ldots,m) \sum_{l=1}^{j} c_{l,i}(1,\ldots,j,i)V_l$$

$$= \sum_{i=1}^{j}\Big(\sum_{k=i}^{m-1} c_{k,m}(1,\ldots,m)\, c_{i,k}(1,\ldots,i,k)\Big)V_i.$$

Consequently comparing in both the expressions for $E(V_m \mid V_1,\ldots,V_j)$ the coefficients of V_j we get (14).

The next identity for the covariance matrix K is another important tool in the proof of Proposition 1.

Lemma 2. *For any* $k = 1,2,\ldots,m$

$$K_{1,k}^{(k)}K^{(m)} = K^{(k)}K_{1,k}^{(m)} - \sum_{i=k+1}^{m} K_{k,i}^{(k,i)}K_{1,i}^{(m)}. \tag{15}$$

Proof. We apply backward induction with respect to k. For $k = m$ this is obvious since the sum after the minus sign on the right hand side of (15) is empty. Take now $k = m-1$ and compute the right hand side with the help of (13)

$$\begin{aligned} K_{1,m-1}^{(m)}K^{(m-1)} - K_{m-1,m}^{(m-1,m)}K_{1,m}^{(m)} &= K_{1,m-1}^{(m)}K_{m,m}^{(m)} - K_{m-1,m}^{(m)}K_{1,m}^{(m)} \\ &= K^{(m)}K_{1,m-1,m,m}^{(m)} = K^{(m)}K_{1,m-1}^{(m-1)}. \end{aligned}$$

Now assume that (15) holds for $k+1, k+2,\ldots,m$. We will prove that (15) is also fulfilled for k. To this end observe that

$$\begin{aligned} K_{1,k}^{(k)}K^{(m)} &= K_{1,k}^{(m)}K^{(k)} - K^{(k)}K^{(m)} \sum_{j=k+1}^{m} \left(\frac{K_{1,k}^{(j)}}{K^{(j)}} - \frac{K_{1,k}^{(j-1)}}{K^{(j-1)}}\right) \\ &= K_{1,k}^{(m)}K^{(k)} - K^{(k)}K^{(m)} \sum_{j=k+1}^{m} \frac{K_{1,k}^{(j)}K_{j,j}^{(j)} - K^{(j)}K_{1,k,j,j}^{(j)}}{K^{(j-1)}K^{(j)}}. \end{aligned}$$

Hence by (13) we have

$$K_{1,k}^{(k)}K^{(m)} = K_{1,k}^{(m)}K^{(k)} - K^{(k)} \sum_{j=k+1}^{m} \frac{K_{k,j}^{(j)}}{K^{(j-1)}K^{(j)}} K_{1,j}^{(j)}K^{(m)}.$$

Apply in the above identity the induction assumption for $K_{1,j}^{(j)}K^{(m)}$, $j = k+1,\ldots,m$. Then

$$K_{1,k}^{(k)}K^{(m)} = K_{1,k}^{(m)}K^{(k)} - K^{(k)} \sum_{j=k+1}^{m} \frac{K_{k,j}^{(j)}}{K^{(j-1)}K^{(j)}}$$

$$\Big(K_{1,j}^{(m)}K^{(j)} - K_{j,j+1}^{(j,j+1)}K_{1,j+1}^{(m)} - K_{j,j+2}^{(j,j+2)}K_{1,j+2}^{(m)} - \cdots - K_{j,m-1}^{(j,m-1)}K_{1,m-1}^{(m)} - K_{j,m}^{(j,m)}K_{1,m}^{(m)}\Big)$$

and after calculation

$$K_{1,k}^{(k)}K^{(m)} = K_{1,k}^{(m)}K^{(k)} -$$
$$- K^{(k)}\sum_{l=k+1}^{m} K_{1,l}^{(m)}\left(\frac{K_{k,l}^{(l)}}{K^{(l-1)}} - \frac{K_{l-1,l}^{(l-1,l)}K_{k,l-1}^{(l-1)}}{K^{(l-2)}K^{(l-1)}} - \dots - \frac{K_{k+1,l}^{(k+1,l)}K_{k,k+1}^{(k+1)}}{K^{(k)}K^{(k+1)}}\right). \quad (16)$$

On the other hand for any $1 \le k < j < l$

$$\frac{K_{k,l}^{(j,l)}}{K^{(j)}} - \frac{K_{j,l}^{(j,l)}K_{k,j}^{(j)}}{K^{(j-1)}K^{(j)}} = \frac{K^{(j,l)}K_{j,j,l,l}^{(j,l)}K_{k,l}^{(j,l)} - K^{(j,l)}K_{k,j,l,l}^{(j,l)}K_{j,l}^{(j,l)}}{K^{(j)}K^{(j-1)}K^{(j,l)}}.$$

Applying in the above formula identity (13) twice we get

$$\frac{K_{k,l}^{(j,l)}}{K^{(j)}} - \frac{K_{j,l}^{(j,l)}K_{k,l}^{(j)}}{K^{(j-1)}K^{(j)}} = \frac{K_{k,l}^{(j,l)}K_{j,j}^{(j,l)} - K_{j,l}^{(j,l)}K_{k,j}^{(j,l)}}{K^{(j-1)}K^{(j,l)}}$$

and now after using once again (13) we obtain

$$\frac{K_{k,l}^{(j,l)}}{K^{(j)}} - \frac{K_{j,l}^{(j,l)}K_{k,j}^{(j)}}{K^{(j-1)}K^{(j)}} = \frac{K_{k,l}^{(j-l,l)}}{K^{(j-1)}}. \quad (17)$$

Hence we have

$$\frac{K_{k,l}^{(l)}}{K^{(l-1)}} - \sum_{j=k+1}^{l-1}\frac{K_{j,l}^{(j,l)}K_{k,j}^{(j)}}{K^{(j-1)}K^{(j)}} = \frac{K_{k,l}^{(k,l)}}{K^{(k)}} \quad (18)$$

by applying (17) successively for $j = l-1, l-2, \dots, k+1$ to the left hand side. Now we end the proof by putting (18) into (16).

Observe that for $k = 1$ formula (15) follows immediately from the Laplace expansion with respect to the first row of the determinant $K^{(m)}$.

We are now ready to prove Proposition 1.

Proof of Proposition 1. Denote by $Z_1, \dots, Z_m$ the linear forms appearing in (6), respectively, and by U_{m-1}, U_m the linear forms from (7), respectively.

We will prove the result in four steps.

First step. It is sufficient to show that

$$U_n = \sum_{j=1}^{n} \alpha_{j,n} Z_j, \qquad n = m-1, m,$$

where

$$\alpha_{j,n} = -\frac{K_{1,j}^{(j)}K_{1,n}^{(n)}}{K^{(j)}K_{1,1}^{(n-1)}}, \qquad n = m-1, m, \quad j = 1, 2, \dots, n-1 \quad (19)$$

and $\alpha_{n,n} = 1$, $n = m-1, m$. Then by (5)

$$K_{1,j}^{(j)}K_{1,m-1}^{(m-1)}K_{1,m}^{(m)} \neq 0, \qquad j = 1, 2, \dots, m-1$$

and consequently each Z_j, $j = 1, \dots, m-1$, is a member of both the linear forms U_{m-1}, U_m. Since Z's are independent then from the Darmois–Skitovitch Theorem we conclude that they are normal. Hence the random vector $(Y_1, \dots, Y_{m-1})$ as a linear transformation of independent normal r.v's has an $(m-1)$-variate Gaussian distribution.

Second step. In this step we will show that for $k = 1, \dots, m$

$$Y_k = \sum_{j=1}^{k} c_{j,k}(1, \dots, j, k) Z_j. \tag{20}$$

Apply induction with respect to k.

For $k = 1, 2$ it is obvious since $a_{j,k} = c_{j,k}(1, \dots, j, k)$. From the definition and the induction assumption for $1, 2, \dots, k-1$ we have

$$Y_k = \sum_{i=1}^{k-1} c_{i,k}(1, \dots, m)\Big(\sum_{j=1}^{i} c_{j,i}(1, \dots, j, i) Z_j\Big) + Z_k =$$
$$= \sum_{j=1}^{k-1} \sum_{i=j}^{k-1} c_{i,k}(1, \dots, k)\, c_{j,i}(1, \dots, j, i) Z_j + Z_k.$$

Now the result follows by Lemma 1.

Third step. In this step we will prove that for any $k = 2, \dots, n-1$, $n = m-1, m$

$$\alpha_{n,k} = \sum_{i=k}^{n-1} c_{k,i}(1, \dots, k, i)\, [c_{i,n}(1, \dots, n) - c_{i,n}(2, \dots, n)]. \tag{21}$$

From (20) and the definition of the U's it follows that

$$U_n = \sum_{j=1}^{n} c_{j,n}(1, \dots, j, n) Z_j - \sum_{i=2}^{n-1} c_{i,n}(2, \dots, n)\Big[\sum_{j=1}^{i} c_{j,i}(1, \dots, j, i) Z_j\Big]$$

and after the calculation we have

$$\begin{aligned} U_n &= Z_n + \sum_{k=2}^{n-1} \Big[c_{k,n}(1, \dots, k, n) - \sum_{i=k}^{n-1} c_{i,n}(2, \dots, n)\, c_{k,i}(1, \dots, k, i)\Big] Z_k + \\ &+ \Big[c_{1,n}(1, n) - \sum_{i=2}^{n-1} c_{i,n}(2, \dots, n)\, c_{1,i}(1, i)\Big] Z_1. \end{aligned}$$

Hence for $k = 2, \dots, n-1$

$$\alpha_{n,k} = c_{k,n}(1, \dots, k, n) - \sum_{i=k}^{n-1} c_{i,n}(2, \dots, n)\, c_{k,i}(1, \dots, k, i).$$

Now apply to $c_{k,n}(1, \dots, k, n)$ in the above formula the identity from Lemma 1 to obtain (21).

Similarly for $k = 1$ we have

$$\alpha_{n,1} = c_{1,n}(1, \dots, n) + \sum_{i=2}^{n-1} c_{1,i}(1, i)\, [c_{i,n}(1, \dots, n) - c_{i,n}(2, \dots, n)]. \tag{22}$$

Fourth step. In this step it will be proved that (19) follows immediately from Lemma 2.

Observe that for any $i = 1, \ldots, n-1$, $n = m-1, m$,

$$c_{i,n}(1, \ldots, n) - c_{i,n}(2, \ldots, n) = \frac{K_{1,1,i,n}^{(n)} K^{(n-1)} - K_{i,n}^{(n)} K_{1,1}^{(n-1)}}{K_{1,1}^{(n-1)} K^{(n-1)}}.$$

Multiplying the numerator and denominator of the expression on the right hand side by $K^{(n)}$ and applying (13) twice we obtain

$$c_{i,n}(1, \ldots, n) - c_{i,n}(2, \ldots, n) = \frac{K_{1,n}^{(n)} [K_{1,n}^{(n)} K_{i,n}^{(n)} - K_{1,i}^{(n)} K_{n,n}^{(n)}]}{K_{1,1}^{(n-1)} K^{(n-1)} K^{(n)}}.$$

Once again applying (13) to the above equation we have

$$c_{i,n}(1, \ldots, n) - c_{i,n}(2, \ldots, n) = -\frac{K_{1,n}^{(n)} K^{(n)} K_{1,i,n,n}^{(n)}}{K_{1,1}^{(n-1)} K^{(n-1)}}.$$

Now (21) or (22) and the above identity give for any $k = 1, \ldots, n-1$

$$\begin{aligned} \alpha_{n,k} &= -\frac{K_{1,n}^{(n)} K_{1,k}^{(n-1)}}{K^{(n-1)} K_{1,1}^{(n-1)}} + \sum_{i=k+1}^{n-1} \frac{K^{(k,i)}}{K^{(k)}} \frac{K_{1,n}^{(n)} K_{1,i}^{(n-1)}}{K^{(n-1)} K_{1,1}^{(n-1)}} = \\ &= -\frac{K_{1,n}^{(n)}}{K^{(k)} K^{(n-1)} K_{1,1}^{(n-1)}} \left[K_{1,k}^{(n-1)} K^{(k)} - \sum_{i=k+1}^{n-1} K_{k,i}^{(k,i)} K_{1,i}^{(n-1)} \right]. \end{aligned}$$

Hence the result follows from Lemma 2.

References

Bryc, W. (1985). Some remarks on random vectors with nice enough behaviour of conditional moments. *Bulletin of Polish Acad, Sci., Math.* **33**, 677–683.

Bryc, W., Plucińska, A. (1985). A characterization of infinite Gaussian sequences by conditional moments. *Sankhyā, Ser. A.* **47**, 166–173.

Bryc, W., Szabłowski, P.J. (1990). Some characteristic of normal distribution by conditional moments. *Bulletin of the Polish Acad. Sci., Math.* **38**, 209–218.

Feller, W. (1971). *An Introduction to Probability Theory and Its Applications.* Vol. II. Wiley, New York.

Kagan, A.M. (1988). New classes of dependent random variables and a generalization of the Darmois–Skitovitch Theorem to the case of several forms. *Teorya Veroyat. Primen.* **33**, 305–315. (In Russian).

Plucińska, A. (1983). On a stochastic process determined by the conditional expectation and the conditional variance. *Stochastics.* **10**, 115–129.

Plucińska, A. (1984). A characterization of the Gaussian distribution based on the Lapunov condition. *Teorya Veroyat. Primen.* **29**, 570–574.

Plucińska, A. (1988). On the property of the memory of stochastic process with the linear regression. *Bulletin of Polish Acad. Sci., Math.* **36**, 175–179.

Plucińska, A. (1994). On a stochastic process with independent linear forms. *Journal of Mathematical Sciences.* (In press).

Wesołowski, J. (1984). A characterization of the Gaussian process based on properties of conditional moments. Demonstrat. Math. **18**, 795–807.

Wesołowski, J. (1991). Gaussian conditional structure of the second order and the Kagan classification of multivariate distributions. *Journal of Multivariate Analysis.* **39**, 79–86.

Exploring Stochastic Laws. pp. 395–408.
Eds. A.V. Skorokhod and Yu.V. Borovskikh

On the irregular-derivative sampling with uniformly dense sample points for bandlimited stochastic processes

TIBOR POGANY

Department of Applied Mathematics Faculty of Maritime Studies, University of Rijeka, 51000 Rijeka, Studentska 2, Croatia

Received January 20, 1995

Abstract. Various sampling formulas are derived for bandlimited functions and weakly stationary stochastic processes using uniformly dense sample points. Sampling restoration is also considered with base functions of an exponential type, when the sampling series uses samples of $r-1$ derivatives of the bandlimited signal, r positive integer, as well as those signals themselves (derivative sampling). The upper bound of the truncation error is derived for deterministic signals, and correspondingly the upper bound of the mean-square truncation error is obtained for processes. Almost sure and mean-square derivative sampling restoration is considered.

1. Introduction

The well known stochastic Kotel'nikov–Shannon formula asserts that for the weakly stationary stochastic process $\{\xi(t);\ t \in \mathbf{R}^1\}$, bandlimited to a certain fixed bandwidth $\tilde{w} > 0$, there holds the *digital-to-analogue* restoration

$$\xi(t) = \sum_{n \in \mathbf{Z}} \xi(n\pi/w) \frac{\sin(wt - n\pi)}{(wt - n\pi)}, \tag{1}$$

in the mean-square sense, and in the almost sure sense (in other words "with probability 1") as well for all $w > \tilde{w}$. Moreover (1) is true for $w = \tilde{w}$ in the mean-square sense, and if the random spectral measure $Z(\{\pm w\}) = 0$ almost surely, Gaposhkin (1977), where $Z(\cdot)$ appears in the spectral representation

$$\xi(t) = \int_{-\tilde{w}}^{\tilde{w}} e^{it\lambda} Z(d\lambda).$$

But it is more realistic to expect that the samples $\xi(t_n)$ of the observed process are perturbed from the uniform spacing, i.e. t_n are not equidistantly spaced.

This perturbation leads us to the so-called *jittered* or *irregular Kotel'nikov–Shannon series*

$$\xi(t) = \sum_{n\in\mathbf{Z}} \xi(t_n) g(t, t_n), \tag{2}$$

of process $\xi(t)$ with respect to the weight-function system $\{g(t,t_n)\}_{n\in Z}$, where the equality is valid in some sense*.

If the weight-function in (2) is replaced with the help of the function

$$G(z) = (z - t_0) \prod_{n=1}^{\infty} \left(1 - \frac{z}{t_n}\right)\left(1 - \frac{z}{t_{-n}}\right), \tag{3}$$

then we get the irregular Kotel'nikov–Shannon formula

$$\xi(t) = \pi \sum_{n\in Z} \frac{\xi(\pi t_n/w) G(t\,w/\pi)}{G'(t_n)(w\,t - t_n\,\pi)}; \quad t \in \mathbf{R}^1. \tag{4}$$

The convergence questions in (4) restrict the deviation $D := \sup_{n\in\mathbf{Z}} |t_n - n|$ to the interval $(0, 1/4)$ (with the help of the well-known Kadec Theorem) for the mean-square convergence, and to $D \in (0, 1/8)$ for the almost sure convergence, Seip (1990b).

It is in our interest to avoid the Kadek 1/4-restriction, since there are some time-sequences $\{t_n\}$ for which the condition $D < 1/4$ is too strong. This was done by Seip, who modified (4) with the help of base functions of exponential decay, Seip (1990a). Namely, he developed certain results in the sampling restoration of bandlimited (BL in the sequel) functions and weakly stationary stochastic processes (WSSP) for uniformly dense sample points in the so-called irregular sampling procedure.

Higgins (1991) established some results for the irregular sampling in restoration BL functions, expanding some earlier results of Seip (1987), to the derivative sampling case, using the contour integration method.

Following Seip and Higgin's approaches, the author summarizes their formulas and generalizes them for the BL WSSP class in the irregular-derivative sampling. Namely, if there is a *guard band* $\varrho \in (0, w)$ and the observed WSSP $\xi(t)$ is BL to $r(w - \varrho)$, it is

$$\xi(t) = G^r(w\,t/\pi) \sum_{n\in\mathbf{Z}} \sum_{k=0}^{r-1} \sum_{j=0}^{r-k-1} \frac{\xi^{(r-k-j-1)}(\pi\,t_n/w) Q_{r\,n\,k}}{k!(r-j-k-1)!(t - \pi\,t_n/w)^{j+1}}, \tag{5}$$

where

$$Q_{r\,n\,k} = \lim_{z\to\infty} \frac{d^k}{d\,z^k} \left(\frac{z - \pi\,t_n/w}{G(z\,w/\pi)}\right)^r, \tag{6}$$

* It seems that this kind of formula has a history going back (just the deterministic case) to Valiron (1925)

and $\xi^{(m)}(t)$ stays for the mth mean-square derivative of process $\xi(t)$. Formula (5) is valid in the mean-square sense if $D < 1/4r$, and it holds in the almost sure sense if $D < 1/8r$, Pogany (1994).

In this paper we consider the irregular-derivative sampling for uniformly dense sample points with base functions of exponential decay. That means, sampling series (4) uses samples of $r-1$ derivatives of the BL signal (function or process). At the same time the sample point set $\{t_n\} \subset \mathbf{R}^1$ (or $\mathbf{C}$) has *uniform density* 1, i.e.

$$\begin{aligned} D &= \sup_{n \in Z} |t_n - n| < \infty; \\ d &= \inf_{n \neq m} |t_n - t_m| > 0. \end{aligned} \tag{7}$$

See for comparison, Duffin and Schaeffer (1952). (For a historical overview consult Seip (1990a) and the references noted there).

As the main result of this paper, the Seip result, the 1-derivative irregular sampling restoration formula (4) will be extended to the general, irregular t^{th}-order derivative sampling restoration procedure, avoiding the 1/4-condition of Kadec. Precisely, our recent approach holds for any real sequence $\{t_n\}_{n \in Z}$ of uniform density 1. With this extension, the frequency spectrum support domain of $\xi(t)$ has to be r times broader then in the conventional, $r = 1$ case (discussed in detail by Seip), however we benefit by introducing the irregular-derivative modification of the Kotel'nikov–Shannon formula (5) with respect to the more general sample point set $\{t_n\}$. The base function with exponential decay enables this generalization, since it neutralizes the "bad behaviour" of the sampling sum (5) if the Kadec $D < 1/4r$ property is not fulfilled (accordingly $D < 1/8r$ for almost sure convergence).

Also, the upper bound of the truncation error is derived in the irregular-derivative sampling restoration of BL deterministic signals. This result (Theorem 1) is the main tool used in our further investigations. In Theorem 2 the deterministic signal is restored by the modified irregular-derivative Kotel'nikov–Shannon series. The upper bound of the mean-square truncation error for a BL WSSP is established in Theorem 3; and our final task, the sampling restoration formula for a BL WSSP with mean-square and almost sure convergence properties are given in Theorem 4. The key to the proofs are some estimates of Levin, Levinson and Seip listed in Levin (1949), Duffin and Schaeffer (1952), Seip (1990a), and Paley–Wiener–Schwartz's Theorem.

Finally, in Chapter 4 we shall show that under some additional conditions on the weight-functions in the sampling restoration sums, we also get a sequence of already known classical results in uniform sampling and derivative sampling.

2. Preliminary definitions. Some results

Let us denote by $\mathcal{S}$ the Schwartz space of infinitely differentiable and rapidly decaying functions. It is well-known that for all $f \in \mathcal{S}$ it is $\mathcal{F}[f] := f^\wedge \in \mathcal{S}$, where $f^\wedge$ denotes the Fourier-transform pair of function f, i.e.

$$f^\wedge(\lambda) = \frac{1}{\sqrt{2\pi}} \int_{\mathbf{R}^1} f(t) e^{-it\lambda} dt, \quad f(t) = \frac{1}{\sqrt{2\pi}} \int_{\mathbf{R}^1} f^\wedge(\lambda) e^{it\lambda} d\lambda.$$

The Banach space $\mathcal{B}_\tau, \tau > 0$ consists of each function f being a Fourier-transform of a distribution with the support $\sup p(f^\wedge) = [-\tau, \tau]$. Using Paley–Wiener–Schwartz's Theorem it follows that $\mathcal{B}_\tau$ is a class of all entire functions $f(z)$ for which there is a non-negative integer L and a positive constant $C_{f,L}$ such that:

$$|f(z)| \leq C_{f,L}(1 + |z|)^L e^{-\tau|\Im\{z\}|}; \quad z \in \mathbf{C}. \tag{8}$$

Consider a real sequence $\{t_n\}_{n\in Z}$ with uniform density 1. By the reasons given in (7), there exist two constants $d > 0, \mathbf{D} \in \mathbf{N} \cup \{0\}$, such that

$$D \geq |t_n - n|; \quad d \leq |t_n - t_m|; \quad n, m \in \mathbf{Z}. \tag{9}$$

It is natural to assume that $\{t_n\}_{n\in Z}$ monotonically increases, without loss of generality.

In proving our main results, the modified irregular-derivative sampling restoration formula, and the corresponding upper bound of the mean-square function error, we need the help of some estimates of function $G(z)$ defined in (3), which can be found in some mathematical literature, e.g. see Seip (1990a). Actually, the sharpest estimates of this kind belong to Levin (1949), with respect to Boas (1952), Duffin and Schaeffer's and Levinson's estimates. Here we use the estimates by Levin and Seip given in Lemma 1, Seip (1990b).

Lemma 1. *There exist positive constants C_1 and C_2 dependent on d and D respectively, such that along all vertical lines with length greater or equal to $d/2$ from $\{t_n\}$, for all $z \in \mathbf{C}$ we have*

$$|G(z)| \geq C_1(1 + |z|)^{-(4D+1)} e^{\pi|\Im\{z\}|}; \tag{10}$$

and for any z with sufficiently large $|\Im\{z\}|$, we have

$$|G(z)| \geq C_2(1 + |z|)^{-(2D+1)} e^{\pi|\Im\{z\}|}. \tag{11}$$

For the proof we also need the contour integration method discussed in Higgins (1991). This method uses the above given estimates. We prove, based on these, the following Lemma.

Lemma 2. *Let $\{t_n\}$ have uniform density 1. Then for all $f \in \mathcal{B}_{r(w-\varrho)}, h \in \mathcal{S} \cap \mathcal{B}_\sigma, 0 < \sigma/r < \varrho < w$ there exist non-negative integers L and M and absolute constants $C_1, C_2, C_{f,L}$ and C_M such that for $|z| < |\zeta = \xi + i\eta|$, $\alpha \geq 1$, $\delta \in (0,1)$, $N \in \mathbf{N}$ we have*

$$\left|\frac{f(z)h(z-\zeta)}{G^r(\zeta\, w/\pi)(\zeta - z)}\right| \leq \frac{C_{f,L}C_1^{-r}C_M\alpha^M}{|\zeta - z|}e^{-(r\,\varrho-\sigma)|\eta|+\sigma|\Im\{z\}|} \times$$
$$\times\ (1+\alpha|z|)^M(1+\alpha|\zeta|)^{r(4D+1)+L-M} \tag{12}$$

on all vertical lines $|\Re\{\zeta\}| = \frac{\pi}{w}(N+\delta)$. Also it holds that

$$\left|\frac{f(z)h(z-\zeta)}{G^r(\zeta\, w/\pi)(\zeta - z)}\right| \leq \frac{C_{f,L}C_2^{-r}C_M\alpha^M}{|\zeta - z|}e^{-(r\,\varrho-\sigma)(N+\delta)\frac{\pi}{w}+\sigma|\Im\{z\}|} \times$$
$$\times\ (1+\alpha|z|)^M(1+\alpha|\zeta|)^{r(2D+1)+L-M} \tag{13}$$

on all horizontal lines $|\Im\{\zeta\}| = \frac{\pi}{w}(N+\delta)$.

Proof. Bearing in mind that $f \in \mathcal{B}_{r(w-\varrho)}$ and $\zeta = \xi + i\eta \in \mathbf{C}$ we get by Paley–Wiener–Schwartz's Theorem estimate (8) for f. Because $h \in \mathcal{S} \cap \mathcal{B}_\sigma$, it follows that $\sup p(h^\wedge) = [-\sigma, \sigma]$. Also by Paley–Wiener–Schwartz's Theorem it follows that there exists an absolute constant C_M for every positive integer M, such that:

$$|h(\zeta)| \leq C_M(1+|\zeta|)^{-M}e^{\sigma|\eta|}\,.\ \zeta \in \mathbf{C}. \tag{14}$$

Since Lemma 1 gives exactly estimate (10) upon the function $G(z)$ if $|\Re\{\zeta\}| = \frac{\pi}{w}(N+\delta)$, the following sequence of inequalities holds:

$$\begin{aligned}
&\left|\frac{f(z)h(z-\zeta)}{G^r(\zeta\, w/\pi)(\zeta - z)}\right| \leq \\
\leq\ & \frac{C_{f,L}C_1^{-r}C_M\alpha^M}{|\zeta - z|}e^{-(r\,\varrho-\sigma)|\eta|+\sigma|\Im\{z\}|}\left(\frac{1+\alpha|\zeta|}{1+|\zeta - z|}\right)^M(1+\alpha|\zeta|)^{r(4D+1)+L-M} \\
\leq\ & \frac{C_{f,L}C_1^{-r}C_M\alpha^M}{|\zeta - z|}e^{-(r\,\varrho-\sigma)|\eta|+\sigma|\Im\{z\}|}\left(\frac{1+\alpha|\zeta|}{1+|\zeta - z|}\right)^M(1+\alpha|\zeta|)^{r(4D+1)+L-M} \\
\leq\ & \frac{C_{f,L}C_1^{-r}C_M\alpha^M}{|\zeta - z|}e^{-(r\,\varrho-\sigma)|\eta|+\sigma|\Im\{z\}|}(1+\alpha|z|)^M(1+\alpha|z|)^{r(4D+1)+L-M}
\end{aligned} \tag{15}$$

This completes the proof of the first part of Lemma 2, given (12).

On the horizontal line $|\eta| = \frac{\pi}{w}(N+\delta)$ combining (8), (11) and (14) and the maximizations given above, we can clearly derive estimate (13). This ends the proof of the Lemma.

3. Main results

We are ready to formulate our main results keeping the notations given above. At first we consider deterministic signals, then with the help of the derived results we prepare the general modified irregular-derivative sampling restoration Kotel'nikov–Shannon formula.

We define the truncated to N irregular r^{th}-derivative sampling restoration sum of function f with respect to the base function h; such that $h(0) = h_0 \neq 0$, as the following term

$$I_N(z;f;G;r;h) := h_0^{-1}G^r(z\,w/\pi) \sum_{|n|\le N} \sum_{j=0}^{r-1} \sum_{k=0}^{r-j-1} \sum_{m=0}^{r-j-k-1} f^{(r-j-k-m-1)}(\pi\,t_n/w) \times \frac{h^{(k)}(z-\pi\,t_n/w)Q_{r\,n\,j}}{j!k!(r-j-k-m-1)!(z-\pi\,t_n/w)^{m+1}}, \tag{16}$$

where $Q_{r\,n\,j}$ is given in (6). The sampling restoration error is given by the quantity $\varepsilon_N(z;f) := f(z) - I_N(z;f;G;r;h)$, it is the so-called *truncation error*. With these notations one formulates our principal tool in the proof of the main results.

Theorem 1. *Let $f \in \mathcal{B}_{r(w-\varrho)}$, and $h \in \mathcal{S} \cap \mathcal{B}_\sigma$ with $h(0) = h_0 \neq 0, 0 < \sigma/r < \varrho < w$. Then there exist positive integers $\mathcal{L}$ and M, and an absolute constant $\kappa = \kappa(f;L;r;M;\alpha)$ such that for all $r \in \mathbf{N}, 0 < \delta < 1$ and for all $N > \frac{w}{\pi}(\max\{|x|,|y|\} + D); z = x + i\,y; \alpha := \max\{1, w/\pi\}$, we have*

$$|\varepsilon_N(z;f)| \le \frac{\kappa}{w|h_0|} e^{\sigma|y|}(1+\alpha|z|)^M |G(z\,w/\pi)|^r \times \left(\frac{w\,S(1)}{\pi(r\,\varrho-\sigma)} + (N+\delta)e^{-(r\,\varrho-\sigma)(N+\sigma)\frac{\pi}{w}} S(2)\right), \tag{17}$$

where

$$S(1) = \sup_{\Re\{\zeta\}|=(\pi/w)(N+\delta)} \left\{\frac{(1+\alpha|\zeta|)^{r(4D+1)+L-M}}{|\zeta - z|}\right\}; \tag{18}$$

$$S(2) = \sup_{\Im\{\zeta\}|=(\pi/w)(N+\delta)} \left\{\frac{(1+\alpha|\zeta|)^{r(2D+1)+L-M}}{|\zeta - z|}\right\}; \tag{19}$$

Proof. Consider the function

$$\Phi(\zeta, z) := \frac{f(\zeta)h(z-\zeta)}{G^r(\zeta\,w/\pi)(\zeta - z)}$$

where z is fixed. The singularities of Φ are poles of order r at the points $\zeta = \pi t_n/w$ which coincide with the zeros of $G^r(\cdot)$. In addition $\zeta = z$ is a pole of the first order. Denote by $\mathrm{Res}_a[\phi(\zeta); m]$ the residual of the function ϕ at pole a with order m.

Let γ_N be the square integration contour with the vertices at the points $\frac{\pi}{w}(N+\delta)(\pm 1 \pm i)$, such that for each $|n| \leq N$ it is $\pi t_n/w \in \mathrm{Int}(\gamma_N)$ and $z \in \mathrm{Int}(\gamma_N)$. By a Theorem on residuals we have

$$\begin{aligned} \varepsilon_N(z;f) &= \frac{G^r(z\,w/\pi)}{h_0}\left(\frac{f(z)h_0}{G^r(\zeta\,w/\pi)} - \sum_{|n|\leq N} \mathrm{Res}_{\frac{\pi}{w}t_n}[\phi(\zeta;z);r]\right) \\ &= \frac{G^r(z\,w/\pi)}{2\pi\,h_0\,i}\oint_{\gamma N}\frac{f(z)h(z_\zeta)}{G^r(\zeta\,w/\pi)(\zeta - z)}\,d\zeta. \end{aligned} \tag{20}$$

The contour integral on γ_N splits into four integrals on the edges of the path γ_N. Let $\int_{\gamma(1)}$ be the integral on the edge between the points $\frac{\pi}{w}(N+\delta)(1-i)$ and $\frac{\pi}{w}(N+\delta)(1+i)$; the other three integrals are indexed successively in the positive direction.

Using these facts we find that by reasons of symmetry it is

$$\left|\int_{\gamma(1)}\right| = \left|\int_{\gamma(3)}\right| := A.$$

With the help of Lemma 2 and estimate (12), we get

$$\begin{aligned} A &\leq 2C_{f,L}C_1^{-r}C_M\alpha^M e^{\sigma|y|}(1+\alpha|z|)^M \int_0^{(\pi/w)(N+\delta)} \frac{(1+\alpha|\zeta|)^{r(4D+1)+L-M}d\eta}{e^{(r\,\varrho-\sigma)\eta|\zeta - z|}} \\ &\leq \frac{C_{f,L}C_1^{-r}C_M\alpha^M}{r\,\varrho - \sigma}e^{\sigma|y|}(1+\alpha|z|)^M \sup_{\zeta\in\gamma(1)}\left\{\frac{(1+\alpha|\zeta|)^{r(4D+1)+L-M}}{|\zeta - z|}\right\} \\ &- \frac{C_{f,L}C_1^{-r}C_M\alpha^M}{r\,\varrho - \sigma}e^{\sigma|y|}(1+\alpha|z|)^M S(1). \end{aligned} \tag{21}$$

Now, we have to evaluate $B := |\int_{\gamma(2)}|$ (at the same time $|\int_{\gamma(4)}|$ as well). Therefore, if $|\eta| = \frac{\pi}{w}(N+\delta)$, with the aid of bound (13) in Lemma 2, we get

$$\begin{aligned} B &\leq \frac{C_{f,L}C_1^{-r}C_M\alpha^M(1+\alpha|z|)^M}{e^{(r\,\varrho-\sigma)(N+\delta)\pi/w+\sigma|y|}}\int_{-(\pi/w)(N+\delta)}^{(\pi/w)(N+\delta)}\frac{(1+\alpha|\zeta|)^{r(2D+1)+L-M}d\varepsilon}{|\zeta - z|} \\ &\leq \frac{2\pi C_{f,L}C_2^{-r}C_M\alpha^M(N+\delta)(1+\alpha|z|)^M}{w(r\,\varrho-\sigma)e^{(r\,\varrho-\sigma)(N+\delta)\pi/w+\sigma|y|}}\sup_{\zeta\in\gamma(2)}\left\{\frac{(1+\alpha|\zeta|)^{r(2D+1)+L-M}}{|\zeta - z|}\right\} \\ &= \frac{2\pi C_{f,L}C_2^{-r}C_M\alpha^M(N+\delta)(1+\alpha|z|)^M}{w(r\,\varrho-\sigma)e^{(r\,\varrho-\sigma)(N+\delta)\pi/w+\sigma|y|}}S(2). \end{aligned} \tag{22}$$

Formulas (21) and (22) will be substituted into (20). This gives

$$|\varepsilon_N(z;f)| = \frac{|G(z\,w/\pi)|^r}{\pi|h_0|}(A+B) \leq$$

$$\leq \frac{2|G(z\,w/\pi)|^r C_{f,L} C_M \alpha^M e^{\sigma|y|}(1+\alpha|z|)^M}{w|h_0|\min\{C_1^r, C_2^r\}} \times$$

$$\times \left(\frac{w\,S(1)}{\pi(r\,\varrho - \sigma)} + \frac{(N+\delta)S(2)}{e^{(r\,\varrho-\sigma)(N+\delta)\pi/w}} \right). \tag{23}$$

Denoting $\kappa = 2C_{f,L}\max\{C_1^r, C_2^r\}C_M\alpha^M$ we complete the proof.

If N increases to infinity, the upper bound of the truncation error (17) vanishes. Thus, we get the modified irregular-derivative sampling restoration Kotel'nikov–Shannon formula for deterministic signals.

Theorem 2. *Let $\{t_n\}_{n\in\mathbf{Z}}$ be a real sequence of uniform density 1, and assume that $0 < \sigma/r < \varrho < w$, where r is a positive integer.*

Then for all $f \in \mathcal{B}_{r(w-\varrho)}$, and for all $h \in \mathcal{S} \cap \mathcal{B}_\sigma$ that satisfy $h(0) = h_0 \neq 0$, we have

$$f(z) = h_0^{-1} G^r(z\,w/\pi) \sum_{n\in\mathbf{Z}} \sum_{j=0}^{r-1} \sum_{k=0}^{r-j-1} \sum_{m=0}^{r-j-k-1} f^{(r-j-k-m-1)}(\pi\,t_n/w)$$

$$\times \frac{h^{(k)}(z-\pi\,t_n/w)Q_{r\,n\,j}}{j!k!(r-j-k-m-1)!(z-\pi\,t_n/w)^{m+1}}, \tag{24}$$

uniformly on all compact subsets of $\mathbf{C}$, where $Q_{r\,n\,j}$ is given in (6).

Proof. It is obvious that our goal is to show that $\varepsilon_N(z;f) \to 0$ as $N \to 0$. The upper bound of the truncation error (17) enables this convergence to be uniform on compact subsets of C. Indeed, because of $h \in \mathcal{S} \cap \mathcal{B}_\sigma$ and $\sigma < r\,\varrho$, Paley–Wiener–Schwartz's Theorem guarantees that for all $M \in \mathbf{N}$ there exists an absolute constant C_M (consequently there is $\kappa(f;L;r;M;\alpha)$) such that (8) is satisfied. If we choose $M > r(4D+1)+L-1$ the suprema $S(1), S(2)$ vanish as $N \to \infty$. Thus, the limit of the evolution (17) proves the Theorem.

Remark 1. Formula (24) differs from that of Valiron not only in the derivative approach, but also in that we have substituted some power of an integer part of $\sin(\psi\,x)/(\psi\,x)$ by a bandlimited Schwartz function h. As can be seen from the above proof, it is not essential that h is in $\mathcal{S}$, but that it has a sufficiently rapid decay to "neutralize" the exponent $r(4D+1)+L-1$ in (17).

At this point we are ready to translate the deterministic signal restoration results to the BL WSSP class. Before we shall do this, we introduce some necessary facts and definitions we need in the sequel.

Let $\{\xi(t); t \in \mathrm{R}^1\}$ be a WSSP, bandlimited to a given $\beta > 0$. That means, the covariance function $K_\xi(t) := \mathbf{E}\,\xi(t)\xi^*(0)$ of process $\xi(t)$ possesses an integral representation

$$K_\xi(t) = \int_{-\beta}^{\beta} e^{it\lambda} F(d\lambda),$$

and accordingly the spectral representation of $\xi(t)$ is

$$\xi(t) = \int_{-\beta}^{\beta} e^{it\lambda} Z(d\lambda),$$

where $F(\Delta) = \mathbf{E}|Z(\Delta)|^2, \quad \Delta \subset \mathrm{R}^1$. The random spectral measure Z is with orthogonal increments, and the spectral function F need to be continuous at $\pm\beta$ for the validity of (1); see the remark after (1) or Gaposhkin (1977), Pogany (1991). For the present approach we do not need this continuity assumption.

Let r be a positive integer. Then all values of the mean-square derivatives $\xi^{(j)}(t); j = \overline{0, r-1}$ shall be measured at the time-values $\{t_n\}_{n\in\mathbf{Z}}$. Here is sequence $\{t_n\}$ with uniform density 1. Since $\xi(t)$ is BL, it is infinitely smooth, therefore $\xi^{(j)}(t)$ exists for all positive integer j, Pogany (1987). Thus, the measuring procedure is relevant for certain fixed $r \in \mathbf{N}$. Also, we have to point out that $\xi^{(j)}(t)$ possesses the spectral representation

$$\xi^{(j)}(t) = \int_{-\beta}^{\beta} (i\lambda)^j e^{it\lambda} Z(d\lambda), \tag{25}$$

where the equality is in the mean-square sense.

Finally, let us introduce the mean-square truncation error

$$\tau_N(t;\xi) := \mathbf{E}|\xi(t) - I_N(t;\xi;G;r;h)|^2, \tag{26}$$

where $I_N(\cdot)$ is defined by (16). Now, we formulate the upper bound of the truncation error result for BL processes and the irregular-derivative sampling restoration result for these processes.

Theorem 3. *Let $\{\xi(t) : t \in \mathrm{R}^1\}$ be a WSSP, BL to $r(w-\varrho), r \in N$, with covariance function of the form*

$$K_\xi(t) = \int_{r(\varrho-w)}^{r(w-\varrho)} e^{it\lambda} F(d\lambda). \tag{27}$$

If $h \in \mathcal{S} \cap \mathcal{B}_\sigma, 0 < \sigma/r < \varrho < w$, then there are positive integers L, M and an absolute constant $\Theta = \Theta(r; M; \alpha)$ such that for $0 < \delta < 1$ and for all $N > \frac{w}{\pi}(|t|+D)$ we have

$$\tau_N(t;\xi) \le \frac{\Theta\, K_\xi(0)}{(w\,h_0)^2} G^{2r}(t\,w/\pi)(1+\alpha|t|)^{2M} \left[\left(\frac{w\,S(1)}{\pi(r\,\varrho-\sigma)}\right)^2 + \frac{(N+\delta)^2 S^2(2)}{e^{2(r\,\varrho-\sigma)(N+\delta)\frac{\pi}{w}}}\right] \tag{28}$$

where $S(1), S(2)$ and α are defined in Theorem 1.

Proof. It is clear that the Fourier-transform kernel $f_\lambda(t) = e^{it\lambda}, t \in \mathrm{R}^1, \|f_\lambda\| = 1$ belongs to $\mathcal{B}_{|\lambda|}$ for all real λ. Since $\xi(t)$ is BL to $r(w-\varrho)$ we shall apply (17) in obtaining the upper bound of the mean-square truncation error (28).

At first, we derive the spectral representation of the truncated sampling sum $I_N(t;\xi;G;r;h)$ by direct computation using spectral representation (25). Thus we find that

$$\begin{aligned} I_N(t;\xi;G;r;h) &= h_0^{-1}G^r(t\,w/\pi)\sum_{n\in\mathbf{Z}}\sum_{j=0}^{r-1}\sum_{k=0}^{r-j-1}\sum_{m=0}^{r-j-k-1}\frac{h^{(k)}(t-\pi\,t_n/w)}{j!k!(r-j-k-m-1)!} \\ &\times \frac{Q_{rnj}}{(t-\pi\,t_n/w)^{m+1}}\int_{r(\varrho-w)}^{r(w-\varrho)}(i\,\lambda)^{r-j-k-m-1}e^{i\pi t_n/w}Z(d\,\lambda) \\ &= \int_{r(\varrho-w)}^{r(w-\varrho)} I_N(t;f_\lambda;G;r;h)Z(d\,\lambda), \end{aligned}$$

where the equality is interpreted in the mean-square sense.

In continuation one calculates the upper bound for $\tau_N(t;\xi)$ with the help of (17), bearing in mind the isometry between the Hilbert spaces $\mathcal{H}(\xi) := \langle \xi(t); t \in \mathrm{R}^1\rangle$ and $L_2(F;r(w-\varrho)) := \langle \phi; \int_{|\lambda|\le r(w-\varrho)} |\phi(\lambda)|^2 F(d\,\lambda) < \infty\rangle$. Therefore, we have

$$\begin{aligned} \tau_N(t;\xi) &= \int_{r(\varrho-w)}^{r(w-\varrho)} |e^{it\lambda} - I_N(t;\xi;G;r;h)|^2 F(d\,\lambda) \\ &= \int_{r(\varrho-w)}^{r(w-\varrho)} |\varepsilon_N(t;f_\lambda)|^2 F(d\,\lambda) \\ &\le \frac{8C_M^2\alpha^{2M}C_{f_\lambda,0}^2}{(w\,h_0)^2\min\{C_1^{2r},C_2^{2r}\}}(1+\alpha|t|)^{2M}G^{2r}(t\,w/\pi) \\ &\times \left[\left(\frac{w\,S(1)}{\pi(r\,\varrho-\sigma)}\right)^2 + \frac{(N+\delta)^2S^2(2)}{e^{2(r\,\varrho-\sigma)(N+\delta)\frac{\pi}{w}}}\right]\int_{r(\varrho-w)}^{r(w-\varrho)} F(d\,\lambda). \end{aligned} \tag{29}$$

Putting $\Theta = 8C_M^2\max\{C_1^{2r},C_2^{2r}\}\alpha^{2M} (= 2\kappa^2(f_\lambda;0;r;L;\alpha))$, remarking that the constant $C_{f,L}$ in (17) becomes $C_{f_\lambda,0} \equiv 1$, by the well-known property of the covariance function

$$\int_{r(\varrho-w)}^{r(w-\varrho)} F(d\,\lambda) = K_\xi(0).$$

The proof of the Theorem is now complete.

Theorem 4. *With the same assumptions as in the previous Theorem we have*

$$\begin{aligned} \xi(t) &= h_0^{-1}G^r(t\,w/\pi)\sum_{n\in\mathbf{Z}}\sum_{j=0}^{r-1}\sum_{k=0}^{r-j-1}\sum_{m=0}^{r-j-k-1}\xi^{(r-j-k-m-1)}(\pi\,t_n/w) \\ &\times \frac{h^{(k)}(t-\pi\,t_n/w)Q_{rnj}}{j!k!(r-j-k-m-1)!(t-\pi\,t_n/w)^{m+1}}, \end{aligned} \tag{30}$$

where the convergence is uniform in the mean-square sense on compact subsets of the real t-axis and also almost sure.

Proof. Combining the proofs of Theorems 1 and 3, we find that (28) explains the uniform mean-square convergence of $I_N(t;\xi;G;r;h)$ on compact subsets of R^1 to the initial BL WSSP $\xi(t)$ as $N \to \infty$ if we choose $M > r(4D+1)+L-1$.

Consider the probability $P_N := P\{\exists n \geq N; |\xi(t) - I_N(t;\xi;G;r;h)| \geq \varepsilon\}$, for some fixed $\varepsilon > 0$. By Chebyshev's inequality we have $P_N \leq \varepsilon^{-2}\tau_N(t;\xi)$. In continuation, since the series

$$\sum_{N>(w/\pi)(|t|+D)} \tau_N(t;\xi) = \sum_{N>(w/\pi)(|t|+D)} O(N^{r(4D+1)+L-M-1})$$

converges as $M > r(4D+1)+L$, we choose such an M, and by Borel–Cantelli's Lemma it implies that almost sure convergence results.

Remark 2. Irregular-derivative Kotel'nikov–Shannon formula (30) generalizes the adequate result of Piranashvili (1967), Corollary 3, with the irregular-derivative approach replacing its sinc-functions in the direction mentioned in Remark 1. Lee constructs a modified sampling restoration formula with a base function without irregular-derivative consideration. So, (30) is an extension of the result by Lee (1977), Theorem 4, in the irregular-derivative manner.

4. Certain special cases

In this part of the paper we shall discuss and report on certain results derived by several authors in irregular sampling, derivative sampling and in equidistantly spaced sampling restoration. It will be shown that these results are special cases of irregular-derivative Kotel'nikov–Shannon formulas, obtained in this paper.

4.1. WSSP case, r = 1

Denote as $\mathcal{X}_A(t)$ the indicator function of the set A. Then, put $h(t) := \mathcal{X}_{[-\sigma,\sigma]}(t)$.

Because $G(t_n) \equiv 0, n \in \mathbf{Z}$ and by considering the $r = 1$ case, it is just $j = 0$ in (6), therefore we have

$$Q_{1n0} = \lim_{z \to \pi t_n/w} \frac{z - \pi t_n/w}{G(zw/\pi)} = \frac{\pi}{w\,G'(t_n)}; \quad n \in \mathbf{Z}. \tag{31}$$

Substituting (31) into (30) together with all the other specified values of indexes, we derive

$$\xi(t) = \pi \sum_{n \in Z} \frac{\xi(\pi t_n/w) G(tw/\pi)}{G'(t_n)(wt - t_n\pi)}; \quad t \in \mathrm{R}^1, \tag{32}$$

which coincides for $w = \pi$ with formula (1) in the paper Seip (1990b). At this point we have to suppose that $D < 1/4$. Then the convergence in (32) is uniform in the mean-square sense. Moreover, if we assume that $D < 1/8$, the convergence in (32) will be with probability 1.

4.2. WSSP case, uniformly dense samples, r = 1

In avoiding the Kadec's 1/4-restriction, supposing the same conditions upon the weight function h as in the introduction of Theorem 4, and by using a sample point set $\{t_n\}_{n\in Z}$ with uniform density 1, the title-assumption concludes with the formula

$$\xi(t) = \pi\, G(t\,w/\pi) \sum_{n\in\mathbf{Z}} \xi(\frac{\pi}{w}t_n) \frac{h(t - \pi\, t_n/w)}{G'(t_n)(w\,t - \pi\, t_n)}.$$

This result is actually formula (20) in Seip (1990a) for $w = \pi$.

4.3. Functions, uniformly spaced general derivative sampling

Assume that $h(t) := \mathcal{X}[-\sigma,\sigma](t)$. Therefore all derivatives $h^{(k)}(z) \equiv 0, k = \overline{0, r-1}$ in (24). Since uniformly spaced sampling means that $D = 0$, i.e. $t_n = n$, we have $G(t) = \sin(\pi\, t)$. By direct computation it is

$$Q_{r\,n\,j} = (-1)^{n\,r} w^{j-r} \lim_{x\to 0} \frac{d^j}{d\,x^j} \left(\frac{1}{\sin c(x)}\right)^r,$$

where the usual notation, $\sin c(x) = x^{-1}\sin(x), \sin c(0) := 1$, is introduced. Denote $\Gamma_{r\,j} := (-1)^{n\,r} w^{r-j} Q_{r\,n\,j}$. Then (24) becomes

$$f(z) = z^r \sin c^r(w\,z) \sum_{n\in\mathbf{Z}} \sum_{j=0}^{r-1} \sum_{m=0}^{r-j-1} \frac{(-1)^{n\,r} w^j f^{(r-j-m-1)}(\frac{\pi}{w}n)\Gamma_{r\,j}}{j!(r-j-m-1)!(z-\frac{\pi}{w}n)^{m+1}}, \tag{33}$$

where the convergence in (34) is uniform, Higgins (1991), because of uniformly spaced sampling satisfies the "uniform density 1" property. The result is already known, it is derived by Linden and Abramson (1960). Compare also Butzer et al. (1988) and Higgins (1991) for further references and for a historical overview.

4. 4. WSSP, uniformly spaced general derivative sampling

According to the previous paragraph, let $h(t)$ be equal to $\mathcal{X}_{[-\sigma,\sigma]}(t)$. Then relation (30) in Theorem 4 becomes

$$\xi(t) = t^r \sin c^r(w\,t) \sum_{n\in\mathbf{Z}} \sum_{j=0}^{r-1} \sum_{m=0}^{r-j-1} \frac{(-1)^{n\,r} w^j \xi^{(r-j-m-1)}(\frac{\pi}{w}n)\Gamma_{r\,j}}{j!(r-j-m-1)!(t-\frac{\pi}{w}n)^{m+1}}. \tag{34}$$

Here (34) is valid in the mean-square and in the almost sure sense, since $D = 0$.

As a special case of relation (34), the case $r = 2$ is precisely the sampling restoration procedure by the Kotel'nikov–Shannon type formula

$$\xi(t) = t^2 \sin c^2(w\,t) \sum_{n\in\mathbf{Z}} \left(\frac{w\,\xi'(\frac{\pi}{w}n)}{w\,t - n\,\pi} + \left(\frac{w}{w\,t - n\,\pi}\right)^2 \xi(\frac{\pi}{w}n)\right), \tag{35}$$

since $\Gamma_{20} = 1$, $\Gamma_{21} = 0$. This formula is well-known from the literature for BL deterministic signals and originates back to the paper Jagerman and Fogel (1956). However, in the BL WSSP case, (35) is a novelty (together with the general r^{th}-order formula (30)).

References

Boas, R.P.Jr. (1952). Integrability along a line for a class of entire functions. *Trans. Am. Math. Soc.* **73**, 191–197.

Butzer, P.L., Splettstösser, W. and Stens, R.L. (1988). The sampling theorem and linear prediction in signal analysis. *Jahresber. Deutsch. Math.-Verein.* **90**, 1–70.

Duffin, R.J. and Schaeffer, A.C. (1952). A class of nonharmonic Fourier series. *Trans. Am. Math. Soc.* **72**, 341–366.

Gaposhkin, V.F. (1977). A theorem on the almost sure convergence of the sequences of measurable functions and its application to the sequences of stochastic integrals. *Mat. Sbornik (N.S.)* 104 (146) **1**(9), 3–21. (In Russian).

Higgins, J.R. (1991). Sampling theorem and the contour integral method. *Applic. Anal.* **4**(1), 155–171.

Jagerman, D.L. and Fogel, L. (1956). Some aspects of the sampling theorem. *IRE Trans. Inform. Theory IT-2,* 139–146.

Lee, A.J. (1977). Approximate interpolation and the sampling theorem. *SIAM J. Appl. Math.* **32**(4), 731–744.

Levin, B.Ya. (1949). On functions of finite degree, bounded on a sequence of points. *Doklady Akad. Nauk SSSR (N.S.)* 65, 265–268. (In Russian).

Linden, D.A. and Abramson, N.M. (1960). A generalization of the sampling theorem. *Inform. Control* 3, 26–31; (see Correction to equation (40), (1961). ibid. 4, pp. 95–96.).

Piranashvili, Z.A. (1967). On the problem of interpolation of random processes. *Teor. Verojat. Primenen. XII,* **4**, 708–717. (In Russian).

Pogany, T. (1987). Singuläre zufällige Prozesse und mittelquadratische Konvergenz. *Publ. Math. Debrecen* **34**, 197–205.

Pogany, T. (1991). Almost sure sampling reconstruction of non-band-limited homogeneous random fields. *Publ. Inst. Math. Belgrade (N.S.)* **49**(63), 221–232.

Pogany, T. (1994). Some irregular-derivative sampling theorems. In: *Proc. Functional Analysis IV*. Butkovic, D. et al. (Eds). Various Publication Series 43, Mathematical Institute, Aarhus.

Seip, K. (1987). An irregular sampling theorem for functions bandlimited in a generalized sense. *SIAM J. Appl. Math.* 47, 1112–1116.

Seip, K. (1990a). On certain irregular sampling formulas for bandlimited functions. *The Univ. Trondheim Preprints. Mathematics,* 4, 1–16.

Seip, K. (1990b). A note on sampling of bandlimited stochastic processes. *IEEE Trans. Inform. Theiry IT-36,* 5, 1186.

Valiron, G. (1925). Sur la formule d'interpolation de Lagrange. *Bull. Sci. Math.* **49**(2), 181–192.

Exploring Stochastic Laws. pp. 409–416.
Eds. A.V. Skorokhod and Yu.V. Borovskikh

Fractal, superfractal and anomalously fractal distributions of random variables with a fixed infinite set of independent n-adic digits

N.V. PRATSEVITYI

Kiev Pedagogical University

Received April 22, 1995

Abstract. We study the fractal properties (determine the Hausdorff measure, find the Hausdorff–Besicovitch dimension) of the spectrum of a random variable with an independent, infinite, fixed set of n-adic ($2 \le n \in N$) digits. The conditions for the set of fixed digits ensuring the distribution to be anomalously fractal or superfractal are obtained.

1. Let

$$\xi = \sum_{k=1}^{\infty} n^{-k} \eta_k = 0.\eta_1 \eta_2 \dots \eta_k \dots$$

be a random variable (r.v.) with independent n-adic (n is a fixed positive integer) digits η_k which are r.v. assuming values $0, 1, \dots, n-1$ with probabilities $P_{0k}, P_{1k}, \dots, P_{(n-1)k}$, respectively, $P_{ik} \ge 0$, $\sum_{i=0}^{n-1} P_{ik} = 1$, $k \in N$.

The matrix $\|P_{ik}\|$, $i = \overline{0, n-1}$, $k \in N$ uniquely determines the distribution of r.v. ξ. The last is a sum of discretely distributed r.v. $n^{-k}\eta_k$ and according to the Jessen–Wintner theorem, it has a pure distribution. The discreteness criterion follows from the Lévy theorem. The singularity criterion has been obtained for $n = 2$ first by Salem and, for general case, by Marsaglia.

A more general model is considered in [1] (details and further references can be found in [2]). The basic problem, which aroused the interest in these researches, can be stated as whether among the distributions of the family of r.v. ξ there are distributions whose spectra are anomalously fractal or superfractal sets. The author studied the fractal properties of distributions of r.v. ξ earlier in [3]–[4].

However, we did not consider all possible cases and, in particular, we did not pay much attention to the case where the matrix $\|P_{ik}\|$ has an infinite number of columns containing 1 and an infinite number of columns not containing 1 (in fact it amounts to an infinite set of n-adic digits being fixed). The present article deals just with this case.

α-dimensional Hausdorff measure of a set $E \subset R^1$ is the number

$$H_\alpha(E) = \lim_{\varepsilon\to 0}\left[\inf_{|E_i|\le\varepsilon}\left\{\sum_i |E_i|^\alpha,\ E\subset \cup E_i\right\}\right].$$

The lower bound (inf) is taken over all possible coverings $\{E_i\}$ of the set E such that the diameter of each set $|E_i|$ is not greater than $\varepsilon > 0$. Regarding $H_\alpha(E)$ as a function of α there is a number $\alpha_0 = \alpha_0(E)$ such that $H_\alpha(E) = 0$ for $\alpha > \alpha_0$ and $H_\alpha(E) = \infty$ for $\alpha < \alpha_0$. This number is called the Hausdorff–Besicovitch dimension of the set E $(0 \le H_{\alpha_0}(E) \le \infty)$. In other words, the Hausdorff–Besicovitch dimension is a point of discontinuity of the function $f(\alpha) = H_\alpha(E)$.

A set $E \subset R^1$ is called fractal if it is continual and $H_{\dim E}(E) = 0$ or ∞. Here $\dim E$ is the topological dimension of the set E. A set E is called anomalously fractal if $\alpha_0(E) = 0$, superfractal if $\alpha_0(E) = 1$ and (properly) fractal if $0 < \alpha_0(E) < 1$. A distribution is called fractal, superfractal or anomalously fractal if its spectrum possesses this property, the spectrum being defined as the set of growth points.

The spectrum S_ξ of the distribution of r.v. ξ can be proved to be the set

$$A\left[\|P_{ik}\|\right] = \left\{x : x\in[0,1],\ P_{\alpha_k(x)k} > 0\ \forall k\in N\right\}$$

where $\alpha_k(x)$ is kth n-adic digit of the number x.

Lemma 1.

(1) *For all $r \in N$ the spectrum S_ξ of the distribution of r.v. ξ consists of*

$$A_r = \prod_{i=1}^{r} m_i$$

disjoint congruent components lying in n-adic intervals of rank r, where m_i is equal to the number of indexes j such that $P_{ji} \ne 0$.

(2) *The entropy dimension* [2] *of S_ξ coincides with the Hausdorff–Besicovitch dimension $\alpha_0(S_\xi)$.*

(3) *Determining the Hausdorff measure $H_\alpha(S_\xi)$ we can confine ourselves to coverings consisting of n-adic intervals of equal rank i*

$$H_\alpha(S_\xi) = \liminf_{r\to\infty}\left[A_r\cdot n^{-\alpha r}\right]$$

where

$$A_r = N_{S_\xi}(n^{-r})$$

is the smallest number of n-adic intervals of rank r which cover S_ξ.

Not giving a detailed proof of Lemma 1 we will only point out that statement (1) follows from the independence of r.v. η_k, $k = 1, 2, \ldots$ and that $S_\xi \cap \Delta_{i_1 \ldots i_r} \neq \emptyset$ if $P_{i_k k} \neq 0$, $k = \overline{1, r}$ and that the following congruence holds

$$S_\xi \cap \Delta_{i_1 \ldots i_r} \cong S_\xi \cap \Delta_{i_1 \ldots i_r}$$

if $P_{i_r r} \cdot P_{i'_r r} \neq 0$ where

$$\Delta_{i_1 \ldots i_r} = \left[\sum_{k=1}^{r} \frac{i_k}{n^k}; \ \sum_{k=1}^{r} \frac{i_k}{n^k} + \frac{1}{n^r} \right].$$

Statements (2) and (3) follow from statement (1), the properties of the Hausdorff measure and Hausdorff–Besicovitch dimension and the argument given in [2, p. 54–55].

2. The case $n = 2$ should be treated separately. We will consider it in detail. In this case the matrix of r.v. ξ satisfies the conditions:

(1) there exists an infinite set $B = \{\beta_k\} \subset N$ such that $P_{0\beta_k} \cdot P_{1\beta_k} = 0$;

(2) $D = N \setminus B = \{1, d_1, d_2, \ldots\}$ is infinite and $P_{0d} \cdot P_{1d} \neq 0$ if $d \in D$.

Obviously B is the set of indices of those columns of the matrix $\|P_{ik}\|$ which contain 1 and D is the set of indices of the columns not containing 1. By the sets B and D we construct the sets (sequences) $\{\gamma_i\}$, $\gamma_i < \gamma_{i+1}$ and $\{c_i\}$, $c_i < c_{i+1}$ by letting γ_1 to be the number of the first consecutive positive integers in D, γ_k to be the length of kth system of consecutive positive integers in D and $1 < c_k$ to be the length of kth system of consecutive positive integers in B. For example, if $D = \{1, 2, 3, 4, 9, 10, 17, 19, 21, 22, 23, 24, 39 \ldots\}$ then $\gamma_1 = 3, c_1 = 5, \gamma_2 = 2, c_2 = 6, \gamma_3 = 1, c_3 = 1, \gamma_4 = 1, c_4 = 1, \gamma_5 = 4, c_5 = 14, \ldots$.

Then r.v. ξ can be written as

$$\xi = 0.(\eta_1 \ldots \eta_{\gamma_1}) u_1^{(1)} \ldots u_{c_1}^{(1)} (\eta_{\gamma_1 + c_1 + 1} \ldots \eta_{\gamma_1 + c_1 + \gamma_2}) u_1^{(2)} \ldots u_{c_2}^{(2)} \ldots$$

$$\ldots (\eta_{\sum_{i=1}^{k-1} (\gamma_i + c_i) + 1} \ldots \eta_{\sum_{i=1}^{k-1} (\gamma_i + c_i) + \gamma_k}) u_1^{(k)} \ldots u_{c_k}^{(k)} \ldots$$

where $u_j^{(k)}$ are fixed binary digits (or digits assuming a given value with probability 1) and r.v. η_j are such that $P_{0j} \cdot P_{1j} \neq 0$.

Lemma 2. *α-dimensional Hausdorff measure of the spectrum S_ξ of distribution of r.v. ξ equals*

$$H_\alpha(S_\xi) = 2^{\liminf_{k \to \infty} S_k(\alpha)}$$

where

$$S_k(\alpha) = \sum_{i=1}^{k} [\gamma_i - \alpha(c_i + \gamma_i)].$$

Proof. Consider the sequence $\{B_i^{(k)}(r_k)\}_{k=1}^{\infty}$ of rank coverings of the set S_ξ with binary intervals of rank r_k where

$$r_k = \sum_{i=1}^{k}(\gamma_i + c_i).$$

The set S_ξ is contained in $2^{\sum_{i=1}^{k}\gamma_i}$ intervals of rank r_k. The "α-volume" of the covering $\{B_i^{(k)}(r_k)\}$ therefore equals

$$\begin{aligned} l_{r_k}^{(\alpha)}[S_\xi] &= 2^{\sum_{i=1}^{k}\gamma_i} \cdot 2^{-\alpha\sum_{i=1}^{k}(\gamma_i+c_i)} \\ &= 2^{\sum_{i=1}^{k}[\gamma_i-\alpha(\gamma_i+c_i)]} \\ &= 2^{S_k(\alpha)}. \end{aligned}$$

By using Lemma 1 we will show that for the "α-volume" $l_r^{(\alpha)}[S_\xi]$ of the covering $\{B_i(r)\}$ of the set S_ξ with binary intervals of rank r there exists k such that

$$l_r^{(\alpha)}[S_\xi] \geq l_{r_k}^{(\alpha)}[S_\xi]$$

where $r_{k-1} < r \leq r_k$.

Let $r = r_k - \nu$. The two complementary cases are possible here: (1) $0 \leq \nu \leq c_k$ and (2) $c_k < \nu < c_k + \gamma_k$.

In case (1) we get

$$\begin{aligned} l_r^{(\alpha)}[S_\xi] &= 2^{\sum_{i=1}^{k}\gamma_i} \cdot 2^{-\alpha[\sum_{i=1}^{k}(\gamma_i+c_i)-\nu]} \\ &= 2^{S_k(\alpha)} \cdot 2^{\alpha\nu} \\ &= l_{r_k}^{(\alpha)}[S_\xi] \cdot 2^{\alpha\nu} \\ &\geq l_{r_k}^{(\alpha)}[S_\xi]. \end{aligned}$$

In case (2) we get

$$\begin{aligned} l_r^{(\alpha)}[S_\xi] &= 2^{c_k-\nu+\sum_{i=1}^{k}\gamma_i} \cdot 2^{-\alpha[\sum_{i=1}^{k}(\gamma_i+c_i)-\nu]} \\ &= 2^{S_k(\alpha)} \cdot 2^{c_k+\nu(\alpha-1)} \\ &= l_{r_k}^{(\alpha)}[S_\xi] \cdot 2^{c_k+\nu(\alpha-1)}. \end{aligned}$$

Consider the function $f(x) = c_k + x(\alpha - 1)$, $x \in [c_k; c_k + \gamma_k]$, c_k and α being the parameters. Since $f'(x) = \alpha - 1 < 0$, the function $f(x)$ decreases and

$$\max_{x\in[c_k;c_k+\gamma_k]} f(x) = f(c_k) = c_k \cdot \alpha > 0. \tag{1}$$

Therefore in this case $l_r^{(\alpha)}[S_\xi] > l_{r_k}^{(\alpha)}[S_\xi]$ holds. Thus we get

$$H_\alpha(S_\xi) = \liminf_{k\to\infty} l_{r_k}^{(\alpha)}[S_\xi] = 2^{\liminf_{k\to\infty} S_k(\alpha)},$$

which completes the proof.

Corollary. *If the limit* $\lim_{i\to\infty} c_i\gamma_i^{-1}$ *exists and for* $\alpha = \alpha_0 = [1 + \lim_{i\to\infty} \frac{c_i}{\gamma_i}]^{-1}$ *the series* $\sum_{i=1}^{\infty}[\gamma_i - \alpha(\gamma_i + c_i)]$ *converges then the Hausdorff–Besicovitch dimension of the spectrum* S_ξ *of distribution of r.v.* ξ *is equal to* α_0, *its* α_0*-dimensional Hausdorff measure satisfying*

$$0 < H_{\alpha_0}(S_\xi) = 2^{\sum_{i=1}^{\infty}[\gamma_i - \alpha(\gamma_i + c_i)]} < \infty.$$

Theorem 1. *If the finite or infinite limit* $\lim_{i\to\infty} c_i\gamma_i^{-1}$ *exists then the Hausdorff–Besicovitch dimension* α_0 *of the spectrum* S_ξ *of distribution of r.v.* ξ *is equal to*

$$\alpha_0(S_\xi) = \left[1 + \lim_{i\to\infty} \frac{c_i}{\gamma_i}\right]^{-1},$$

and its α_0*-dimensional Hausdorff measure satisfies*

$$0 < H_{\alpha_0}(S_\xi) < \infty \tag{2}$$

if $0 < S(\alpha_0) = \liminf_{k\to\infty} \sum_{i=1}^{k}[\gamma_i - \alpha_0(\gamma_i + c_i)] < \infty$.

Proof. Let $S(\alpha_0) = \liminf_{k\to\infty} S_k(\alpha_0) = \lim_{i\to\infty} S_{k_i}(\alpha_0)$. We will consider three possible cases: (1) $-\infty < S(\alpha_0) < \infty$, (2) $S(\alpha_0) = \infty$ and (3) $S(\alpha_0) = -\infty$. In case (1) Lemma 2 shows that (2) holds, which means that α_0 is the Hausdorff–Besicovitch dimension.

In case (2) we can write $S(\alpha_0) = \lim_{k\to\infty} S_k(\alpha_0) = \sum_{i=1}^{\infty}[\gamma_i - \alpha_0(1 + \frac{c_i}{\gamma_i})]$. Given $\alpha \neq \alpha_0$ we will first consider the case where $\alpha > \alpha_0$. Writing $\alpha = \alpha_0 + \varepsilon$, $\varepsilon > 0$ we obtain

$$S_k(\alpha) = \sum_{i=1}^{k} \gamma_i \left[1 - \alpha\left(1 + \frac{c_i}{\gamma_i}\right)\right] = \sum_{i-1}^{k} \gamma_i \left[\left(1 - \alpha_0\left(1 + \frac{c_i}{\gamma_i}\right)\right) - \varepsilon\left(1 + \frac{c_i}{\gamma_i}\right)\right].$$

Since $1 - \alpha_0(1 + \frac{c_i}{\gamma_i}) \to 0$ when $i \to \infty$, the terms in the last sum are negative and their absolute values are greater than $\frac{\varepsilon}{2}$ for sufficiently large i. Therefore $S(\alpha) = \liminf_{k\to\infty} S_k(\alpha) = -\infty$. Lemma 2 now yields $H_\alpha(E) = 0$. Considering the case where $\alpha < \alpha_0$ and writing $\alpha = \alpha_0 - \varepsilon$, $\varepsilon > 0$ we obtain

$$S_k(\alpha) = \sum_{i=1}^{k} \gamma_i \left[\left(1 - \alpha_0\left(1 + \frac{c_i}{\gamma_i}\right)\right) + \varepsilon\left(1 + \frac{c_i}{\gamma_i}\right)\right].$$

Proceeding as before we see that the terms in the last sum are positive and greater than $\varepsilon/2$ for large enough i. Therefore $S(\alpha) = +\infty$, which yields $H_\alpha(E) = \infty$. This proves Theorem 1 for case (2).

Similar reasoning shows that Theorem 2 is true for case (3), which completes the proof.

Corollary 1. *If $c_i = a\gamma_i$ for all $i > i_0$ then $\alpha_0(S_\xi) = (1+a)^{-1}$.*

Corollary 2. *If $c_i\gamma_i^{-1} \to 0$ when $i \to \infty$ then $\alpha_0(S_\xi) = 1$ and the distribution of r.v. ξ is superfractal.*

Corollary 3. *If $c_i\gamma_i^{-1} \to \infty$ when $i \to \infty$ then $\alpha_0(S_\xi) = 0$ and the distribution of r.v. ξ is anomalously fractal.*

Remark 1. Since the sequences $\{\gamma_i\}$ and $\{c_i\}$ are 'independent', there is a rather broad spectrum of various possible cases and the fraction $c_i\gamma_i^{-1}$ might not tend to a limit (finite or infinite). In that case Theorem 1 does not help to determine the Hausdorff–Besicovitch dimension $\alpha_0(S_\xi)$. According to Lemma 2 the dimension α_0 however meets the condition

$$\liminf_{k\to\infty} \sum_{i=1}^{k} [\gamma_i - \alpha(\gamma_i + c_i)] = \begin{cases} -\infty & \text{if } \alpha > \alpha_0, \\ \infty & \text{if } \alpha < \alpha_0. \end{cases}$$

For instance, if $c_{2m} = \gamma_{2m-1} = 2$, $c_{2m-1} = \gamma_{2m} = 1$ we have

$$S_k(\alpha) = \sum_{i=1}^{k} [\gamma_i - \alpha(\gamma_i + c_i)] = \begin{cases} 3m(1-2\alpha) - (1-3\alpha) & \text{if } k = 2m-1, \\ 3m(1-2\alpha) & \text{if } k = 2m. \end{cases}$$

Therefore $S(\alpha) = \frac{1}{2}$ for $\alpha = \frac{1}{2}$ and $H_{1/2}(S_\xi) = \sqrt{2}$.

Remark 2. Theorem 1 and Remark 1 allow us to determine the Hausdorff–Besicovitch dimension of the spectrum S_ξ of r.v. $\xi = \sum_{i\in D} 2^{-i}\eta_i$ where η_i are Bernoulli random variables, i. e. they satisfy $P_{0i} = P_{1i} = \frac{1}{2}$, and the sets D and $N \setminus D$ are infinite. To this end the sequences $\{\gamma_i\}$ and $\{c_i\}$ are constructed by using the sets D and $N \setminus D$.

For $n = 2$ the matrices $\|P_{ik}\|$ have the following property **B**: a column not containing 1 does not contain 0. For $n > 2$ a column not containing 1 might contain 0. If $n > 2$ and property **B** is fulfilled then the above argument yields the same result. The results however are different for the general case. For example, if $n = 3$, $P_{od} \neq P_{1d} = 0 \neq P_{2d}$ and the limit exists then by using rank coverings we obtain

$$l_{r_k}^{(\alpha)}[S_\xi] = 3^{\log_3 2 \cdot \sum_{i=1}^{k} \gamma_i - \alpha \sum_{i=1}^{k} (\gamma_i + c_i)}$$

for all $d \in D$. Whence it follows that

$$\alpha_0(S_\xi) = \frac{\log_3 2}{1 + \lim_{i\to\infty} \frac{c_i}{\gamma_i}}.$$

In general the following theorem is true.

Theorem 2. *α-dimensional Hausdorff measure of the spectrum S_ξ of distribution of r.v. ξ equals*

$$H_\alpha(S_\xi) = n^{\liminf_{k\to\infty}[\log_n \prod_{\nu=1}^{r_k} m_\nu - \alpha\sum_{i=1}^{k}(\gamma_i+c_i)]}$$

and the Hausdorff-Besicovitch dimension $\alpha_0 = \alpha_0(S_\xi)$ satisfies the condition

$$\liminf_{k\to\infty}[\log_n \prod_{\nu=1}^{r_k} m_\nu - \alpha\sum_{i=1}^{k}(\gamma_i + c_i)] = \begin{cases} -\infty & \text{if } \alpha < \alpha_0, \\ \infty & \text{if } \alpha > \alpha_0, \end{cases}$$

where m_ν is the number of indexes j such that $P_{j\nu} \neq 0$.

Proof. The set S_ξ is contained in

$$\prod_{\nu=1}^{r_k} m_\nu = \prod_{i=1}^{k-1}\prod_{s=1}^{\gamma_i+1} m_{r_i+s}$$

intervals of rank r_k, each having a non-empty intersection with this set. Hence the α-volume of the covering with these intervals equals

$$l_{r_k}^{(\alpha)}[S_\xi] = \prod_{\nu=1}^{r_k} m_\nu \cdot n^{-\alpha r_k} = n^{\log_n \prod_{\nu=1}^{r_k} m_\nu - \alpha r_k}.$$

We will show that the following inequality

$$l_r^{(\alpha)}[S_\xi] \geq l_{r_k}^{(\alpha)}[S_\xi], \quad r_{k-1} < r \leq r_k \tag{3}$$

takes place for the 'α volume' of the covering of spectrum S_ξ with n-adic intervals of rank r. Let $r = r_k - \tau$. Consider the two cases: (1) $0 \leq \nu \leq c_k$ and (2) $c_k < \nu < c_k + \gamma_k$. In case (1) we get

$$\begin{aligned} l_r^{(\alpha)}[S_\xi] &= \prod_{\nu=1}^{r_k} m_\nu \cdot n^{-\alpha[r_k-\tau]} \\ &= l_{r_k}^{(\alpha)}[S_\xi] \cdot n^{\alpha\tau} \\ &\geq l_{r_k}^{(\alpha)}[S_\xi]. \end{aligned}$$

In case (2) we obtain

$$l_r^{(\alpha)}[S_\xi] = \prod_{\nu=1}^{r} m_\nu \cdot \prod_{j=1}^{\tau} m_{r_k-j}^{-1} \cdot n^{-\alpha[r_k-\tau]} \geq l_{r_k}^{(\alpha)}[S_\xi] \cdot n^{c_k+\tau(\alpha-1)}$$

since $m_{r_k-j} = 1$ for $j = \overline{1, c_k}$ and $m_{r_k-(c_k+j)} \leq n$ for $j = \overline{1, \gamma_k}$.

Taking into account (1), we arrive at (3), which in view of Lemma 1 results in the statement of Theorem 2.

References

1. **Pratsevityi, N.V.** (1990). Distributions of random variables with independent Q-symbols. In: *Asymptotic and Applied Problems of Random Evolutions.* Kiev, Inst. Math., Ukr. SSR Acad. Sci., 92–101. (In Russian).

2. **Turbin, A.F. and Pratsevityi, N.V.** (1992). *Fractal Sets, Functions, Distributions.* Kiev, Naukova Dumka. (In Russian).

3. **Pratsevityi, N.V.** (1985). *Random Variables with Fractal Supports of Cantor Type.* Prepr. 85.93, Kiev, Inst. Math., Ukr. SSR Acad. Sci. (In Russian).

4. **Pratsevityi, N.V.** (1988). *Singular Distributions of Cantor and Salem types.* Prepr. 88.6, Kiev, Inst. Math., Ukr. SSR Acad. Sci. (In Russian).

Exploring Stochastic Laws. pp. 417–442.
Eds. A.V. Skorokhod and Yu.V. Borovskikh

Semi-Markov random evolutions: some ideas, methods and results

A.V. SWISHCHUK*

National Academy of Sciences of Ukraine, International Mathematical Center, Kiev, Ukraine

Received March 27, 1995

Abstract. We consider some ideas, methods and results from the theory of semi-Markov random evolutions. It includes averaging and diffusion approximations of random evolutions, rates of convergence in limit theorems, normal deviations, stability and control of random evolutions, and evolutionary models of financial mathematics.

Introduction

Many physical processes in nature appear in random media. The evolution of systems in random media are models of dynamical systems perturbed by the random disturbances of an outer medium. Abstract mathematical models of this evolution are called *random evolutions* (RE), namely, stochastic operator integral equations in a separable Banach space [3]. Investigation of RE is usually carried out by using asymptotic perturbation theory, which is based on asymptotical analysis of Markov renewal equations for expectations of RE, and also on the theory of operators perturbed on a spectrum [1].

These methods allow us to obtain averaging and diffusion approximations for RE in a series scheme. But an especially effective approach to the study of RE turns out to be martingale methods.

It allows us to obtain the solution to the martingale problem, rates of convergence in the limit theorems for RE, stability, control, hedging strategies, etc. [3, 4].

The results of RE are applied to evolutionary stochastic systems in random media, that has almost certain methodological advantages. These applications concern traffic, storage, branching, wave, diffusion and switching systems in random media, integral functionals and U-statistical processes, motions on Lie groups, evolutionary models in financial mathematics, etc. [3, 4].

* The research described in this publication was made possible in part by Grant no. K43100 from the Joint Fund of the Government of Ukraine and International Science Foundation

In this article we consider a survey of ideas, methods and results from the theory of semi-Markov random evolutions (SMRE), namely, operator evolutionary dynamical systems in a separable Banach space in a semi-Markov random media.

In Section 1 we give a definition of SMRE and its classification. Section 2 is devoted to the application of SMRE. The main idea and method from the theory of SMRE are described in Section 3. Averaging and diffusion approximations of SMRE in series schemes in ergodic media are considered in Sections 4 and 5 respectively, and in reducible media — in Sections 7 and 8 respectively. Uniqueness of the limiting *RE* is described in Section 6.

Normal deviations of SMRE are investigated in Section 9, and rates of convergence in limit theorems — in Section 10.

Stability and control of SMRE are studied in Sections 11 and 12 respectively.

Applications of SMRE in a series scheme are considered in Section 13. We consider SMRE in financial mathematics in Section 14.

Article [19] is the most similar to this article.

1. Definition and classification of semi-Markov random evolutions [3, 4, 15]

Semi-Markov random evolutions are defined by the solution of stochastic operator equations in a separable Banach space $(B, \mathcal{B}, \|\cdot\|)$:

$$V(t)f = f + \int_0^t \Gamma(x(s))\, V(s)f\, ds + \sum_{k=1}^{\nu(t)} [D(x_{k-1}, x_k) - I] V(\tau_k^-) f, \quad \forall f \in B. \tag{1}$$

Here: $\{x(s); s \in R_+\}$ is a semi-Markov process in a measurable phase space $(X, \mathcal{X})$; τ_k, $k = 0, 1, 2, 3, \ldots$, are the jump moments of $x(t)$;

$$\tau_k - = \tau_k - 0, \quad x_k := x(\tau_k), \quad \nu(t) := \max\{n: \tau_n \le t\};$$

$$\tau_n := \sum_{k=1}^{n} \theta_k,$$

$(x_n, \theta_n; n \ge 0)$ is a Markov renewal process [1, 2]; $\{\Gamma(x); x \in X\}$ is a family of operators on B, in general non-commuting and unbounded, $\{D(x,y); x_1 y \in X\}$ is a family of bounded operators on B.

If $D(x,y) \equiv I$ is an identity operator on B, $\forall x, y \in X$, then $V(t) \equiv V^c(t)$ is a *continuous SMRE* in (1). If $\Gamma(x) \equiv 0$ is a zero operator on B, $\forall x \in X$, then $V(t) = V^d(t)$ is *a jump SMRE* in (1).

The SMRE $V_n := V(\tau_n)$ is called *a discrete SMRE.*

Operators $\Gamma(x)$ describe a continuous component $V^c(t)$ of SMRE $V(t)$, and operators $D(x,y)$ describe a jump component $V^d(t)$ of SMRE $V(t)$ in (1).

In such a way, SMRE can be described by the following:

1) semi-Markov process $x(t)$;

2) operator dynamical system $V(t)$.

We note, that is turns out to be

$$V(t) = \Gamma_{x(t)}(t - \tau_{\nu(t)}) \prod_{k=1}^{\nu(t)} D(x_{k-1}, x_k)\, \Gamma_{x_{k-1}}(\theta_k), \tag{2}$$

where $\Gamma_x(t)$ are the semigroup of operators generated by the operators $\Gamma(x)$, $\forall x \in X$.

We also note that SMRE in (1) is usually called a *discontinuous SMRE.*

2. Realizations of SMRE [3, 4, 19]

The connection of *RE* with applied problems is explained by the generality of definition (1). It includes any homogeneous linear evolutionary system. For example, if

$$\Gamma(x) := \vartheta(x) \cdot \frac{d}{dz}, \quad D(x,y) \equiv I, \quad B = C^1(R),$$

then equation (1) is a *transport equation* that describes a motion of a particle with random speed $\vartheta(x(t))$. In such a way, various interpretations of operators $\Gamma(x)$ and $D(x,y)$ gives us many realizations of SMRE.

Now we show by *the storage process* (as an example) how *RE*'s are the realization of SMRE $V(t)$ in (1).

Let $B = C^1(R)$ and operators $\Gamma(x)$ and $D(x,y)$ be defined in the following way:

$$\begin{aligned} \Gamma(x)f(z) &:= \vartheta(z,x)\frac{d}{dz}f(z), \\ D(x,y)f(z) &:= f(z + a(x,y)), \end{aligned} \tag{3}$$

where functions $\vartheta(z,x)$ and $a(x,y)$ are continuous and bounded,

$$z \in R, \quad x, y \in X, \quad f(z) \in C^1(R).$$

Then relation (1) takes the form:

$$\begin{aligned} f(z_t) &= f(z) + \int_0^t \vartheta(z_s, x(s))\frac{d}{dz}f(z_s)\,ds + \\ &\quad + \sum_{k=1}^{\nu(t)}[f(z_{\tau_k -} + a(x_{k-1}, x_k)) - f(z_{\tau_k -})], \end{aligned} \tag{4}$$

and SMRE $V(t)$ is defined by the relation:

$$V(t)f(z) := f(z_t), \quad z_0 = z.$$

It is a functional equation for *storage process* z_t that satisfies equation:

$$\begin{aligned} z_t &= z + \int_0^t \vartheta(z_s, x(s))\, ds + \\ &+ \sum_{k=1}^{\nu(t)} a(x_{k-1}, x_k). \end{aligned} \tag{5}$$

If $a(x,y) \equiv 0$, $\forall x, y \in X$, then equation (4) takes the form:

$$f(z_t) = f(z) + \int_0^t \vartheta(z_s, x(s)) \frac{d}{dz} f(z_s)\, ds \tag{6}$$

which is a functional equation for *traffic process* Z_t:

$$\begin{cases} dz_t/dt &= \vartheta(z_t, x(t)) \\ z_0 &= z. \end{cases} \tag{7}$$

The above equation is the *transport equation.* Hence, defining the operators $\Gamma(x)$ and $D(x,y)$ by the expression in (3) we obtain traffic and storage processes in (5) and (7) as continuous and discontinuous SMRE.

Now let $\vartheta(z,x) \equiv 0$ in (5), $\forall z \in R$, $\forall x \in X$. Then the process

$$z_t = z + \sum_{k=1}^{\nu(t)} a(x_{k-1}, x_k)$$

is a summation on a Markov chain $(x_n; n \geq 0)$, and therefore it is a realization of a jump SMRE.

3. The main idea and method from the theory of SMRE [13, 14, 15, 19]

The main approach to the investigation of SMRE is a martingale method.

The martingale method of obtaining limit theorems (averaging and diffusion approximations) for sequences of SMRE is based on the solutions of the following problems:

1) weak compactness of a family of measures generated by a sequence of SMRE;

2) any limiting point of this family of measures is the solution of the martingale problem;

3) the solution of the martingale problem is unique.

Conditions 1)–2) guarantee the existence of a weakly converging subsequence, and condition 3) gives the uniqueness of a weak limit.

From 1)–3) it follows that a sequence of a SMRE converges weakly to the unique solution of the martingale problem.

We obtain a *weak convergence of SMRE* in a series scheme from the criterion of weak compactness of the process with values in a separable Banach space [22]. We obtain the *limit SMRE* from the solution of some martingale problem in a form of some integral operator equations in a Banach space B.

The main idea is that process

$$Mn := V_n - I - \sum_{k=0}^{n-1} E\left[V_{k+1} - V_k/\mathcal{F}_k\right], \quad V_0 = I, \tag{8}$$

is an $\mathcal{F}_n$-*martingale* on B, where

$$\mathcal{F}_n := \sigma\{x_k, \tau_k; 0 \le k \le n\}, \quad V_n := V(\tau_n),$$

E is an expectation with probability $\mathcal{P}$ on a probability space $(\Omega, \mathcal{F}, \mathcal{P})$.

Representation of the martingale M_n in the form of *martingale-differences*

$$Mn = \sum_{k=0}^{n-1} [V_{k+1} - E(V_{k+1}/\mathcal{F}_k)] \tag{9}$$

gives us the possibility to calculate the *weak quadratic variation*:

$$<l(Mnf)> := \sum_{k=0}^{n-1} E\left[l^2((V_{k+1} - V_k)f)/\mathcal{F}_k\right], \tag{10}$$

where $l \in B^*$, and B^* is the dual space of B, dividing the points of B.

From (2) it follows that

$$V_{k+1} - V_k = [\Gamma_{x_k}(\theta_{k+1}) D(x_k, x_{k+1}) - I] \cdot V_k. \tag{11}$$

We note that the following expression for semigroup of operators $\Gamma_x(t)$ is satisfied:

$$\begin{aligned} \Gamma_x(t)f &= I + \sum_{k=1}^{n-1} \frac{t^k}{k!} \Gamma^k(x) f \\ &\quad + \frac{1}{(n-1)!} \int_0^t (t-s)^n \Gamma_x(s) \Gamma^n(x) f \, ds, \\ &\quad \forall x \in X, \quad \forall f \in \bigcap_x \mathrm{Dom}(\Gamma^n(x)). \end{aligned} \tag{12}$$

Taking into account (8)–(12) we obtain the above mentioned results.

Everywhere we suppose that the following conditions are satisfied:

A) there exist Hilbert spaces H and H^* compactly imbedded in Banach spaces B and B^* respectively, $H \subset B$, $H^* \subset B^*$, where B^* is a dual space to B, that divides points of B;

B) operators $\Gamma(x)$ and $(\Gamma(x))^*$ are dissipative on any Hilbert space H and H^* respectively;

C) operators $D(x, y)$ and $D^*(x, y)$ are contractive on any Hilbert space H and H^* respectively;

D) $(x_n;\ n \geq 0)$ is a uniformly ergodic Markov chain with stationary distribution $\rho(A)$, $A \in \mathcal{X}$;

E) $m_i(x) := \int_0^\infty t^i\, G_x(dt)$ are uniformly integrable, $\forall i = 1, 2, 3$, where

$$G_x(t) := \mathcal{P}\{\omega: \theta_{n+1} \leq t / x_n = x\}; \tag{13}$$

F)

$$\begin{gathered}\int_X \rho(dx) \|\Gamma(x) f\|^k < +\infty; \quad \int_X \rho(dx) \|P D_j(x,\cdot) f\|^k < +\infty; \\ \int_X \rho(dx) \|\Gamma(x) f\|^{k-1} \cdot \|P D_j(x,\cdot) f\|^{k-1} < +\infty; \\ \forall k = 1,2,3,4, \ f \in B,\end{gathered} \tag{14}$$

where P is an operator generated by transition probabilities $P(x, A)$ of Markov chain $(x_n;\ n \geq 0)$:

$$P(x, A) := \mathcal{P}\{\omega: x_{n+1} \in A / x_n = x\}, \tag{15}$$

and $\{D_j(x, y);\ x, y \in X,\ j = 1, 2\}$ is a family of some closed operators.

If $B := C_0(R)$, then $H := W^{l,2}(R)$ is a Sobolev space, and $W^{l,2}(R) \subset C_0(R)$ and this imbedding is compact. For the spaces $B := L_2(R)$ and $H := W^{e,2}(R)$ it is the same.

It follows from conditions A)–B) that operators $\Gamma(x)$ and $(\Gamma(x))^*$ generate a strongly continuous contractive semigroup of operators $\Gamma_x(t)$ and $\Gamma_x^*(t)$, $\forall x \in X$, in H and H^* respectively. From conditions A–C it follows that SMRE $V(t)$ in (1) is a contractive operator in H, $\forall t \in R_+$, and $\|V(t)f\|_H$ is a semimartingale $\forall f \in H$. In such a way, conditions A)–C) imply the following result:

SMRE $V(t)f$ is a tight process in B, namely, $\forall \Delta > 0$ there exists a compact set K_Δ:

$$\mathcal{P}\{V(t)f \in K_\Delta; 0 \leq t \leq T\} \geq 1 - \Delta. \tag{16}$$

This result follows from Kolmogorov–Doob's inequality for semimartingale $\|V(t)f\|_H$.

Condition (16) is the main step in the proofs of limit theorems and rates of convergence for the sequence of SMRE in a series scheme.

4. Averaging of SMRE in a series scheme [3, 4, 15, 19]

Now consider a SMRE in a series scheme:

$$\begin{aligned} V_\varepsilon(t) &= f + \int_0^t \Gamma(x(s/\varepsilon))\, V_\varepsilon(s) f\, ds \\ &\quad + \sum_{k=1}^{\nu(t/\varepsilon)} [D^\varepsilon(x_{k-1}, x_k) - I]\, V_\varepsilon(\varepsilon\tau_k -) f, \end{aligned} \tag{17}$$

where

$$D^\varepsilon(x,y) = I + \varepsilon D_1(x,y) + 0(\varepsilon), \tag{18}$$

$\{D_1(x,y); x,y \in X\}$ is a family of closed linear operators, $\|0(\varepsilon)f\|/\varepsilon \to 0$, as $\varepsilon \to 0$, ε is a small parameter,

$$f \in B_0 := \bigcap_{x,y\in X} \mathrm{Dom}(\Gamma^2(x)) \cap \mathrm{Dom}(D_1^2(x,y)). \tag{19}$$

Another form for $V_\varepsilon(t)$ in (17) is:

$$V_\varepsilon(t) = \Gamma_{x(t/\varepsilon)}(t - \varepsilon\tau_{\nu(t/\varepsilon)}) \prod_{k=1}^{\nu(t/\varepsilon)} D^\varepsilon(x_{k-1}, x_k)\, \Gamma_{k-1}(\varepsilon\theta_k). \tag{20}$$

Under conditions A)–C) the sequence of SMRE $V_\varepsilon(t)f$ is tight (see (16)). ρ-a.s.

Under conditions D), E), $i=2$, F), $k=2$, $j=1$, the sequence of SMRE $V_\varepsilon(t)f$ is weakly compact ρ-a.s. in $D_B[0,+\infty)$ with limit points in $C_B[0,+\infty)$, $f \in B_0$.

Now consider the following process in $D_B[0,+\infty)$:

$$M^\varepsilon_{\nu(t/\varepsilon)} f^\varepsilon := V^\varepsilon_{\nu(t/\varepsilon)} f^\varepsilon - f^\varepsilon - \sum_{k=0}^{\nu(t/\varepsilon)-1} E_\rho[V^\varepsilon_{k+1} f^\varepsilon_{k+1} - V^\varepsilon_k f^\varepsilon_k / \mathcal{F}_k], \tag{21}$$

where $V_n^\varepsilon := V_\varepsilon(\varepsilon\tau_n)$ (see (20)),

$$f^\varepsilon := f + \varepsilon f_1(x(t/\varepsilon)),$$

$$f_k^\varepsilon := f^\varepsilon(x_k),$$

where function $f_1(x)$ is defined from the equation

$$\begin{gathered}(P-I)f_1(x) = \left[(\hat\Gamma + \hat D) - (m(x)\Gamma(x) + PD_1(x,\cdot))\right] f,\\ \hat\Gamma := \int_X \rho(dx)\, m(x)\,\Gamma(x), \quad \hat D := \int_X \rho(dx)\, PD_1(x,\cdot),\\ m(x) := m_1(x)\end{gathered} \tag{22}$$

(see E), $f \in B_0$.

The process $M^\varepsilon_{\nu(t/\varepsilon)} f^\varepsilon$ is an $\mathcal{F}_t^\varepsilon$-martingale with respect to the σ-algebra $\mathcal{F}_t^\varepsilon := \sigma\{x(s/\varepsilon); 0 \le s \le t\}$.

The martingale $M^\varepsilon_{\nu(t/\varepsilon)} f^\varepsilon$ in (21) has the asymptotic representation:

$$M^\varepsilon_{\nu(t/\varepsilon)} f^\varepsilon = V^\varepsilon_{\nu(t/\varepsilon)} f - f - \varepsilon \sum_{k=0}^{\nu(t/\varepsilon)} (\hat\Gamma + \hat D)\, V_k^\varepsilon f + 0_f(\varepsilon), \tag{23}$$

where $\hat\Gamma$, $\hat D$, f, f^ε are defined in (21)–(22) and

$$\|0_f(\varepsilon)\|/\varepsilon \to \text{const} \quad \text{as } \varepsilon \to 0,\ \forall f \in B_0.$$

We have used (11) and (12) with $n = 2$, and representation (18) and (21) in (23).

The families $l(M^{\varepsilon}_{\nu(t/\varepsilon)} f^{\varepsilon})$ and

$$l\Big(\sum_{k=0}^{\nu(t/\varepsilon)} E_{\rho}[(V^{\varepsilon}_{k+1} f^{\varepsilon}_{k+1} - V^{\varepsilon}_{k} f^{\varepsilon}_{k})/\mathcal{F}_k]\Big)$$

are weakly compact for all $l \in B^*_0$ where B^*_0 is some dense subset from B^*. Let $V_0(t)$ be a limit process for $V_\varepsilon(t)$ as $\varepsilon \to 0$.

Since (see (20))

$$[V_\varepsilon(t) - V^{\varepsilon}_{\nu(t/\varepsilon)}] = [\Gamma_{x(t/\varepsilon)}(t - \varepsilon\tau_{\nu(t/\varepsilon)}) - I] \cdot V^{\varepsilon}_{\nu(t/\varepsilon)} \tag{24}$$

and the right-hand side in (24) tends to zero as $\varepsilon \to 0$, then it is clear that the limits for $V_\varepsilon(t)$ and $V^{\varepsilon}_{\nu(t/\varepsilon)}$ are the same, namely, $V_0(t)$ ρ-a.s.

The sum $\varepsilon \cdot \sum_{k=0}^{\nu(t/\varepsilon)} (\hat{\Gamma} + \hat{D})\, V^{\varepsilon}_{k} f$ converges strongly as $\varepsilon \to 0$ to the integral

$$m^{-1} \cdot \int_0^t (\hat{\Gamma} + \hat{D})\, V_0(s) f\, ds.$$

The quadratic variation of the martingale $l(M^{\varepsilon}_{\nu(t/\varepsilon)} f^{\varepsilon})$ tends to zero, and, hence,

$$M^{\varepsilon}_{\nu(t/\varepsilon)} f^{\varepsilon} \to 0 \quad as \quad \varepsilon \to 0, \quad \forall f \in B_0, \quad \forall e \in B^*_0.$$

Passing to the limit in (23) as $\varepsilon \to 0$ and taking into account all previous reasonings we can show that the limit process $V_0(t)$ satisfies the equation:

$$0 = V_0(t) f - f - m^{-1} \int_0^t (\hat{\Gamma} + \hat{D})\, V_0(s) f\, ds, \tag{25}$$

where

$$m := \int_X \rho(dx) m(x), \quad f \in B_0, \quad t \in [0, T].$$

5. Diffusion approximation of SMRE in a series scheme [3, 4, 15, 19]

Now consider SMRE $V_\varepsilon(t/\varepsilon)$, where $V_\varepsilon(t)$ is defined in (17) or (20), with the operators

$$D^{\varepsilon}(x, y) := I + \varepsilon D_1(x, y) + \varepsilon^2 D_2(x, y) + 0(\varepsilon^2), \tag{26}$$

$\{D_i(x, y);\ x, y \in X,\ i = 1, 2\}$ are closed linear operators and $\|0(\varepsilon^2) f\| / \varepsilon^2 \to 0$, $\varepsilon \to 0$

$$\begin{gathered} \forall f \in B_0 := \bigcap_{x,y \in X} \mathrm{Dom}(\Gamma^4(x)) \bigcap \mathrm{Dom}(D_2(x, y)), \\ \mathrm{Dom}(D_2(x, y)) \subseteq \mathrm{Dom}(D_1(x, y)); \quad D_1(x, y) \subseteq \mathrm{Dom}(D_1(x, y)), \\ \forall x, y \in X, \quad \Gamma^i(x) \subset \mathrm{Dom}(D_2(x, y)), \quad i = \overline{1, 3}. \end{gathered} \tag{27}$$

In such a way

$$V_\varepsilon(t/\varepsilon) = \Gamma_{x(t/\varepsilon^2)}(t/\varepsilon - \varepsilon\tau_{\nu(t/\varepsilon^2)}) \prod_{k=1}^{\nu(t/\varepsilon^2)} D^\varepsilon(x_{k-1}, x_k)\Gamma_{x_{k-1}}(\varepsilon\theta_k), \tag{28}$$

where $D^\varepsilon(x,y)$ is defined in (26).

Under conditions A)–C) the sequence of SMRE $V_\varepsilon(t/\varepsilon)f$ is tight (see (16)) ρ-a.s.

Under conditions D), E), $i = 3$, F), $k = 4$, the sequence of SMRE $V_\varepsilon(t/\varepsilon)f$ is weakly compact ρ-a.s. in $D_B[0, +\infty)$ with limit points in $C_B[0, +\infty)$, $f \in B_0$.

Let the *balance condition* be satisfied:

$$\int_X \rho(dx)[m(x)\Gamma(x) + PD_1(x,\cdot)]f = 0, \quad \forall f \in B_0 \tag{29}$$

Now consider the following process in $D_B[0, +\infty)$:

$$M^\varepsilon_{\nu(t/\varepsilon^2)} f^\varepsilon := V^\varepsilon_{\nu(t/\varepsilon^2)} f^\varepsilon - f^\varepsilon - \sum_{k=0}^{\nu(t/\varepsilon^2)-1} E_\rho[V^\varepsilon_{k+1} f^\varepsilon_{k+1} - V^\varepsilon_k f^\varepsilon_k / \mathcal{F}_k], \tag{30}$$

where $f^\varepsilon := f + \varepsilon f_1(x(t/\varepsilon^2)) + \varepsilon^2 f_2(x(t/\varepsilon^2))$, and functions f_1 and f_2 are defined from the following equations:

$$\begin{aligned} (P - I)f_1(x) &= -[m(x)\Gamma(x) + PD_1(x,\cdot)]f, \\ (P - I)f_2(x) &= [\hat{L} - L(x)]f, \\ \hat{L} : &= \int_X \rho(dx)L(x), \end{aligned} \tag{31}$$

$$\begin{aligned} L(x) \;:=\; & (m(x)\Gamma(x) + PD_1(x,\cdot))(R_0 - I)(m(x)\Gamma(x) + PD_1(x,\cdot)) + \\ & + m_2(x)\Gamma^2(x)/2 + m(x)PD_1(x,\cdot)\Gamma(x) + PD_2(x,\cdot), \end{aligned}$$

R_0 is a potential operator of $(x_n;\ n \geq 0)$.

Balance condition (29) and condition $\prod(\hat{L} - L(x)) = 0$ enable the equations in (31) to be solved.

Process $M^\varepsilon_{\nu(t/\varepsilon^2)} f^\varepsilon$ is an $\mathcal{F}^\varepsilon_t$-martingale with respect to the σ-algebra $\mathcal{F}^\varepsilon_t := \sigma\{x(s/\varepsilon^2); 0 \leq s \leq t\}$.

This martingale has the asymptotic representation:

$$M^\varepsilon_{\nu(t/\varepsilon^2)} f^\varepsilon = V^\varepsilon_{\nu(t/\varepsilon^2)} f - f - \varepsilon^2 \sum_{k=0}^{\nu(t/\varepsilon^2)-1} \hat{L} V^\varepsilon_k f - 0_f(\varepsilon t), \tag{32}$$

where $\hat{L}$ is defined in (31) and

$$\|0_f(\varepsilon)\|/\varepsilon \to \text{const}, \quad \varepsilon \to 0, \quad \forall f \in B_0.$$

We have used (11) and (12) with $n = 3$, and representation (26) and (31) in (32).

The families $l(M^{\varepsilon}_{\nu(t/\varepsilon^2)}f^{\varepsilon})$ and $l(\sum_{k=0}^{\nu(t/\varepsilon^2)} E_{\rho}[(V^{\varepsilon}_{k+1}f^{\varepsilon}_{k+1} - V^{\varepsilon}_{k}f^{\varepsilon}_{k})/\mathcal{F}_k])$ are weakly compact for all $l \in B_0^*$, $f \in B_0$.

Set $V^0(t)$ for the limit process for $V_\varepsilon(t/\varepsilon)$ as $\varepsilon \to 0$. From (20) we can show that the limits for $V_\varepsilon(t/\varepsilon)$ and $V^{\varepsilon}_{\nu(t/\varepsilon^2)}$ are the same, namely, $V^0(t)$.

The sum $\varepsilon^2 \sum_{k=0}^{\nu(t/\varepsilon^2)} \hat{L}V^{\varepsilon}_{k}f$ converges strongly as $\varepsilon \to 0$ to the integral

$$m^{-1}\int_0^t \hat{L}V^0(s)f\,ds.$$

Let $M^0(t)f$ be a limit martingale for $M^{\varepsilon}_{\nu(t/\varepsilon^2)}f^{\varepsilon}$ as $\varepsilon \to 0$.

Then, from (30)–(32) and previous reasonings we have as $\varepsilon \to 0$:

$$M^0(t)f = V^0(t)f - f - m^{-1}\cdot\int_0^t \hat{L}V^0(s)f\,ds. \tag{33}$$

The quadratic variation of the martingale $M^0(t)f$ has the form:

$$<l(M^0(t)f)> = \int_0^t \int_X l^2(\sigma(x)\,\Gamma(x)\,V^0(s)f)\,\sigma(dx)\,ds, \tag{34}$$

where

$$\sigma^2(x) := [m_2(x) - m^2(x)]/m.$$

The solution of the martingale problem for $M^0(t)$ (namely, to find the representation of $M^0(t)$ with quadratic variation (34)) is expressed by the integral over Wiener orthogonal martingale measure $W(dx,ds)$ with quadratic variation $\rho(dx)\cdot ds$:

$$M^0(t)f = \int_0^t \int_X \sigma(x)\,\Gamma(x)V^0(s)f\,W(dx,ds). \tag{35}$$

In such a way, the limit process $V^0(t)$ satisfies the following equation (see (33) and (35)):

$$V^0(t)f = f + m^{-1}\cdot\int_0^t \hat{L}\cdot V^0(s)f\,ds + \int_0^t\int_X \sigma(x)\,\Gamma(x)\,V^0(s)f\,W(dx,ds). \tag{36}$$

If the operator $\hat{L}$ generates the semigroup $U(t)$ then the process $V^0(t)f$ in (36) satisfies equation:

$$V^0(t)f = U(t)f + \int_0^t\int_X \sigma(x)\,U(t-s)\,\Gamma(x)\,V^0(s)f\,W(dx,ds). \tag{37}$$

6. The uniqueness of limit random evolutions. Dual SMRE [16, 17]

The *uniqueness* of the limit evolution $V_0(t)f$ in an *averaging* scheme follows from equation (25) and the fact that if the operator $\hat{\Gamma}+\hat{D}$ (see (22)) generates a semigroup, then $V_0(t)f = \exp\{(\hat{\Gamma}+\hat{D})\cdot t\}f$ and this representation is unique.

The *uniqueness* of the limit evolution $V^0(t)f$ in a *diffusion approximation* scheme follows from the uniqueness of the solution of the martingale problem for $V^0(t)f$ (see (33)–(35)). The latter is proved by *dual SMRE* in a series scheme by constructing the limit equation in diffusion approximation and by using a dual identity [15].

7. Averaging of SMRE in reducible phase space. Merged *RE* [3, 18]

Suppose that the following conditions hold true:

a) *decomposition* of phase space X (*reducible* phase space):

$$X = \bigcup_{u \in U} X_u, \quad X_u \bigcap X_{u'} = \emptyset, \quad u \neq u' : \tag{38}$$

where $(U, \mathcal{U})$ is a measurable phase space (*merged* phase space);

b) Markov renewal process $(x_n^\varepsilon, \theta_n; n \geq 0)$ on $(X, \mathcal{X})$ has the *semi-Markov kernel*:

$$Q_\varepsilon(x, A, t) := P_\varepsilon(x, A) G_x(t), \tag{39}$$

where $P_\varepsilon(x, A) = P(x, A) - \varepsilon^l P_1(x, A)$, $x \in X$, $A \in \mathcal{X}$, $l = 1, 2$; $P(x, A)$ are the transition probabilities of the *supporting non-perturbed* Markov chain $(x_n; n \geq 0)$;

c) the stochastic kernel $P(x, A)$ is adapted to decomposition (38) in the following form:

$$P(x, X_u) = \begin{cases} 1, & x \in X_u \\ 0, & x \overline{\in} X_u, \quad u \in U; \end{cases}$$

d) the Markov chain $(x_n; n \geq 0)$ is uniformly ergodic with stationary distributions $\rho_u(B)$:

$$\rho_u(B) = \int_{X_u} P(x, B)\, \rho_u(dx), \quad \forall u \in U, \quad \forall B \in \mathcal{X}. \tag{40}$$

e) there is a family $\{\rho_u^\varepsilon(A); u \in U, A \in \chi, \varepsilon > 0\}$ of stationary distributions of perturbed Markov chain $(x_n^\varepsilon; n \geq 0)$;

f)

$$\begin{aligned} b(u) &:= \int_{X_u} \rho_u(dx)\, P_1(x, X_u) > 0, \quad \forall u \in U, \\ b(u, \Delta) &:= -\int_{X_u} \rho_u(dx)\, P_1(x, X_\Delta) > 0, \quad \forall u \overline{\in} \Delta, \quad \Delta \in U; \end{aligned} \tag{41}$$

g) the operators $\Gamma(u) := \int_{X_u} \rho_u(dx)\, m(x)\, \Gamma(x)$ and

$$\hat{D}(u) := \int_{X_u} \rho_u(dx) \int_{X_u} P(x, dy)\, D_1(x, y) \tag{42}$$

are closed $\forall u \in U$ with common domain B_0, and operators $\hat{\Gamma}(u) + \hat{D}(u)$ generate the semigroup of operators $\forall u \in U$.

Decomposition (38) in *a*) defines the *merging* function

$$u(x) = u \quad \forall x \in Xu, \quad u \in U. \tag{43}$$

We note that σ-algebras $\mathcal{X}$ and $\mathcal{U}$ are coordinated such that

$$X_\Delta = \bigcup_{u\in\Delta} Xu, \quad \forall u \in U, \quad \Delta \in \mathcal{U}. \tag{44}$$

We set $\prod_u f(u) := \int_{X_u} \rho_u(dx)\, f(x)$ and $x^\varepsilon(t) := x^\varepsilon_{\nu(t/\varepsilon^2)}$.

SMRE in a reducible phase space X is defined by the solution of the equation:

$$\begin{aligned} V_\varepsilon(t) &= I + \int_0^t \Gamma(x^\varepsilon(s/\varepsilon))\, V_\varepsilon(s)\, ds \\ &+ \sum_{k=0}^{\nu(t/\varepsilon)} [D^\varepsilon(x^\varepsilon_{k-1}, x^\varepsilon_k) - I]\, V_\varepsilon(\varepsilon\tau_k^-), \end{aligned} \tag{45}$$

where $D^\varepsilon(x, y)$ is defined in (18).

Now consider the martingale

$$\begin{aligned} M^\varepsilon_{\nu(t/\varepsilon)} f^\varepsilon(x^\varepsilon(t/\varepsilon)) &:= V^\varepsilon_{\nu(t/\varepsilon)} f^\varepsilon(x^\varepsilon(t/\varepsilon)) - f^\varepsilon(x) \\ &- \sum_{k=0}^{\nu(t/\varepsilon)-1} E_{\rho^\varepsilon_u}[V^\varepsilon_{k+1} f^\varepsilon_{k+1} - V^\varepsilon_k f^\varepsilon_k / \mathcal{F}^\varepsilon_k], \end{aligned} \tag{46}$$

where

$$\begin{gathered} \mathcal{F}^\varepsilon_n := \sigma\{x^\varepsilon_k, \theta_k; 0 \le k \le n\}, \\ f^\varepsilon(x) := \hat{f}(u(x)) + \varepsilon f^1(x), \quad \hat{f}(u) := \int_{X_u} \rho_u(dx) f(x), \end{gathered} \tag{47}$$

$$\begin{aligned} (P - I) f_1(x) &= [-(m(x)\Gamma(x) + P D_1(x, \cdot)) + \hat{\Gamma}(u) \\ &+ \hat{D}(u) + (\Pi_u - I) P_1] \hat{f}(u), \end{aligned} \tag{48}$$

$$f^\varepsilon_k := f^\varepsilon(x^\varepsilon_k), \quad V^\varepsilon_n := V_\varepsilon(\varepsilon\tau_n),$$

and $V_\varepsilon(t)$ is defined in (45), P_1 is an operator generated by $P_1(x, A)$ (see (38)).

Under conditions A)–C) the sequence of SMRE $V_\varepsilon(t)f$ in (45), $f \in B_0$ (see (19)), is tight (see (16)) ρ_u-a.s., $\forall u \in U$.

Under conditions D), E), $i = 2$, F), $k = 2$, $j = 1$, the sequence of SMRE $V_\varepsilon(t)f$ is weakly compact ρ_u-a.s., $\forall u \in U$, in $D_B[0, +\infty)$ with limit points in $C_B[0, +\infty)$.

We note that $u(x^\varepsilon(t/\varepsilon)) \to \hat{x}(t)$, as $\varepsilon \to 0$, where $\hat{x}(t)$ is a *merged* jump Markov process in $(U, \mathcal{U})$ with an *infinitesimal operator* $\Lambda(\hat{P} - I)$,

$$\begin{aligned} \Lambda \hat{f}(u) &:= [b(u)/m(u)]\, \hat{f}(u), \\ \hat{P} \hat{f}(u) &:= \int_U [b(u, du')/b(u)]\, \hat{f}(u), \\ m(u) &:= \int_{X_u} \rho_u(dx)\, m(x), \end{aligned} \tag{49}$$

$b(u)$ and $b(u, \Delta)$ are defined in (41). We also note that

$$\Pi_u P_1 = \Lambda(\hat{P} - I), \tag{50}$$

where Π_u is defined in (44), P_1 — in (39), Λ and $\hat{P}$ — in (49).

Using (46)–(48) and (50), we can show that the limit evolution $\hat{V}_0(t)$ satisfies equation:

$$\hat{V}_0(t)\,\hat{f}(\hat{x}(t)) = \hat{f}(u) + \int_0^t [\Lambda(\hat{P} - I) + \hat{\Gamma}(\hat{x}(s)) + \hat{D}(\hat{x}(s))]\,\hat{V}_0(s)\,\hat{f}(\hat{x}(s))\,ds. \quad (51)$$

RE $\hat{V}_0(t)$ is called a *merged RE* in an averaging scheme.

8. Diffusion approximation of SMRE in reducible phase space [18]

Now consider SMRE $V_\varepsilon(t/\varepsilon)$ with expansion (26), where $V_\varepsilon(t)$ is defined in (45), and conditions $A)$–$F)$ (when $i = 3$, $k = 4$, $j = 1, 2$) and conditions $a)$–$f)$ ($e = 2$) are satisfied.

Let the balance condition

$$\int_{X_u} \rho_u(dx)\,[m(x)\Gamma(x) + PD_1(x,\cdot)]f = 0, \quad \forall u \in U, \quad (52)$$

also be satisfied and let operator

$$L(u) := \int_{X_u} \rho_u(dx)\,L(x)/m(u), \quad (53)$$

generate the semigroup of operators, where $L(x)$ is defined in (31) and $m(u)$ in (49).

Now also consider the martingale

$$\begin{aligned} M^\varepsilon_{\nu(t/\varepsilon^2)} f^\varepsilon(x^\varepsilon(t/\varepsilon^2)) &= V^\varepsilon_{\nu(t/\varepsilon^2)} f^\varepsilon(x^\varepsilon(t/\varepsilon^2)) - f^\varepsilon(x) \\ &\quad - \sum_{k=0}^{\nu(t/\varepsilon^2)} E\rho^\varepsilon_u[V^\varepsilon_{k+1} f^\varepsilon_{k+1} - V^\varepsilon_k f^\varepsilon_k / \mathcal{F}^\varepsilon_k], \end{aligned} \quad (54)$$

where

$$\begin{aligned} f^\varepsilon(x) &:= \hat{f}(u(x)) + \varepsilon f^1(x) + \varepsilon^2 f^2(x), \\ (P - I)f^1(x) &= [m(x)\Gamma(x) + PD_1(x,\cdot)]\,\hat{f}(u), \\ (P \quad I)f^2(x) &= [m(u)L(u) - L(x) + (\Pi_u - I)P_1]\,\hat{f}(u), \end{aligned} \quad (55)$$

where $L(u)$ is defined in (53).

Let $\hat{M}^0(t)f$ be a limit martingale for

$$M^\varepsilon_{\nu(t/\varepsilon^2)} f^\varepsilon(x^\varepsilon(t/\varepsilon^2)) \quad \text{as } \varepsilon \to 0.$$

In such a way from (53)–(58) we have the following equation as $\varepsilon \to 0$:

$$\begin{aligned} \hat{M}^0(t)\hat{f}(\hat{x}(t)) &= \hat{V}^0(t)\hat{f}(\hat{x}(t)) - \hat{f}(u) \\ &\quad - \int_0^t [\Lambda(\hat{P} - I) + L(\hat{x}(s))]\,\hat{V}^0(s)\,\hat{f}(\hat{x}(s))\,ds. \end{aligned} \quad (56)$$

The quadratic variation of the martingale $\hat{M}^0(t)$ has the form:

$$<l(\hat{M}^0(t)\hat{f}(u))> = \int_0^t \int_{X_u} l^2(\sigma(x,u)\,\Gamma(x)\,\hat{V}^0(s)\,\hat{f}(u))\,\rho_u(dx)\,ds, \tag{57}$$

where

$$\sigma^2(x,u) := [m_2(x) - m^2(x)]/m(u).$$

The solution of the martingale problem for $\hat{M}^0(t)$ is expressed by the integral:

$$\hat{M}^0(t)\hat{f}(\hat{x}(t)) = \int_0^t \hat{W}(ds, \hat{x}(s))\,\hat{V}^0(s)\,\hat{f}(\hat{x}(s)), \tag{58}$$

where

$$\hat{W}(t,u)f := \int_{X_u} W_{\rho_u}(t,dx)\,\sigma(x,u)\,\Gamma(x)f.$$

Finally, from (56)–(58) it follows that the limit process $\hat{V}^0(t)$ satisfies the following equation:

$$\begin{aligned}\hat{V}^0(t)\hat{f}(\hat{x}(t)) &= \hat{f}(u) + \int_0^t [\Lambda(\hat{P} - I) + L(\hat{x}(s))]\,\hat{V}^0(s)\,\hat{f}(\hat{x}(s))\,ds \\ &\quad + \int_0^t \hat{W}(ds, \hat{x}(s))\,\hat{V}^0(s)\,\hat{f}(\hat{x}(s)). \end{aligned}\tag{59}$$

RE $\hat{V}^0(t)$ in (59) is called a *merged RE* in a diffusion approximation scheme. If operator $\hat{U}^0(t)$ is a solution of the Cauchy problem:

$$\begin{cases} d\hat{U}^0(t)\,dt &= \hat{U}^0(t)\,L(\hat{x}(t)) \\ \hat{U}^0(0) &= I, \end{cases}$$

then the operator process $\hat{V}^0\hat{f}(\hat{x}(t))$ satisfies equation:

$$\begin{aligned}\hat{V}^0(t)\hat{f}(\hat{x}(t)) &= \hat{U}^0(t)\hat{f}(u) + \int_0^t \hat{U}^0(t-s)\,\Lambda(\hat{P} - I)\,\hat{V}^0(s)\,\hat{f}(\hat{x}(s))\,ds \\ &\quad + \int_0^t \hat{U}^0(t-s)\,\hat{W}(ds, \hat{x}(s))\,\hat{V}^0(s)\,\hat{f}(\hat{x}(s)). \end{aligned}\tag{60}$$

The uniqueness of the limit *RE* $\hat{V}^0(t)$ is established with the help of dual SMRE (see Section 6).

9. Normal deviations of SMRE [3, 4, 10]

The averaged evolution obtained in Sections 4 and 7 can be considered as the first approximation to the initial evolution. The diffusion approximation of the SMRE described in Sections 5 and 8 determines the second approximation to the initial evolution, since the first approximation under the balance condition, the averaged evolution, appears to be trivial.

Here we consider the *double approximation* to the SMRE — the averaged and the diffusion approximation — provided that the balance condition fails. We introduce the *deviation process* as the normalized difference between the initial and averaged evolutions. In the limit we obtain the *normal deviations* of the initial SMRE from the averaged one.

Now consider the SMRE $V_\varepsilon(t)$ in (17) and the averaged evolution $V_0(t)$ in (25). Also consider the deviation of the initial evolution $V_\varepsilon(t)f$ from the averaged one $V_0(t)f$:

$$W_\varepsilon(t)f := \varepsilon^{-1/2} \cdot [V_\varepsilon(t) - V_0(t)]f, \quad \forall f \in B_0 \tag{61}$$

(see (19)).

Taking into account equations (17) and (25) we obtain for $W_\varepsilon(t)$ the relation:

$$\begin{aligned} W_\varepsilon(t)f &= \varepsilon^{-1/2} \int_0^t (\Gamma(x(s/\varepsilon)) - \hat{\Gamma})\, V_\varepsilon(s)f\, ds \\ &+ \int_0^t \hat{\Gamma}\, W_\varepsilon(s)f\, ds + \\ &+ \varepsilon^{-1/2}\Big[V_\varepsilon^d(t) - \int_0^t \hat{D} \cdot V_0(s)\, ds\Big] f, \quad \forall f \in B_0, \end{aligned} \tag{62}$$

where

$$V_\varepsilon^d(t)f := \sum_{k=1}^{\nu(t/\varepsilon)} [D^\varepsilon(x_{k-1}, x_k) - I]\, V_\varepsilon(\varepsilon\tau_k^-)f,$$

and $\hat{\Gamma}, \hat{D}$ are defined in (22).

The passage to the limit as $\varepsilon \to 0$ in formula (62) by encountering (61) arrives at the equation for $W_0(t)f$:

$$W_0(t)f = \int_0^t \hat{\Gamma}\, W_0(s)f\, ds + \sigma f w(t), \tag{63}$$

where the variance operator σ is determined from the relation:

$$l^2(\sigma f) := l^2(\sigma_1 f) + l^2(\sigma_2 f), \quad \forall l \in B_0, \quad \forall l \in B_0^*, \tag{64}$$

where operators σ_1 and σ_2 are defined by:

$$\begin{aligned} l^2(\sigma_2 f) &:= \int_X \rho(dx)\, l((PD_1(x,\cdot) - \hat{D})f)(R_0 - I) \cdot l((PD_1(x,\cdot) - \hat{D})f), \\ l^2(\sigma_1 f) &= \int_X \rho(dx)\, [m(x)\, l((\Gamma(x) - \hat{\Gamma})f)(R_0 - I)\, m(x)\, l((\Gamma(x) - \hat{\Gamma})f) \\ &\quad + 2^{-1} \cdot m_2(x)\, l^2((\Gamma(x) - \hat{\Gamma})f)]/m, \end{aligned}$$

$\forall l \in B_0$, $w(t)$ is a standard Wiener process.

Double approximation of the SMRE has the form:

$$V_\varepsilon(t)f \approx V_0(t)f + \sqrt{\varepsilon}\, W_0(t)f$$

for small ε, which perfectly fits the standard form of the *CLT* with a non-zero limiting mean value.

10. Rates of convergence in the limit theorems for SMRE [4, 11, 12]

The rates of convergence of the averaging and a diffusion approximation scheme for the sequence of SMRE are considered in this section.

Averaging scheme. The problem is to estimate the value

$$\|E_\rho[V_\varepsilon(t)f^\varepsilon(x(t/\varepsilon)) - V_0(t)f]\|, \quad \forall f \in B_0, \tag{65}$$

where $V_0(t)$, $V_\varepsilon(t)$, f^ε, f and B_0 are defined in (17), (19), (21), (25) respectively.

We use the following representation

$$\begin{aligned}\|E_\rho[V_\varepsilon(t)f^\varepsilon(x(t/\varepsilon)) - V_0(t)f]\| &\le \|E_\rho[V_\varepsilon(t)f - V_\varepsilon(\tau_{\nu(t/\varepsilon)})f]\| + \\ &+ \|E_\rho[V_\varepsilon(\tau_{\nu(t/\varepsilon)})f - V_0(t)f]\| + \varepsilon\|E_\rho V_\varepsilon(t)f_1(x(t/\varepsilon))\|\end{aligned}$$

that follows from (17), (21), (25) and (65).

The estimate of the value in (70), namely, the *rate of convergence in an averaging scheme for SMRE* has the form:

$$\|E_\rho[V_\varepsilon(t)f^\varepsilon(x(t/\varepsilon)) - V_0(t)f]\| \le \varepsilon \cdot C(T,f), \tag{66}$$

where constant $C(T,f)$ is expressed by $C_i(T,f)$, $i = \overline{1,3}$,

$$\begin{aligned}C_1(T,f) &:= \int_X \rho(dx)\,[C_0(T,x,f) + C_0^2(T,x,f)],\\ C_0(T,x,f) &:= T \cdot m_2(x)\|\Gamma(x)f\|/2m, \quad \forall f \in B_0.\end{aligned}$$

and constant $C_2(T,f)$ is expressed by the algebraic sum of $\int_X m_i(x)\,\|\Gamma^i(x)f\|\,\rho(dx)$ and

$$\int_X m_i(x)\,\|PD_1(x,\cdot)\cdot\Gamma^i(x)f\|\,\rho(dx), \quad i = 1,2, \quad f \in B_0,$$

and $\|R_0\|$, R_0 is a potential of Markov chain $(x_n; n \ge 0)$;

$$C_3(f) := \|R_0\| \cdot \int_X p(dx)\,[m(x)\|\Gamma(x)f\| + \|PD_1(x,\cdot)f\|].$$

Diffusion approximation. The problem is to estimate the value:

$$\|E_\rho[V_\varepsilon(t/\varepsilon)f^\varepsilon(x(t/\varepsilon^2)) - V^0(t)f]\|, \quad \forall f \in B_0, \tag{67}$$

where $V_\varepsilon(t/\varepsilon)$, f^ε, $V^0(t)$, f, B_0 are defined in (28), (30), (36), (27) respectively.

Here, we use the following representation:

$$\begin{aligned}&\|E_\rho[V_\varepsilon(t/\varepsilon)f^\varepsilon(x(t/\varepsilon^2)) - V^0(t)f]\| \le\\ &\le \|E_\rho[V_\varepsilon(t/\varepsilon)f - V_\varepsilon(\tau_{\nu(t/\varepsilon^2)})f]\| + \|E_\rho[V_\varepsilon(\tau_{\nu(t/\varepsilon^2)})f - V^0(t)f]\| +\\ &\quad + \varepsilon\|E_\rho[V_\varepsilon(t/\varepsilon)f_1(x(t/\varepsilon^2))]\| + \varepsilon^2\|E_\rho[V_\varepsilon(t/\varepsilon)f_2(x(t/\varepsilon^2))]\|,\end{aligned} \tag{68}$$

$$2\|R_0\| \cdot \int_X \rho(dx)\|L(x)f\| := \varepsilon^2 d_1(f),$$

where $L(x)$ is defined in (31),

$$d_2(f) := 2\|R_0\| \cdot \int_X \rho(dx)\,[m(x)\|\Gamma(x)f\| + \|PD_1(x,\cdot)f\|], \quad f \in B_0.$$

constant $d_3(f)$ is expressed by the algebraic sum of

$$\int_X m_i(x)\,\|\Gamma^j(x)PD_e(x,\cdot)f\|\,\rho(dx), \quad i=\overline{1,3},\ j=\overline{0,3},\ e=\overline{1,2}.$$

We note that

$$\Big\|E_\rho\Big[\sum_{k=0}^{\nu(t/\varepsilon^2)-1} \hat{L}V_k^\varepsilon f - \varepsilon^{-2}m^{-1}\int_0^t \hat{L}V^0(s)f\,ds\Big]\Big\| \le d_4(T,f). \tag{69}$$

The estimate of the value in (67), namely, the *rate of convergence in a diffusion approximation scheme for SMRE* has the form:

$$\|E_\rho[V_\varepsilon(t/\varepsilon)f^\varepsilon(x(t/\varepsilon^2)) - V^0(t)f]\| \le \varepsilon \cdot d(T,f), \tag{70}$$

where constant $d(T,f)$ is expressed by d_i, $i=\overline{1,4}$, and $C_1(T,f)$, $f \in B_0$.

$$2\|R_0\| \cdot \int_X \rho(dx)\,\|L(x)f\| := \varepsilon^2 d_1(f),$$

where $L(x)$ is defined in (31),

$$d_2(f) := 2\|R_0\| \cdot \int_X \rho(dx)\,[m(x)\|\Gamma(x)f\| + \|PD_1(x,\cdot)f\|], \quad f \in B_0,$$

constant $d_3(f)$ is expressed by the algebraic sum of

$$\int_X m_i(x)\,\|\Gamma^j(x)PD_e(x,\cdot)f\|\,\rho(dx), \quad i=\overline{1,3},\ j=\overline{0,3},\ e=\overline{1,2}$$

11. Stability of SMRE [4, 6–8]

Now we have a SMRE $V(t)f$ in (1). A solution of equation (1) is said to be the solution with property $\mathcal{L}(\nu,N)$ if any solution $V(t)$ of equation (1) with initial value $f \in S_\rho := \{y : y \in B : \|y\| \le \rho\}$ satisfies the inequality:

$$\|V(t)f\| \le N \cdot e^{-\nu(t-s)} \cdot \|V(s)f\|, \quad a.s.\ \ \forall \nu, N > 0,\ 0 < s < t. \tag{71}$$

We note that another form of SMRE in (1) is (see (20)):

$$V(t) = \Gamma_{x(t)}(t-\tau_{\nu(t)}) \cdot \prod_{k=1}^{\nu(t)} D(x_{k-1},x_k)\Gamma_{x_{k-1}}(\theta_k). \tag{72}$$

Since the operators $\Gamma_x(t)$ and $D(x,y)$ are contractive then it follows from (86) that

$$\|V(t)f\| \le \|V(s)f\|, \quad \forall f \in S_{\rho 1}\ a.s. \tag{73}$$

It follows from (73) that SMRE in (72) has the property $\mathcal{L}(\nu, N)$ with $\nu = 0$, $N = 1$, namely, $\mathcal{L}(0,1)$.

A solution $V(t)f \equiv f$ of equation (1) is said to be *a weakly stable solution in probability* (as $t \geq s$), if for any $\varepsilon > 0$, $\delta > 0$ there is $\rho > 0$ such that as $t \geq s$, $\|f\| < \rho$, then

$$\mathcal{P}\{\|V(t,s)f\| > \varepsilon\} < \delta,$$

where $V(t,s)$ is a SMRE on the interval $[s,t]$ (see (1)).

The solution of equation (1), namely, SMRE $V(t)$ is weakly stable in probability.

12. Control of SMRE [4, 9]

Let $(U, \mathcal{U})$ be some measurable control space, let x_t^u be a semi-Markov process on $(X, \mathcal{X})$ which is a metric space, $u \in U$. Let $V^u(t)$ be a family of SMRE:

$$dV^u(t)\,dt = V^u(t)\,\Gamma(x_t^u, U_t), \quad V^u(0) = I, \quad u_t \in U. \tag{74}$$

The *optimal control problem* in SMRE is to choose u such that $V^u(t)f \to S$ as $t \to \tau_u$, where

$$\tau_u := \inf\{t := x_t^u \in \mathcal{Y}\}, \quad \mathcal{Y} \in X, \quad S \subset B, \quad \forall f \in B.$$

For the solution of this problem we introduce the *cost function*:

$$C^u(f,x) := E_x^u\, b(V^u(\tau_u)f, x_{\tau_u}^u) + E_x^u \int_0^{\tau_u} k(V^u(t)f, x_t^u, U_t)\,dt, \tag{75}$$

where $b(f,x)$ and $k(f,x,u)$ are some functions on the spaces $B \times X$ and $B \times X \times U$ respectively, $\forall f \in B$, $\forall x \in X$, $\forall u \in U$.

Let $h(f,x)$ be a minimal value of cost function $C^u(f,x)$, i. e.,

$$h(f,x) = \min_u C^u(f,x), \quad h \leq C^u, \quad \forall u \in U.$$

Function h satisfies the boundary condition $h = b(f,x)$ for those $(f,x) \in \partial s \cdot \partial Y$, $Y \subset X$.

We note that process $\hat{x}_t^u := (x_t^u, t - \tau_{\nu(t)})$ is a Markov process on $X \cdot R_+$ with infinitesimal operator:

$$\hat{Q}^u f(x,t) := \frac{d}{dt} f(x,t) + \frac{g_x^u(t)}{\overline{G_x^u}(t)_u}[P^u f(x,0) - f(x,t)], \tag{76}$$

where $g_x^u(t) := dG_x^u(t)/dt$, P^u is an operator corresponding to kernel

$$P^u(x,A) := \mathcal{P}\{\omega : x_{u+1}^u \in A / x_n^u = x\}, \quad G_x^u(t) := \mathcal{P}\{\theta_{n+1} \leq t / x_n^u = x\}.$$

In such a way, process $\hat{Z}_t^u := (V^u(t)f, \hat{x}_t^u)$ is a Markov process on $B \times X \times R_+$ with infinitesimal operator

$$\hat{L}^u := \hat{Q}^u + \Gamma(x, u),$$

where $\hat{Q}^u$ and $\Gamma(x, u)$ are defined in (75) and (76) respectively.

Optimal control of SMRE is defined by the following equation

$$\min_u [\hat{Q}^u h(f, x, t) + \Gamma(x) h(f, x, t) + k(f, x, u, t)] = 0, \tag{77}$$

where k is defined on $B \times X \times U \times R_+$, h on $B \times X \times R_+$, $h \in \mathrm{Dom}(\hat{Q}^u) \bigcap B_0$.

The main idea in the investigation of the control of SMRE is to use the analogue of Dynkin's formula [21] for *RE*:

$$E\,[V(\tau) f(x_\tau)] - f(x) = E_x \int_0^\tau [\hat{Q} + \Gamma(x)]\, V(s) f(xs)\, ds,$$

where τ is the stopping time.

13. SMRE in financial mathematics [20]

RE have many applications (see Section 2). The latest applications of *RE* are connected with models of financial mathematics. We will describe one of them.

In a famous Black-Scholes model, that is used for the evolution of option prices, it is supposed that the movement of stock prices is set by the linear stochastic differential equation:

$$dS_t = a \cdot S_t\, dt + \sigma S_t\, dw_t, \quad S_0 = Z, \tag{78}$$

where a and σ are deterministic functions (in the simplest case — constants).

We suppose that in our model the coefficients a and σ, appreciation rate and volatility respectively, are defined from semi-Markov process x_t that does not depend on standard Wiener process W_t, i. e.,

$$dS_t = a(x_t)\, S_t\, dt + \sigma(x_t)\, S_t\, dw_t, \quad S_0 = Z. \tag{79}$$

We consider the hedging problem of the European call option with terminal payment $H = f(S_t)$, $T > 0$.

Since an additional source of randomness exists (namely, semi-Markov process x_t) besides Wiener process W_t, the market is not complete and perfect hedging is not possible. We will find a strategy that locally minimizes the risk.

The approach is based on the idea of hedging by using the mean-square criterion. The semi-Markov *RE* approach is also used. Namely, *RE* is defined in such a way:

$$V(t) f(z) := E_z[f(S_t)/x(s);\ 0 \le s \le t]. \tag{80}$$

The optimal strategy is defined from Kunita–Watanabe's expansion for terminal payment and martingale properties of *RE* in (80).

14. Applications of SMRE [3, 4, 19]

Here we consider the applications of SMRE in a series scheme for storage processes as an example (see Section 2).

These applications include averaging, diffusion approximation, normal deviations, stability and control.

14.1. Averaging of storage processes

Let us have a storage process z_t^ε in a series scheme (see Section 2, (5)):

$$z_t^\varepsilon = z + \varepsilon \sum_{k=1}^{\nu(t/\varepsilon)} a(x_{k-1}, x_k) - \int_0^t v(z_s^\varepsilon, x(s/\varepsilon))\, ds. \tag{81}$$

Here

$$\begin{aligned} \Gamma(x) f(z) &:= -v(z,x) \frac{d}{dz} f(z), \\ D^\varepsilon(x,y) f(z) &:= f(z + \varepsilon a(x,y)), \quad \forall f(z) \in C^1(R), \end{aligned}$$

(see Section 4, (17)-(18)).

Under some conditions the *averaged storage process* is defined by the equation:

$$z_t = z + \hat{a} \cdot t - \int_0^t \hat{v}(z_s)\, ds, \tag{82}$$

where

$$\begin{aligned} \hat{a} &:= \int_X \rho(dx) \int_X P(x, dy)\, a(x,y)/m, \\ \hat{v}_z &:= \int_X \rho(dx)\, m(x)\, v(z,x)/m. \end{aligned}$$

Process z_t is a weak limit for z_t^ε as $\varepsilon \to 0$.

The *merged storage process* under the averaging of storage process z_t^ε in a reducible phase space is defined by the equation (see Section 7):

$$\hat{z}_t = z + \int_0^t [\hat{a}(\hat{x}(s)) - \hat{v}(\hat{z}_s, \hat{x}(s))]\, ds, \tag{83}$$

where

$$\begin{aligned} \hat{a}(u) &:= \int_{X_u} \rho_u(dx) \int_{X_u} P(x,dy)\, a(x,y)/m(u), \\ \hat{v}(z,u) &:= \int_{X_u} \rho_u(dx)\, m(x)\, v(z,x)/m(u), \end{aligned}$$

$\hat{x}(t)$ is a merged Markov process in $(U,\mathcal{U})$ with infinitesimal operator $\Lambda(\hat{P} - I)$ (see (48)).

14.2. Diffusion approximation of a storage process

Let us have a storage process $z^{\varepsilon}_{t/\varepsilon}$ in a series scheme (see (87), Section 2, (5)):

$$z^{\varepsilon}_{t/\varepsilon} = z + \varepsilon \sum_{k=1}^{\nu(t/\varepsilon^2)} a(x_{k-1}, x_k) - \int_0^{t/\varepsilon} v(z^{\varepsilon}_s, x(s/\varepsilon))\, ds \tag{84}$$

Under some conditions the process $z^{\varepsilon}_{t/\varepsilon}$ converges weakly as $\varepsilon \to 0$ to the *diffusion process* z^0_t:

$$dz^0_t = \alpha(z^0_t)\, dt + \beta(z^0_t)\, dw_t, \tag{85}$$

where w_t is a standard Wiener process,

$$\begin{aligned} \alpha(z) &:= \int_X \rho(dx)\, [(-m(x)\, v(z,x) + Pa(x,\cdot))(-R_0 + I)\, m(x) \times \\ &\quad \times v'_z(z,x) + m_2(x)\, v(z,x)\, v'_z(z,x)/2]/m, \\ \beta^2(z) &:= 2\int_X \rho(dx)\, [(-m(x)v(z,x) + Pa(x,\cdot))(R_0 - I) \times \\ &\quad \times(-m(x)\, v(z,x) + Pa(x,\cdot)) - m(x)v(z,x)Pa(x,\cdot) + \\ &\quad + Pa^2(x,\cdot)/2 + m_2(x)v^2(z,x)/2]/m. \end{aligned} \tag{86}$$

The main condition is the *balance condition*:

$$\int_X \rho(dx)\, [Pa(x,\cdot) - m(x)\, v(z,x)] = 0. \tag{87}$$

14.3. Normal deviations of storage process

Let us have a storage process z^{ε}_t in a series scheme in (81) and an averaged storage process z_t in (83). Now we define a *deviation process* for a storage process, namely:

$$\eta^{\varepsilon}_t := [z^{\varepsilon}_t - z_t]/\sqrt{\varepsilon}. \tag{88}$$

Under some conditions the *deviation process* η^{ε}_t in (88) converges weakly as $\varepsilon \to 0$ to the *diffusion process* η^0_t:

$$\eta^0_t = \int_0^t a^0(z_s) \cdot \eta^0_s\, ds + \int_0^t [\upsilon^0(z_s) + \sigma]\, dw_s, \tag{89}$$

where

$$\begin{aligned} \alpha^0(z) &:= \int_x \rho(dx)\, m(x)\, v'_z(z,x)/m, \\ (\sigma^0(z))^2 &:= 2\int_X \rho(dx)\, [m(x)(v(z,x) - \hat{v}(z))(R_0 - I)m(x) \\ &\quad + (v(z,x) - \hat{v}(z)) + 2^{-1} m_2(x)(v(z,x) - \hat{v}(z))^2]/m; \\ \sigma^2 &:= \int_X \rho(dx)\, [(Pa(x,\cdot) - \hat{a})(R_0 - I)(Pa(x,\cdot) - \hat{a}) \\ &\quad + (Pa(x,\cdot) - \hat{a})^2/2]/m. \end{aligned}$$

It follows from (88) that we have a representation for small ε:

$$\eta_t^\varepsilon \approx z_t + \sqrt{\varepsilon}\,\eta_t^0. \tag{90}$$

Since η_t^0 is a Gaussian process then we have in (90) the *normal deviations* of storage process z_t^ε from averaged process z_t.

14.4. Rates of convergence of storage processes in limit theorems

Rates of convergence of storage process z_t^ε in an averaging scheme to the averaged process z_t have the form (see Section 14.1):

$$|E_\rho[z_t^\varepsilon - z_t]| \leq \varepsilon \cdot C(T), \quad t \in [0, T], \tag{91}$$

where

$$\begin{aligned} C(T) \;:=\; & 2 \cdot \|R_0\| \cdot \sup_z \int_X \rho(dx)\,[m(x)\,|v(z,x)| + |Pa(x,\cdot)|]/m + \\ & + 2 \cdot (T/m) \cdot \sup_z \int_X \rho(dx)\,[m(x)\,|v(z,x)| + |Pa(x\cdot)|] + \\ & + (T/2m) \cdot \sup_z \int_X \rho(dx)\,[m_2(x)\,|v(z,x)| + m_2^2\,|v(z,x)|^2/2m]. \end{aligned}$$

The rate of convergence of storage process $z_{t/\varepsilon}^\varepsilon$ in a diffusion approximation scheme to the diffusion process z_t^0 has the form (see Section 14.2):

$$|E_\rho[z_{t/\varepsilon}^\varepsilon - z_t^0]| \leq \varepsilon \cdot d(T), \quad t \in [0, T], \tag{92}$$

where

$$\begin{aligned} d(T) \;:=\; & 2 \cdot \|R_0\| \cdot \sup_z \int_X \rho(dx)\,[m(x)\,|v(z,x)| + |Pa(x,\cdot)|] + \\ & + 2(\|R_0\|^2 + \|R_0\|) \sup_z \int_X \rho(dx)\,[m^2(x)\,|v(z,x) \cdot v_z'(z,x)| + \\ & + \frac{1}{2} m_2(x)\,|v(z,x) v_z'(z,x)| bigr] + \\ & + 2(T/m) \sup_z \int_X \rho(dx)\,[m^2(x)\,|v(z,x)\theta_z'(z,x)| + \\ & + \frac{1}{2} m_2(x)\,|v(z,x) v'(z,x)|] + \\ & + (T/2m) \sup_z \int_X \rho(dx)\,[m_2(x)\,|v(z,x)| + \\ & + T \cdot m_2^2(x) |v(z,x) \cdot v_z'(z,x)|/2m)]. \end{aligned}$$

14.5. Stability of storage processes [7, 8]

We consider the stability of storage processes in averaging and diffusion approximation schemes.

Stability of storage processes in an averaging scheme.

Let us have a storage process z_t^ε in (81) and an averaged process z_t. Let there be a function $W(z)$:

$$[\hat{a} + \hat{v}(z)]\, W_z'(z) \le -\beta W(z), \tag{93}$$

where $\hat{a}$, $\hat{v}(z)$ are defined in (81), $\beta > 0$, and $G_x(t)$ is a smooth function on t. Therefore process z_t^ε in (87) is a uniformly stochastically asymptotically stable process, namely, $\forall \Delta_1 > 0$ and $\Delta_2 > 0$, $\exists \delta_1 > 0$ such that if $|z_0^\varepsilon| < \delta_1$ then

$$\mathcal{P}_{z,x}\{|z_t^\varepsilon| \le \Delta_2 \cdot e^{-\hat{\beta} t}; t \ge 0\} \ge 1 - \Delta_1 \tag{94}$$

$\forall x \in X$, $\forall z \in R_+$, $\hat{\beta} > 0$ is a constant. Moreover,

$$\mathcal{P}_{z,x}\{\lim_{z \to \infty} |z_t^\varepsilon| = 0\} = 1.$$

Stability of storage processes in a diffusion approximation scheme [7, 8]

Let us have a storage process $z_{t/\varepsilon}^\varepsilon$ in (84) and limit diffusion process z_t^0 in (85). Let there be a function $V(z)$:

$$\alpha(z) V_z'(z) + \frac{1}{2}\beta^2(z) V_{zz}''(z) \le -\gamma \cdot V(z), \tag{95}$$

where $\alpha(z)$, $\beta(z)$ are defined in (86), $\gamma > 0$, $\forall z \in R$, and G_x is a smooth function on t.

Therefore the process $z_{t/\varepsilon}^\varepsilon$ is a uniformly stochastically asymptotically stable process (see (94)) and, moreover,

$$\mathcal{P}_{z,x}\{\lim_{t \to +\infty} |z_{t/\varepsilon}^\varepsilon| = 0\} = 1.$$

The functions $W(z)$ and $V(z)$ in (92) and (95) are called *Liapunov functions.*

14.6. Control of storage process [9]

Now let us consider the storage process z_t^u:

$$z_t^u = z + \sum_{k=1}^{\nu(t)} a(x_{k-1}^u, x_k^u) + \int_0^t v(z_s^u, x^u(s), u_t)\, ds, \tag{96}$$

where $u \in (U, \mathcal{U})$ is a control space (see Section 12).

Define the following operators:

$$\begin{aligned} Q_1^u &:= \frac{d}{dt} + \frac{g_x^u(t)}{\overline{G}_x^u(t)}[P^u - I], \\ Q_2^u &:= v(z,x,u)\frac{d}{dz} f(z,x) + \int_X P^u(x,dy)\,[f(z + a(x,y), x) - f(z,x)], \end{aligned}$$

$$\forall f(z,x) \in C^1(R_1 \times X), \quad g_x^u(t) = dG_x^u(t)/dt.$$

We note that the process $(Z_t^u, x_t^u, \gamma(t))$ in $(R \times X \times R_+)$ is a Markov process with infinitesimal operator

$$Q_1^u + Q_2^u, \quad u \in U, \quad \gamma(t) := t - \tau_{\nu(t)}.$$

Optimal stochastic control of storage processes. z_t^u is defined by the following equation:

$$\min_u [Q_1^u \cdot h(z,x,t) + Q_2^u h(z,x,t) + k(z,x,u,t)] = 0. \tag{97}$$

Function $h(z,x,t)$ satisfies the boundary condition $h = b(z,x,t)$ for $(z,x,t) \in \delta S \times \delta Y \times \delta R_+$, where δS, δY, δR_+ are boundaries of the sets S, Y, R_+ from R, X, R_+ respectively.

Equation (97) has the solution:

$$h(z,x,t) = E_{z,x}^u b(z_{\tau_u}^u, x_{\tau_u}^u, \tau_u) + E_{z,x}^u \int_0^{\tau_u} k(z_s^u, x_s^u, u_s, s)\, ds.$$

The results follow from the analogue of Dynkin's formula for random evolutions [21].

References

1. **Korolyuk, V.S. and Turbin, A.F.** (1993). *Mathematical groups of phase merging of the compound system.* Kluwer Acad. Publ., The Netherlands.

2. **Korolyuk, V.S. and Turbin, A.F.** (1976). *Semi-Markov process and their applications.* Nauk. dumka publ., Kiev. (In Russian).

3. **Korolyuk, V.S. and Swishchuk, A.V.** (1995). *Semi-Markov Random Evolutions.* Kluwer Acad. Publ.

4. **Korolyuk, V.S. and Swishchuk, A.V.** (1995). *Evolution of Systems in Random Media.* CRC Press, USA. (In press).

5. **Korolyuk, V.S.** (1987). Semi-Markov systems in Random Media. *Cybernetics.*

6. **Korolyuk, V.S.** (1991). Stability of autonomous dynamical systems with rapid Markov switchings. *Ukrain. Math. J.* **33** (12), 1176–1181.

7. **Swishchuk, A.V.** (1994). Stability of semi-Markov evolutionary stochastic systems in averaging and diffusion approximation schemes. In: *Asymptotic analysis of random evolutions.* Inst. of Math., Ukr. Ac. Sci., Kiev, 255–269.

8. **Swishchuk, A.V.** (1993). Stability of semi-Markov evolutionary stochastic systems in the limit theorems. *Proceed.* 6th *Intern. Vilnius Conf. on Prob. Th. Math. Stat., Abstr. of Commun.*, **2.** Vilnius, p. 149.

9. **Swishchuk, A.V.** (1994). Control of Evolutionary stochastic Systems. *Proceed.* 12th *Conf. on Stoch. Proc.*, Prague, 285–288.

10. **Swishchuk, A.V.** (1992). Double approximation of SMRE. *Ukrain Math. J.* **44** (3), 400–408. (In Russian).

11. **Swishchuk, A.V.** (1990). On the rates of convergence in the limit theorems for SMRE. *Proceed.* 11th *Prague Conf. on inf. th., stat. des. funct., rand. proc.*, Prague, 27–31 August 1990, p. 170.

12. **Swishchuk, A.V.** (1990). Rates of Convergence in the limit theorems for SMRE. In: *Stochastic systems and their applications.* Inst. of Math., Ukr. Ac. Sci., Kiev, 86–92. (In Russian).

13. **Swishchuk, A.V.** (1990). Solution of martingale problems for SMRE. In: *Asymptotic and applied problems of the theory of random evolutions.* Inst. of Math., Ukr. Ac. Sci., Kiev, 98–107. (In Russian).

14. **Swishchuk, A.V.** (1990). The solution of a martingale problem for random evolution. In: 2nd *World Congr. Berh. Soc. for Math. Stat. and Prob.* Uppsala, Sweden, 13–18 Aug. 1990. Abstr., NY, p. 190.

15. **Swishchuk, A.V.** (1994). The Martingale problem and stochastic integral equations in a Banach space for the limit semi-Markov random evolutions. I. *Random Operators and Stochastic Equations.* **2** (3), 277–301. (II. *ROSE* **2** (4), 302–327).

16. **Swishchuk, A.V.** (1991). Uniqueness of the solution of the martingale problem for SMRE. In: *Asymptot. meth. in the problem of random evolutions,* Inst. of Math., Ukr. Ac. Sci., Kiev, 92–107. (In Russian).

17. **Swishchuk, A.V.** (1990). Uniqueness of solution of the martingale problem for SMRE. In: *II Donetsk conf. "Probab. models of proc. in control and reliability"* Donetsk, 23–25 May 1990, Inst. Appl. Math. Mech., Donetsk, p. 58.

18. **Swishchuk, A.V.** (1992). Stochastic integral operator equations for the limiting SMRE in lumping scheme. In: *Prob. Th. J. Math. Stat.* Proceed. VI USSR–Japan symp., Kiev, Ukraine, 5–10 August 1991, WSP, Singapore, 382–391.

19. Swishchuk, A.V. (1992). SMRE. A survey of recent results. *Proceed.* 11th *Prague Conf.*, Prague, 27–30 August 1990, Academia, Prague, 403–414.

20. Swishchuk, A.V. (1995). Hedging of options with mean-square criterion and semi-Markov Volatility. *Ukrain. Math. J.* (In press).

21. Swishchuk, A.V. (1993). *The Analogue of Dynkin's formula and boundary value problems for multiplicative operator functionals of Markov processes.* Preprint 93.44, Inst. of Math., Ukr. Ac. Sci., Kiev.

22. Ethier, S. and Kurtz, T. (1986). *Markov process: characterization and convergence.* Wiley and Sons, N.Y.

Exploring Stochastic Laws. pp. 443–455.
Eds. A.V. Skorokhod and Yu.V. Borovskikh

On the limit distribution of random permanents

G.J. SZÉKELY[1]*AND L.SZEIDL[2]

[1] *Technical University Department of Mathematics Stoczek u.2. Bldg. H.V.7. 1521 Budapest, HUNGARY*

[2] *Eötvös Loránd University Department of Probability Theory Múzeum krt. 6–8. 1088 Budapest, HUNGARY*

Received February 1, 1995

Abstract. In this paper, limit distributions of permanents of random block matrices will be discussed where each row is a series of independent infinitesimal random variables. The rows are identical within each block, and different blocks are independent. It is proved that the distribution of the permanent in the limit is the same as that of the product of the individual block permanents. As a corollary we find the limit distribution of permanents of block matrices where the entries are functions of the above mentioned random variables. These functions may vary from block to block and are supposed to be Peano differentiable of order 2 at the point 0. This corollary is used to compute the limit distribution of generating functions for elementary symmetric polynomials of triangular arrays of independent random variables.

1. Introduction

Korolyuk and Borovskikh (1993) proved several general theorems on the limiting behavior of random permanents. In this paper we are going to generalize some of their results. Instead of i.i.d. random variables within each row or martingale differences with finite second moment type conditions we consider independent infinitesimal series, and in the results where functions of entries are considered we shall include a broader class of functions than Korolyuk and Borovskikh did.

Let $k(n), \quad n = 1, 2, \ldots$ be a sequence of integers tending to infinity and let

$$\{X_{nj}^{(b)}, 1 \leq j \leq k(n)\}, \quad b = 1, 2, \ldots, B$$

be independent random variables. Let $m_1, \ldots, m_B$ be fixed positive integers and define a random block matrix $A_{m_1,\ldots,m_B}^{(n)}$ of size $(m_1 + \ldots + m_B) \times k(n)$ as follows:

* Research supported by the Hungarian National Foundation for Scientific Research, Grant 1405

$$A^{(n)}_{m_1,\dots,m_B} = \begin{pmatrix} A^{(n)}_{m_1} \\ A^{(n)}_{m_2} \\ \vdots \\ A^{(n)}_{m_B} \end{pmatrix}, \quad n = 1, 2, \dots \tag{1}$$

where the block $A^{(n)}_{m_b}$, $b = 1, \dots, B$ is of size $m_b \times k(n)$, and within any given block the rows are identical, and in the b-th block the rows are the following:

$$X^{(b)}_{n1}, \dots, X^{(b)}_{nk(n)}, \quad b = 1, 2, \dots, B.$$

In this paper the limiting behavior of the permanent

$$\mathrm{Per}(A^{(n)}_{m_1,\dots,m_B}) = \sum_{1 \le i_1 \ne i_2 \ne \dots \ne i_B \le n} a^{(1)}_{ni_1} a^{(2)}_{ni_2} \dots a^{(m_1+m_2+\dots+m_B)}_{ni_B}$$

of $A^{(n)}_{m_1,\dots,m_B}$ will be discussed (lower-case letters $a^{(b)}_{nj}$ denote the entries of the underlying matrix) where the number of columns $k(n)$ is supposed to be bigger or equal to the number of rows: $m_1 + m_2 + \dots + m_b \le k(n)$ (which surely holds if n is big enough).

The special case of the permanent when the number of blocks is 1 ($B = 1$) is in fact the elementary symmetric polynomial of order $b = b_1$ of the row elements $X^{(1)}_{n1}, \dots, X^{(1)}_{nk(n)}$, multiplied by $b!$. On this special case for i.i.d. random variables see Székely (1975), (1982), Halász and Székely (1974), Móri (1981), Móri and Székely (1982), Zolotarev (1985), (1986), Szeidl (1986), (1988), Korolyuk and Borovskikh (1989), Yanushkevichiene (1993), and for independent triangular arrays see Avram and Taqqu (1986).

2. Main results

Throughout this paper we shall suppose the following conditions for the random variables $\{X^{(b)}_{nj},\ 1 \le j \le k(n)\}$, $b = 1, 2, \dots, B$.

(A) For any given positive integer n the random variables $X^{(b)}_{nj}$ are independent for $b = 1, 2, \dots, B$, $j = 1, 2, \dots, k(n)$, and for any given b the sequence $\{X^{(b)}_{nj}, 1 \le j \le k(n)\}$ is infinitesimal, i.e. for any $\varepsilon > 0$ $\max_{1 \le j \le k(n)} P(|X^{(b)}_{nj}| \ge \varepsilon) \to 0$ if $n \to \infty$.

(B) The probability distribution of the sum $\sum_{j=1}^{k(n)} X^{(b)}_{nj}$ converges weakly to a non-degenerate probability distribution as $n \to \infty$.

(C) There exists a $C > 0$ such that for the numbers $a^{(b)}_{nj}(C) = EX^{(b)}_{nj} I(|X^{(b)}_{nj}| < C)$

we have

$$\sum_{j=1}^{k(n)} (a_{nj}^{(b)}(C))^2 \to 0 \quad \text{if} \quad n \to \infty$$

(I denotes the indicator function).

Under these conditions Avram and Taqqu (1986) describe the limit distribution of the elementary symmetric polynomials, i.e. they settle the case $B = 1$. Our first result is a generalization of this result for the case $B > 1$.

Theorem. *Under conditions* (A), (B), *and* (C) *we have*

$$\mathrm{Per}(A^{(n)}_{m_1,\dots,m_B}) - \prod_{b=1}^{B} \mathrm{Per}(A^{(n)}_{m_b}) \xrightarrow{P} 0 \quad \text{if } n \to \infty.$$

Corollary 1. *The limit distribution of the random permanent* $\mathrm{Per}(A^{(n)}_{m_1,\dots,m_B})$ *is the same as that of the product of the permanents*

$$\prod_{b=1}^{B} \mathrm{Per}(A^{(n)}_{m_b}).$$

Remark 1. In this product the factors are independent and the limit distribution of each factor is known from Avram and Taqqu (1986) thus we immediately get the limit distribution of $\mathrm{Per}(A^{(n)}_{m_1,\dots,m_B})$.

Corollary 2. *The Theorem remains true if we apply it to permanents of block matrices where the entries are functions of the original entries, these functions* g_b, $b = 1, 2, \dots, B$ *may vary from block to block and satisfy the following conditions*

(i) *they are continuous, and take the value* 0 *at the point* 0*;*

(ii) *they belong to the class of Peano differentiable functions of order* 2 *at the point 0.**

The validity of Corollary 2 follows from the fact that we can easily check that under conditions (i) and (ii) the basic conditions (A), (B) and (C) remain true for the functional block matrices. It is interesting to note that LeCam (1986) p.443, Proposition 5 gives the characteristic function of the limit distribution of the functional infinitesimal series $\sum_{j=1}^{k(n)} g_b(X_{nj}^{(b)})$ in the following explicit form:

$$\varphi_{g_b}(\lambda) = \exp\{ i(\gamma_b(\tau) g_b' + \frac{1}{2}\sigma_b^2 g_b'')\lambda - \frac{1}{2}\sigma_b^2 (g_b')^2 \lambda^2 + \int_{-\infty}^{\infty} (e^{i\lambda g_b(x)} - 1 - i\lambda g_b' x I(|x| < \tau)) G_b(dx) \},$$

* a function g is called Peano differentiable of order 2 at the point 0 if the following condition is satisfied: $x^{-2}|g(x) - g(0) - xg' - x^2 2^{-1} g''| \to 0$ as $x \to 0$.

where $\gamma_b(\tau), \sigma_b$ and G_b are the parameters in canonical representation (7) of infinitely divisible limit distribution of sequence $\sum_{j=1}^{k(n)} X_{nj}^{(b)}$, $n = 1, 2, \ldots$.

Finally we note that the condition of continuity in (i) can be replaced by Borel measurability if the Lévy measure of the points of discontinuity is 0 (see Michaletzky and Szeidl (1994), Remark 2).

3. Proof of the Theorem

Before starting the proof of the Theorem let us introduce some notations and recall some well-known results. Conditions (A), (B) and (C) imply (see Gnedenko and Kolmogorov (1954), §25, Theorem 1) the existence of a non-negative constant σ_b and Lévy measures L_b, R_b $(b = 1, \ldots, B)$ on $(-\infty, 0)$ and $(0, +\infty)$ respectively such that

$$L_b(-\infty) = R_b(+\infty) = 0, \tag{2}$$

$$\int_{-\varepsilon}^{0} x^2 dL_b(x) + \int_{0}^{\varepsilon} x^2 dR_b(x) < \infty \quad \text{for every } \varepsilon > 0, \tag{3}$$

(i) for the points of continuity of L_b and R_b we have

$$L_b^{(n)}(x) = \sum_{j=1}^{k(n)} P(X_{nj}^{(b)} < x) \to L_b(x), \quad x < 0, \tag{4}$$

$$R_b^{(n)}(x) = \sum_{j=1}^{k(n)} (1 - P(X_{nj}^{(b)} < x)) \to R_b(x), \quad x < 0, \tag{5}$$

(ii)

$$\lim_{\varepsilon \to 0} \liminf_{n \to \infty} \sum_{j=1}^{k(n)} D^2(X_{nj}^{b(n)} I(|X_{nj}^{(b)}| < \varepsilon)) =$$

$$= \lim_{\varepsilon \to 0} \limsup_{n \to \infty} \sum_{j=1}^{k(n)} D^2(X_{nj}^{b(n)} I(|X_{nj}^{(b)}| < \varepsilon)) =$$

$$= \sigma_b^2. \tag{6}$$

Then for the characteristic function $\varphi_b(\lambda)$ of the limit distribution of $\sum_{j=1}^{k(n)} X_{nj}^{(b)}$

$$\log \varphi_b(\lambda) = i\gamma_b(\tau)\lambda - \frac{\sigma_b^2}{2}\lambda^2 + \int_{-\infty}^{0} (e^{i\lambda x} - 1 - i\lambda x I(|x| < \tau)) dL_b(x)$$
$$+ \int_{0}^{+\infty} (e^{i\lambda x} - 1 - i\lambda x I(|x| < \tau)) dR_b(x), \tag{7}$$

where $\gamma_b(\tau) = \lim_{n\to\infty} \sum_{j=1}^{k(n)} a_{nj}^{(b)}(\tau)$, $-\tau$ and τ are points of continuity of L_b and R_b (we may suppose that τ is the same for all values of b since the points of discontinuity is a countable set).

If conditions (A), (B) and (C) hold for all values of b then in (C) we may choose an arbitrary C since if (C) holds for one single C then it holds for all positive C (see Michaletzky and Szeidl (1994), Lemma 1).

For the proof of the Theorem we need two Lemmas.

Lemma 1. *Under conditions* (A), (B) *and* (C) *we have*

$$\sum_{j=1}^{k(n)} X_{nj}^{(1)} X_{nj}^{(2)} \xrightarrow{d} 0 \quad \text{if } n \to \infty.$$

Proof of Lemma 1. We are going to prove that for all $z > 0$

$$\sum_{j=1}^{k(n)} P(|X_{nj}^{(1)} X_{nj}^{(2)}| > z) \to 0 \quad \text{as } n \to \infty, \tag{8}$$

$$\lim_{\varepsilon\to 0} \limsup_{n\to\infty} \sum_{j=1}^{k(n)} D^2(X_{nj}^{(1)} X_{nj}^{(2)} I(|X_{nj}^{(1)} X_{nj}^{(2)}| < \varepsilon)) = 0, \tag{9}$$

$$h_n = \sum_{j=1}^{k(n)} E(X_{nj}^{(1)} X_{nj}^{(2)} I(|X_{nj}^{(1)} X_{nj}^{(2)}| < \tau)) \to 0 \quad \text{as } n \to \infty \text{ and } \tau \to \infty. \tag{10}$$

These relations show that $X_{nj}^{(1)} X_{nj}^{(2)}$, $1 \le j \le k(n)$, $n = 1, 2, \ldots$ are infinitesimal in the neighborhood of 0 with Gaussian, Poisson and constant components in the limit distribution, thus this limit distribution is degenerate at 0.

Proof of (8). For all positive z, ε_1 and ε_2 we have

$$P(|X_{nj}^{(1)} X_{nj}^{(2)}| > z) = P(|X_{nj}^{(1)} X_{nj}^{(2)}| > z, |X_{nj}^{(2)}| < \varepsilon_1) + P(|X_{nj}^{(1)} X_{nj}^{(2)}| > z, |X_{nj}^{(2)}| \ge \varepsilon_1)$$

$$\le P(|X_{nj}^{(1)}| > \frac{z}{\varepsilon_1}) + P(|X_{nj}^{(1)} X_{nj}^{(2)}| > z, |X_{nj}^{(2)}| \ge \varepsilon_1)$$

$$= P(|X_{nj}^{(1)}| > \frac{z}{\varepsilon_1}) + P(|X_{nj}^{(1)} X_{nj}^{(2)}| > z, |X_{nj}^{(1)}| < \varepsilon_2, |X_{nj}^{(2)}| \ge \varepsilon_1) +$$

$$+ P(|X_{nj}^{(1)} X_{nj}^{(2)}| > z, |X_{nj}^{(1)}| \ge \varepsilon_2, |X_{nj}^{(2)}| \ge \varepsilon_1)$$

$$\le P(|X_{nj}^{(1)}| > \frac{z}{\varepsilon_1}) + P(|X_{nj}^{(2)}| > \frac{z}{\varepsilon_2}) + P(|X_{nj}^{(1)}| \ge \varepsilon_1) P(|X_{nj}^{(2)}| \ge \varepsilon_2). \tag{11}$$

Denote by G_b, $b = 1, \dots, B$ the Lévy measure on $(-\infty, 0) \cup (0, +\infty)$ determined by L_b and R_b. Choose the points $z_i = z_i(\varepsilon_i), z/(2\varepsilon_i) < z_i \leq z/\varepsilon_i$, $i = 1, 2$ such that both $-z_i$ and z_i are the points of continuity of G_i.

Using (4) and (5) we get the upper estimate from (11)

$$\limsup_{n\to\infty} \sum_{j=1}^{k(n)} P(|X_{nj}^{(1)} X_{nj}^{(2)}| > z) \leq G_1(|x| > z_1) + G_2(|x| > z_2) +$$

$$+ \limsup_{n\to\infty} \left[\max_{1\leq j \leq k(n)} P(|X_{nj}^{(1)}| \geq \varepsilon_1) \sum_{j=1}^{k(n)} P(|X_{nj}^{(2)}| \geq \varepsilon_2) \right]$$

$$= G_1(|x| > z_1) + G_2(|x| > z_2).$$

Here the left-hand side does not depend on ε_1 and ε_1, thus on the right-hand side we can take the limit $\varepsilon_1 \to 0$ and $\varepsilon_2 \to 0$ implying (8).

Proof of (9). For any $0 < \varepsilon < 1$ and fix $1 \leq j \leq k(n)$ we can easily get the following

$$D^2(X_{nj}^{(1)} X_{nj}^{(2)} I(|X_{nj}^{(1)} X_{nj}^{(2)}| < \varepsilon)) \leq$$

$$\leq 3\{D^2(X_{nj}^{(1)} X_{nj}^{(2)} I(|X_{nj}^{(1)} X_{nj}^{(2)}| < \varepsilon, |X_{nj}^{(1)}| < \varepsilon, |X_{nj}^{(2)}| < \varepsilon)) +$$

$$+ D^2(X_{nj}^{(1)} X_{nj}^{(2)} I(|X_{nj}^{(1)} X_{nj}^{(2)}| < \varepsilon, |X_{nj}^{(1)}| < \varepsilon, |X_{nj}^{(2)}| \geq \varepsilon)) +$$

$$+ D^2(X_{nj}^{(1)} X_{nj}^{(2)} I(|X_{nj}^{(1)} X_{nj}^{(2)}| < \varepsilon, |X_{nj}^{(1)}| \geq \varepsilon))\} \leq$$

$$\leq 3\{D^2(X_{nj}^{(1)} I(|X_{nj}^{(1)}| < \varepsilon) X_{nj}^{(2)} I(|X_{nj}^{(2)}| < \varepsilon)) +$$

$$+ \varepsilon^2 P(|X_{nj}^{(1)}| \geq \varepsilon) + \varepsilon^2 P(|X_{nj}^{(2)}| \geq \varepsilon)\}. \tag{12}$$

Using the notation $X_{nj}^{(i)}(\varepsilon) = X_{nj}^{(i)} I(|X_{nj}^{(i)}| < \varepsilon)$, $i = 1, 2$ we get

$$D^2(X_{nj}^{(1)} I(|X_{nj}^{(1)}| < \varepsilon) X_{nj}^{(2)} I(|X_{nj}^{(2)}| < \varepsilon)) =$$

$$= E\left[X_{nj}^{(1)}(\varepsilon)(X_{nj}^{(2)}(\varepsilon) - EX_{nj}^{(2)}(\varepsilon)) + E[X_{nj}^{(2)}(\varepsilon)(X_{nj}^{(1)}(\varepsilon) - EX_{nj}^{(1)}(\varepsilon))\right]^2 \leq$$

$$\leq 2\varepsilon^2 \left[D^2(X_{nj}^{(1)}(\varepsilon)) + D^2(X_{nj}^{(2)}(\varepsilon))\right]. \tag{13}$$

By condition (C) we get from (6) that $\limsup_{n\to\infty}\sum_{j=1}^{k(n)} D^2(X_{nj}^{(i)}(\varepsilon))$, $i = 1,2$ are bounded for sufficiently small ε and thus

$$\lim_{\varepsilon\to 0}\limsup_{n\to\infty}\sum_{j=1}^{k(n)} D^2(X_{nj}^{(1)}(\varepsilon)X_{nj}^{(2)}(\varepsilon)) = 0. \tag{14}$$

On the other hand for every $\varepsilon > 0$ one can find an ε_*, $\frac{1}{2}\varepsilon < \varepsilon_* < \varepsilon$, such that $-\varepsilon_*$ and ε_* are points of continuity of G_i, $i = 1,2$ thus

$$\limsup_{n\to\infty}\left[\varepsilon^2\sum_{j=1}^{k(n)} P(|X_{nj}^{(1)}| \geq \varepsilon) + \varepsilon^2\sum_{j=1}^{k(n)} P(|X_{nj}^{(2)}| \geq \varepsilon)\right] \leq$$

$$\leq 4\varepsilon_*^2\left[\limsup_{n\to\infty}\sum_{j=1}^{k(n)} P(|X_{nj}^{(1)}| \geq \varepsilon) + \limsup_{n\to\infty}\sum_{j=1}^{k(n)} P(|X_{nj}^{(2)}| \geq \varepsilon)\right] =$$

$$= 4\varepsilon_*^2\,[G_1(|x| \geq \varepsilon_*) + G_2(|x| \geq \varepsilon_*)] \to 0 \quad \text{as} \quad \varepsilon_* \to 0,$$

and this relation, together with (11), (12) and (13) gives (9).

Proof of (10). Suppose that $-\sqrt{\tau}$ and $\sqrt{\tau}$ are points of continuity of G_i, $i = 1,2$. Then

$$\begin{aligned}\overline{\lim_{n\to\infty}}\,|h_n| \;\leq\; & \overline{\lim_{n\to\infty}}\,|\sum_{j=1}^{k(n)} E(X_{nj}^{(1)}X_{nj}^{(2)}[I(|X_{nj}^{(1)}X_{nj}^{(2)}| < \tau, |X_{nj}^{(1)}| < \sqrt{\tau}, |X_{nj}^{(2)}| < \sqrt{\tau}) + \\ & +I(|X_{nj}^{(1)}X_{nj}^{(2)}| < \tau, |X_{nj}^{(1)}| < \sqrt{\tau}, |X_{nj}^{(2)}| \geq \sqrt{\tau}) + \\ & +I(|X_{nj}^{(1)}X_{nj}^{(2)}| < \tau, |X_{nj}^{(1)}| \geq \sqrt{\tau})]| \\ \leq\; & \overline{\lim_{n\to\infty}}\,\{\sum_{j=1}^{k(n)} \frac{1}{2}[(EX_{nj}^{(1)}(\sqrt{\tau}))^2 + (EX_{nj}^{(2)}(\sqrt{\tau}))^2] + \\ & +\sum_{j=1}^{k(n)}(\tau P(|X_{nj}^{(1)}| \geq \sqrt{\tau}) + P(|X_{nj}^{(1)}| \geq \sqrt{\tau}))\} \\ =\; & \tau(G_1(\Gamma_{\sqrt{\tau}}) + G_2(\Gamma_{\sqrt{\tau}})) \to 0 \quad \text{as} \quad \tau \to 0,\end{aligned}$$

where $\Gamma_u = \{x : |x| \geq u\}$, $u > 0$.

Now let $L \geq 1$ and $1 \leq r \leq L$ be arbitrary integers, $k_r = (k_{r1}, \ldots, k_{rB}) \in N^B$, $k_{r1} + \ldots + k_{rB} \geq 1$. For $n = 1,2,\ldots$ put

$$Z_{nj}^{(r)} = \left(X_{nj}^{(r)}\right)^{k_{r1}} \ldots \left(X_{nj}^{(r)}\right)^{k_{rB}}, \quad 1 \leq j \leq k(n)$$

and

$$Z_n(L) = \begin{pmatrix} Z_{n1}^{(1)} & \dots & Z_{nk(n)}^{(1)} \\ & \dots & \\ Z_{n1}^{(L)} & \dots & Z_{nk(n)}^{(L)} \end{pmatrix}.$$

Lemma 2. *Under conditions* (A), (B) *and* (C) *the limit distribution of the permanent* $\mathrm{Per}(Z_n(L))$ *exists. If, further, for some* r, $1 \le r \le L$

$$I(k_{r1} \ge 1) + I(k_{r2} \ge 1) + \ldots + I(k_{rB} \ge 1) \ge 2, \tag{15}$$

then

$$\mathrm{Per}(Z_n(L)) \xrightarrow{P} 0, \quad n \to \infty. \tag{16}$$

Proof. For arbitrary $i_b \ge 1$, $1 \le b \le B$ the sequences $\{(X_{nj}^{(b)})^{i_b},\ 1 \le j \le k(n)\}$, $1 \le b \le B$, $n = 1, 2, \ldots$ satisfy conditions (A), (B) and (C). The sums

$$\sum_{j=1}^{k(n)} (X_{nj}^{(b)})^{i_b}, \quad 1 \le b \le B$$

have a joint limit distribution (its characteristic function and representation by Poisson integrals can be found in Michaletzky and Szeidl (1994) for stable attraction of domain, see Szeidl (1994)).

Thus $\sum_{j=1}^{k(n)} Z_{nj}^{(r)}$ has a limit distribution which may differ from the degenerate distribution at 0 only if condition (15) does not hold (see Lemma 1), i.e. if $Z_{nj}^{(r)} = (X_{nj}^{(b)})^i$ for some $i \ge 1$ and $1 \le b \le B$.

In the proof we are going to apply induction. For $L = 1$ the statement is obvious. Now suppose that $L \ge 2$. If condition (15) holds for some r then we can suppose that it holds for $r = L$ (the permanent is invariant under permutation). Clearly we have

$$\begin{aligned} \mathrm{Per}(Z_n(L)) &= \sum_{1 \le i_1 \ne i_2 \ne \ldots \ne i_{L-1} \le n} Z_{ni_1}^{(1)} Z_{ni_2}^{(2)} \ldots Z_{ni_{L-1}}^{(L-1)} \left(\sum_{j=1}^{k(n)} Z_{nj}^{(L)} - Z_{ni_1}^{(L)} - \ldots - Z_{ni_{L-1}}^{(L)} \right) \\ &= \sum_{j=1}^{k(n)} Z_{nj}^{(L)} \sum_{1 \le i_1 \ne i_2 \ne \ldots \ne i_{L-1} \le n} Z_{ni_1}^{(1)} Z_{ni_2}^{(2)} \ldots Z_{ni_{L-1}}^{(L-1)} - \\ &\quad - \sum_{1 \le i_1 \ne i_2 \ne \ldots \ne i_{L-1} \le n} (Z_{ni_1}^{(1)} Z_{ni_1}^{(L)}) Z_{ni_2}^{(2)} \ldots Z_{ni_{L-1}}^{(L-1)} - \ldots \\ &\quad - \sum_{1 \le i_1 \ne i_2 \ne \ldots \ne i_{L-1} \le n} Z_{ni_1}^{(1)} Z_{ni_2}^{(2)} \ldots (Z_{ni_{L-1}}^{(L-1)} Z_{ni_{L-1}}^{(L)}) . \end{aligned} \tag{17}$$

Here, under condition (15) the limit distribution of the first factor of the first term is degenerate at 0. The second factor of the first term is the permanent of the matrix formed from the first $(L-1)$ rows of $Z_n(L)$. All other terms form a permanent of a matrix with the same structure as $Z_n(L)$ having only $(L-1)$ rows and satisfying condition (15). Thus we can apply induction.

Proof of the Theorem. The proof of the Theorem follows from Lemma 2 since

$$\mathrm{Per}(A^{(n)}_{m_1,\dots,m_B}) - \prod_{b=1}^{B} \mathrm{Per}(A^{(n)}_{m_b})$$

can be expanded to the sum of permanents that appear in Lemma 2 (see below for the case $B=2$). These permanents satisfy condition (15), thus they tend to 0 stochastically.

The above mentioned expansion for the case $B=2$ is the following (for a similar expansion see Theorem 1.1.1 in Korolyuk and Borovskikh (1993))

$$\frac{1}{m_1!\, m_2!}\left(\mathrm{Per}(A^{(n)}_{m_1,m_2}) - \prod_{b=1}^{2} \mathrm{Per}(A^{(n)}_{m_b})\right) =$$

$$= \sum_{q\in Q_{m_1}} \left\{\prod_{s=1}^{m_1} X^{(1)}_{nq_s} \sum_{q'\in Q_{m_2}(q)} \prod_{t=1}^{m_2} X^{(2)}_{nq'_t}\right\} - \left(\sum_{q\in Q_{m_1}} \prod_{s=1}^{m_1} X^{(1)}_{nq_s}\right)\left(\sum_{q'\in Q_{m_2}} \prod_{t=1}^{m_2} X^{(2)}_{nq'_t}\right) =$$

$$= -\sum_{q\in Q_{m_1}} \left\{\prod_{s=1}^{m_1} X^{(1)}_{nq_s} \sum_{q''\in Q_{m_2}(q)} \prod_{t=1}^{m_2} X^{(2)}_{nq''_t}\right\},$$

where

$$\begin{aligned} Q_{m_b} &= \{q=(q_1,\dots,q_{m_b}) \in N^{m_b} :\ 1 \le q_1 < \dots < q_{m_b} \le n\}, \\ Q_{m_2}(q) &= \{q' \in Q_{m_2} : q \text{ and } q' \text{ have no joint components}\}, \text{ and} \\ R_{m_2}(q) &= \{q' \in Q_{m_2} : q \text{ and } q' \text{ have joint components}\}. \end{aligned}$$

4. Generating functions

In this final section of our paper we are going to determine the generating function $P^{(n)}(z)$ for elementary symmetric polynomials $P^{(n)}_m$ of order $m = 0,1,\dots,k(n)$ of $X^{(1)}_{nj}$, $j=1,2,\dots,k(n)$

$$P^{(n)}(z) = \sum_{m=0}^{k(n)} P^{(n)}_m z^m = \prod_{j=1}^{k(n)} (1 + zX^{(1)}_{nj}),$$

where $z > 0$. Generating functions of this type were already considered in Korolyuk and Borovskikh (1993) Chapter 3 for random variables $X^{(1)}_{nj}$ obtained from a sequence of i.i.d. random variables by linear normalization. Here we drop the condition that the X variables are identically distributed, and include infinitesimal triangular arrays of independent random variables.

Using the notations

$$\begin{aligned} U_n(z) &= \sum_{j=1}^{k(n)} \log|1 + zX_{nj}^{(1)} I(X_{nj}^{(1)} > -1/z)|, \\ V_n(z) &= \sum_{j=1}^{k(n)} \log|1 + zX_{nj}^{(1)} I(X_{nj}^{(1)} < -1/z)|, \\ W_n(z) &= \sum_{j=1}^{k(n)} I(X_{nj}^{(1)} < -1/z), \\ Y_n(z) &= I\left(\sum_{j=1}^{k(n)} I(X_{nj}^{(1)} = -1/z) > 0\right), \end{aligned}$$

we have

$$P^{(n)}(z) = \prod_{j=1}^{k(n)} (1 + zX_{nj}^{(1)}) = (-1)^{W_n(z)} \exp\{U_n(z) + V_n(z)\}(1 - Y_n(z)), \; n = 1, 2, \ldots$$

Proposition. *Suppose that conditions* (A), (B), (C) *hold with* $B = 1$. *If* $-1/z$ *is a point of continuity of the Lévy measure* G_1 *defined in Corollary* 2, *then*

$$Y_n \xrightarrow{P} 0, \quad \text{as } n \to \infty, \tag{18}$$

and the joint limit distribution of $(U_n(z), V_n(z), W_n(z))$ *exists. For their joint characteristic function we have*

$$\begin{aligned} &\lim_{n\to\infty} E \exp\{i\lambda_1 U_n(z) + i\lambda_2 V_n(z) + i\lambda_3 W_n(z)\} = \\ &= \exp\{i(\gamma_1(\tau) z - \frac{1}{2}\sigma_1^2 z^2)\lambda_1 - \frac{1}{2}\sigma_1^2 z^2 \lambda_1^2 \\ &\quad + \int_{(-\infty,-1/z)} [\exp\{i\lambda_2 \log|1 + zx| + i\lambda_3\} - 1] G_1(dx) \\ &\quad + \int_{(-1/z,\infty)} [\exp\{i\lambda_1 \log|1 + zx|\} - 1 - i\lambda_1 zx I(|x| < \tau)] G_1(dx)\}. \end{aligned} \tag{19}$$

Proof. Since $-1/z$ is a point of continuity of our Lévy measure, we have

$$E(Y_n) = E\left(\sum_{j=1}^{k(n)} I(X_{nj}^{(1)} = -1/z)\right) = \sum_{j=1}^{k(n)} P(X_{nj}^{(1)} = -1/z) \to 0 \quad \text{as } n \to \infty.$$

This relation together with $Y_n \geq 0$ implies (18).

To see (19) apply Corollary 2 for

$$g(x) = \lambda_1 \log|1 + zxI(x > -1/z)| + \lambda_2 \log|1 + zxI(x < -1/z)| + \lambda_3 I(x < -1/z),$$

where $\lambda_i, i = 1,2,3$ are arbitrary real numbers. Observe that where $\lambda_i, \quad i = 1,2,3$ are arbitrary real numbers. Observe that

$$\lambda_1 U_n + \lambda_2 V_n + \lambda_3 W_n = \sum_{j=1}^{k(n)} g(X_{nj}^{(1)}),$$

and $g(0) = 0, \; g'(0) = \lambda_1 z, \; g''(0) = -\lambda_1 z^2$. Thus Corollary 2 implies our Proposition.

Corollary 3. *If the random vector $(U(z), V(z), W(z))$ represents the limit distribution of $(U_n(z), V_n(z), W_n(z))$ then the characteristic function above shows that $U(z)$ is independent of $(V(z), W(z))$, the Gaussian and Poissonian part of $U(z)$ is independent, the marginal distribution of $W(z)$ is Poissonian whose parameter is the Lévy measure of the interval $(-\infty, -1/z)$. Thus*

$$P^{(n)}(z) \xrightarrow{d} (-1)^{W(z)} \exp\{U(z) + V(z)\}, \quad \text{as } n \to \infty.$$

Remark 2. Suppose the Lévy measure turns out to be identically 0, $\sigma_1 = 1, \gamma_1(\tau) = 0$. This is the case e.g. when the triangular array of our random variables is constructed by the usual linear transformation of i.i.d. random variables with finite variance (here conditions (A), (B) clearly hold and (C) can easily be verified). This case was discussed in Korolyuk and Borovskikh (1993). Under the above mentioned conditions the representation of the limit distribution of $P^{(n)}(z)$ is $\exp\{zN - z^2/2\}$ where N is a standard normal random variable. The m-th derivative of this formula with respect to z at 0 gives a representation of the limit distribution of the elementary symmetric polynomial $P_m^{(n)}$. Thus a representation of this limit distribution is $H_m(N)$ where H_m is the m-th Hermite polynomial.

Remark 3. If $X_{nj}^{(1)} = Y_j/b_n$, where the sequence $Y_1, \; ,Y_2, \ldots$ is i.i.d., from the domain of attraction of a positive one-sided stable distribution with characteristic exponent $0 < \gamma < 1$ (b_n is the usual norming factor, and the centering constant can be supposed to be 0), then $V(z)$ and $W(z)$ are degenerate, and the characteristic function of $U(z)$ is

$$\begin{aligned} E\exp\{i\lambda U(z)\} &= \exp\left\{\int_0^\infty (\exp\{i\lambda \log(1+zx)\} - 1) d(-x^{-\gamma})\right\} \\ &= \exp\left\{z^\gamma \int_0^\infty (\exp\{i\lambda x\} - 1) d(-(\exp x - 1)^{-\gamma})\right\} \end{aligned}$$

since in our general formula the Lévy measure corresponds to the Lévy measure of the stable distribution with characteristic exponent γ.

References

Avram, F. and Taqqu, M.S. (1986). Symmetric polynomials of random variables attracted to an infinitely divisible law. *Probab. Th. Rel. Fields* **71**(4), 491–500.

Borovskikh, Yu.V. and Korolyuk, V.S. (1994). Random permanents and symmetric statistics. *Acta Appl. Math.* **36**(3), 227–288.

Gnedenko, B.V. and Kolmogorov, A.N. (1964). *Limit distributions for sums of independent random variables.* Addoson–Wesley, Cambridge.

Halász, G. and Székely, G.J. (1976). On the elementary symmetric polynomials of independent random variables. *Acta Math. Acad. Sci. Hungar* 28, 397–400.

Korolyuk, V.S. and Borovskikh, Yu.V. (1989). *Theory of U-statistics.* Naukova Dumka, Kiev. (In Russian).

Korolyuk, V.S. and Borovskikh, Yu.V. (1993). *Random permanents.* Inst. Math., Kiev. (In Russian).

Le Cam, L. (1986). *Asymptotic methods in statistical decision theory.* Springer–Verlag, New-York, Berlin, Heidelberg.

Michaletzky, Gy. and Szeidl, L. (1994). Lévy processes with values in a space of continuous functions arising as limits of independent triangular arrays. *Proceedings of XVI-th Seminar on Stability Problems of Stochastic Models*, Eger (Hungary), Plenum Press. (In press).

Móri, T.F. (1981). On a Hoeffding-type problem. *Lecture Notes in Statistics* **8**, 174–181.

Móri, T.F. and Székely, G.J. (1982). Asymptotic behaviour of symmetric polynomial statistics. *Ann. Probab.* **10**(1), 124–131.

Szeidl, L. (1986). On the limit distribution of random symmetric functions. *Theor. Probab. Appl.* **31**(3), 590–603.

Szeidl, L. (1988). Limit distribution of random symmetric polynomials. *Theor. Probab. Appl.* **33**(2), 248–259.

Szeidl, L. (1994). On power processes defined by independent identically distributed random variables having regularly varying distribution at the infinity. *Annales Univ., Sci. Budapest., Sect. Comp.* 14, 189–217.

Székely, G.J. (1975). On the polynomials of independent random variables. In: *Coll. Math. Soc. J. Bolyai: Limit Theorems of Probability Theory,* Révész P. (ed.), Keszthely, 1974, (Hungary), North Holland, Amsterdam, London, 365–371.

Székely, G.J. (1982). A limit theorem for elementary symmetric polynomials of independent random variables. *Z. Wahrscheinlichkeitstheorie verw. Geb.* 59, 355–360.

Yanushkevichiene, O.L. (1993) On the convergence of random symmetric polynomials. *Lecture Notes in Mathematics* **1546**, 184–188.

Zolotarev, V.M. (1985). Limit theorems for random symmetric polynomials. *Theor. Probab. Appl.* **30**(3), 598–599.

Zolotarev, V.M. (1986). On random symmetric polynomials. In: *Probability Distributions and Mathematical Statistics,* FAN, Tashkent, 170–188.

Exploring Stochastic Laws. pp. 457–468.
Eds. A.V. Skorokhod and Yu.V. Borovskikh

A simple approach to classical extreme value theory

JOZEF L. TEUGELS AND JAN BEIRLANT

Department of Mathematics Katholieke Universiteit Leuven Celestijnenlaan 200B, 3001 Heverlee, Belgium

Received June 24, 1995

Abstract. It is possible to solve the domain of attraction problem for the maximum directly, without recourse to the three extremal laws separately. The natural criterion turns out to be the so-called C_γ-condition. The versatility of the latter condition is further illustrated by a number of examples dealing with functions of large order statistics.

1. Introduction

Assume $\{X_1, X_2, \dots, X_n\}$ is a sequence of i.i.d. random variables with common distribution F. The corresponding order statistics of this sequence will be denoted by

$$\{X_1^* \le X_2^* \le \dots \le X_n^*\}.$$

We give an alternative method for the weak limit theorem for higher order statistics when the sample size n tends to infinity. For standard treatises on this question we refer to De Haan [4], Galambos [6], Leadbetter *et al.* [10] and Resnick [12].

In this section we collect a couple of items that will prove to be useful later. We start with classes of distributions that are crucial in the domain of attration problem for extremes.

1.1. Regularly varying classes

A class of distributions that has been thoroughly studied is the class of *regularly varying distributions.* The concept of regular variation is by now standard.

Definition 1. A distribution function F on $[0, \infty)$ is called *of regular variation* with *index* $\alpha \ge 0$ or *α-varying* if it is of the form $1 - F(x) = x^{-\alpha} l(x)$ where l is slowly varying, i.e. for all $\lambda > 0$ the ratio $l(\lambda x)/l(x)$ tends to 1 when $x \uparrow \infty$. We will write that $F \in \mathcal{P}_\alpha$. The class of all such distributions will be denoted by $\mathcal{P}$. Alternatively we might call such distributions *of Pareto-type* with *Pareto-index* α.

For information on regular variation and its impact in probability theory and other fields, see Bingham *et al.* [3] and Resnick [12].

1.2. Classes $\mathcal{C}_\gamma$

The most versatile class of distributions for our purposes is defined in the following fashion.

Definition 2. Let F be a distribution with inverse (the quantile function) $F^{\leftarrow}(u) = \inf\{x : F(x) \geq u\}$ and *tail quantile function* $U(t) = F^{\leftarrow}(1 - \frac{1}{t})$. The distribution F belongs to the *extremal-type* class $\mathcal{C}_\gamma(g)$ if there exists a real valued constant γ and an ultimately positive *auxiliary function* g such that

$$\{U(xu) - U(x)\}/g(x) \to h_\gamma(u) := \int_1^u v^{\gamma-1} dv \tag{1}$$

for all $u \geq 1$ as $x \to \infty$. Condition (1) will be called the $\mathcal{C}_\gamma$-*condition.*

It follows from the general theory of regularly varying functions that we do not need to specify the explicit form of the function h_γ on the right hand side above. See for example Bingham *et al.* [3], p. 128, where it is also shown that $h_0(u) = \log u$. Furthermore it can be shown that g is automatically of regular variation with the same index γ as the one appearing in h_γ. The restriction $u \geq 1$ can be broadened to $u > 0$.

Notice also that for the case when $\gamma = 0$ alternatively

$$U \in \Pi(g) \Longleftrightarrow F \in \mathcal{C}_0(g)$$

linking the class $\mathcal{C}_0$ with the so-called *de Haan class* Π familiar in regular variation.

It has been shown — for example in Bingham *et al.* [3] — that the above definition can be equivalently stated in terms of the original distribution. The alternative condition is ($x \to \infty$),

$$\frac{1 - F\{x + uh(x)\}}{1 - F(x)} \to \eta_\gamma(u) := (1 + \gamma u)^{-\frac{1}{\gamma}}, \tag{2}$$

for all $u > -\gamma^{-1}$. Here $h \circ U = g$ provides the link between conditions (1) and (2).

The following additional information is available for different values of γ.

(i) For $\gamma > 0$, condition (2) is equivalent to $F \in \mathcal{P}_{\gamma^{-1}}$. As such the more or less elaborate statistical literature on the Pareto-index is immediately relavant to this case. Furthermore the ratio $g(x)/U(x) \to \gamma$ as $x \to \infty$. This last result shows that if $\gamma > 0$ then U is of the same regular variation as the auxiliary function g.

(ii) For $\gamma = 0$ we should read $(1+\gamma u)^{-1/\gamma} = e^{-u}$. Again $g(x)/U(x) \to 0$ as $x \to \infty$, the statement is improved by the relation $U \in \Pi(g)$ which is more informative.

(iii) If $\gamma < 0$ then $\sup\{x : F(x) < 1\} := x_+$, the right bound of the carrier of F, is finite. The regular variation condition can again be recovered by writing that $x_+ - U(1/x)$ is γ-varying or that $1 - F(x_+ - (1/x))$ is $1/\gamma$-varying. Moreover as $x \to \infty$, $g(x)/\{x_+ - U(x)\} \to -\gamma$.

2. The domain of attraction for the maximum

Here is the *raison d'être* of the class $\mathcal{C}_\gamma$.

Proposition 1. *Assume $F \in \mathcal{C}_\gamma$ for some γ. Then there exist positive constants $\{a_n\}$ and real constants $\{b_n\}$ such that for all bounded, continuous functions k*

$$E\{k\left(a_n^{-1}(X_n^* - b_n)\right)\} \to \int_0^\infty k\left(h_\gamma(1/v)\right) e^{-v} dv.$$

Proof. Let k be any bounded continuous function over the domain of F. Then since $P\{X_n^* < x\} = F^n(x)$ we find that

$$E\left\{k\left(a_n^{-1}(X_n^* - b_n)\right)\right\} = n\int_{-\infty}^\infty k\left(\frac{x - b_n}{a_n}\right) F^{n-1}(x)\, dF(x).$$

Put $F(x) = 1 - n^{-1}v$. Then the right hand side equals

$$\int_0^\infty k\left(\frac{U(n/v) - b_n}{a_n}\right)(1 - (v/n))^{n-1}\, I\{0 \le v \le n\} dv.$$

Take $a_n = g(n)$ and $b_n = U(n)$ and apply Lebesgue's dominated convergence theorem.

It is easy to derive the three standard extremal types from this. Since clearly

$$h_\gamma\left(\frac{1}{v}\right) = y \Leftrightarrow v = \eta_\gamma(y)$$

we find

- that if $\gamma > 0$

$$E\left\{k\left(a_n^{-1}(X_n^* - b_n)\right)\right\} \to \int_{-\gamma^{-1}}^\infty k(u)\, d(\exp -(1 + \gamma u)^{-1/\gamma});$$

- if $\gamma = 0$, then a similar calculation leads to

$$E\left\{k\left(a_n^{-1}(X_n^* - b_n)\right)\right\} \to \int_{-\infty}^\infty k(u)\, d(\exp(-e^{-u}));$$

- finally if $\gamma < 0$, then one obtains

$$E\left\{k\left(a_n^{-1}(X_n^* - b_n)\right)\right\} \to \int_{-\infty}^{-\gamma^{-1}} k(u)\, d(\exp -(1+\gamma u)^{-1/\gamma}).$$

Since the convergence in the above proposition holds for all real bounded continuous functions, the classical Helly-Bray Theorem gives us that

$$F \in \mathcal{C}_\gamma \Rightarrow \frac{X_n^* - b_n}{a_n} \xrightarrow{\mathcal{D}} Y_\gamma$$

where Y_γ has extreme value law G_γ, i.e. $G_\gamma(y) = \exp -\{\eta_\gamma(y)\}$ on the appropriate domain.

In establishing the converse, let us put

$$\psi_n(u) := \frac{U(n/u) - b_n}{a_n}.$$

We have to show that if the normalised maximum converges in distribution to a nondegenerate limit, then $F \in \mathcal{C}_\gamma$ for some γ. We will first show that if for all bounded, continuous functions k

$$E\left\{k\left(a_n^{-1}(X_n^* - b_n)\right)\right\} \to \int_0^\infty k(\psi(v))e^{-v} dv$$

then $\psi_n(v) \to \psi(v)$. We apply the relation above to the function $k(u) = e^{itu}$ so that we can direct ourselves to the convergence of characteristic functions. Then

$$\int_0^\infty e^{it\psi_n(v)}\,(1-(v/n))^{n-1}\, I\{0 \le v \le n\}\, dv \to \int_0^\infty \exp(it\psi(v))e^{-v}\, dv.$$

Let us show that we can replace $(1-(v/n))^{n-1} I\{0 \le v \le n\}$ by e^{-v}. As shown for example in Pfeiffer [11], we have that for n a positive integer and $x \in [0, \sqrt{2n-1})$ the following inequalities are valid

$$\frac{x^2}{2n} e^{-x} \le e^{-x} - \left(1 - \frac{x}{n}\right)^n \le \frac{x^2}{2n-1} e^{-x}.$$

From these inequalities it is immediate that there exist constants C and A such that

$$\left|\left(1-\frac{v}{n}\right)^{n-1} I\{0 \le v \le n\} - e^{-v}\right| \le Ce^{-v}\frac{v^2}{n}$$

for $0 \le v \le A\sqrt{n}$. Hence

$$\left|\int_0^{A\sqrt{n}} e^{it\psi_n(v)} \left\{\left(1-\frac{v}{n}\right)^{n-1} I\{0 \le v \le n\} - e^{-v}\right\} dv\right| \le C\int_0^{A\sqrt{n}} e^{-v}\frac{v^2}{n}\, dv \le 2\frac{C}{n}.$$

Furthermore

$$\left|\int_{A\sqrt{n}}^\infty e^{it\psi_n(v)} \left\{\left(1-\frac{v}{n}\right)^{n-1} I\{0 \le v \le n\} - e^{-v}\right\} dv\right| \le$$

$$\leq \int_{A\sqrt{n}}^{n} \left(1-\frac{v}{n}\right)^{n-1} dv + \int_{A\sqrt{n}}^{\infty} e^{-v} dv$$

which can be made smaller than $C_2 \exp\{-A\sqrt{n}\}$ for an appropriate constant C_2. Hence we see that our assumption yields that for all real t

$$\int_0^\infty e^{it\psi_n(v)} e^{-v}\, dv \to \int_0^\infty e^{it\psi(v)} e^{-v} dv.$$

Now replace $\psi_n(v)$ by y on the left and $\psi(v)$ by y on the right. By the Lévy-Cramér Theorem it follows that $\psi_n(v) \to \psi(v)$ for all non-negative v. The classical theory of regular variation now shows that the condition

$$\psi_n(v) = \frac{U(n/v) - b_n}{a_n} \to \psi(v)$$

is equivalent to the definition of $F \in C_\gamma$. The explicit form of the possible functions $\psi(v)$ follows from this analysis as well.

We conclude with a classical formulation of the domain of attraction problem for the extremes.

Theorem 1. *There exist sequences of positive constants $\{a_n\}$ and real constants $\{b_n\}$ respectively such that $a_n^{-1}\{X_n^* - b_n\}$ converges in distribution to a non-degenerate distribution G if $F \in C_\gamma$ for some $\gamma \in \Re$. Moreover the possible limit distributions are given by $G_\gamma(v) = \exp(-\eta_\gamma(v))$.*

The above theorem contains a statement about the possible limit laws. In this full generality is originally due to Fisher–Tippett [5]. But the theorem also gives a domain of attraction condition. For the non-zero γ *cases* these results are due to Gnedenko [7]. For the case $\gamma = 0$, see De Haan [4], where concise treatment is given for the other cases as well.

3. Some applications

In Beirlant *et al.* [1, 2] it is shown how the C_γ-condition was useful in the discussion of the asymptotic behaviour of the Hill-estimator. To illustrate the usefulness of the C_γ-approach even further, we derive the limiting distributions of some less standard examples. They are taken from the classical treatise of Gumbel [9]. The last example refers to a statistic introduced by Galton, the first two refer (for simplicity) to symmetric distributions. Note that for the latter $F^{\leftarrow}(1-a) = -F^{\leftarrow}(a)$.

3.1. The extremal quotient

The *extremal quotient* is defined for symmetric distributions (only) by the expression $Q_n := X_n^*/(-X_1^*)$. To find its distribution, recall from the theory of order statistics that for arbitrary pairs

$$E\{g(X_k^*, X_m^*)\} =$$
$$c_{k,m}^{(n)} \int_{-\infty}^{\infty} dF(x) \int_x^{\infty} g(x,y) F^{k-1}(x)[F(y) - F(x)]^{m-k-1}[1 - F(y)]^{n-m} dF(y)$$

where

$$c_{k,m}^{(n)} = \frac{n!}{(k-1)!(m-k-1)!(n-m)!}.$$

First replace in the general expression for the joint expectation $g(x, y)$ by $h(-y/x)$. Then substitute in the integrating quantities $F(x)$ and $F(y)$, by $1 - a$ and $1 - ab$ respectively. One easily finds the general expression

$$E\{h(Q_n)\} = n(n-1) \int_0^1 a^{n-1} da \int_0^1 (1-b)^{n-2} h\left(\frac{F^{\leftarrow}(1-ab)}{F^{\leftarrow}(a)}\right) db. \qquad (3)$$

From this formula it is rather transparant that the asymptotic behaviour of Q_n will heavily depend upon the tailbehaviour of F. Recall that we had put $F^{\leftarrow}(1 - v) := U(1/v)$, where U is the tail quantile function of F. So, we replace a by $1 - \frac{u}{n}$ and b by $\frac{v}{n}$. Further, we probably need centering and/or normalisation for Q_n. So,

$$P\{Q_n \le b_n + a_n w\} =$$
$$\frac{n-1}{n} \int_0^n (1 - \frac{u}{n})^{n-1} du \int_0^n (1 - \frac{v}{n})^{n-2} I\left\{U\Big(\frac{n}{v}(1 - \frac{u}{n})^{-1}\Big) \Big/ U(\frac{n}{u}) \le b_n + a_n w\right\} dv.$$

So, we have to guarantee the inequality

$$U\Big(\frac{n}{v}(1 - \frac{u}{n})^{-1}\Big) - b_n U(\frac{n}{u}) \le a_n U(\frac{n}{u}) w.$$

The choice $b_n = 1$ is obvious, also in view of the definition of Q_n and the assumed symmetry of F. For a_n we seemingly need to fall back on the class C_γ. Hence $a_n U(n) = g(n)$ seems appropriate. With these two choices the above inequality turns into

$$\frac{1}{g(n)} \left\{U\Big(\frac{n}{v}(1 - \frac{u}{n})^{-1}\Big) - U\Big(\frac{n}{u}\Big)\right\} \le \frac{U(n/u)}{U(n)} w$$

which can now be handled. The ratio on the left converges to $h_\gamma(v^{-1}) - h_\gamma(u^{-1})$ while on the right the remaining ratio leads to $u^{-\gamma_+}$ which equals $u^{-\gamma}$ if $\gamma > 0$ while the expression equals 1 for $\gamma \le 0$.

In the limit for $n \uparrow \infty$ the inequality reads

$$h_\gamma(v^{-1}) - h_\gamma(u^{-1}) \le u^{-\gamma_+} w$$

and we get the limiting result

$$P\left\{\frac{U(n)}{g(n)}\left\{\left|\frac{X_n^*}{X_1^*}\right| - 1\right\} \le w\right\} \Rightarrow \int_0^\infty e^{-u} du \int_0^\infty e^{-v} I\left\{\frac{v^{-\gamma} - u^{-\gamma}}{\gamma} \le u^{-\gamma_+} w\right\} dv$$

which should now be evaluated.

- Let us look at the case $\gamma = 0$ first. Then the inequality is satisfied when $-\log v + \log u \leq w$ or when $v \geq ue^{-w}$. After the required integrations we obtain $(1+e^{-w})^{-1}$ as the limit, which is known as the *logistic distribution.*

- For $\gamma > 0$ the v-integration is again easy. We ultimately find for the limit on the right $\left(1+[1+\gamma w]_+^{-\frac{1}{\gamma}}\right)^{-1}$ which can be interpreted as a one-parameter extension of the logistic distribution.

- Finally, for $\gamma < 0$. Here the remaining calculations do not seem to simplify any further. First for $w \geq 0$ we find that $u^{-\gamma} \leq v^{-\gamma} - \gamma w$. For $w \leq 0$, $v \geq (u^{-\gamma} + \gamma w)^{-1/\gamma}$. Hence we find

$$P\left\{\frac{U(n)}{g(n)}\left\{\left|\frac{X_n^*}{X_1^*}\right| - 1\right\} \leq w\right\} \Rightarrow \begin{cases} 1 - \int_0^\infty \exp -v\{1 + (1-\gamma w v^\gamma)^{-1/\gamma}\}dv, & w \geq 0, \\ \\ \int_0^\infty \exp -v\{1 + (1+\gamma w v^\gamma)^{-1/\gamma}\}dv, & w \leq 0. \end{cases}$$

We formulate the result only for non-negative γ:

Theorem 2. *Assume F is symmetric and belongs to C_γ, $\gamma \geq 0$. Then*

$$P\left\{\frac{U(n)}{g(n)}\left\{\left|\frac{X_n^*}{X_1^*}\right| - 1\right\} \leq w\right\} \Rightarrow \left(1+[1+\gamma w]_+^{-\frac{1}{\gamma}}\right)^{-1}.$$

3.2. The geometric range

As another interesting quantity allied to the two extremes consider the *geometric range* which is again defined for a symmetric distribution by the relation $GR_n = \sqrt{X_n^*(-X_1^*)}$. With the choice $g(x,y) = h(\sqrt{-xy})$ in the formula above we find similarly that

$$E\{h(GR_n)\} = n(n-1)\int_0^1 a^{n-1}da \int_0^1 (1-b)^{n-2} h\left(\sqrt{F^{\leftarrow}(1-ab)F^{\leftarrow}(a)}\right) db.$$

To get the weak limit we again replace a by $1-u/n$ and b by v/n. We further introduce centering and normalisation constants to be determined later. We find

$$P\{GR_n \leq b_n + a_n w\} =$$

$$= \frac{n-1}{n}\int_0^n (1-\frac{u}{n})^{n-1}du \int_0^n (1-\frac{v}{n})^{n-2} I\left\{\sqrt{U\left(\frac{n}{v}(1-\frac{u}{n})^{-1}\right)U(\frac{n}{u})} \leq b_n + a_n w\right\} dv.$$

Let us assume again that $F \in C_\gamma$. Then

$$U(nz) - U(n) \sim g(n)h_\gamma(z)$$

for all $z > 0$. In the current case we have to satisfy the inequality

$$U\left(\frac{n}{v}(1-\frac{u}{n})^{-1}\right)U(\frac{n}{u}) \leq (b_n + a_n w)^2.$$

But then using the approximation for $U(n/u)$ and $U(n/v)$ we find that approximately we need

$$U^2(n) + g(n)U(n)[h_\gamma(v^{-1}) + h_\gamma(u^{-1})] + g^2(n)h_\gamma(v^{-1})h_\gamma(u^{-1}) + \ldots \leq$$
$$\leq b_n^2 + 2a_n b_n w + a_n^2 w^2.$$

Now the choice $b_n = U(n)$ is quite obvious. Further, it seems appropriate to take $a_n = g(n)$. The remaining inequality reads then

$$h_\gamma(v^{-1}) + h_\gamma(u^{-1}) + \frac{g(n)}{U(n)}h_\gamma(v^{-1})h_\gamma(u^{-1}) + \ldots \leq 2w + \frac{g(n)}{U(n)}w^2.$$

Now recall that for $n \uparrow \infty$, $g(n)/U(n) \to \gamma_+$. Hence asymptotically we end up with the result

$$P\left\{\frac{\sqrt{|X_n^* X_1^*|} - U(n)}{g(n)} \leq w\right\} \Rightarrow$$

$$\int_0^\infty e^{-u}du \int_0^\infty e^{-v} I\{h_\gamma(v^{-1}) + h_\gamma(u^{-1}) + \gamma_+ h_\gamma(v^{-1})h_\gamma(u^{-1}) \leq 2w + \gamma_+ w^2\}dv.$$

- For $\gamma = 0$ the remaining calculations are easy and lead to the inequality $vu > \exp(-2w)$. However for the next case the calculations are only slightly more complicated and lead to a similar result.

- For $\gamma > 0$ we find similarly $uv \geq (1+\gamma w)^{-2/\gamma}$. The resulting limit distribution can be expressed in terms of the *McDonald function* wich is often refered to as a (*modified*) *Bessel function* (*of the third kind*). Here is one possible definition as given in Gradhsteyn *et al.* [8], 8.432.7. For $\nu \in \Re$ and complex z

$$K_\nu(z) = \frac{1}{2}\left(\frac{z}{2}\right)^\nu \int_0^\infty t^{-\nu-1} \exp\left\{-t - \frac{1}{t}\left(\frac{z}{2}\right)^2\right\} dt. \tag{4}$$

- Finally, for $\gamma < 0$ again the calculations are a bit messy. For the region of integration is now determined by $v^{-\gamma} + u^\gamma \geq 2(1+\gamma w)$. For $w \geq -\gamma^{-1}$ this is the entire first quadrant. For $w < -\gamma^{-1}$ however we end up with

$$\exp -g(w) + \int_0^{g(w)} \exp -u\left\{1 + [(ug(w))^\gamma - 1]^{-1/\gamma}\right\} du$$

where $g(w) := [2(1+\gamma w)]^{-1/\gamma}$.

We only state the result for $\gamma \geq 0$.

Theorem 3. *Assume F is symmetric and belongs to C_γ with $\gamma \geq 0$. Then*

$$P\left\{\frac{\sqrt{|X_n^* X_1^*|} - U(n)}{g(n)} \leq w\right\} \Rightarrow zK_{-1}(z)$$

where $z = 2\{h_\gamma^{\leftarrow}(w)\}^{-1} = 2\eta_\gamma(w)$.

3.3. The Galton statistic

Let $\kappa < \lambda < \mu$ be positive integers. From the general theory of order statistics one easily gets the distribution of three order statistics. Taking the distributions as the variables of integration one gets for appropriate function g

$$E\{g(X^*_{n-\mu+1}, X^*_{n-\lambda+1}, X^*_{n-\kappa+1})\} =$$

$$= c_3 \int_0^1 v^{\mu-1}(1-v)^{n-\mu} dv \int_0^1 w^{\lambda-1}(1-w)^{\mu-\lambda-1} dw \int_0^1 z^{\kappa-1}(1-z)^{\lambda-\kappa-1} g_3 dz$$

where

$$c_3 := c(n, \mu, \lambda, \kappa) = n!\{(n-\mu)!(\mu-\lambda-1)!(\lambda-\kappa-1)!(\kappa-1)!\}^{-1}$$

and

$$g_3 = g[F^{\leftarrow}(1-v), F^{\leftarrow}(1-vw), F^{\leftarrow}(1-vwz)].$$

The ratio

$$G_n := \frac{X^*_{n-1} - X^*_{n-2}}{X^*_n - X^*_{n-2}}$$

(or rather its reciprocal) is known as *Galton's ratio.* Its properties can be readily obtained by putting $\mu = 3, \lambda = 2$ and $\kappa = 1$ in the above. We get the following simplification

$$Eg\{X^*_{n-2}, X^*_{n-1}, X^*_n\} = \frac{n!}{(n-3)!} \int_0^1 v^2(1-v)^{n-3} dv \int_0^1 w dw \int_0^1 g_3 dz$$

where g_3 is as above.

If we now specialize to a function of G_n it is reasonable to take

$$g(a, b, c) = k\left(\frac{b-a}{c-a}\right).$$

We then obtain that

$$E\{h(G_n)\} = \frac{n!}{(n-3)!} \int_0^1 v^2(1-v)^{n-3} dv \int_0^1 w dw \int_0^1 h(v, w, z) dz$$

with

$$h(v, w, z) = k\left(\frac{F^{\leftarrow}(1-vw) - F^{\leftarrow}(1-v)}{F^{\leftarrow}(1-vwz) - F^{\leftarrow}(1-v)}\right).$$

To get information about the asymptotic properties of G_n for large n we need to replace v by a/n. We immediately get

$$E\{h(G_n)\} = \frac{n!n^{-3}}{(n-3)!} \int_0^n a^2(1-\frac{a}{n})^{n-3} da \int_0^1 w dw \int_0^1 h(\frac{a}{n}, w, z) dz$$

with

$$h(\frac{a}{n}, w, z) = k\left(\frac{U(\frac{n}{aw}) - U(\frac{n}{a})}{U(\frac{n}{awz}) - U(\frac{n}{a})}\right).$$

Dividing the U differences in the numerator and denominator by $g(n/a)$, we can apply the C_γ-condition to find that for any continuous function k

$$E\{k(G_n)\} \to 2\int_0^1 w\,dw \int_0^1 k\left(\frac{h_\gamma(1/w)}{h_\gamma(1/wz)}\right) dz.$$

By a classical Berry–Esseen argument we can obtain the limit in distribution. Let $0 \le y \le 1$. For the particular case where $k(u) = I\{-\infty < y \le u\}$ we still need a couple of extra steps to derive the limit distribution. We have in general

$$P\{G_n \le y\} \Rightarrow 2\int_0^1 w \int_0^1 I\left\{h_\gamma(1/w) \le y h_\gamma(1/wz)\right\} dz\,dw. \tag{5}$$

Case 1: $\gamma > 0$.
The inequality $w^{-\gamma} - 1 \le y\{(wz)^{-\gamma} - 1\}$ readily leads to

$$z \le (1/w)(1 + (1/y)(w^{-\gamma} - 1))^{-1/\gamma}$$

and hence to the remaining integral

$$2\int_0^1 \left(1 + \frac{1}{y}(w^{-\gamma} - 1)\right)^{-1/\gamma} dw.$$

The substitution $w = (1 + xy)^{-1/\gamma}$ turns the integral on the right-hand side of (1) into a standard form to which 3.197.1 in Gradhsteyn *et al.* [8] applies. One finds as the limit $yF(\frac{1}{\gamma} + 1, 1, \frac{2}{\gamma} + 1; 1 - y)$ where F is the *hypergeometric function*.

Case 2: $\gamma = 0$.
Here the calculations are very simple and lead to the limit $2y(1 + y)^{-1}$.

Case 3: $\gamma < 0$.
This case is only slightly more cumbersome. The region of integration is now determined by $w^{-\gamma}(1 - yz^{-\gamma}) \ge 1 - y$ and $w \le 1$. Integration with respect to w can be carried out. The remaining integration is then

$$\int_0^1 \left\{1 - \left(\frac{1 - yz^{-\gamma}}{1 - y}\right)^{2/\gamma}\right\} dz.$$

However we can now apply 3.197.3 in Gradhsteyn *et al.* [8]. After a couple of reductions using 9.193.1 in Gradhsteyn *et al.* [8] we find again a hypergeometric limit. Let us summarize our findings in a theorem that extends a former result by Gumbel to the case where the underlying distributions are in an arbitrary domain of attraction of an extreme value law.

Theorem 4. *Assume that F belongs to $\mathcal{C}_\gamma$. Then for $0 \le y \le 1$*

$$P\left\{\frac{X^*_{n-1} - X^*_{n-2}}{X^*_n - X^*_{n-2}} \le y\right\} \Rightarrow \begin{cases} yF(\frac{1}{\gamma}+1,\ 1,\ \frac{2}{\gamma}+1;\ 1-y) & \text{if } \gamma > 0 \\ 2y/(1+y) & \text{if } \gamma = 0 \\ 1-(1-y)F(\frac{1}{\gamma}+1,\ 1,\ 1-\frac{1}{\gamma};\ y) & \text{if } \gamma < 0. \end{cases}$$

Acknowlegements

We are proud to present this modest contribution to Academician V. Korolyuk on the occasion of his seventieth birthday.

We also gladly acknowledge Petra Vynckier for helpful discussions in connection with this paper.

References

1. **Beirlant, J. and Teugels, J.L.** (1986). Asymptotics of Hill's estimator. *Theory Probabil. Appl.* **31**, 463–469.

2. **Beirlant, J. and Teugels, J.L.** (1989). Asymptotic normality of Hill's estimator. In: *Extreme Value Theory,* J. Hüsler and R.D. Reiss (Eds). Lecture Notes in Statistics, **51**, 148–155.

3. **Bingham, N.H., Goldie, C.M. and Teugels, J.L.** (1987). *Regular Variation.* Cambridge University Press.

4. **De Haan, L.** (1970). *On Regular Variation and its Applications to the Weak Convergence of Sample Extremes.* Math. Center Tract 32, Amsterdam.

5. **Fisher, R.A. and Tippett, L.H.C.** (1928). Limiting form of the frequency distribution of the largest or the smallest member of a sample. *Proc. Cambridge Philosophic. Soc.* **24**, 180–190.

6. **Galambos, J.** (1978). *The Asymptotic Theory of Extreme Order Statistics.* J.Wiley and Sons, New York.

7. **Gnedenko, B.V.** (1943). Sur la distribution limite du terme maximum d'une série aléatoire. *Ann. Math.* **44**, 423–453.

8. **Gradhsteyn, I.S. and Ryzhik, I.M.** (1980). *Table of Integrals, Series and Products.* Academic Press, Orlando.

9. **Gumbel, E.J.** (1958). *Statistics of Extremes.* Columbia University Press, New York.

10. **Leadbetter, M.R., Lindgren, G. and Rootzén, H.** (1982). *Extremal and Related Properties of Stationary Processes.* Springer-Verlag, New York.

11. **Pfeiffer, D.** (1989). *Einführung in die Extremwertstatistik.* B.G. Teubner, Stuttgart.

12. **Resnick, S.I.** (1987). *Extreme Values, Regular Variation and Point Processes.* Springer-Verlag, New York.

Exploring Stochastic Laws. pp. 469–479.
Eds. A.V. Skorokhod and Yu.V. Borovskikh

Averaging and stability of cocycles under dynamical systems with rapid Markov switching

YE.F. TSARKOV

Applied Mathematics Department, Riga Technical University, Ausekla St. 9, Riga LV - 1010, Latvia

Received February 1, 1995

Abstract. A linear differential equation in $\mathbb{R}^n$ with coefficients dependent on the trajectories of dynamical systems, directed by a Markov process with rapid Markov switching, is studied. It will be proved that under the hypothesis of ergodicity the averaged or enlarged by the phase lumping method dynamical system can be used for the mean square stability analysis of this differential equation.

Let $\zeta_\varepsilon(t)$ be a homogeneous Markov process in the metric phase space $\mathbb{U}$ with C-infinitesimal operator $Q_\varepsilon = \frac{1}{\varepsilon}(Q + \varepsilon Q_1)$, where $\varepsilon \in (0, \delta)$, and operators Q and Q_1 are given by the formulas

$$(Qv)(z) = a(z) \int_{\mathbb{U}} [v(u) - v(z)]\, p(z, du), \tag{1}$$

$$(Q_1 v)(z) = a(z) \int_{\mathbb{U}} [v(u) - v(z)]\, p_1(z, du), \tag{2}$$

for any measurable function on $\mathbb{U}$. It is a well-known piecewise constant process (see, for example, Skorokhod (1991)) with intensity of switching $a(z)/\varepsilon$ and embedded Markov chain with transition probability $p(z, du) + \varepsilon p_1(z, du)$. We shall assume that $a(y)$ is a positive continuous function, $p(z, du)$ is the transition probability of some Markov chain with Feller property, $p_1(z, A)$ is a function continuous on z for any measurable subset A and countably additive on A measure for any z, and the number δ is sufficiently small.

Let us suppose that the above mentioned Markov process $\zeta_\varepsilon(t)$ is a switching process for a dynamical system, given by the differential equation in $\mathbb{R}^d$:

$$\frac{dy_\varepsilon}{dt} = \varphi(y_\varepsilon, \zeta_\varepsilon(t)), \tag{3}$$

where $\varphi(y,z)$ is a continuous and bounded function with two continuous and bounded y-derivatives, i.e., there exists a positive number c, such that

$$\sup_{y,u}\Big\{|\varphi(y,u)| + \|D\varphi(y,u)\| + \|D^2\varphi(y,u)\|\Big\} \le c,$$

where D and D^2 are the first and second Frechet derivatives with respect to the argument y. System (3) can be analyzed by the *averaging method* (see, for example, Korolyuk (1991)): if $\mu(du)$ is a unique invariant measure corresponding to operator Q, then the family of processes $y_\varepsilon(t)$ weakly converge as $\varepsilon \to 0$ to the solution of the *averaged equation*

$$\frac{dy}{dt} = \bar{\varphi}(y), \tag{4}$$

where

$$\bar{\varphi}(y) = \int_{\mathbb{U}} \varphi(y,u)\,\mu(du),$$

and both systems (3) and (4) have the same properties of equilibrium stability. The object, we shall deal with, is a linear differential equation in $\mathbb{R}^n$ directed by the Markov process $\zeta_\varepsilon(t)$ and dynamical system (3):

$$\frac{dx_\varepsilon}{dt} = A(y_\varepsilon(t),\zeta_\varepsilon(t))\,x_\varepsilon, \tag{5}$$

where the matrix-valued function $A(y,z)$ is bounded and continuous with two continuous y-derivatives, i.e., there exists a positive number c, such that

$$\sup_{y,u}\Big\{\|A(y,u)\| + \|DA(y,u)\| + \|D^2A(y,u)\|\Big\} \le c.$$

Differential equation (5) defines the multiplicative family of Cauchy operators $X_\varepsilon(t,s,y)$ in $\mathbb{R}^n$, depending on the trajectories of dynamical system (3) with initial condition $y_\varepsilon(s) = y$ and having an evolutionary property. This family and equation (5), following the paper of Oseledec (1968), we shall name *cocycle under dynamical system* (3). Cocycle (5) is named *exponentially stable in the mean square*, if there exist such positive constants M and γ, that for any $t \ge s$, $y \in \mathbb{R}^d$ and $u \in \mathbb{U}$ the inequality

$$\mathbf{E}_{s,u}\|X_\varepsilon(t,s,y)\|^2 \le Me^{-\gamma(t-s)} \tag{6}$$

is fulfilled. Here and further the subindices of expectation denote the condition $\zeta(s) = u$. Side by side with cocycle (5) we shall consider the *averaged cocycle* defined by the differential equation

$$\frac{dx}{dt} = \bar{A}(y(t))\,x, \tag{7}$$

where $y(t)$ is a solution of the averaged equation (4), and $\bar{A}(y)$ is the averaged matrix

$$\bar{A}(y) = \int_{\mathbb{U}} A(y,u)\,\mu(du). \tag{8}$$

The Cauchy operators of (7), corresponding to the solution of averaged equation (4) with initial condition $y(s) = y$, we shall denote as $X(t, s, y)$. Cocycle (7) is named *exponentially stable*, if there exist such positive constants C and ρ, that for any $t \geq s$, and $y \in \mathbb{R}^d$ the following inequality can be written:

$$\|X(t,s,y)\| \leq Ce^{-\rho(t-s)}. \tag{9}$$

According to what has been proved in Korolyuk (1991) the solutions of the systems (3)–(5) converge to corresponding solutions of averaging systems (4)–(7). We will analyze the stability of cocycle (7) by means of cocycle (5).

Let us denote $P(t, z, du)$ the transition probability of the Markov process with C-infinitesimal operator Q. We shall say that *operator Q satisfies the condition of uniform ergodicity*, if

$$\sup_{\|v\|=1} \left| \int_{\mathbb{U}} [P(t,z,du) - \mu(du)]\, v(u) \right| \leq e^{-\alpha t} \tag{10}$$

for some positive number α and any $t \geq 0$, $z \in \mathbb{U}$. Under this condition we can define the linear continuous operator $\Pi \in \mathbf{L}(C(\mathbb{U}))$ by the equality

$$(\Pi v)(z) := \int_0^\infty \int_{\mathbb{U}} [P(t,z,du) - \mu(du)]\, v(u)\, dt.$$

It is clearly seen that one can write

$$Q\,\Pi v = -v + \bar{v},$$

for any $v \in C(\mathbb{U})$. Here and further the line over the function of the argument $u \in \mathbb{U}$ denotes the averaging with respect to measure μ. The above operator Π we shall name as *the potential.*

By using the averaged cocycle one can define the quadratic form of argument x:

$$V(x,y) := (q(y)\,x, x) := \int_0^T |X(t,0,y)\,x|^2\, dt, \tag{11}$$

where $(\cdot,\cdot)$ denotes the scalar product in the space $\mathbb{R}^n$. The matrix $q(y)$ of this quadratic form depends on the number T also, but in our further considerations this number will be fixed.

Theorem 1. *If operator Q is uniformly ergodic and the averaged cocycle is exponentially stable, then there exists a positive number ε_0, such that cocycle* (5) *is exponentially stable in the mean square for any $\varepsilon \in (0, \varepsilon_0)$.*

Proof. Let us define the function

$$V_\varepsilon(x,y,z) = V(x,y) + \varepsilon\, W(x,y,z),$$

where

$$W(x,y,z) = \Pi\Big\{\Big(\nabla_y V(x,y), \varphi(y,z)\Big) + \Big([A^*(y,z)\,q(y) + q(y)\,A(y,z)]\,x, x\Big)\Big\},$$

∇_y denotes the operator gradient in $\mathbb{R}^d$ and upper index * denotes a transposition. We put $T = -(\ln 2 + 2\ln C)/2\rho$ in definition (11) of the function $V(x,y)$, where ρ and C are constants from inequality (9). It is clearly seen, that due to the assumptions of function φ and matrix A one can write the inequalities:

$$\begin{aligned} \sup_y \Big\{|\bar\varphi(y)| + \|D\bar\varphi(y)\| + \|D^2\bar\varphi(u)\|\Big\} &\leq c, \\ \sup_y \Big\{\|\bar A(y)\| + \|D\bar A(y)\| + \|D^2\bar A(u)\|\Big\} &\leq c. \end{aligned}$$

Hence, there are constants c_1 and c_2, such that for any $x \in \mathbb{R}^n$ and $y \in \mathbb{R}^d$ the inequalities

$$\begin{gathered} c_1\,|x|^2 \leq V(x,y) \leq c_2\,|x|^2, \quad |\nabla_y V(x,y)| \leq c_2\,|x|^2, \quad |\nabla_x V(x,y)| \leq c_2\,|x|, \\ c_1 \leq \|q(y)\| \leq c_2, \quad \|Dq(y)\| \leq c_2, \quad \|D^2 q(y)\| \leq c_2, \end{gathered} \tag{12}$$

are fulfilled. Here and further ∇_x is the operator-gradient in $\mathbb{R}^n$. Let us introduce *the Lyapunov operator according to the system* (4)–(7):

$$(\mathcal{L}_0 G)(x,y) := (\nabla_y G(x,y), \bar\varphi(y)) + (\nabla_x G(x,y), \bar A(y)x), \tag{13}$$

and *weak infinitesimal operator* $\mathcal{L}_\varepsilon$ *of the Markov process* $\{x_\varepsilon(t), y_\varepsilon(t), \zeta_\varepsilon(t)\}$, *defined by* $\zeta_\varepsilon(t)$ *and the systems* (3)–(5) (see, for example, Skorokhod (1991)):

$$\begin{aligned} \mathcal{L}_\varepsilon H(x,y,z) &= (\nabla_y H(x,y,z), \varphi(y,z)) + Q_\varepsilon H(x,y,z) + \\ &\quad + (\nabla_x H(x,y,z), A(y,z)x), \end{aligned} \tag{14}$$

where functions $G(x,y)$ and $H(x,y,z)$ are sufficiently smooth. By definition

$$\begin{aligned} \mathcal{L}_0 V(x,y) &= \lim_{t\to 0} \frac{1}{t}\Big[V(X(t,0,y)\,x, y(t,0,y)) - V(x,y)\Big] = \\ &= \lim_{t\to 0} \frac{1}{t}\Big[\int_0^T \Big|X(t+s,t,y(t,0,y))\,X(t,0,y)\,x\Big|^2 ds - \\ &\qquad - \int_0^T |X(s,0,y)\,x|^2\,ds\Big] = \\ &= |X(T,0,y)\,x|^2 - |x|^2, \end{aligned}$$

where $y(t,0,y)$ is a solution of the averaged dynamical system with initial condition $y(0) = y$. Hence, in view of the value of T, the inequality

$$\mathcal{L}_0 V(x,y) \leq -\frac{1}{2}\,|x|^2 \tag{15}$$

follows from (12). Since the potential Π is a bounded linear operator and $|\varphi(y,z)|$, $\|A(y,z)\|$ are uniformly bounded by number c, one can deduce from inequalities (12) the inequalities:

$$\sup_{y,z} |W(x,y,z)| \leq 3\,\|\Pi\|\, cc_2\, |x|^2,$$

for any $x \in \mathbb{R}^n$, and

$$(c_1 - 3\varepsilon\,\|\Pi\|\, cc_2)\,|x|^2 \leq V_\varepsilon(x,y,z) \leq (c_2 + 3\varepsilon\,\|\Pi\|\, cc_2)\,|x|^2$$

for any $x \in \mathbb{R}^n$, $y \in \mathbb{R}^d$, $z \in \mathbb{U}$, $\varepsilon \in (0,\delta)$. Hence, there exists a sufficiently small number $\varepsilon_1 \leq \delta$, such that

$$\frac{1}{2} c_1\,|x|^2 \leq V_\varepsilon(x,y,z) \leq 2c_2\,|x|^2 \tag{16}$$

for any $x \in \mathbb{R}^n$, $y \in \mathbb{R}^d$, $z \in \mathbb{U}$ and $\varepsilon \in (0,\varepsilon_1)$. By construction, the function $V_\varepsilon(x,y,z)$ is sufficiently smooth and we may use the weak infinitesimal operator $\mathcal{L}_\varepsilon$ from (14) of the Markov process $\{x_\varepsilon(t), y_\varepsilon(t), \zeta_\varepsilon(t)\}$. By the definition of this function, we have

$$\begin{aligned}
(\mathcal{L}_\varepsilon V_\varepsilon)(x,y,z) &= (\nabla_y V_\varepsilon(x,y,z), \varphi(y,z)) + \\
&\quad + Q_\varepsilon V_\varepsilon(x,y,z) + (\nabla_x V_\varepsilon(x,y,z), A(y,z)x) = \\
&= \frac{1}{\varepsilon}(Q + \varepsilon Q_1)\, V(x,y) + Q\, W(x,y,z) + \\
&\quad + (\nabla_y V(x,y), \varphi(y,z)) + (\nabla_x V(x,y), A(y,z)\, x) + \\
&\quad + \varepsilon \left[(\nabla_y W(x,y,z), \varphi(y,z)) + (\nabla_x W(x,y,z), A(y,z)\, x)\right] + \\
&\quad + \varepsilon Q_1 W(x,y,z) = \\
&= -\left\{(\nabla_y V(x,y), \varphi(y,z)) + ([A^*(y,z)\, q(y) + q(y)\, A(y,z)]\, x, x)\right\} + \\
&\quad + (\nabla_y V(x,y), \bar{\varphi}(y)) + ([\bar{A}^*(y)\, q(y) + q(y)\, \bar{A}(y)]\, x, x) + \\
&\quad + (\nabla_y V(x,y), \varphi(y,z)) + (\nabla_x V(x,y), A(y,z)x) + \\
&\quad + \varepsilon \left[(\nabla_y W(x,y,z), \varphi(y,z)) + (\nabla_x W(x,y,z), A(y,z)x)\right] + \\
&\quad + \varepsilon Q_1 W(x,y,z),
\end{aligned} \tag{17}$$

or

$$(\mathcal{L}_\varepsilon V_\varepsilon)(x,y,z) = (\mathcal{L}_0 V)(x,y) + \varepsilon\, r(x,y,z), \tag{18}$$

where

$$\begin{aligned}
&r(x,y,z) = \\
&\quad = \left[(\nabla_y W(x,y,z), \varphi(y,z)) + (\nabla_x W(x,y,z), A(y,z)\, x)\right] + Q_1 W(x,y,z).
\end{aligned}$$

Now we use the uniform boundedness of $\varphi(y,z)$ and $A(y,z)$, continuity of the linear operator Π and inequalities (12) for the estimation of

$$\sup_{y,z} |r(x,y,z)| \leq c_3\, |x|^2$$

with some positive constant c_3. Hence, from (15) and (16) we have the inequality:

$$(\mathcal{L}_\varepsilon V_\varepsilon)(x,y,z) \le \left(-\frac{1}{2} + \varepsilon c_3\right) |x|^2 \le \\ \le \left(-\frac{1}{2} + \varepsilon c_3\right) \frac{1}{2c_2} V_\varepsilon(x,y,z) \le -\frac{1}{8c_2} V_\varepsilon(x,y,z) \tag{19}$$

for any $x \in \mathbb{R}^n$, $y \in \mathbb{R}^d$, $z \in \mathbb{U}$, $\varepsilon \in (0, \varepsilon_2)$ and sufficiently small positive number ε_2. Now we may use the well-known formula for Markov semigroups from Dynkin (1965):

$$\mathbf{E}_{s,z} V_\varepsilon(X_\varepsilon(t,s,y)\,x, y_\varepsilon(t,s,y), \zeta_\varepsilon(t)) = \\ = V_\varepsilon(x,y,z) + \int_s^t \mathbf{E}_{s,z}(\mathcal{L}_\varepsilon V_\varepsilon)(X_\varepsilon(\tau,s,y)\,x, y_\varepsilon(\tau,s,y), \zeta_\varepsilon(\tau))\,d\tau,$$

and deduce from (19) the inequality

$$\mathbf{E}_{s,z} V_\varepsilon(X_\varepsilon(t,s,y)\,x, y_\varepsilon(t,s,y), \zeta_\varepsilon(t)) \le \\ \le V_\varepsilon(x,y,z) - \frac{1}{8c_2} \int_s^t \mathbf{E}_{s,z} V_\varepsilon(X_\varepsilon(\tau,s,y)\,x, y_\varepsilon(\tau t,s,y), \zeta_\varepsilon(\tau))\,d\tau,$$

where $y_\varepsilon(t,s,y)$ is the solution of (3) with initial condition $y_\varepsilon(s) = y$. From (16) and the above inequality it follows, that

$$\frac{1}{2} c_1\, \mathbf{E}_{s,z} |X_\varepsilon(t,s,y)\,x|^2 \le \mathbf{E}_{s,z} V_\varepsilon(X_\varepsilon(t,s,y)\,x, y_\varepsilon(t,s,y), \zeta_\varepsilon(t)) \le \\ \le V_\varepsilon(x,y,z) \exp\left\{-\frac{1}{8c_2}(t-s)\right\} \le 2\,c_2\,|x|^2 \exp\left\{-\frac{1}{8c_2}(t-s)\right\}$$

for any $x \in \mathbb{R}^n$, $y \in \mathbb{R}^d$, $z \in \mathbb{U}$ and $\varepsilon \in (0, \varepsilon_0)$, where $\varepsilon_0 = \min\{\varepsilon_1, \varepsilon_2\}$. The proof of the Theorem is now complete.

Now we shall use Q from (1), which has eigenvalue 0 of multiplicity $m > 1$, and a Markov process, corresponding to the infinitesimal operator Q, which has m invariant measures μ_j with disjoint supports, $j = \overline{1,m}$. According to the phase lumping method of Korolyuk (1991), we shall suppose that *there exists a decomposition* $\mathbb{U} = \bigcup_{j=1}^m \mathbb{U}_j$ *of the phase space, such that*

$$p(z, \mathbb{U}_j) = \begin{cases} 1 & \textit{for } z \in \mathbb{U}_j, \\ 0 & \textit{for } z \notin \mathbb{U}_j, \end{cases} \tag{20}$$

and there exists a positive number α, *such that the transition probability of the Markov process with infinitesimal operator* Q *satisfies the inequality:*

$$\sup_{\substack{\|v\|=1 \\ z \in \mathbb{U}_j}} \left| \int_{\mathbb{U}_j} [P(t,z,du) - \mu_j(du)]\, v(u) \right| \le e^{-\alpha t} \tag{21}$$

for any $j = \overline{1,m}$.

Under this assumption we shall consider:
the lumped Markov process $\zeta(t)$ with phase space $\bar{\mathbb{U}} = \{1, 2, \dots, m\}$ and infinitesimal matrix

$$Q_{ij} = \begin{cases} \int_{\mathbb{U}_i} a(z)\, p_1(z, \mathbb{U}_j)\, \mu_i(dz) & \text{for} \quad i \neq j, \\ -\sum_{j \neq i} \int_{\mathbb{U}_i} a(z)\, p_1(z, \mathbb{U}_j)\, \mu_i(dz) & \text{for} \quad i = j, \end{cases}$$

where $p_1(z, du)$ is defined by (2);
the lumped dynamical system

$$\frac{d\tilde{y}}{dt} = \tilde{\varphi}(\tilde{y}, \zeta(t)), \tag{22}$$

where

$$\tilde{\varphi}(y, j) = \int_{\mathbb{U}_j} \varphi(y, z)\, \mu_j(dz), \qquad j = \overline{1, m};$$

and *the lumped cocycle*

$$\frac{dx}{dt} = \tilde{A}(\tilde{y}(t), \zeta(t))\, x, \tag{23}$$

where

$$\tilde{A}(y, j) = \int_{\mathbb{U}_j} A(y, u)\, \mu_j(du), \qquad j = \overline{1, m},$$

and $\tilde{y}(t)$ satisfies equation (22). Equation (23) defines the Cauchy operator $\tilde{X}(t, s, y)$, corresponding to the solution of (22) with the initial condition $\tilde{y}(s) = y$. The conditional expectations, generated by the Markov process $\zeta(t)$ with initial condition $\zeta(s) = j$ we shall denote as $\tilde{\mathbf{E}}_{s,j}$.

From the paper of Korolyuk (1991) it follows that the solutions of systems (3)–(5) weakly converge (as $\varepsilon \to 0$) to the corresponding solutions of lumped systems (22)–(23). Besides, in this paper it is proved that the mean square stability of the equilibrium of (5) guarantees the stochastic stability of the corresponding equilibrium of (3). We will analyze the mean square stability of cocycle (5) by means of cocycle (23).

Theorem 2. *If operator Q satisfies the above conditions (20)–(21) and lumped cocycle (23) is exponentially stable in the mean square, then there exists a positive number ε_0, such that the lumped cocycle is exponentially stable in the mean square for any $\varepsilon \in (0, \varepsilon_0)$.*

Proof. Let P be the projective operator in the kernel of the extension of operator Q on the space $\mathbf{B}\mathbb{U}$ of measurable functions, i.e.

$$(Pv)(z) = \int_{\mathbb{U}_j} v(u)\, \mu_j(du), \qquad z \in \mathbb{U}_j,$$

for any $j = \overline{1,m}$ and $v \in \mathbf{B}(\mathbb{U})$. *The potential* $\tilde{\Pi}$, corresponding to the above extension of the operator Q can also be defined for all $v \in \mathbf{B}(\mathbb{U})$ by the equality:

$$(\tilde{\Pi}v)(z) = \int_0^\infty \int_{\mathbb{U}_j} [P(t,z,du) - \mu_j(du)]\, v(u)\, dt, \qquad z \in \mathbb{U}_j,$$

for any $j = \overline{1,m}$. It is clearly seen, that $\tilde{\Pi}$ is a linear continuous operator on the space $\mathbf{B}(\mathbb{U})$ and for any $v \in \mathbf{B}(\mathbb{U})$ the equality

$$Q\,\tilde{\Pi}v = -v + Pv. \tag{24}$$

can be written. Lumped cocycle (23), lumped dynamical system (22) and Markov process $\zeta(t)$ define on the phase space $\mathbb{R}^n \times \mathbb{R}^d \times \tilde{\mathbb{U}}$ some new Markov process with *weak infinitesimal operator* $\tilde{\mathcal{L}}$, which is defined on sufficiently smooth (according to x and y) functions by the equality (see, for example, Dynkin (1965)):

$$\begin{aligned}(\tilde{\mathcal{L}}v)(x,y,j) &:= \lim_{t\to 0}\frac{1}{t}\left[\tilde{\mathbf{E}}_{0,j}v(\tilde{X}(t,0,y)\,x, \tilde{y}(t,0,y), \zeta(t)) - v(x,y,j)\right]\\ &= (\nabla_x v(x,y,j), \tilde{A}(y,j)\,x) + (\nabla_y v(x,y,j), \tilde{\varphi}(y,j)) +\\ &\quad + \sum_{i=1}^m Q_{ji}\, v(x,y,i),\end{aligned} \tag{25}$$

for any $x \in \mathbb{R}^n$, $y \in \mathbb{R}^d$ and $j = \overline{1,m}$. Here and elsewhere $\tilde{y}(t,s,y)$ denotes the solution of (22) with initial condition $\tilde{y}(s) = y$. Let us consider the quadratic form in $\mathbb{R}^n$, defined by equality

$$\tilde{V}(x,y,j) := (\tilde{q}(y,j)\,x, x) := \int_0^T \tilde{\mathbf{E}}_{0,j}|\tilde{X}(s,0,y)\,x|^2\, ds.$$

One can prove that in view of the exponential mean square stability of the lumping cocycle, this function and its derivatives have the estimations:

$$\begin{gathered} c_1\,|x|^2 \le \tilde{V}(x,y,j) \le c_2\,|x|^2, \quad |\nabla_y \tilde{V}(x,y,j)| \le c_2\,|x|^2,\\ |\nabla_x \tilde{V}(x,y,j)| \le c_2\,|x|, \quad c_1 \le \|\tilde{q}(y,j)\| \le c_2,\\ \|D_y\,\tilde{q}(y,j)\| \le c_2, \quad \|D^2\,\tilde{q}(y,j)\| \le c_2 \end{gathered} \tag{26}$$

with some positive numbers c_1 and c_2 uniform on $T > 0$, $y \in \mathbb{R}^d$, $x \in \mathbb{R}^n$ and $j = \overline{1,m}$. It is clearly seen, that due to the homogeneity of the Markov process, we may use the identity

$$\tilde{\mathbf{E}}_{t,j}|\tilde{X}(s+t,t,y)\,x|^2 \equiv \tilde{\mathbf{E}}_{0,j}|\tilde{X}(s,0,y)\,x|^2.$$

Hence,

$$\tilde{\mathbf{E}}_{0,j}\tilde{\mathbf{E}}_{t,\zeta(t)}|\tilde{X}(s+t,t,\tilde{y}(t,0,y))\,\tilde{X}(t,0,y)\,x|^2 = \tilde{\mathbf{E}}_{0,j}|\tilde{X}(s+t,0,y)\,x|^2,$$

and, by definition of $\tilde{V}(x,y,j)$, we can choose a number T so large, that the inequality:

$$(\tilde{\mathcal{L}}\tilde{V})(x,y,j) = \tilde{\mathbf{E}}_{0,j}|\tilde{X}(T,0,y)\,x|^2 - |x|^2 \le -\frac{1}{2}\,|x|^2 \tag{27}$$

is fulfilled uniformly on $y \in \mathbb{R}^d$, $x \in \mathbb{R}^n$ and $j = \overline{1,m}$. This value of T will be used in the definition of the Lyapunov function:

$$\tilde{V}_\varepsilon(x,y,z) = U(x,y,z) + \varepsilon\,\tilde{W}(x,y,z),$$

where

$$\begin{aligned}
\tilde{W}(x,y,z) &= \tilde{\Pi}[G(x,y,z) + Q_1 U(x,y,z)],\\
G(x,y,z) &= (\nabla_y U(x,y,z),\varphi(y,z)) + (\nabla_x U(x,y,z), A(y,z)\,x),\\
U(x,y,z) &= \tilde{V}(x,y,j), \qquad z \in \mathbb{U}_j,\ j = \overline{1,m}.
\end{aligned}$$

By construction, the function $U(x,y,z)$ is of a quadratic form with matrix $h(y,z)$, defined by equalities

$$h(y,z) = \tilde{q}(y,j), \qquad z \in \mathbb{U}_j,\ j = \overline{1,m}.$$

These functions and matrices satisfy condition (26). Hence, by definition of the projective operator P, one can write $PU = U$, and, according to equality (24),

$$Q\,U(x,y,z) = 0. \tag{28}$$

Due to the boundedness of operators $\tilde{\Pi}$ and Q_1, the inequalities

$$\|\nabla_x \tilde{W}(x,y,z)\| \le k_1\,|x|, \quad \|\nabla_y \tilde{W}(x,y,z)\| + |\tilde{W}(x,y,z)| \le k_1\,|x|^2 \tag{29}$$

with some positive k_1 follows immediately from (25) and (26). Then, there exists a sufficiently small number ε_1, such that

$$\frac{1}{2}c_1\,|x|^2 \le \tilde{V}_\varepsilon(x,y,z) \le 2c_2\,|x|^2 \tag{30}$$

for any $x \in \mathbb{R}^n$, $y \in \mathbb{R}^d$, $z \in \mathbb{U}$ and $\varepsilon \in (0,\varepsilon_1)$. By construction, the function $\tilde{V}_\varepsilon(x,y,z)$ is sufficiently smooth and we may use the weak infinitesimal operator $\mathcal{L}_\varepsilon$ from (14) of the Markov process $\{x_\varepsilon(t), y_\varepsilon(t), \zeta_\varepsilon(t)\}$. By definition of this function we have the equality:

$$\begin{aligned}
(\mathcal{L}_\varepsilon \tilde{V}_\varepsilon)(x,y,z) &= (\nabla_y \tilde{V}_\varepsilon(x,y,z),\varphi(y,z)) +\\
&\quad + Q_\varepsilon \tilde{V}_\varepsilon(x,y,z) + (\nabla_x \tilde{V}_\varepsilon(x,y,z), A(y,z)\,x) =\\
&= \frac{1}{\varepsilon}Q\,U(x,y,z) + Q_1 U(x,y,z) + Q\,\tilde{W}(x,y,z) +\\
&\quad + (\nabla_y U(x,y,z),\varphi(y,z)) +\\
&\quad + (\nabla_x U(x,y,z), A(y,z)\,x) + \varepsilon R(x,y,z),
\end{aligned}$$

where

$$R(x,y,z) = (\nabla_y \tilde{W}(x,y,z), \varphi(y,z)) + \\ + (\nabla_x \tilde{W}(x,y,z), A(y,z)\,x) + Q_1 \tilde{W}(x,y,z).$$

According to formulas (23) and (25) one can write

$$\begin{aligned} Q\tilde{W}(x,y,z) &= -G(x,y,z) - Q_1U(x,y,z) + P\,G(x,y,z) + P\,Q_1U(x,y,z) = \\ &= -G(x,y,z) - Q_1U(x,y,z) + \\ &\quad + \int_{\mathbf{U}_j} \Big\{ G(x,y,u) + \\ &\qquad + a(u) \int_{\mathbf{U}} [U(x,y,s) - U(x,y,u)]\, p_1(u,ds) \Big\} \mu_j(du) = \\ &= -G(x,y,z) - Q_1U(x,y,z) + \\ &\quad + (\nabla_x U(x,y,j), \tilde{A}(y,j)\,x) + (\nabla_y U(x,y,j), \tilde{\varphi}(y,j)) + \\ &\quad + \sum_{k=1}^{m} \int_{\mathbf{U}_j} a(u)\, p_1(u,\mathbf{U}_k)\, [\tilde{V}(x,y,k) - \tilde{V}(x,y,j)]\, \mu_j(du) = \\ &= -G(x,y,z) - Q_1U(x,y,z) + (\tilde{\mathcal{L}}\tilde{V})(x,y,j), \end{aligned}$$

if $z \in \mathbf{U}_j$, $j = \overline{1,m}$. Hence, in view of the above equalities and formula (28), it is clearly seen, that

$$(\mathcal{L}_\varepsilon \tilde{V}_\varepsilon)(x,y,z) = (\tilde{\mathcal{L}}\tilde{V})(x,y,j) + \varepsilon\, R(x,y,z),$$

for all $z \in \mathbf{U}_j$, $j = \overline{1,m}$, and therefore, the inequality

$$(\mathcal{L}_\varepsilon \tilde{V}_\varepsilon)(x,y,z) \le -\frac{1}{2}|x|^2 + \varepsilon k_1 |x|^2 \le -\frac{1}{4}|x|^2 \le -\frac{1}{8c_2}\tilde{V}_\varepsilon(x,y,z)$$

follows immediately from (27), (29) and (30) for any $x \in \mathbf{R}^n$, $y \in \mathbf{R}^d$, $z \in \mathbf{U}$ and $\varepsilon \in (0,\varepsilon_0)$, where ε_0 is a sufficiently small positive number, less than the above ε_1. Now we may use the well-known formula for the Markov semigroups from Dynkin (1965)

$$\mathbf{E}_{s,z}\tilde{V}_\varepsilon(X_\varepsilon(t,s,y)\,x, y_\varepsilon(t,s,y), \zeta(t)) = \\ = \tilde{V}_\varepsilon(x,y,z) + \int_s^t \mathbf{E}_{s,z}(\mathcal{L}_\varepsilon \tilde{V}_\varepsilon)(X_\varepsilon(\tau,s,y)\,x, y_\varepsilon(\tau,s,y), \zeta_\varepsilon(\tau))\, d\tau$$

and deduce from (19) the inequality

$$\mathbf{E}_{s,z}\tilde{V}_\varepsilon(X_\varepsilon(t,s,y)\,x, y_\varepsilon(t,s,y), \zeta(t)) \le \\ = \tilde{V}_\varepsilon(x,y,z) - \frac{1}{8c_2}\int_s^t \mathbf{E}_{s,z}\tilde{V}_\varepsilon(X_\varepsilon(\tau,s,y)\,x, y_\varepsilon(\tau,s,y), \zeta_\varepsilon(\tau))\, d\tau.$$

From (16) and the above inequality it follows that

$$\frac{1}{2}c_1\,\mathbf{E}_{s,z}|X_\varepsilon(t,s,y)\,x|^2 \le \mathbf{E}_{s,z}\tilde{V}_\varepsilon(X_\varepsilon(t,s,y)\,x, y_\varepsilon(t,s,y), \zeta_\varepsilon(t)) \le \\ \le \tilde{V}_\varepsilon(x,y,z)\exp\Big\{-\frac{1}{8c_2}(t-s)\Big\} \le 2c_2|x|^2\exp\Big\{-\frac{1}{8c_2}(t-s)\Big\}$$

for any $x \in \mathbf{R}^n$, $y \in \mathbf{R}^d$, $z \in \mathbf{U}$ and $\varepsilon \in (0,\varepsilon_0)$.

The proof of Theorem 2 is now complete.

References

Dynkin, E.B. (1965). *Markov processes.* Academic Press, New York.

Korolyuk, V.S. (1991). *Averaging and stability of dynamical systems with rapid Markov switching.* UmeåUniv. Press, Umeå, Sweden.

Oseledec, V.I. (1968). A multiplicative ergodic theorem. Lyapunov characteristic numbers for dynamical systems. *Transactions of Moscow Mathematical Society.* **19**, 197–231.

Skorokhod, A.V. (1991). *Asymptotic methods of stochastic differential equations.* AMS, Providence.

Exploring Stochastic Laws. pp. 481–490.
Eds. A.V. Skorokhod and Yu.V. Borovskikh

The bootstrap estimator for the asymptotic variance of *U*-quantiles

NOËL VERAVERBEKE

Limburgs Universitair Centrum, Universitaire Campus B 3590 Diepenbeek, Belgium

Received January 18, 1995

Abstract. U-quantiles are of the form $H_F^{-1}(p)$ where $0 < p < 1$, and where H_F is the distribution function of $h(X_1, \ldots, X_m)$, with h some kernel function and $X_1, \ldots, X_n$ a random sample from a distribution function F. These generalized quantiles are naturally estimated by $H_n^{-1}(p)$, where H_n is the empirical distribution function of a U-statistic structure. The main result in this paper is the strong consistency of the bootstrap variance estimator, thereby generalizing a well known result of Ghosh, Parr, Singh and Babu (1984). As a related result, the pointwise consistency is shown for the Swanepoel (1986) non-parametric estimator for the density function of $h(X_1, \ldots, X_m)$.

1. Introduction

Let $h(x_1, \ldots, x_m)$ be a kernel of degree m, i. e. a real-valued measurable function, symmetric in its arguments, and let $X_1, \ldots, X_n$ be a random sample of size $n \geq m$ from a distribution function F. The distribution function of $h(X_1, \ldots, X_m)$ is denoted by

$$H_F(t) = P(h(X_1, \ldots, X_m) \leq t), \quad t \in \mathrm{R},$$

and its natural estimator is the so called empirical distribution function of U-statistic structure

$$H_n(t) = \binom{n}{m}^{-1} \sum_{C_{nm}} I(h(X_1, \ldots X_{i_m}) \leq t), \quad t \in \mathrm{R}$$

where $C_{nm} = \{(i_1, \ldots, i_m) : 1 \leq i_1 < i_2 < \ldots < i_m \leq n\}$. For the theory of U-statistics, we refer the reader to the recent book by Korolyuk and Borovskikh (1994).

For $0 < p < 1$, we denote the p-th quantiles of H_F and H_n as $\xi_p = H_F^{-1}(p)$ and $\xi_{pn} = H_n^{-1}(p)$. Asymptotic theory for these U-quantiles has been obtained by

Choudhury and Serfling (1988). If H_F is differentiable at ξ_p with $h_F(\xi_p) = H'_F(\xi_p) > 0$, then the following asymptotic normality result holds

$$n^{1/2}(\xi_{pn} - \xi_p) \xrightarrow{d} N(0;\sigma_p^2) \tag{1}$$

where

$$\sigma_p^2 = m^2 \frac{\zeta_p}{h_F^2(\xi_p)} \tag{2}$$

with

$$\zeta_p = \text{Var}(g_p(X_1)) > 0 \tag{3}$$

and

$$g_p(X_1) = E[I(h(X_1,\ldots,X_m) \leq \xi_p)|X_1] - p.$$

In this paper we consider the following bootstrap procedure: a random sample $X_1^*,\ldots,X_n^*$ is drawn from F_n, the empirical distribution function of the original sample $X_1,\ldots,X_n$. Then consider

$$\xi_{pn}^* = H_n^{*^{-1}}(p)$$

where H_n^* denotes the empirical distribution function of a U-statistic structure, based on the bootstrap sample $X_1^*,\ldots,X_n^*$, i. e.

$$H_n^*(t) = \binom{n}{m}^{-1} \sum_{C_{nm}} I(h(X_{i_1}^*,\ldots,X_{i_m}^*) \leq t), \quad t \in \mathrm{R}. \tag{4}$$

In Helmers, Janssen and Veraverbeke (1990) it was shown that for almost all sample sequences $X_1, X_2, \ldots$ the limiting distribution of $n^{1/2}(\xi_{pn}^* - \xi_{pn})$ is the same $N(0;\sigma_p^2)$ distribution as in (1), provided H_F is continuously differentiable in a neighborhood of ξ_p with $h_F(\xi_p) > 0$. We prove here that, if in addition a moment condition on the kernel h is satisfied, also the variance (under F_n) of $n^{1/2}(\xi_{pn}^* - \xi_{pn})$ converges a.s. to σ_p^2. This result generalizes that of Ghosh, Parr, Singh and Babu (1984) who proved this for usual quantiles (i.e. the case $m = 1$ and $h(x) = x$). The key tool is to prove uniform integrability of the sequence $\{n(\xi_{pn}^* - \xi_{pn})^2\}, \quad n \geq m$. This is done in Section 2 in a slightly more general context than needed, namely with p replaced by p_n where $\{p_n\}$ is a sequence of random variables, $0 < p_n < 1$, converging almost surely to some fixed $0 < p < 1$. This generality will be useful in Section 3, where we use the uniform integrability to prove a strong consistency result for an estimator of the density h_F of $h(X_1, \ldots X_m)$. This generalizes a result of Swanepoel (1986).

2. Strong consistency of the bootstrap variance estimator

The result of Helmers, Janssen and Veraverbeke (1990) on the limiting distribution of $n^{1/2}(\xi^*_{pn} - \xi_{pn})$ is now supplemented by the almost sure convergence to σ_p^2 of the bootstrap estimator

$$\hat{\sigma}_n^2 = \mathrm{Var}^*(\mathrm{n}^{1/2}(\xi^*_{\mathrm{pn}} - \xi_{\mathrm{pn}})).$$

Note that "starred" notations such as $P^*, E^*, \mathrm{Var}^*, \ldots$ denote probability, expectation, variance, ... under F_n.

Theorem 1. *Suppose that H_F is continuously differentiable in a neighborhood of ξ_p with $h_F(\xi_p) = H'_F(\xi_p) > 0$. Assume moreover that, for some $\alpha > 0$,*

$$E|h(X_{i_1}, \ldots, X_{i_m})|^\alpha < \infty,$$

for all $1 \leq i_1, i_2, \ldots, i_m \leq n$.

Then, as $n \to \infty$,

$$\hat{\sigma}_n^2 = Var^*(n^{1/2}(\xi^*_{pn} - \xi_{pn})) \to \sigma_p^2 \quad \text{a.s.}$$

with σ_p^2 as in (2).

Because of the convergence result of Theorem 2.1 in Helmers, Janssen and Veraverbeke (1990), it is sufficient to prove the uniform integrability of the sequence $\{n(\xi^*_{pn} - \xi_{pn})\}, n \geq m$. In view of a further application in Section 3, we prove in Lemma 1 below, a slightly more general result.

Lemma 1. *Assume the conditions of Theorem 1. Let $\{p_n\}$ be a sequence of random variables, $0 < p_n < 1$, such that $p_n \to p$ a.s.*

Then the sequence $\{n(\xi^*_{p_n n} - \xi_{p_n n})\}, n \geq m$ is uniformly integrable.

Proof. It is sufficient to show that $\sup_{n \geq m} E^*|n^{1/2}(\xi^*_{p_n n} - \xi_{p_n n})|^{2+\delta} < \infty$ a.s. for some $\delta > 0$, or that

$$\sup_{n \geq m} \int_0^\infty t^{1+\delta} P^*(n^{1/2}|\xi^*_{p_n n} - \xi_{p_n n}| > t)dt < \infty \quad \text{a.s.} \tag{5}$$

We will prove (5) with $n^{1/2}|\xi^*_{p_n n} - \xi_{p_n n}|$ replaced by $n^{1/2}(\xi^*_{p_n n} - \xi_{p_n n})$. The other case works similarly. We have, with H_n^* as in (4) and with

$$\overline{H}_n(t) = n^{-m} \sum_{i_1=1}^{n} \cdots \sum_{i_m=1}^{n} I(h(X_{i_1}, \ldots, X_{i_m}) \leq t), \quad t \in \mathrm{R} \tag{6}$$

and

$$\overline{D}_n = n^{1/2}\{\overline{H}_n(\xi_{p_n n} + tn^{-1/2}) - p_n\} \tag{7}$$

that

$$P^*(n^{1/2}(\xi^*_{p_n n} - \xi_{p_n n}) > t) =$$

$$= P^*(\xi^*_{p_n n} \geq \xi_{p_n n} + tn^{-1/2}) = P^*(p_n \geq H^*_n(\xi_{p_n n} + tn^{-1/2})) =$$

$$= P^*\left(H^*_n(\xi_{p_n n} + tn^{-1/2}) - \overline{H}_n(\xi_{p_n n} + tn^{-1/2}) \geq p_n - \overline{H}_n(\xi_{p_n n} + tn^{-1/2})\right) \leq$$

$$\leq 2\exp\{-2[\frac{n}{m}][\overline{H}_n(\xi_{p_n n} + tn^{-1/2}) - p_n]^2\} =$$

$$= 2\exp\{-2[\frac{n}{m}]n^{-1}\overline{D}^2_n\}$$

by Hoeffding's probability inequality (see e.g. Serfling (1980), p. 201). We need an a.s. order bound for the exponent in this last expression. Since the bound has to be integrated out, the dependence on t is important. First note that we may replace $\overline{D}_n$ by D_n, where

$$D_n = n^{1/2}\{H_n(\xi_{p_n n} + tn^{-1/2}) - p_n\}$$

since $\sup_t |\overline{H}_n(t) - H_n(t)| = O(n^{-1})$ a.s., by the well known relation between U-and V-statistics (see e.g. Serfling (1980), p.206). We therefore have that for all $t > 0$,

$$P^*(n^{1/2}(\xi_{p_n n} - \xi_{p_n n}) > t) \leq c_1 e^{-c_2 D^2_n} \tag{8}$$

where c_1 and c_2 are positive constants. Now write

$$D_n = D_{n1} + D_{n2} + D_{n3}$$

where

$$D_{n1} = n^{1/2}\{[H_n(\xi_{p_n n} + tn^{-1/2}) - H_F(\xi_{p_n n} + tn^{-1/2})] -$$

$$-[H_n(\xi_{p_n n}) - H_F(\xi_{p_n n})]\}, \tag{9}$$

$$D_{n2} = n^{1/2}\{H_F(\xi_{p_n n} + tn^{-1/2}) - H_F(\xi_{p_n n})\}, \tag{10}$$

$$D_{n3} = n^{1/2}\{H_n(\xi_{p_n n}) - p_n\}. \tag{11}$$

From Lemma 3.1 of Choudhury and Serfling (1988), we have that $\xi_{p_n n} - \xi_p = O(n^{-1/2}(\log n)^{1/2})$ a.s. Hence, for a constant $c_3 > 0$ (to be specified later) and a constant $c_4 > 0$ we have: if

$$t \leq c_3(\log n)^{1/2}, \tag{12}$$

then

$$|D_{n1}| \leq 2n^{1/2} \sup_{|y| \leq c_4 n^{-1/2}(\log n)^{1/2}} |[H_n(\xi_p + y) - H_F(\xi_p + y)] -$$

$$-[H_n(\xi_p) - H_F(\xi_p)]| = O(n^{-1/4}(\log n)^{3/4}) \quad \text{a.s} \tag{13}$$

by Lemma 3.2 of Choudhury and Serfling (1988), since h_F is bounded in some neighborhood of ξ_p with $h_F(\xi_p) > 0$. For D_{n2} we have that $D_{n2} = th_F(\theta_{tn})$ where θ_{tn} is between $\xi_{p_n n}$ and $\xi_{p_n n} + tn^{-1/2}$. Hence, with some constant $c_5 > 0$, we have for all large n,

$$D_{2n} > c_5 t \quad \text{a.s.} \tag{14}$$

since h_F is continuous in some neighborhood of ξ_p and $h_F(\xi_p) > 0$. Finally

$$D_{3n} = O(n^{-1/2}) \quad \text{a.s.} \tag{15}$$

It follows from (13)–(15) that, with some constant $c_6 > 0$, we have for all large n,

$$D_n^2 > c_6 t^2,$$

which gives with (8) and (11) that for $t \leq c_3(\log n)^{1/2}$

$$P^*(n^{1/2}(\xi^*_{p_n n} - \xi_{p_n n}) > t) \leq c_1 e^{-c_7 t^2} \quad \text{a.s.}$$

where $c_7 = c_2 \cdot c_6$, and hence,

$$\int_0^{c_3(\log n)^{1/2}} t^{1+\delta} P^*(n^{1/2}(\xi_{p_n n} - \xi_{p_n n}) > t)dt < \infty \quad \text{a.s.}$$

If $t > c_3(\log n)^{1/2}$, we have

$$P^*(n^{1/2}(\xi^*_{p_n n} - \xi_{p_n n}) > t) \leq P^*(n^{1/2}(\xi^*_{p_n n} - \xi_{p_n n}) > c_3(\log n)^{1/2}) \leq c_1 n^{-c_7 c_3^2},$$

and hence,

$$\int_{c_3(\log n)^{1/2}}^{n^{\frac{1}{2}+\frac{m}{\alpha}}} t^{1+\delta} P^*(n^{1/2}(\xi^*_{p_n n} - \xi_{p_n n}) > t)dt \leq c_1 n^{-c_7 c_3^2 + (2+\delta)(\frac{1}{2}+\frac{m}{\alpha})} = O(1) \quad \text{a.s.}$$

by choice of c_3. Finally,

$$\int_{n^{\frac{1}{2}+\frac{m}{\alpha}}}^{\infty} t^{1+\delta} P^*(n^{1/2}(\xi^*_{p_n n} - \xi_{p_n n}) > t)dt \leq$$

$$\leq \int_{n^{\frac{1}{2}+\frac{m}{\alpha}}}^{\infty} t^{1+\delta} P^*(\xi^*_{p_n n} - \xi_{p_n n} > n^{\frac{m}{\alpha}})dt = 0 \quad \text{a.s.}$$

for all large n. Indeed, with $V_{1:n} \leq V_{2:n} \leq \ldots \leq V_{n^m:n}$ the ordered values of $h(X_{i_1}, \ldots, X_{i_m})(1 \leq i_1, \ldots, i_m \leq n)$, we have

$$P^*(\xi^*_{p_n n} - \xi_{p_n n} > n^{\frac{m}{\alpha}}) \leq I(V_{n^m:n} - V_{1:n} > n^{\frac{m}{\alpha}}) = 0 \quad \text{a.s.}$$

since, by Lemma 2 below, the moment condition on the kernel h implies that $|V_{n^m:n}| + |V_{1:n}| = 0(n^{\frac{m}{\alpha}})$ a.s. as $n \to \infty$. This proves Lemma 1.

Let $V_{1:n} \leq V_{2:n} \leq \ldots \leq V_{n^m:n}$ denote the ordered values of $h(X_{i_1}, \ldots, X_{i_m})$, taken over all m-tuples $(i_1, \ldots, i_m)$ of (not necessarily distinct) elements from $\{1, \ldots, n\}$. In the previous proof we used the following property on the a.s. behaviour of the extremes of these generalized order statistics.

Lemma 2. *If for some* $\alpha > 0, E|h(X_{i_1}, \ldots, X_{i_m})|^\alpha < \infty$ *for all* $1 \leq i_1, \ldots, i_m \leq n$, *then as* $n \to \infty$,

$$n^{-\frac{m}{\alpha}}\{|V_{1:n}| + |V_{n^m:n}|\} \to 0 \quad \text{a.s.}$$

Proof. It is sufficient to show that, as $n \to \infty$,

$$n^{-\frac{m}{\alpha}} \max_{1 \leq i_1, \ldots, i_m \leq n} \{|h(X_{i_1}, \ldots, X_{i_m})|\} \to 0 \quad \text{a.s.} \tag{16}$$

Now

$$\max_{1 \leq i_1, \ldots, i_m \leq n} \{|h(X_{i_1}, \ldots, X_{i_m})|\} = \max\{T_{n1}, \ldots, T_{nm}\}$$

with, for $j = 1, \ldots, m$,

$$T_{nj} = \max_{(j)}\{|h(X_{i_1}, \ldots, X_{i_m})|\}$$

where $\max_{(j)}$ represents the maximum taken over the m-tuples $(i_1, \ldots, i_m)$ from $\{1, \ldots, n\}$, for which exactly j among $i_1, \ldots, i_m$ are distinct.

In view of the identity

$$n^m = \sum_{j=1}^{m} S_m^{(j)} n_{(j)}$$

where $n_{(j)} = n(n-1)\ldots(n-j+1)$ and $S_m^{(j)}$ is the Stirling number of the second kind, we have that for each $j = 1, \ldots, m, \max_{(j)}$ is taken over $S_m^{(j)} n_{(j)} = O(n^j)$ elements. The validity of (16) is established by showing that, for each $j = 1, \ldots, m$

$$n^{-\frac{j}{\alpha}} T_{nj} \to 0 \quad \text{a.s.} \tag{17}$$

For $j = 1$, this is known (see e.g. Lemma 3 in Ghosh, Parr, Singh and Babu (1984)). We give a proof for $j = m$. For this we relabel $h(X_{i_1}, \ldots, X_{i_m}), 1 \leq i_1 \neq i_2 \neq \ldots \neq i_m \leq n$, by one single index $H_1, H_2, \ldots, H_N$, where $N = n_{(m)}$. For every $\epsilon > 0$, we have

$$\begin{aligned} \sum_{n=1}^{\infty} P(|H_n| \geq \epsilon n^{\frac{1}{\alpha}}) &= \sum_{n=1}^{\infty} P(|h(X_1, \ldots, X_m)| \geq \epsilon n^{\frac{1}{\alpha}}) \leq \\ &\leq \epsilon^{-\alpha} E|h(X_1, \ldots, X_m)|^\alpha < \infty \end{aligned}$$

(by the Lemma in Serfling (1980), p. 12). By the Borel–Cantelli Lemma, we have

$$P(|H_n| \geq \epsilon n^{1/\alpha} \quad i.o.) = 0,$$

and hence also

$$P(|H_N| \geq \epsilon N^{1/\alpha} \quad i.o.) = 0.$$

By Theorem 4.2.1 in Galambos (1978), this implies that

$$P(\max(H_1, \ldots, H_N) \geq \epsilon N^{1/\alpha} \quad i.o.) = 0$$

and since $N^{1/\alpha} = O(n^{m/\alpha})$, we have that $T_{nm} = o(n^{m/\alpha})$ a.s. It is now clear that for T_{nj} with $1 < j < m$, the proof works along the same lines.

3. Strong consistency of a density estimator

The uniform integrability result in Lemma 1 is also an important tool in establishing strong consistency of a non-parametric estimator for the density function h_F of the random variable $h(X_1, \ldots, X_m)$. One possible estimator for $h_F(t)$ could be the obvious generalization of the kernel estimator

$$\frac{1}{a_n} \int_{-\infty}^{\infty} K(\frac{t-x}{a_n}) dH_n(x)$$

where $a_n > 0$ and K is a known probability density function. A totally different idea (due to Swanepoel (1986) in the case $m = 1$ and $h(x) = x$) is as follows: since by Theorem 1,

$$Var^*(n^{1/2}(\xi^*_{pn} - \xi_{pn})) \rightarrow \sigma_p^2 = m^2 \frac{\zeta_p}{h_F^2(\xi_p)} \quad \text{a.s.}$$

we could propose as estimator for $h_F(t)$

$$\hat{h}_n(t) = \left\{ \frac{m^2 \hat{\zeta}_{p_n n}}{Var^*(n^{1/2}(\xi^*_{p_n n} - \xi_{p_n n}))} \right\}^{1/2},$$

where $p_n = H_n(t)$ and $\hat{\zeta}_{pn}$ is the estimator for the parameter ζ_p in (3), due to Choudhury and Serfling (1988).

Theorem 2. *Let $t \in R$ be such that $0 < H_F(t) < 1$. Suppose that H_F is continuously differentiable in a neighborhood of t with $h_F(t) = H_F'(t) > 0$. Assume moreover that for some $\alpha > 0$, $E|h(X_{i_1}, \ldots, X_{i_m})|^\alpha < \infty$, for all $1 \leq i_1, \ldots, i_m \leq n$.*

Then, as $n \rightarrow \infty$,

$$\hat{h}_n(t) \rightarrow h(t) \quad \text{a.s.}$$

Proof. First note that $\hat{\zeta}_{p_n n} \to \zeta_p$ a.s. as $n \to \infty$, where $p_n = H_n(t)$ and $p = H(t)$. This follows from a slight modification of the proof of Theorem 5.1 in Choudhury and Serfling (1988), together with the fact that $\xi_{p_n n} \to \xi_p$, a.s. The latter follows from Lemma 3.1 in Choudhury and Serfling (1988) and the fact that $p_n = H_n(t)$ converges a.s. to $p = H(t)$ by the strong law for U-statistics. It remains to show that

$$\mathrm{Var}^*(\mathrm{n}^{1/2}(\xi^*_{\mathrm{p_n n}} - \xi_{\mathrm{p_n n}})) \to \sigma^2_{\mathrm{p}} = \sigma^2_{\mathrm{H(t)}} \quad \text{a.s.}$$

This requires that

$$n^{1/2}(\xi^*_{p_n n} - \xi_{p_n n}) \xrightarrow{d^*} N(0; \sigma^2_p) \tag{18}$$

and that the sequence $\{n(\xi^*_{p_n n} - \xi_{p_n n})\}$, $n \geq m$ is uniformly integrable. The latter holds by Lemma 1. The convergence result in (18) is the content of the next lemma, which is a generalization from fixed p to random p_n of Theorem 2.1 in Helmers, Janssen and Veraverbeke (1990).

Lemma 3. *Suppose that H_F is continuously differentiable in a neighborhood of ξ_p with $h_F(\xi_p) = H'_F(\xi_p) > 0$. Let $\{p_n\}$ be a sequence of random variables, $0 < p_n < 1$, such that $p_n \to p$ a.s.*

Then, for almost every sample sequence $X_1, X_2, \ldots$

$$n^{1/2}(\xi^*_{p_n n} - \xi_{p_n n}) \xrightarrow{d^*} N(0; \sigma^2_p)$$

with σ^2_p as in (2).

Proof. As in Helmers, Janssen and Veraverbeke (1990), we have

$$P^*(n^{1/2}(\xi^*_{p_n n} - \xi_{p_n n}) \leq t) = P^*(W^*_n \geq -\overline{D}_n) \tag{19}$$

where $W^*_n = n^{1/2}\{H^*_n(\xi_{p_n n} + tn^{-1/2}) - \overline{H}_n(\xi_{p_n n} + tn^{-1/2})\}$ and $H^*_n, \overline{H}_n$ and $\overline{D}_n$ are given by (4), (6) and (7) respectively. As in the proof of Lemma 1, $\overline{D}_n = D_n + O(n^{-1/2})$ a.s. and $D_n = D_{n1} + D_{n2} + D_{n3}$ with the D_{ni} given in (9)–(11). Since $\xi_{p_n n} \to \xi_p$ a.s. (use e.g. Lemma 3.1 in Choudhury and Serfling (1988)) we can use exactly the same arguments as in Helmers, Janssen and Veraverbeke (1990) to find that

$$\overline{D}_n \to th_F(\xi_p) \quad \text{a.s.}$$

As to the limit behaviour of $W^*_n, n \geq m$, we have given $X_1, \ldots, X_n$, W^*_n is a properly normalized U-statistic with bounded kernel, depending on n, and given by

$$h_n(y_1, \ldots, y_m) = I(h(y_1, \ldots, y_m) \leq \xi_{p_n n} + tn^{-1/2}) - \overline{H}_n(\xi_{p_n n} + tn^{-1/2}).$$

We have $E^*[h_n(X_1^*,\ldots,X_m^*)] = 0$ and

$$E^*[h_n^2(X_1^*,\ldots,X_m^*)] = \overline{H}_n(\xi_{p_n n} + tn^{-1/2})(1 - \overline{H}_n(\xi_{p_n n} + tn^{-1/2})).$$

Also with

$$g_n(y_1) = E^*[h_n(y_1, X_2^*,\ldots,X_m^*)]$$

we have

$$\begin{aligned}
\sigma_{g_n}^2 &= E^*[g_n^2(X_1^*)] \\
&= \frac{1}{n}\sum_{i=1}^{n} g_n^2(X_i) \\
&= n^{-2m+1}\sum_{i_1=1}^{n}\cdots\sum_{i_{2m-1}=1}^{n} h_n(X_{i_1},\ldots,X_{i_m})h_n(X_{i_1},X_{i_{m+1}},\ldots,X_{i_{2m-1}}).
\end{aligned}$$

Using Theorem 2.9 of Iverson and Randles (1989) on the strong law for *U*-statistics with estimated parameters, gives

$$E^*[h_n^2(X_1^*,\ldots,X_m^*)] \to p(1-p) \quad \text{a.s.}$$

and

$$\sigma_{g_n}^2 \to \mathrm{Cov}[I(h(X_1,\ldots,X_m) \le \xi_p),\ I(h(X_1,X_{m+1},\ldots,X_{2m-1}) \le \xi_p)] = \zeta_p \quad \text{a.s.}$$

with ζ_p as in (3). The central limit theorem for *U*- statistics with the kernel depending on n, as given in Theorem 1 of Weber (1983) now applies, so that

$$\sup_y |P^*(W_n^* \le y) - \Phi(y^{-1}\zeta_p^{-1/2})| \to 0 \quad \text{a.s.}$$

Hence, from (19),

$$\begin{aligned}
P^*(n^{1/2}(\xi_{p_n n}^* - \xi_{p_n n}) \le t) &= \Phi(-\overline{D}_n m^{-1}\zeta_p^{-1/2}) + o(1) \\
&= \Phi(t\sigma^{-1}) + o(1)
\end{aligned}$$

a.s. as $n \to \infty$. This proves the lemma.

References

Choudhury, J. and Serfling, R.J. (1988). Generalized order statistics, Bahadur representations, and sequential non-parametric fixed-width confidence intervals. *J. Statist. Planning Inf.* **19**, 269–282.

Galambos, J. (1978). *The asymptotic theory of extreme order statistics.* John Wiley, New York.

Ghosh, M.W., Singh K. and Babu, G.J. (1984). A note on bootstrapping the sample median. *Ann. Statist.* **12**, 1130–1135.

Helmers, R., Janssen, P. and Veraverbeke, N. (1990). Bootstrapping *U*-quantiles. In: *Exploring the limits of the bootstrap,* E. Le Page and L. Billard (Eds). Wiley, New York, pp. 145–155.

Iverson, H. and Randles, R. (1989). The effect on convergence of substituting parameter estimates into *U*-statistics and other families of statistics. *Probab. Th. Rel. Fields* **81**, 453–471.

Koroljuk, V.S. and Borovskich, Yu.V. (1994). *Theory of U-statistics.* Kluwer Academic Publishers, Dordrecht.

Serfling, R.J. (1980). *Approximation theorems of mathematical statistics.* John Wiley, New York.

Swanepoel, J.W.H. (1986). On the construction of non-parametric density function estimators using the bootstrap. *Commun. Statist. - Theor. Meth.* **15**, 1399–1415.

Weber N.C. (1983). Central limit theorems for a class of symmetric statistics. *Math. Proc. Camb. Phil. Soc.* **94**, 307–313.

Bei Fragen zur Produktsicherheit wenden Sie sich bitte an:
If you have any questions regarding product safety,
please contact:

Walter de Gruyter GmbH
Genthiner Straße 13
10785 Berlin
productsafety@degruyterbrill.com